计算流体力学

原理和应用

（第3版）

Computational Fluid Dynamics

Principles and Applications

Third Edition

[德]吉瑞·布拉泽克（Jiri Blazek）　著

张耀冰　陈江涛　陈喜兰　周乃春　等译

国防工業出版社

·北京·

著作权合同登记 图字:01-2023-2622号

图书在版编目(CIP)数据

计算流体力学原理和应用:第3版/(德)吉瑞·布拉泽克(Jiri Blazek)著;张耀冰等译. —北京:国防工业出版社,2024.2重印

书名原文:Computational Fluid Dynamics:Principles and Applications(Third Edition)

ISBN 978-7-118-13039-3

Ⅰ.①计… Ⅱ.①吉… ②张… Ⅲ.①计算流体力学 Ⅳ.①O35

中国国家版本馆CIP数据核字(2023)第116088号

Computational Fluid Dynamics:Principles and Applications,Third Edition
Jiri Blazek
ISBN:9780080999951

※

国防工业出版社出版发行
(北京市海淀区紫竹院南路23号 邮政编码100048)
北京凌奇印刷有限责任公司印刷
新华书店经售

*

开本710×1000 1/16 印张26¼ 字数460千字
2024年2月第3版第2次印刷 印数1501—2500册 定价186.00元

(本书如有印装错误,我社负责调换)

国防书店:(010)88540777 书店传真:(010)88540776
发行业务:(010)88540717 发行传真:(010)88540762

译者序

计算流体力学(computational fluid dynamics,CFD)是一门利用计算机和数值方法对流体力学问题进行数值模拟和分析的交叉学科,它是流体力学、计算机科学和计算科学相结合的产物。随着现代计算机技术的飞速发展和数值计算方法的日益成熟,CFD 得到了长足的发展,在航空、航天和航海等领域都得到了广泛的应用。

目前,CFD 方面的教材很多,但是多数都是讲解 CFD 基础知识或理论研究方面的,介绍工程实用的 CFD 方法,特别是非结构网格方法方面的书籍很少。译者所在课题组的主要工作是 CFD 软件的开发及应用,2005 年译者的工作开始由结构网格方法研究转向非结构混合网格流场求解方法研究时,对非结构混合网格方法一片茫然,无意中读到本书,惊为“神书”,读过本书后,一种醍醐灌顶之感油然而生,之后本书就成为课题组的必读书目,人手一本,多年来在日常工作中发挥了重要作用。

本书的特点是理论联系实际,重点介绍工程实用的 CFD 方法,侧重于具体算法的实现。在介绍各种数值方法的同时讨论方法的优缺点,每个理论部分都会有相应的实践部分来增强对知识的理解,并展示这些方法如何应用到实际 CFD 程序中,本书最后一章提供了各种 CFD 源代码,帮助读者实现从理论到实践的过渡。本书还有一个重要特点是参考文献广泛,读者可以本书的参考文献为基础开展进一步的研究工作。

鉴于初次从事 CFD 工作的很多人都是刚从大学本科或硕士研究生毕业的,专业知识的英文水平不高,本书中文版的面世,将对读者正确理解和掌握 CFD 知识有所帮助,对他们快速进入 CFD 行业领域起到促进作用。

本书翻译工作获得国家数值风洞(National Numerical Wind-tunnel,NNW)工程资助。参加本书翻译工作的还有龚小权、唐静、李明、王建涛、赵辉、崔鹏程、李欢、周桂宇、张健、牟斌、吴晓军、张培红、李彬、贾洪印、赵凡、程锋、付云峰。本书翻译工作还得到了马明生研究员和邓有奇研究员的指导,并得到译者所在单位中国空气动力研究与发展中心计算空气动力研究所的大力支持，在此一并致谢。

由于译者水平所限，书中难免有错误或不妥之处，敬请广大读者及时给予指正。

译　者

2022年5月于四川绵阳

符号列表

$\boldsymbol{A}_{\mathrm{c}}$	对流通量雅可比矩阵
$\boldsymbol{A}_{\mathrm{v}}$	黏性通量雅可比矩阵
b	二维控制体的纵深常数
c	声速
c_p	比定压热容
c_V	比定容热容
$\boldsymbol{C}$	特征矢量
C_m	组分 m 的摩尔浓度($=\rho Y_m/W_m$)
C_{S}	Smagorinsky 常数
curl $\boldsymbol{v}$	速度 $\boldsymbol{v}$ 的旋度$\left(=\nabla\times\boldsymbol{v}=\left[\frac{\partial w}{\partial y}-\frac{\partial v}{\partial z},\frac{\partial u}{\partial z}-\frac{\partial w}{\partial x},\frac{\partial v}{\partial x}-\frac{\partial u}{\partial y}\right]\right)$
d	距离
$\mathbf{D}$	隐式算子的对角部分
$\boldsymbol{D}$	人工耗散
D_m	组分 m 的有效二元扩散系数
div $\boldsymbol{v}$	速度 $\boldsymbol{v}$ 的散度$\left(=\nabla\cdot\boldsymbol{v}=\frac{\partial u}{\partial x}+\frac{\partial v}{\partial y}+\frac{\partial w}{\partial z}\right)$
e	单位质量的内能
E	单位质量的总能
f	时间推进算子的傅里叶记号
$\boldsymbol{f}_{\mathrm{e}}$	体积外力矢量
$\boldsymbol{F}$	矢通量;通量张量
g	放大因子
$\boldsymbol{g}$	网格速度
h	焓
Δh	当地网格(单元)尺寸
H	总(滞止)焓

续表

$\boldsymbol{H}$	Hessian 矩阵(二阶导数矩阵)
I	虚数单位($\mathrm{I}=\sqrt{-1}$)
$\boldsymbol{I}$	单位矩阵;单位张量
$\hat{I}_h^{2h}$	插值算子
I_h^{2h}	限制算子
I_{2h}^h	延拓算子
$\boldsymbol{J}$	系统矩阵(隐式算子)
J^{-1}	坐标变换雅可比矩阵行列式的倒数
k	热传导系数
K	湍动能
$K_\mathrm{f}, K_\mathrm{b}$	正向和逆向反应率常数
l_T	湍流长度尺度
$\mathbf{L}$	隐式算子的严格下三角部分
L_{ij}	Leonard 应力张量分量
M	马赫数
$\boldsymbol{M}$	质量矩阵
$\boldsymbol{n}$	控制体面的单位法向矢量(向外为正)
n_x, n_y, n_z	单位法矢在 x, y, z 方向的分量
N	网格点数,单元数或控制体的数量
N_A	相邻控制体的数量
N_F	控制体的面数
p	静压
P	湍动能的生成项
$\boldsymbol{P}$	从原始变量到守恒变量的转换矩阵
$\boldsymbol{P}_\mathrm{L}, \boldsymbol{P}_\mathrm{R}$	(Krylov 子空间方法的)左,右预处理矩阵
Pr	普朗特数
$\dot{q}_\mathrm{h}$	辐射、化学反应等引起的热通量
Q	源项
$\boldsymbol{r}$	(笛卡儿坐标的)位矢;(GMRES 的)残差
$\boldsymbol{r}_{ij}$	从点 i 指向点 j 的矢量
R	气体常数

续表

R_u	普适气体常数(=8314. 34 J/(kmol · K))
$\boldsymbol{R}$	残差;右端项;旋转矩阵
$\boldsymbol{R}^*$	光顺后的残差
Re	雷诺数
$\dot{s}_m$	组分 m 的化学反应变化率
$\boldsymbol{S}$	面积矢量($=\boldsymbol{n}\Delta S$)
S_{ij}	应变率张量的分量
S_x, S_y, S_z	面矢量的笛卡儿坐标分量
$\mathrm{d}S$	面微元
ΔS	控制体面的长度/面积
t	时间
t_T	湍流时间尺度
Δt	时间步长
T	静温
$\boldsymbol{T}$	右特征矢量矩阵
$\boldsymbol{T}^{-1}$	左特征矢量矩阵
u, v, w	速度的笛卡儿坐标分量
u_τ	表面摩擦速度($=\sqrt{\tau_w/\rho}$)
U	一般(标量形式的)流场变量
$\mathbf{U}$	隐式算子的严格上三角部分
$\boldsymbol{U}$	一般流场变量的矢量形式
$\boldsymbol{v}$	速度矢量,分量为 u,v 和 w
V	逆变速度
V_r	相对于运动网格的逆变速度
V_t	控制体的面的逆变速度
W_m	组分 m 的分子量
$\boldsymbol{W}$	守恒变量矢量($=[\rho, \rho u, \rho v, \rho w, \rho E]^T$)
$\boldsymbol{W}_p$	原始变量矢量($=[p, u, v, w, T]^T$)
x, y, z	笛卡儿坐标系
Δx	网格单元在 x 方向的尺寸
y^+	无量纲壁面坐标($=\rho y u_\tau/\mu_w$)

续表

Y_m	组分 m 的质量分数
z	空间算子的傅里叶记号
α	迎角，入口角
α_m	龙格-库塔法（第 m 步的）系数
β	隐式格式中控制时间精度的参数
β_m	（龙格-库塔法第 m 步的）混合系数
γ	比热比（比定压热容与比定容热容之比）
Γ	环量
$\boldsymbol{\Gamma}$	（低马赫数流动的）预处理矩阵
δ_{ij}	Kronecker 符号
ϵ	湍能耗散率；（隐式残差光顺的）光滑系数；参数
κ	热扩散系数
λ	第二黏性系数
Λ_c	对流通量雅可比矩阵的特征值
$\boldsymbol{\Lambda}_c$	对流通量雅可比矩阵的特征值对角矩阵
$\hat{\Lambda}_c$	对流通量雅可比矩阵的谱半径
$\hat{\Lambda}_v$	黏性通量雅可比矩阵的谱半径
μ	动力学黏性系数
ν	运动学黏性系数（$=\mu/\rho$）
ξ,η,ζ	曲线坐标系
ρ	密度
σ	CFL 数
σ^*	使用残值光顺的 CFL 数
τ	黏性应力
τ_w	壁面剪切应力
$\boldsymbol{\tau}$	黏性应力张量（正应力和剪切应力）
τ_{ij}	黏性应力张量的分量
τ_{ij}^F	Favre 平均雷诺应力张量的分量
τ_{ij}^R	雷诺应力张量的分量
τ_{ij}^S	亚格子尺度应力张量的分量

续表

τ_{ij}^{SF}	Favre 滤波的亚格子尺度应力张量的分量
τ_{ij}^{SR}	亚格子尺度雷诺应力张量的分量
ω	单位湍动能的耗散率($=\epsilon/K$)
Υ	压力探测器
Ω	控制体
Ω_{ij}	旋转率张量的分量
$\partial\Omega$	控制体的边界
Ψ	限制器函数
∇U	标量 U 的梯度$\left(=\left[\frac{\partial U}{\partial x},\frac{\partial U}{\partial y},\frac{\partial U}{\partial z}\right]\right)$
$\nabla^2 U$	标量 U 的拉普拉斯算子$\left(=\frac{\partial^2 U}{\partial x^2}+\frac{\partial^2 U}{\partial y^2}+\frac{\partial^2 U}{\partial z^2}\right)$
$\lVert \boldsymbol{U} \rVert_2$	矢量 $\boldsymbol{U}$ 的二范数($=\sqrt{\boldsymbol{U}\cdot\boldsymbol{U}}$)

下标

C	对流部分
c	与对流相关
D	扩散部分
i,j,k	节点的索引
I,J,K	控制体的索引
L	层流;左
m	控制体面的索引;组分
R	右
T	湍流
v	黏性部分
V	体积相关的
w	壁面
x,y,z	在 x,y,z 方向的分量
∞	在无穷远处(远场)

上标

I,J,K	计算空间中的方向
n	前一时间层
$n+1$	新时间层
T	转置
~	Favre 平均的平均值;(LES 的)Favre 滤波值
″	Favre 分解的波动部分;(LES 的)亚格子尺度
–	雷诺平均的平均值;(LES 的)滤波值
′	雷诺分解的波动部分;(LES 的)亚格子尺度

缩略语

AGARD	(北约)航空航天研究与发展顾问组
AIAA	美国航空航天学会
ARC	英国航空研究委员会
ASME	美国机械工程师协会
CERCA	加拿大蒙特利尔计算及应用研究中心
CERFACS	(法国)欧洲科学计算研究中心
DFVLR	德国航空航天研究试验院(现为德国航空航天中心)
DLR	德国航空航天中心
ERCOFTAC	欧洲流动、湍流和燃烧研究共同体
ESA	欧洲空间局
FFA	瑞典航空研究所
GAMM	德国应用数学和力学学会
ICASE	美国太空总署兰利研究中心科学与工程计算机应用研究所
INRIA	法国国家信息与自动化研究所
ISABE	国际吸气式发动机协会
MAE	美国普林斯顿大学机械与航空航天工程系
NACA	美国国家航空咨询委员会(现为美国太空总署)
NASA	美国太空总署
NLR	荷兰国家航空航天实验室
ONERA	法国宇航院
SIAM	美国工业与应用数学学会
VKI	比利时冯·卡门流体动力学研究所
ZAMM	德国应用数学与力学学报
ZFW	德国航空与空间研究杂志
1D	一维
2D	二维
3D	三维

目　录

第1章 引 言

计算流体力学(computational fluid dynamics,CFD)的历史始于20世纪70年代初,大约自那时起,CFD成为了物理学、计算数学以及某种程度上的计算机科学组合的缩略词——这些学科都被用来模拟流体流动。CFD的产生是由于越来越强大的大型计算机的出现,而CFD的发展也与计算机技术的发展密切相关。CFD的第一个应用是基于求解非线性位势方程的跨声速流动模拟。20世纪80年代初,二维(2D)欧拉(Euler)方程组和三维(3D)欧拉方程组求解先后具备可行性。由于超级计算机速度的迅速提高,以及多种数值加速技术(如多重网格)的发展,飞机全机无黏绕流和涡轮机械内流的无黏流计算成为可能。到了20世纪80年代中期,研究的重点开始转向由纳维-斯托克斯(Navier-Stokes)方程组控制的要求更苛刻的黏性流动模拟。与此同时,具备不同数值复杂程度和精度的众多湍流模型也发展起来,其先进代表是直接数值模拟(direct numerical simulation,DNS)和大涡模拟(large eddy simulation,LES)技术。

随着数值方法,特别是隐式格式的发展,到20世纪80年代末,考虑真实气体(real gas)模型的流动问题求解也变得可行。在第一次大规模应用中,3D再入飞行器的高超声速绕流,如欧洲HERMES航天飞机,先后使用平衡化学模型和非平衡化学模型进行了计算。一直以来,许多研究活动致力于燃烧,特别是火焰模型的数值模拟,这些工作对于低排放燃气轮机和发动机的发展非常重要。此外,蒸汽的建模,尤其是蒸汽冷凝成为设计高效汽轮机的关键因素。

由于流动模拟的复杂性和对精度要求的不断提高,网格生成方法变得越来越复杂。网格生成技术的发展始于相对简单的结构网格,其由代数方法或偏微分方程求解方法构建。随着外形几何复杂性的增加,网格被划分成许多拓扑上更简单的块(多块方法)。随后发展出允许非对接关系的多块网格,以减轻对单块网格生成的限制。最后则发展出了重叠网格处理方法(Chimera技术),例如用于模拟完整的航天飞机与外部燃料箱和助推器连接的绕流流动问题。尽管如此,生成一个复杂几何外形的结构多块网格可能仍然需要数周的时间,因此,研究也集中到非结构网格生成器和流场解算器的发展上,这将显著减少生成时间,且只需很少的用户干预。非结构方法的另一个非常重要的特性是基于解的网格自适应的可能性。第一种非结构网格只由各向同性的四面体组成,它完全满足由欧拉方程控制的无黏流计算,然而,求解高雷诺数(Re)纳维-斯托克斯方程组时在剪切层需要使用极度拉伸的网格,虽然这种网格也可以由四面体单元构成,但通常建议在黏性流区域使用三棱柱或六面体单元,而在外部使用四面体单元。这样不仅能提高解的精度,而且还节省了单元、面和边的数量。因此,数值模拟的内存和运行时间需求也显著降低。

时至今日,CFD 在飞行器、涡轮机械、汽车和船舶设计领域的使用已司空见惯。此外,CFD 还应用于气象学、海洋学、天体物理学、生物学、石油开采和建筑设计等领域。许多为 CFD 开发的数值技术也被用于求解麦克斯韦方程组或气动声学中。因此,CFD 已经成为工程项目中重要的设计工具,也是各种科学中不可缺少的研究工具。由于数值求解方法和计算机技术的进步,几何和物理上复杂的项目甚至可以在 PC 或 PC 集群上运行。在如今的超级计算机上,由数千万单元组成的网格上的黏性流动的大规模模拟只需几个小时就能完成。然而,如果认为如今的 CFD 代表着一种成熟的技术,像固体力学中的有限元方法那样,那就大错特错了。其实,仍然还有许多有待研究的问题,如湍流和燃烧建模、传热、黏性流动的高效求解技术、鲁棒而又精确的离散方法、自动网格生成器等。CFD 与其他学科(如固体力学)的耦合也需要进一步的研究。利用 CFD 进行设计优化也带来了新的机遇。

本书的目标是为大学生的学习提供一个坚实的基础,以理解当今 CFD 中使用的数值方法,并通过实践经验使他们熟悉现代 CFD 程序。本书也适合于即将在 CFD 领域开展工作,或将开始使用 CFD 程序的工程师和科学家。本书所使用的数学方法总是与基础物理知识联系在一起,以促进对问题的理解。本书也可作为参考手册,每一章都有广泛的参考书目,可以作为进一步研究的基础。

CFD 关注流体运动方程的解以及流体与固体之间的相互作用。第 2 章推导了无黏流体(欧拉方程组)和黏性流体(纳维-斯托克斯方程组)运动的控制方程组,另外还讨论了理想气体和真实气体的热力学关系。第 3 章讨论了控制方程组的求解原理,简要介绍了其中最重要的方法,并提供了相应的参考文献。第 3 章可以与第 2 章结合使用,以了解 CFD 的基本原理。

以往人们发展了许多欧拉方程组和纳维-斯托克斯方程组的空间离散格式。本书的一个独特之处在于,既涉及了结构的(第 4 章),也涉及了非结构的

(第5章)有限体积法,因为它们都具有广泛的应用可能性,特别是在处理工业环境中经常遇到的复杂流体问题时。本书特别关注了对流和黏性通量空间离散方法使用的各种类型控制体的定义,还详细介绍了最流行的中心格式和迎风格式的3D有限体积形式。

控制方程组的时间离散方法主要分为两类,一类是显式时间推进方法(6.1节),另一类是隐式时间推进方法(6.2节)。为了提供一个更完整的概述,本书不但讨论了标准求解技术,如显式龙格-库塔(Runge-Kutta)法,还讨论了最近发展的基于牛顿迭代(Newton-iteration)的求解方法。

通常我们会遇到:层流和湍流两种性质不同的黏性流体流动。在层流情况下,纳维-斯托克斯方程组的解不会出现任何根本性的困难,然而,湍流的模拟仍然像以前一样是一个重大的挑战。湍流的一种相对简单的模拟方法是求解雷诺平均纳维-斯托克斯(Reynolds-averaged Navier-Stokes,RANS)方程组,另外,雷诺应力模型或者LES能相对更准确地预测湍流。第7章详细介绍了各种经过验证并广泛应用的复杂程度不同的湍流模型。

为了考虑特定问题的特定特征,求得控制方程组的唯一解,有必要指定适当的边界条件。从根本上来说,存在两类边界条件:物理边界条件和数值边界条件。第8章讨论了这两类边界条件在不同的情况下的应用,例如固壁、入口、出口、入射和远场边界。对称面、周期和块边界也做了介绍。

为了减少求解复杂流动问题的控制方程组所需的计算时间,采用数值加速技术是非常必要的。第9章讨论了隐式残差光顺和多重网格等技术,还描述了另外一个重要的方法,即控制方程组的预处理,当流动马赫数在接近零和跨声速或者更高值之间变化时,它允许应用单一的数值格式。9.6节介绍了使用不同方法对数值计算程序进行并行化。

控制方程组的每种离散都会引入一定的误差——离散误差。为了确保离散方程的解足够近似于原始方程的解,离散格式必须满足几个相容性要求。这个问题在第10章的前两部分讨论。在实施一种特定的数值求解方法之前,至少大致了解该方法将如何影响CFD程序的稳定性和收敛性是很重要的。经常被证实的是,冯·诺依曼(John von Neumann)稳定性分析可以为数值格式的特性提供一个很好的评估。因此,10.3节针对各种模型方程进行了稳定性分析。

CFD中一个具有挑战性的任务是在复杂几何外形周围生成结构或非结构的贴体网格。网格被用于在空间中离散控制方程组,因此,流动解的精度与网格的质量密切相关。第11章深入讨论了生成结构和非结构网格的最主要的方法。

为了演示不同数值求解方法的实际特性,本书给出了多个源代码供下载使用,包括准一维(1D)欧拉,以及2D欧拉和纳维-斯托克斯方程组的结构和非结构流场解算器源代码,还包括2D结构代数和椭圆网格生成器的源程序,以及结构网格到非结构网格的转换器源程序,此外,还给出了显式和隐式时间格式的线性稳定性分析程序。源程序由一组求解出的示例完成,包括网格、输入文件和计算结果,程序包还包括几个演示并行化技术的程序。第12章描述了源代码文件目录的内容、特定程序的功能,并提供了它们的用法示例。

本书附录包含了微分形式的控制方程组及其特性,讨论了用于运动网格的旋转坐标系下的控制方程组公式,并给出了控制方程组的一些简化形式。此外,给出了2D和3D雅可比(Jacobian)矩阵以及守恒变量到特征变量的转换矩阵,还介绍了求解线性方程组的GMRES共轭梯度法,附录A.13提供了张量符号的简要说明。

第2章
控制方程

2.1 流动及其数学描述

在开始推导描述流体特性的基本方程之前,最好先阐明一下"流体动力学"一词的含义。事实上,它是对大量单个粒子相互作用运动的研究,在我们的研究对象中,这些粒子是分子或原子。这意味着,如果假设流体的密度足够高,它就可以近似为一个连续体,即使流体的一个无限小的(在微分学意义上的)单元仍然包含足够多的粒子,从而我们可以为其指定平均速度和平均动能。这样,我们就可以定义流体中每一点的速度、压力、温度、密度和其他重要的量。

流体动力学基本方程组的推导基于以下事实,即流体的动力学行为由下列守恒定律决定:

(1) 质量守恒;

(2) 动量守恒;

(3) 能量守恒。

某一流动变量的守恒意味着它在任意体积内的总变化量可以表示为流过边界的量、任何内力和源,以及作用于体积的外力的净效应。流过边界的量称为通量,一般可以分解为两个不同的部分:一部分与对流输运相关,另一部分则是由于静止流体中存在的分子运动造成的。第二部分具有扩散性质——它与所考虑的量的梯度成正比,因而对于均匀分布的量这一部分将消失。

对守恒定律的讨论很自然地使我们想到把流场分成若干个体积,并集中于一个有限的区域内流体行为的模拟。为此,我们将定义有限控制体,并试图发展对其物理性质的数学描述。

2.1.1 有限控制体

考虑图 2.1 中以流线表示的一般流场，由封闭曲面 $\partial\Omega$ 为边界并固定在空间中的流动的一个任意有限区域，定义了控制体 Ω。我们再引入面单元 dS 及其相关的指向外的单位法向矢量（法矢）$\boldsymbol{n}$。将守恒定律应用于单位体积的标量 U，它在 Ω 内随时间的变化表示为

$$\frac{\partial}{\partial t}\int_{\Omega} U \mathrm{d}\Omega$$

对流通量是 U 以速度 $\boldsymbol{v}$ 通过边界面进入控制体的量为

$$-\oint_{\partial\Omega} U(\boldsymbol{v}\cdot\boldsymbol{n})\,\mathrm{d}S$$

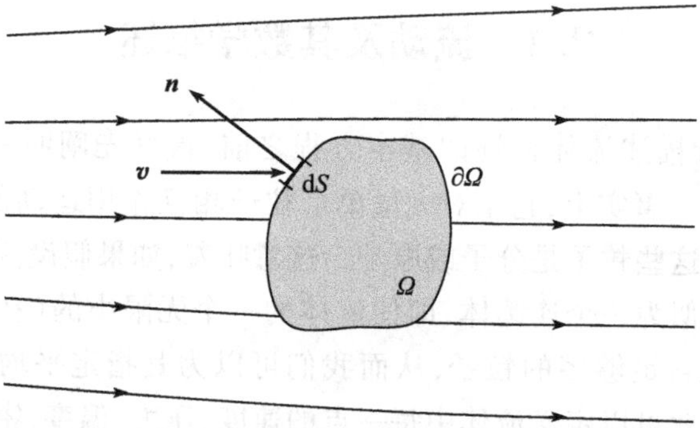

图 2.1 （固定在空间中的）有限控制体的定义

用广义菲克（Fick）梯度定律表示的扩散通量为

$$\oint_{\partial\Omega} \kappa\rho[\nabla(U/\rho)\cdot\boldsymbol{n}]\,\mathrm{d}S$$

其中，κ 是热扩散系数。最后，体积源项 Q_V 以及面源项 $\boldsymbol{Q}_S$ 的贡献为

$$\int_{\Omega} Q_V \mathrm{d}\Omega + \oint_{\partial\Omega}\cdot(\boldsymbol{Q}_S\cdot\boldsymbol{n})\,\mathrm{d}S$$

对上述贡献进行求和，得到如下标量 U 的守恒定律的一般形式

$$\frac{\partial}{\partial t}\int_{\Omega} U\mathrm{d}\Omega + \oint_{\partial\Omega}[U(\boldsymbol{v}\cdot\boldsymbol{n}) - \kappa\rho(\nabla U^{*}\cdot\boldsymbol{n})]\,\mathrm{d}S = \int_{\Omega} Q_V\mathrm{d}\Omega + \oint_{\partial\Omega}(\boldsymbol{Q}_S\cdot\boldsymbol{n})\,\mathrm{d}S \tag{2.1}$$

其中，U^{*} 表示单位质量的 U，即 U/ρ。

值得注意的是，如果守恒量是矢量而不是标量，在形式上，式（2.1）仍然有效。但不同的是，对流通量和扩散通量会由矢量变为张量——$\boldsymbol{F}_C$ 表示对流通量张量，$\boldsymbol{F}_D$ 表示扩散通量张量。体积源项变为矢量 $\boldsymbol{Q}_V$，面源项将变为张量 $\boldsymbol{Q}_S$。

因此可以写出一般矢量 $\boldsymbol{U}$ 的守恒定律为

$$\frac{\partial}{\partial t}\int_{\Omega}\boldsymbol{U}\mathrm{d}\Omega+\oint_{\partial\Omega}[(\boldsymbol{F}_{\mathrm{C}}-\boldsymbol{F}_{\mathrm{D}})\cdot\boldsymbol{n}]\mathrm{d}S=\int_{\Omega}\boldsymbol{Q}_{\mathrm{V}}\mathrm{d}\Omega+\oint_{\partial\Omega}(\boldsymbol{Q}_{\mathrm{S}}\cdot\boldsymbol{n})\mathrm{d}S \tag{2.2}$$

式(2.1)或式(2.2)给出了守恒定律的积分形式,它具有两个非常重要且十分有用的特性:

(1) 如果没有体积源项存在,U 的变化仅取决于穿过边界 $\partial\Omega$ 的通量,而不依赖于控制体积 Ω 内部的任何通量。

(2) 在存在间断的流场,比如激波或接触间断,这种形式仍然有效[1]。

由于守恒定律积分形式的通用性,以及上述非常有用的特性,如今大多数 CFD 程序都基于控制方程的积分形式也就不奇怪了。

在 2.2 节中,将利用上述积分形式来推导流体力学中三个守恒定律的相应表达式。

2.2 守恒定律

2.2.1 连续性方程

如果将关注点聚焦于单相流体上,质量守恒定律表明,在这样的流体系统中,质量不可能产生,也不可能消失。因为对于静止的流体,质量的任何变化都意味着流体粒子的位移,因此扩散通量对连续性方程没有贡献。

为了推导连续性方程,我们考虑空间位置固定的有限控制体模型,如图 2.1 所示。在控制面上的某一点,流动速度为 $\boldsymbol{v}$,单位法向矢量为 $\boldsymbol{n}$,$\mathrm{d}S$ 为微元面的面积。此时守恒变量是密度 ρ。对于有限体积 Ω 内总质量的时间变化率,有

$$\frac{\partial}{\partial t}\int_{\Omega}\rho\mathrm{d}\Omega$$

流体通过空间中固定面的质量流量=密度×面积×垂直于该面的速度分量。因此,通过单元面 $\mathrm{d}S$ 的对流通量为

$$\rho(\boldsymbol{v}\cdot\boldsymbol{n})\mathrm{d}S$$

通常,$\boldsymbol{n}$ 总是指向控制体之外,如果$(\boldsymbol{v}\cdot\boldsymbol{n})$的乘积为负,我们称为入流,如果为正,则为出流,即质量离开控制体。

如上所述,在连续性方程中没有体积或面的源项存在。因此,根据式(2.1)的一般公式,可得

$$\frac{\partial}{\partial t}\int_{\Omega}\rho \mathrm{d}\Omega + \oint_{\partial\Omega}\rho(\boldsymbol{v}\cdot\boldsymbol{n})\mathrm{d}S = 0 \tag{2.3}$$

此即为连续性方程——质量守恒定律的积分形式。

2.2.2 动量方程

我们通过回顾牛顿第二定律的特殊形式来开始动量方程的推导，牛顿第二定律指出，动量的变化是由作用在质量单元上的合力引起的。控制体 Ω（图 2.1）内无限小的微元的动量为

$$\rho\boldsymbol{v}\mathrm{d}\Omega$$

则控制体内动量随时间的变化等于

$$\frac{\partial}{\partial t}\int_{\Omega}\rho\boldsymbol{v}\mathrm{d}\Omega$$

因而，在此守恒变量是密度和速度的乘积，即

$$\rho\boldsymbol{v}=[\rho u,\rho v,\rho w]^{\mathrm{T}}$$

描述动量越过控制体边界转移的对流通量张量，在笛卡儿坐标系中包含以下三个分量：

$$x\text{ 方向分量}:\rho u\boldsymbol{v}$$
$$y\text{ 方向分量}:\rho v\boldsymbol{v}$$
$$z\text{ 方向分量}:\rho w\boldsymbol{v}$$

对流通量张量对动量守恒的贡献为

$$-\oint_{\partial\Omega}\rho\boldsymbol{v}(\boldsymbol{v}\cdot\boldsymbol{n})\mathrm{d}S$$

因为静止的流体中不可能有动量扩散，因此扩散通量为零。现在剩下的问题是，流体单元受到的力有哪些？我们可以确定作用在控制体上有两种力：

(1) 体积外力，直接作用于控制体上，例如，重力、浮力、科里奥利（Coriolis）力（科氏力）或离心力，在某些情况下，也可能存在电磁力。

(2) 面力，直接作用于控制体表面，只来源于以下两个方面：

①控制体周围的流体施加的压力分布；

②流体与控制体表面摩擦产生的切向应力和法向应力。

由上述可知，单位体积的体积力对应式（2.2）中的体积源项，表示为 $\rho\boldsymbol{f}_{\mathrm{e}}$。因此，体积（外）力对动量守恒的贡献为

$$\int_{\Omega}\rho\boldsymbol{f}_{\mathrm{e}}\mathrm{d}\Omega$$

面源项由两部分组成——各向同性压力分量和黏性应力张量 $\boldsymbol{\tau}$，即

$$\boldsymbol{Q}_{\mathrm{S}}=-p\boldsymbol{I}+\boldsymbol{\tau} \tag{2.4}$$

$\boldsymbol{I}$ 是单位张量（关于张量，可参阅文献[2]）。面源项对控制体的影响如图 2.2 所示。在 2.3 节中，将更详细地阐述应力张量的形式，并重点介绍法向应力和切向应力与流动速度的关系。

因此，根据一般守恒定律（式(2.2)），将上述所有量相加，最终得到固定于空间的任意控制体 Ω 的动量守恒表达式为

$$\frac{\partial}{\partial t}\int_{\Omega}\rho\boldsymbol{v}\mathrm{d}\Omega+\oint_{\partial\Omega}\rho\boldsymbol{v}(\boldsymbol{v}\cdot\boldsymbol{n})\mathrm{d}S=\int_{\Omega}\rho\boldsymbol{f}_{\mathrm{e}}\mathrm{d}\Omega-\oint_{\partial\Omega}p\boldsymbol{n}\mathrm{d}S+\oint_{\partial\Omega}(\boldsymbol{\tau}\cdot\boldsymbol{n})\mathrm{d}S \tag{2.5}$$

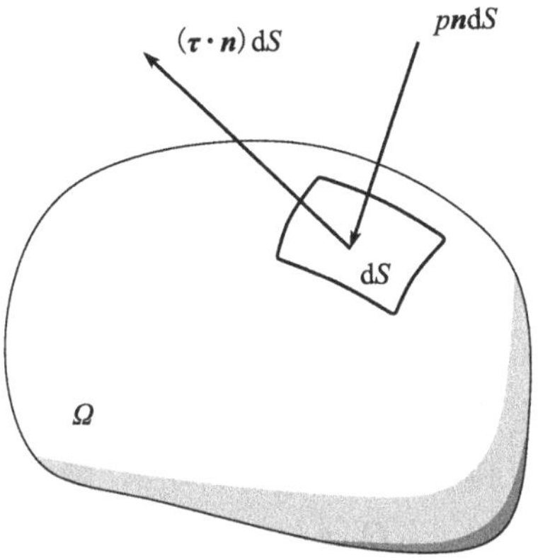

图 2.2　作用于控制体面元的面力

2.2.3　能量方程

推导能量方程应用的基本原理是热力学第一定律。对于图 2.1 所示的控制体，它表明控制体内总能量随时间的变化是由作用于控制体的力做功的功率和进入体积的净热通量引起的。流体的单位质量的总能 E 由它的单位质量的内能 e 加上单位质量的动能 $|\boldsymbol{v}|^2/2$ 得到，因此，总能为

$$E=e+\frac{|\boldsymbol{v}|^2}{2}=e+\frac{u^2+v^2+w^2}{2} \tag{2.6}$$

在此，守恒变量是单位体积的总能，即 ρE。控制体 Ω 内 ρE 随时间的变化可以表示为

$$\frac{\partial}{\partial t}\int_{\Omega}\rho E\mathrm{d}\Omega$$

根据推导一般守恒定律（式(2.1)）过程中的讨论，可以很容易地将对流通量的贡献表示为

$$-\oint_{\partial\Omega}\rho E(\boldsymbol{v}\cdot\boldsymbol{n})\mathrm{d}S$$

与连续性方程和动量方程相反，能量方程存在扩散通量，正如之前所说，它与单位质量的守恒变量的梯度成正比（菲克定律）。由于扩散通量 $\boldsymbol{F}_{\mathrm{D}}$ 是针对静止流体定义的，从而只有内能对其有影响，因此可以得到

$$\boldsymbol{F}_{\mathrm{D}}=-\gamma\rho\kappa\nabla e \tag{2.7}$$

式中：$\gamma=c_p/c_V$ 为比热比；κ 为热扩散系数。扩散通量代表进入控制体的一部分热通量，即由分子热传导引起的热扩散——由温度梯度引起的热量传递。因此，式(2.7)一般写成傅里叶(Fourier)热传导定律的形式，即

$$\boldsymbol{F}_{\mathrm{D}}=-k\nabla T \tag{2.8}$$

其中，k 是热传导系数，T 是绝对静温。

另一部分进入有限控制体的净热通量由辐射的吸收或发射，或者化学反应而产生的体积加热构成。用 $\dot{q}_{\mathrm{h}}$ 表示热源——单位质量的传热率，再加上在动量方程中引入的体积力 $\boldsymbol{f}_{\mathrm{e}}$ 做功的功率，得到完整的体积源项

$$Q_{\mathrm{V}}=\rho\boldsymbol{f}_{\mathrm{e}}\cdot\boldsymbol{v}+\dot{q}_{\mathrm{h}} \tag{2.9}$$

能量守恒还没有确定的最后一个贡献是面源项 $\boldsymbol{Q}_{\mathrm{S}}$，它对应于作用在流体单元上的压力以及切向应力和法向应力做功的功率（图2.2），即

$$\boldsymbol{Q}_{\mathrm{S}}=-p\boldsymbol{v}+\boldsymbol{\tau}\cdot\boldsymbol{v} \tag{2.10}$$

整理上述所有贡献项，得到能量守恒方程为

$$\frac{\partial}{\partial t}\int_{\Omega}\rho E\mathrm{d}\Omega+\oint_{\partial\Omega}\rho E(\boldsymbol{v}\cdot\boldsymbol{n})\mathrm{d}S=\oint_{\partial\Omega}k(\nabla T\cdot\boldsymbol{n})\mathrm{d}S+\int_{\Omega}(\rho\boldsymbol{f}_{\mathrm{e}}\cdot\boldsymbol{v}+\dot{q}_{\mathrm{h}})\mathrm{d}\Omega$$
$$-\oint_{\partial\Omega}p(\boldsymbol{v}\cdot\boldsymbol{n})\mathrm{d}S+\oint_{\partial\Omega}(\boldsymbol{\tau}\cdot\boldsymbol{v})\cdot\boldsymbol{n}\mathrm{d}S \tag{2.11}$$

能量方程(2.11)通常被写成稍微不同的形式。为此目的，利用下面总焓、总能和压力之间的一般关系：

$$H=h+\frac{|\boldsymbol{v}|^2}{2}=E+\frac{p}{\rho} \tag{2.12}$$

把能量方程(2.11)中的对流项($\rho E\boldsymbol{v}$)和压力项($p\boldsymbol{v}$)组合起来，并利用式(2.12)，就得到以下能量方程的最终形式：

$$\frac{\partial}{\partial t}\int_{\Omega}\rho E\mathrm{d}\Omega+\oint_{\partial\Omega}\rho H(\boldsymbol{v}\cdot\boldsymbol{n})\mathrm{d}S=\oint_{\partial\Omega}k(\nabla T\cdot\boldsymbol{n})\mathrm{d}S$$
$$+\int_{\Omega}(\rho\boldsymbol{f}_{\mathrm{e}}\cdot\boldsymbol{v}+\dot{q}_{\mathrm{h}})\mathrm{d}\Omega+\oint_{\partial\Omega}(\boldsymbol{\tau}\cdot\boldsymbol{v})\cdot\boldsymbol{n}\mathrm{d}S \tag{2.13}$$

从而,推导出了三个守恒定律的积分公式:质量守恒方程(2.3)、动量守恒方程(2.5)和能量守恒方程(2.13)。2.3 节中,将更详细地推导法向应力和切向应力的公式。

2.3 黏性应力

黏性应力来源于流体和单元表面之间的摩擦,用应力张量 $\boldsymbol{\tau}$ 来表示。在笛卡儿坐标系中它的一般形式为

$$\boldsymbol{\tau}=\begin{bmatrix}\tau_{xx} & \tau_{xy} & \tau_{xz}\\ \tau_{yx} & \tau_{yy} & \tau_{yz}\\ \tau_{zx} & \tau_{zy} & \tau_{zz}\end{bmatrix} \tag{2.14}$$

通常,符号 τ_{ij} 表示作用于垂直 j 轴的平面的应力在 i 方向的分量。τ_{xx}、τ_{yy} 和 τ_{zz} 代表法向应力,$\boldsymbol{\tau}$ 的其他分量代表切向应力。图 2.3 显示了四边形流体单元的应力,可以看到,法向应力(图 2.3(a))试图在三个相互垂直的方向上移动单元的面,而切向应力(图 2.3(b))则试图剪切单元。

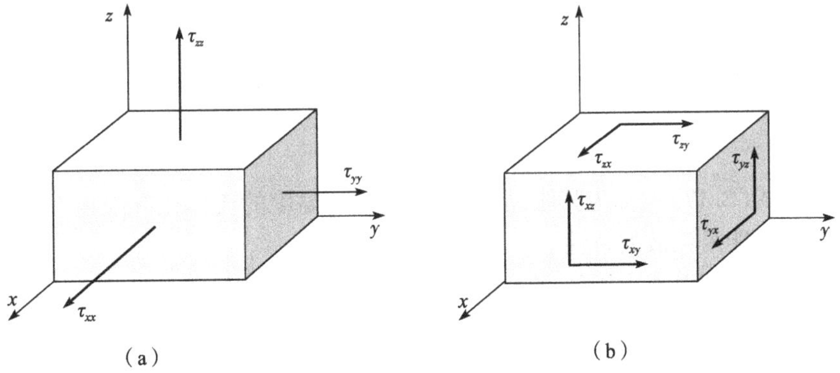

图 2.3 作用于一个有限流体单元的应力

(a) 法向应力;(b) 切向应力。

现在,你可能会问,黏性应力是如何计算的呢?首先,它依赖于介质的动力学特性,对于像空气或水这样的流体,牛顿认为切向应力与速度梯度成正比,因此,这类介质被称为牛顿流体。另一类流体,如熔化的塑料或血液,表现出不同的性质——它们是非牛顿流体。在绝大多数的实际问题中,流体可以假定为牛顿流体,黏性应力张量的分量由下列关系定义[3-4]:

$$
\begin{aligned}
\tau_{xx} &= \lambda\left(\frac{\partial u}{\partial x}+\frac{\partial v}{\partial y}+\frac{\partial w}{\partial z}\right)+2\mu\frac{\partial u}{\partial x}\\
\tau_{yy} &= \lambda\left(\frac{\partial u}{\partial x}+\frac{\partial v}{\partial y}+\frac{\partial w}{\partial z}\right)+2\mu\frac{\partial v}{\partial y}\\
\tau_{zz} &= \lambda\left(\frac{\partial u}{\partial x}+\frac{\partial v}{\partial y}+\frac{\partial w}{\partial z}\right)+2\mu\frac{\partial w}{\partial z}\\
\tau_{xy} &= \tau_{yx}=\mu\left(\frac{\partial u}{\partial y}+\frac{\partial v}{\partial x}\right)\\
\tau_{xz} &= \tau_{zx}=\mu\left(\frac{\partial u}{\partial z}+\frac{\partial w}{\partial x}\right)\\
\tau_{yz} &= \tau_{zy}=\mu\left(\frac{\partial v}{\partial z}+\frac{\partial w}{\partial y}\right)
\end{aligned}
\tag{2.15}
$$

其中，λ 为第二黏性系数，μ 为动力学黏性系数。为方便起见，还可以定义运动学黏性系数

$$
\nu=\mu/\rho \tag{2.16}
$$

式(2.15)是英国人乔治·斯托克斯(George Stokes)在19世纪中叶推导出来的。法向应力 $\mu(\partial u/\partial x)$ 等项代表线性膨胀的速率——形状的变化。另外，式(2.15)中 $\lambda(\mathrm{div}\boldsymbol{v})$ 项表示体积膨胀——体积变化率，实质上是密度的变化。

为了封闭法向应力的表达式，斯托克斯引入了假设[5]：

$$
\lambda+\frac{2}{3}\mu=0 \tag{2.17}
$$

关系式(2.17)被称为膨胀黏性，它代表均匀温度的流体在有限速率的体积变化过程中能量耗散的特性。

除了极端高温或高压外，迄今为止还没有实验证据证明式(2.17)的斯托克斯假设不成立(参见文献[6]中的讨论)。因此，通常使用式(2.17)将式(2.15)中的 λ 消去，从而得到法向黏性应力公式为

$$
\begin{aligned}
\tau_{xx} &= 2\mu\left(\frac{\partial u}{\partial x}-\frac{1}{3}\mathrm{div}\boldsymbol{v}\right)\\
\tau_{yy} &= 2\mu\left(\frac{\partial v}{\partial y}-\frac{1}{3}\mathrm{div}\boldsymbol{v}\right)\\
\tau_{zz} &= 2\mu\left(\frac{\partial w}{\partial z}-\frac{1}{3}\mathrm{div}\boldsymbol{v}\right)
\end{aligned}
\tag{2.18}
$$

需要注意的是，对于不可压缩流体(密度不变)，由于 $\mathrm{div}\boldsymbol{v}=0$(连续性方程)，式(2.18)中的法向应力表达式可以简化。

尚待确定的还有作为流体状态函数的黏性系数 μ 和热传导系数 k，在连续

介质力学的框架内它们仅仅建立于经验假设的基础上，下面章节中将对这个问题进行讨论。

2.4 纳维-斯托克斯方程组的完整形式

前几节分别推导了质量守恒定律、动量守恒定律和能量守恒定律，现在，把它们集合到一个方程组中，以便于更好地了解所涉及的各项。为此，我们回到一般矢量守恒定律，如式(2.2)所示。由于一些将在后面解释的原因，在此引入两个通量矢量 $\boldsymbol{F}_{\mathrm{c}}$ 和 $\boldsymbol{F}_{\mathrm{v}}$。其中 $\boldsymbol{F}_{\mathrm{c}}$ 与流体中量的对流输运相关，它通常被称为对流通量的矢量，尽管对于动量和能量方程，它还分别包括压力项 $p\boldsymbol{n}$（式(2.5)）和 $p(\boldsymbol{v}\cdot\boldsymbol{n})$（式(2.11)）。第二个通量矢量 $\boldsymbol{F}_{\mathrm{v}}$ 称为黏性通量矢量，包含黏性应力以及热扩散。此外，还定义一个源项 $\boldsymbol{Q}$，它包括由体积力和体积加热而产生的所有体积源。考虑到所有这些，并针对单位法矢 $\boldsymbol{n}$ 求标量积，然后将式(2.2)和式(2.3)、式(2.5)、式(2.13)组合在一起，得到

$$\frac{\partial}{\partial t}\int_{\Omega}\boldsymbol{W}\mathrm{d}\Omega+\oint_{\partial\Omega}(\boldsymbol{F}_{\mathrm{c}}-\boldsymbol{F}_{\mathrm{v}})\mathrm{d}S=\int_{\Omega}\boldsymbol{Q}\mathrm{d}\Omega \tag{2.19}$$

守恒变量 $\boldsymbol{W}$ 在3D情况下由下面的五个分量组成：

$$\boldsymbol{W}=\begin{bmatrix}\rho\\ \rho u\\ \rho v\\ \rho w\\ \rho E\end{bmatrix} \tag{2.20}$$

对流通量为

$$\boldsymbol{F}_{\mathrm{c}}=\begin{bmatrix}\rho V\\ \rho uV+n_xp\\ \rho vV+n_yp\\ \rho wV+n_zp\\ \rho HV\end{bmatrix} \tag{2.21}$$

垂直于面元 dS 的速度 V 称为逆变速度，它定义为速度矢量与单位法矢的标量积，即

$$V\equiv\boldsymbol{v}\cdot\boldsymbol{n}=n_xu+n_yv+n_zw \tag{2.22}$$

式(2.21)中的总焓 H 由式(2.12)给出。对于黏性通量，由式(2.14)可知

$$\boldsymbol{F}_{\mathrm{v}}=\begin{bmatrix}0\\ n_x\tau_{xx}+n_y\tau_{xy}+n_z\tau_{xz}\\ n_x\tau_{yx}+n_y\tau_{yy}+n_z\tau_{yz}\\ n_x\tau_{zx}+n_y\tau_{zy}+n_z\tau_{zz}\\ n_x\Theta_x+n_y\Theta_y+n_z\Theta_z\end{bmatrix} \tag{2.23}$$

其中

$$\begin{aligned}\Theta_x&=u\tau_{xx}+v\tau_{xy}+w\tau_{xz}+k\frac{\partial T}{\partial x}\\ \Theta_y&=u\tau_{yx}+v\tau_{yy}+w\tau_{yz}+k\frac{\partial T}{\partial y}\\ \Theta_z&=u\tau_{zx}+v\tau_{zy}+w\tau_{zz}+k\frac{\partial T}{\partial z}\end{aligned} \tag{2.24}$$

描述了流体中黏性应力和热传导所做的功。最后，源项为

$$\boldsymbol{Q}=\begin{bmatrix}0\\ \rho f_{\mathrm{e},x}\\ \rho f_{\mathrm{e},y}\\ \rho f_{\mathrm{e},z}\\ \rho\boldsymbol{f}_{\mathrm{e}}\cdot\boldsymbol{v}+\dot{q}_{\mathrm{h}}\end{bmatrix} \tag{2.25}$$

如果所考虑的流体为牛顿流体，即黏性应力的关系式(2.15)成立，方程组(2.19)称为纳维-斯托克斯方程组，它描述了通过固定在空间中的控制体 Ω(见图 2.1)的边界 $\partial\Omega$ 的质量、动量和能量的交换(通量)。这样，我们按照守恒定律推导出了积分形式的纳维-斯托克斯方程组，应用高斯定理，式(2.19)可以改写成微分形式[7]。微分形式的控制方程组在文献中经常看到，为了完整起见，将它放在附录 A.1 节中。

在某些情况下，例如叶轮机械应用或地球物理中，控制体围绕某一轴(通常是有规律地)旋转，在这种情况下，可以将纳维-斯托克斯方程组转换到旋转参考系下。对此，源项 $\boldsymbol{Q}$ 中必须增加科氏力和离心力的影响[8]，旋转坐标系下的纳维-斯托克斯方程组的形式见附录 A.4 节。在有些情况下，控制体还可能出现平移或者变形，例如研究流体与结构的相互作用时，就可能发生这种情况，对此，必须对纳维-斯托克斯方程组(2.19)增加一项，以描述面元 $\mathrm{d}S$ 相对于固定坐标系的相对运动[9]，此外，还必须满足几何守恒律(geometric conservation law, GCL)[10-12]，我们在附录 A.5 节中给出了相应的公式。

纳维-斯托克斯方程组在 3D 情况下是由五个守恒变量 ρ，ρu，ρv，ρw 和 ρE 的

五个方程组成的方程组，但它包含七个未知的流场变量，即 ρ, u, v, w, E, p 和 T，因此，我们必须提供两个额外的方程，来描述热状态变量之间的热力学关系。例如，压力表示为密度和温度的函数，内能或焓表示为压力和温度的函数。除此之外，我们还必须提供黏性系数 μ 和热传导系数 k 作为流体状态的函数，以封闭整个方程组。显然，这些关系取决于被研究的流体对象的类型，下面针对两种常见情况说明封闭方程组的方法。

2.4.1　完全气体形式

在理论空气动力学中，通常可以合理地假定工作流体的性质类似于量热完全气体，其状态方程为[13-14]

$$p=\rho RT \tag{2.26}$$

其中，R 为气体常数。焓的形式为

$$h=c_p T \tag{2.27}$$

通常使用守恒变量来表达压力比较方便，为此，需要将总焓与总能的关系式(2.12)与状态方程(2.26)结合，用式(2.27)表示焓，并使用下面的定义

$$R=c_p-c_V, \quad \gamma=\frac{c_p}{c_V} \tag{2.28}$$

最终得到压力为

$$p=(\gamma-1)\rho\left[E-\frac{u^2+v^2+w^2}{2}\right] \tag{2.29}$$

然后借助状态方程(2.26)可以计算出温度。对于完全气体，动力学黏性系数 μ 强烈依赖于温度，而对压力的依赖性较弱，通常应用萨瑟兰(Sutherland)公式计算 μ。对于空气，(国际单位制下的)形式为

$$\mu=\frac{1.45T^{3/2}}{T+110}\times10^{-6} \tag{2.30}$$

其中，温度 T 的单位为 K，在 $T=288\text{K}$ 时，得到 $\mu=1.78\times10^{-5}\text{kg/ms}$。在气体中，热传导系数 k 的温度依赖性类似于 μ，与之相比，在液体中 k 实际上是常数，对于空气，k 的关系式为

$$k=c_p\frac{\mu}{Pr} \tag{2.31}$$

一般假定普朗特数(Pr)在整个流场中为常数，对于空气，其值为 $Pr=0.72$。

2.4.2　真实气体形式

当我们必须处理真实气体时，事情就变得非常复杂，因为除了流体力学外，

现在还必须为热力学过程和化学反应建立模型。真实气体流动的例子包括燃烧、再入飞行器的高超声速流动、蒸汽轮机内的流动等。

在理论上可以采用两种不同的方法来解决这个问题。第一种方法适用于气体处于化学平衡和热力学平衡的情况，这意味着存在一个统一的状态方程，控制方程组(2.19)保持不变，只有压力、温度、黏性等变量的值需要使用查表数据通过曲线拟合得到[15-18]。然而，在实际情况下，气体更多地处于化学和/或热力学的非平衡状态，因此必须进行相应的模拟。

举个例子，考虑由 N 种不同组分组成的气体混合物，对于有限的达姆科勒(Damkohler)数(其定义为流动停留时间与化学反应时间之比)，必须将有限速率化学反应纳入到我们的模型中，以描述由于化学反应而产生/消失的组分。接下来，进一步假设流体力学和化学反应的时间和空间尺度比热力学的尺度大得多，因此，可以假定气体处于热力学平衡而化学非平衡的状态。为了模拟这种气体混合物的特性，纳维-斯托克斯方程组必须增加 N 种组分的 $N-1$ 个附加输运方程[19-27]。因此，我们得到了形式上与式(2.19)相同的方程组，但现在使用的是由 $N-1$ 个组分方程扩展了的守恒变量 $\boldsymbol{W}$、通量 $\boldsymbol{F}_c$ 和 $\boldsymbol{F}_v$，以及源项 $\boldsymbol{Q}$。回顾式(2.20)~式(2.25)，守恒变量变为

$$\boldsymbol{W}=\begin{bmatrix}\rho\\ \rho u\\ \rho v\\ \rho w\\ \rho E\\ \rho Y_1\\ \vdots\\ \rho Y_{N-1}\end{bmatrix} \tag{2.32}$$

对流通量和黏性通量转换为

$$\boldsymbol{F}_c=\begin{bmatrix}\rho V\\ \rho uV+n_x p\\ \rho vV+n_y p\\ \rho wV+n_z p\\ \rho HV\\ \rho Y_1 V\\ \vdots\\ \rho Y_{N-1}V\end{bmatrix},\quad \boldsymbol{F}_v=\begin{bmatrix}0\\ n_x\tau_{xx}+n_y\tau_{xy}+n_z\tau_{xz}\\ n_x\tau_{yx}+n_y\tau_{yy}+n_z\tau_{yz}\\ n_x\tau_{zx}+n_y\tau_{zy}+n_z\tau_{zz}\\ n_x\Theta_x+n_y\Theta_y+n_z\Theta_z\\ n_x\Phi_{x,1}+n_y\Phi_{y,1}+n_z\Phi_{z,1}\\ \vdots\\ n_x\Phi_{x,N-1}+n_y\Phi_{y,N-1}+n_z\Phi_{z,N-1}\end{bmatrix} \tag{2.33}$$

其中

$$
\begin{aligned}
\Theta_x &= u\tau_{xx}+v\tau_{xy}+w\tau_{xz}+k\frac{\partial T}{\partial x}+\rho\sum_{m=1}^{N}h_m D_m\frac{\partial Y_m}{\partial x}\\
\Theta_y &= u\tau_{yx}+v\tau_{yy}+w\tau_{yz}+k\frac{\partial T}{\partial y}+\rho\sum_{m=1}^{N}h_m D_m\frac{\partial Y_m}{\partial y}\\
\Theta_z &= u\tau_{zx}+v\tau_{zy}+w\tau_{zz}+k\frac{\partial T}{\partial z}+\rho\sum_{m=1}^{N}h_m D_m\frac{\partial Y_m}{\partial z}\\
\Phi_{x,m} &= \rho D_m\frac{\partial Y_m}{\partial x}\\
\Phi_{y,m} &= \rho D_m\frac{\partial Y_m}{\partial y}\\
\Phi_{z,m} &= \rho D_m\frac{\partial Y_m}{\partial z}
\end{aligned}
\tag{2.34}
$$

最后,源项变为

$$
\boldsymbol{Q}=\begin{bmatrix} 0\\ \rho f_{\mathrm{e},x}\\ \rho f_{\mathrm{e},y}\\ \rho f_{\mathrm{e},z}\\ \rho\boldsymbol{f}_{\mathrm{e}}\cdot\boldsymbol{v}+\dot{q}_{\mathrm{h}}\\ \dot{s}_1\\ \vdots\\ \dot{s}_{N-1}\end{bmatrix}
\tag{2.35}
$$

在式(2.32)~式(2.35)中,Y_m,h_m 和 D_m 分别表示组分 m 的质量分数、焓和有效二元扩散系数,$\dot{s}_m$ 是组分 m 在化学反应中的变化率。我们注意到混合物的总密度 ρ 等于组分密度 ρY_m 之和,由于总密度被视为一个独立的量,因此只存在 $N-1$ 个独立的分密度 ρY_m,剩余的一个质量分数 Y_N 为

$$
Y_N = 1 - \sum_{m=1}^{N-1} Y_m \tag{2.36}
$$

为了找到压力 p 的表达式,我们首先假设单独组分的性质类似于理想气体,即

$$
p_m=\rho Y_m\frac{R_{\mathrm{u}}}{W_m}T \tag{2.37}
$$

其中，R_u 为普适气体常数，W_m 为分子量。根据道尔顿(Dalton)定律，有

$$p = \sum_{m=1}^{N} p_m \tag{2.38}$$

因此

$$p = \rho R_u T \sum_{m=1}^{N} \frac{Y_m}{W_m} \tag{2.39}$$

需要注意的是，因为气体处于热力学平衡状态，所以所有的组分都具有相同的温度 T，温度必须从下式通过迭代计算得到[22,28]

$$e = \sum_{m=1}^{N} \left[Y_m \left(h_{f,m}^0 + \int_{T_{ref}}^{T} c_{p,m} \mathrm{d}T \right) \right] - \frac{p}{\rho} \tag{2.40}$$

气体混合物的内能 e 由式(2.6)得到。式(2.40)中的 $h_{f,m}^0$，$c_{p,m}$，T_{ref} 分别表示组分 m 的生成热、比定压热容和参考温度。上述量的值以及组分的热传导系数 k 和动力学黏性系数 μ 的值都通过曲线拟合来确定[20,22,24]。

需要建模的最后一部分是式(2.35)中的化学源项 $\dot{s}_m$。N 种组分，N_R 个基元反应方程式可以写成如下一般形式：

$$\sum_{m=1}^{N} \nu'_{lm} C_m \underset{K_{bl}}{\overset{K_{fl}}{\rightleftharpoons}} \sum_{m=1}^{N} \nu''_{lm} C_m, \quad l = 1,2,\cdots,N_R \tag{2.41}$$

其中，ν'_{lm} 和 ν''_{lm} 分别为组分 m 在第 l 种基元反应中正向和逆向反应的化学当量系数。其中 C_m 为组分 m 的摩尔浓度($C_m = \rho Y_m / W_m$)，K_{fl} 和 K_{bl} 分别为第 l 反应步的正向和逆向反应速率常数，它们是由经验阿伦尼乌斯(Arrhenius)公式给出的：

$$K_f = A_f T^{B_f} \exp(-E_f / R_u T)$$

$$K_b = A_b T^{B_b} \exp(-E_b / R_u T)$$

式中：A_f 和 A_b 为阿伦尼乌斯系数；E_f 和 E_b 为活化能；B_f 和 B_b 为常数。组分 m 的摩尔浓度随第 l 个反应的变化率为

$$\dot{C}_{lm} = (\nu''_{lm} - \nu'_{lm}) \left(K_{fl} \prod_{n=1}^{N} C_n^{\nu'_{ln}} - K_{bl} \prod_{n=1}^{N} C_n^{\nu''_{ln}} \right) \tag{2.42}$$

因此，结合式(2.42)可以计算出组分 m 的总变化率为

$$\dot{s}_m = W_m \sum_{l=1}^{N_R} \dot{C}_{lm} \tag{2.43}$$

更多的细节可以在上面引用的文献中找到，关于控制化学反应流动的方程的详细综述，以及通量的雅可比矩阵及其特征值，也可以在文献[28]中找到。

真实气体的另一个实际例子是在涡轮机械应用中的蒸汽模拟或要求更高的

湿蒸汽模拟[29-39]，在后一种情况中，蒸汽与水滴混合，因此我们称之为多相流，它或者求解额外的一组输运方程，或者沿着若干条流线追踪水滴。这些模拟在现代汽轮机叶栅设计中具有重要的应用价值，例如，通过对涡轮叶片流动的分析，可以帮助我们理解冷凝产生的超临界激波和流动不稳定的发生，它们对叶片造成额外的动态负荷，是导致效率损失的原因。

2.4.3　纳维–斯托克斯方程组的简化

下面介绍纳维–斯托克斯方程组(2.19)的三种常见简化，在此重点关注每一种近似背后的物理原理，附录中提供了前两种简化形式的方程组。

1. 薄层近似

当模拟高 *Re* 数(即边界层相对于特征尺寸较薄)绕流时，可以简化纳维–斯托克斯方程组(2.19)。这种简化的一个必要条件是不存在大范围的边界层分离，由此可以预料，只有流场变量的梯度沿物体表面法向的分量(图 2.4 中的 η 方向)对黏性应力有贡献[40-41]，另外，计算切向应力张量(式(2.14)和式(2.15))时忽略梯度在其他坐标方向上的分量(图 2.4 中的 ξ 方向)，这样就得到了纳维–斯托克斯方程组的薄层近似(thin shear layer，TSL)。进行 TSL 修改的动机是使黏性项的数值计算变得更廉价，但在此假设条件下，解仍然能够保持足够的精度。TSL 也可以从实践的角度进行论证，在高 *Re* 数流动中，为了正确地解析边界层，网格沿壁面法向必须非常精细，由于有限的计算机内存和速度，在其他方向上的网格则很粗，这反过来导致梯度求解的数值精度明显低于法线方向。TSL 方程组的完整形式在附录 A.6 中给出。由于二次流动(如在叶栅中)不能被恰当地解析，TSL 通常只应用于外流空气动力学。

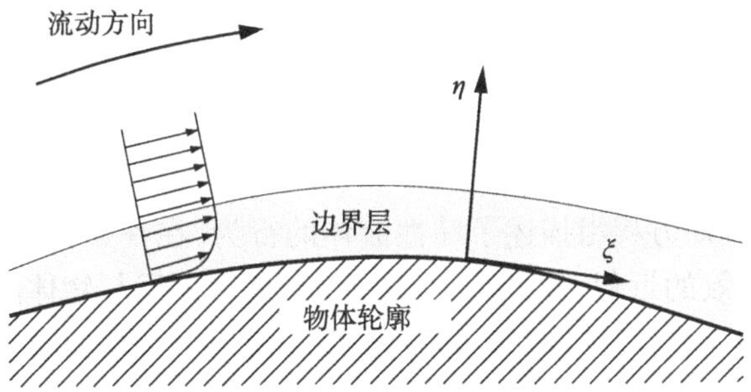

图 2.4　薄边界层示意图

2. 抛物型纳维–斯托克斯方程组

在满足以下三个条件的情况下，控制方程组(2.19)可以简化为抛物型纳维-斯托克斯(Parabolized Navier-Stokes,PNS)方程组[8,42-44]。

(1) 流动是定常的(即 $\partial \boldsymbol{W}/\partial t=0$);

(2) 流体主要向一个方向运动(例如,不能出现边界层分离);

(3) 横流分量可以忽略不计。

如果满足以上条件,我们可以将黏性应力项(式(2.15))中的 u,v 和 w 关于流向的导数取为零。此外,黏性应力张量 $\boldsymbol{\tau}$ 做的功($\boldsymbol{\tau}\cdot\boldsymbol{v}$)以及热传导 $k\nabla T$ 在流向方向的分量都可以从式(2.23)的黏性通量中去掉,而连续性方程和对流通量(式(2.21))则保持不变,详细情况请参阅附录 A.7。如图 2.5 所示,主流方向与 x 轴方向一致,可以看出 PNS 近似产生了一个抛物型/椭圆型混合方程组,即沿流动方向的动量方程和能量方程一起变成抛物型,因此它们可以沿 x 方向推进求解,而在 y 和 z 方向上动量方程是椭圆型的,它们必须在每个 x 平面上迭代求解。PNS 方法的主要好处是在很大程度上降低了流动求解的复杂度——从一个完全的 3D 流场问题变为一系列 2D 问题。PNS 方程组的典型应用是计算风道和管道的内部流动,以及利用空间推进法模拟定常超声速流动[45-48]。

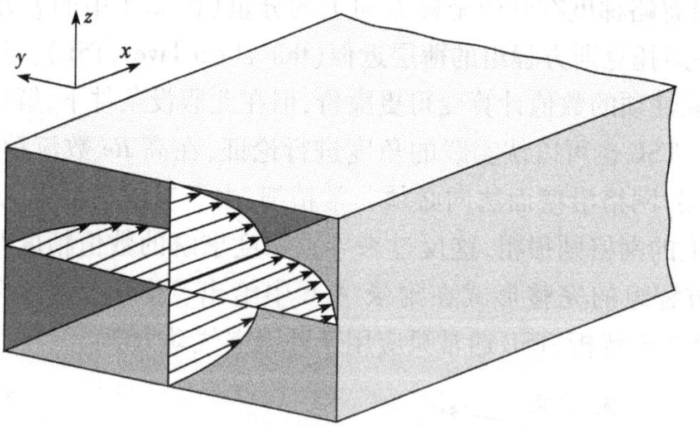

图 2.5　管道内部流动——抛物型纳维-斯托克斯方程组

3. 欧拉方程组

纳维-斯托克斯方程组描述了黏性流体的行为,在许多情形下,完全忽略黏性效应是一种有效的近似,例如高 Re 数流动,其边界层与物体的尺寸相比非常薄,在这种情况下,可以忽略式(2.19)中的黏性通量 $\boldsymbol{F}_\mathrm{v}$,从而得到

$$\frac{\partial}{\partial t}\int_{\Omega}\boldsymbol{W}\mathrm{d}\Omega+\oint_{\partial\Omega}\boldsymbol{F}_\mathrm{c}\mathrm{d}S=\int_{\Omega}\boldsymbol{Q}\mathrm{d}\Omega \tag{2.44}$$

其余的项由同样的关系式(2.20)~式(2.22)和式(2.25)给出。这种控制方程组的简化形式称为欧拉方程组,它描述了无黏流体中流动变量的纯对流情况。如果欧拉方程组以守恒形式表述(如式(2.44)所示),它能精确地描述诸如

激波、膨胀波和(尖前缘)三角翼上的涡流等重要现象。此外,欧拉方程组在过去和现在都是发展离散方法和边界条件的基础。

然而,应该指出的是,如今,即使个人计算机也具有强大的计算能力,人们对模拟质量的要求也不断增加,欧拉方程组已经很少被用于流动模拟。

参考文献

[1] Lax PD. Weak solutions of nonlinear hyperbolic equations and their numerical computation. Comm Pure Appl Math 1954;7:159-93.

[2] Aris R. Vectors, tensors and the basic equations of fluid mechanics. New York: Dover Publ. Inc.;1989.

[3] Schlichting H. Boundary layer theory. 7th ed. New York: McGraw-Hill;1979.

[4] White FM. Viscous fluid flow. New York: McGraw-Hill;1991.

[5] Stokes GG. On the theories of internal friction of fluids in motion. Trans Camb Phil Soc 1845;8:287-305.

[6] Gad-el-Hak M. Questions in fluid mechanics: Stokes' hypothesis for a Newtonian, isotropic fluid. J Fluids Eng 1995;117:3-5.

[7] Vinokur M. Conservation equations of gas dynamics in curvilinear coordinate systems. J Comput Phys 1974;14:105-25.

[8] Hirsch C. Numerical computation of internal and external flows, vols. 1 and 2, Chichester, UK: John Wiley and Sons;1988.

[9] Pulliam TH, Steger JL. Recent improvements in efficiency, accuracy, and convergence for implicit approximate factorization algorithms. AIAA Paper 85-0360;1985.

[10] Thomas PD, Lombard CK. Geometric conservation law and its application to flow computations on moving grids. AIAA J 1979;17:1030-37.

[11] Lesoinne M, Farhat C. Geometric conservation laws for flow problems with moving boundaries and deformable meshes and their impact on aeroelastic computations. AIAA Paper 95-1709; 1995 [also in Comp Meth Appl Mech Eng 1996;134:71-90].

[12] Guillard H, Farhat C. On the significance of the GCL for flow computations on moving meshes. AIAA Paper 99-0793;1999.

[13] Zierep J. Vorlesungen über theoretische Gasdynamik (Lectures on theoretical gas dynamics). Karlsruhe: G. Braun Verlag;1963.

[14] Liepmann HW, Roshko A. Elements of gas dynamics. Mineola, NY: Dover Publications;2002.

[15] Srinivasan S, Weilmuenster KJ. Simplified curve fits for the thermodynamic properties of equilibrium air. NASA RP-1181;1987.

[16] Schmatz MA. Hypersonic three-dimensional Navier-Stokes calculations for equilibrium gas.

AIAA Paper 89-2183;1989.

[17] Mundt Ch, Keraus R, Fischer J. New, accurate, vectorized approximations of state surfaces for the thermodynamic and transport properties of equilibrium air. ZFW 1991;15:179-84.

[18] Rinaldi E, Pecnik R, Colonna P. Exact Jacobians for implicit Navier-Stokes simulations of equilibrium real gas flows. J Comput Phys 2014;270:459-77.

[19] Bussing TRA, Murman EM. Finite-volume method for the calculation of compressible chemically reacting flows. AIAA J 1988;26:1070-78.

[20] Molvik GA, Merkle CL. A set of strongly coupled, upwind algorithms for computing flows in chemical non-equilibrium. AIAA Paper 89-0199;1989.

[21] Slomski JF, Anderson JD, Gorski JJ. Effectiveness of multigrid in accelerating convergence of multidimensional flows in chemical non-equilibrium. AIAA Paper 90-1575;1990.

[22] Shuen JS, Liou MS, van Leer B. Inviscid flux-splitting algorithms for real gases with non-equilibrium chemistry. J Comput Phys 1990;90:371-95.

[23] Li CP. Computational aspects of chemically reacting flows. AIAA Paper 91-1574;1991.

[24] Shuen J-S, Chen K-H, Choi Y. A coupled implicit method for chemical non-equilibrium flows at all speeds. J Comput Phys 1993;106:306-18.

[25] Raman V, Fox RO, Harvey AD. Hybrid finite-volume/transported PDF simulations of a partially premixed methane-air flame. Combust Flame 2004;136:327-50.

[26] Selle L, Lartigue G, Poinsot T, Koch R, Schildmacher K-U, Krebs W, et al. Compressible large eddy simulation of turbulent combustion in complex geometry on unstructured meshes. Combust Flame 2004;137:489-505.

[27] Ajmani K, Mongia H, Lee P. CFD computations of emissions for LDI-2 combustors with simplex and airblast injectors. AIAA Paper 2014-3529;2014.

[28] Yu S-T, Chang S-C, Jorgenson PCE, Park S-J, Lai M-C. Basic equations of chemically reactive flow for computational fluid dynamics. AIAA Paper 98-1051;1998.

[29] Bakhtar F, Tochai MTM. An investigation of two-dimensional flows of nucleating and wet stream by the time-marching method. Int J Heat Fluid Flow 1980;2:5-18.

[30] Young JB, Snoeck J. In: Aerothermodynamics of Low Pressure Steam Turbines and Condensers. Moore MJ, Sieverding CH, editors. New York: Springer Verlag; 1987. p. 87-133.

[31] Bakhtar F, So KS. A study of nucleating flow of steam in a cascade of supersonic blading by the time-marching method. Int J Heat Fluid Flow 1991;12:54-62.

[32] Young JB. Two-dimensional non-equilibrium wet-steam calculations for nozzles and turbine cascades. Trans ASME J Turbomach 1992;114:569-79.

[33] White AJ, Young JB. Time-marching method for the prediction of two-dimensional, unsteady flows of condensing steam. AIAA J Propul Power 1993;9:579-87.

[34] Liberson A, Kosolapov Y, Rieger N, Hesler S. Calculation of 3-D condensing flows in nozzles and turbine stages. In: EPRI Nucleation Workshop, Rochester, New York; October 24-

26,1995.
[35] Bakhtar F,Mahpeykar MR,Abbas KK. An investigation of nucleating flows of steam in a cascade of turbine blading—theoretical treatment. Trans ASME 1995;117:138-44.
[36] White AJ,Young JB,Walters PT. Experimental validation of condensing flow theory for a stationary cascade of steam turbine blades. Phil Trans R Soc Lond A 1996;354:59-88.
[37] Fakhari K. Development of a two-phase Eulerian/Lagrangian algorithm for condensing steam flow. AIAA Paper 2006-597;2006.
[38] Fakhari K. Unsteady phenomena in the condensing steam flow of an industrial steam turbine stage. AIAA Paper 2008-1449;2008.
[39] Giordano M,Congedo P,Cinnella P. Nozzle shape optimization for wet-steam flows. AIAA Paper 2009-4157;2009.
[40] Steger JL. Implicit finite difference simulation of flows about two-dimensional arbitrary geometries. AIAA J 1978;17:679-86.
[41] Pulliam TH,Steger JL. Implicit finite difference simulations of three-dimensional compressible flows. AIAA J 1980;18:159-67.
[42] Patankar SV,Spalding DB. A calculation procedure for heat, mass and momentum transfer in three-dimensional parabolic flows. Int J Heat Mass Transfer 1972;15:1787-806.
[43] Aslan AR,Grundmann R. Computation of three-dimensional subsonic flows in ducts using the PNS approach. ZFW 1990;14:373-80.
[44] Kirtley KR,Lakshminarayana B. A multiple passspace-marching method for three-dimensional incompressible viscous flow. ZFW 1992;16:49-59.
[45] Hollenbäck DM,Blom GA. Application of a parabolized Navier-Stokes code to an HSCT configuration and comparison to wind tunnel test data. AIAA Paper 93-3537;1993.
[46] Krishnan RR, Eidson TM. An efficient, parallel space-marching Euler solver for HSCT research. AIAA Paper 95-1749;1995.
[47] Nakahashi K,Saitoh E. Space-marching method on unstructured grid for supersonic flows with embedded subsonic regions. AIAA Paper 96-0418;1996.
[48] Yamaleev NK,Ballmann J. Space-marching method for calculating steady supersonic flows on a grid adapted to the solution. J Comput Phy 1998;146:436-63.

第3章
控制方程求解原理

第2章,得到了完整的纳维-斯托克斯/欧拉方程组,引入了完全气体附加的热力学关系,并定义了化学反应气体附加的输运方程,因此,我们现在可以求解整个控制方程组的流动变量了。可以想见,存在大量的求解方法,如果不考虑只适用于简单流动问题的解析方法,那么几乎所有的求解策略都遵循相同的途径。

首先,计算流动的空间(即物理空间)被划分成大量的几何单元,这些单元被称作网格单元,这个过程被称为网格生成。它也可以看作首先在物理空间中放置网格点(也被称为节点或顶点),其次用直线(网格线)将它们连接起来。2D网格通常由三角形和/或四边形组成,在3D空间中,它通常由四面体、六面体、三棱柱或金字塔组成。对于网格生成工具来说,最重要的要求是网格单元之间既不能有缝隙,也不能重叠,另外,网格应该是光滑的,即网格单元的体积或拉伸比不应该有突变,单元形状应该尽可能地接近标准形状。此外,如果网格由四边形或六面体组成,网格线不应存在较大的扭结,否则,数值误差将显著增加。

一方面,网格可以紧贴物理空间的边界生成,我们称这种网格为贴体网格(图3.1(a)),它的主要优点是可以非常精确地在边界处解析流动,这对于沿着固壁的剪切层至关重要。生成贴体网格需要付出的代价是网格生成工具高度复杂,特别是对于真实的几何外形而言。另一方面,我们可以很容易地生成笛卡儿(Cartesian)网格[1-8],其网格单元的边平行于笛卡儿坐标,它的优点是在式(2.19)中通量的求解比贴体网格要简单得多。但是,当考虑图3.1(b)或(c)时,可以明显看出,它很难实现对边界的一般而精确处理[3],由于这一缺点,通常更倾向于采用贴体或组合的方法(见11.2.4节),特别是在模拟的外形几何复杂度通常非常高的工业环境中。

目前,绝大多数求解欧拉和纳维-斯托克斯方程组的数值方法都采用在空间和时间上分别离散的方法——所谓的直线法(method of lines)[9]。根据所选择的特定算法,网格被用于构建控制体和计算通量积分,或者用于近似流场变量的空间导数。接着,从已知的初始解出发,借助于合适的方法,将时间相关方程组的解沿时间进行推进。另一种可能性是,当流场变量不随时间变化时,通过迭代过程得到控制方程组的定常解。

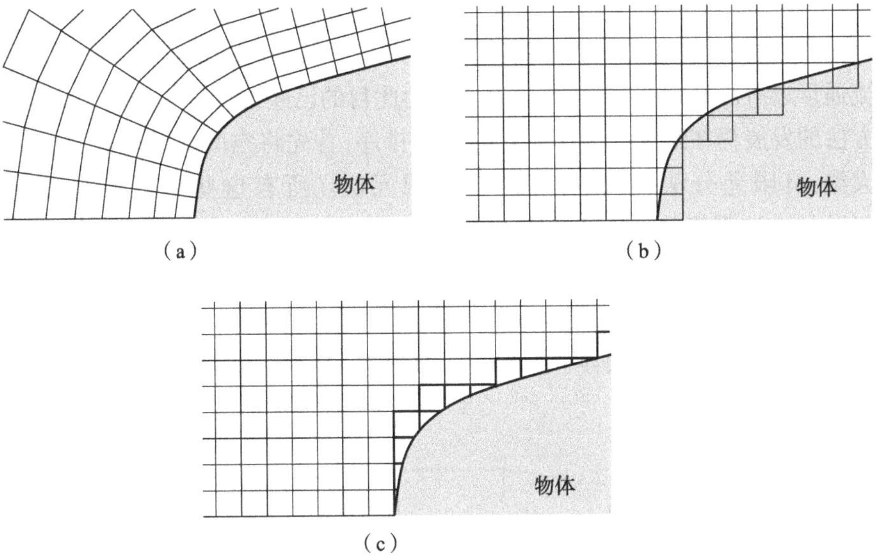

图 3.1

(a) 固壁附加的 2D 贴体网格;(b) 浸入式边界的笛卡儿网格;

(c) 多边形边界单元的笛卡儿网格(在(c)中边界单元用粗线标出)。

在推导控制方程组(2.19)的过程中,连续性方程(2.3)中包含了密度的时间导数,由于使用密度作为自变量来计算压力(式(2.29)),因此动量方程组中密度与压力的时间演化存在耦合关系,由于这个原因,采用控制方程组(2.19)离散的求解方法被称为密度基方法。这种公式存在的问题是,对于不可压缩流体,因为密度的时间导数从连续性方程中消失,压力不再由任何自变量推得。另一个难题是随着马赫数的减小,声波与对流波之间的速度差距越来越大,使得控制方程的刚性增加,从而很难求解[10]。大致说来,目前发展了三种方法来解决上述问题:第一种是求解从动量方程组推导出来的基于压力的泊松(Poisson)方程[11-17],这种方法被称为压力基方法;第二种称为人工可压缩性方法,其思想是将连续性方程中密度的时间导数替换为压力的时间导数[18-19],这样,速度场和压力场就直接耦合了;第三种,也是最普遍的一种,是基于控制方程组的预处理[20-31],这种方法允许对马赫数非常低的流动使用与高马赫数流动相同的数值格式,我们将在 9.5 节更详细地讨论这种方法。

下面各节将介绍各种求解方法的基本原理,包括控制方程组在空间和时间上的数值逼近、湍流建模以及边界处理等。

3.1 空间离散

首先,我们将注意力转向第一步——纳维-斯托克斯方程组的空间离散,即对流通量、黏性通量和源项的数值近似。为此目的已经设计了各种不同的方法,且方法的发展仍在继续。为了对它们进行排序,首先将空间离散格式分为以下三大类:有限差分法、有限体积法和有限元法。所有这些离散控制方程组(2.19)的方法都依赖于某种网格,从根本上说,存在两种不同类型的网格:

(1) 结构网格(图3.2):每个网格点(顶点、节点)分别由唯一的索引 i,j 和 k 标识,对应的笛卡儿坐标分别为 $x_{i,j,k}$,$y_{i,j,k}$ 和 $z_{i,j,k}$。网格单元在2D情况下为四边形,在3D情况下为六面体。如果网格为图3.1(a)所示的贴体网格,那么我们也可以称之为曲线网格。

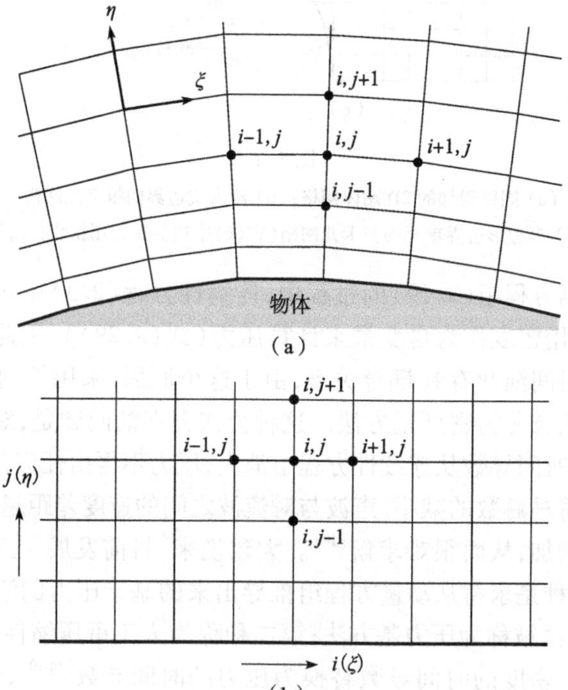

图3.2 2D结构贴体网格方法

(a)物理空间;(b)计算空间;ξ 和 η 表示曲线坐标系。

(2) 非结构网格(图 3.3):网格单元和网格点没有特定的排序,即相邻的单元或网格点不能通过其索引直接识别(如单元 6 与单元 119 相邻)。以前,网格单元在 2D 中是三角形,在 3D 中是四面体。如今,为了正确地解析边界层,非结构网格通常由 2D 的四边形和三角形,3D 的六面体、四面体、三棱柱和金字塔组成。因此,在这种情况下我们称之为混合网格。

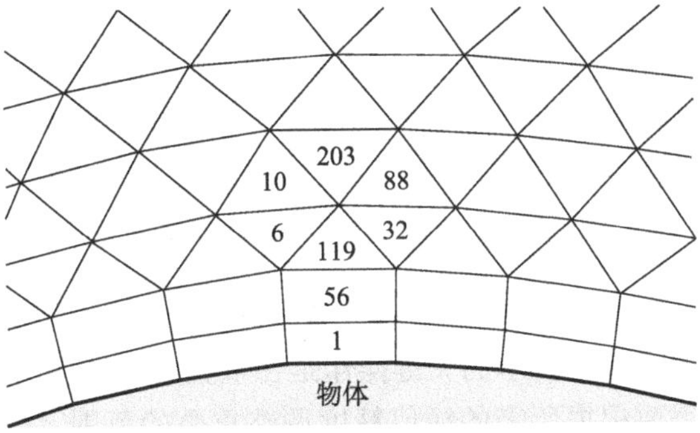

图 3.3　2D 非结构混合网格方法,数字标记单独的单元

结构网格的主要优点来自由索引 i,j,k 代表的线性地址空间——也称为计算空间,因为它直接对应于流场变量在计算机内存中的存储方式。这个属性使得只需要在相应的索引中增加或减去一个整数值(例如$(i+1)$,$(k-3)$,…,(见图 3.2)),即可非常快速和方便地访问相邻的网格节点,可以想象,这一特性使得梯度和通量的求解,以及边界条件的处理大大简化,对于隐式格式的实现也一样,因为它具有有序的带状通量雅可比矩阵。但是结构网格也存在缺点,对于复杂的几何外形,结构网格的生成是一件很困难的事情,如图 3.4 所示,一种可能的改进方法是将物理空间划分为若干拓扑上更简单的部分——块——在这些部分上更容易进行网格生成,我们称之为多块网格方法[32-36]。当然,由于块之间交换物理量或通量需要特殊的逻辑,因此流场解算器的复杂性会增加。如果边界两侧的网格点可以彼此独立放置,即网格线不需要在块边界处对接(如图 3.4 中的 C 或 F 内的网格块),则会增加更多的灵活性,只位于块边界一侧的网格点称为悬挂节点,这种方法的优点非常明显——可以根据需要为每个块分别选择网格线的数量,为增强灵活性所付出的代价是增加了针对悬挂节点的守恒处理的开销[37-38]。多块方法还提供了一项极为诱人的可能性,可以在并行计算机上通过域分解来执行流场解算器。然而,在复杂外形的情况下,网格生成仍然需要很长的时间(通常是几周或几个月)。

另一种与多块结构网格相关的方法是嵌套技术(Chimera technique)[39-46],它的基本思想是首先围绕域内的每个几何外形分别生成网格,其次,使这些网格

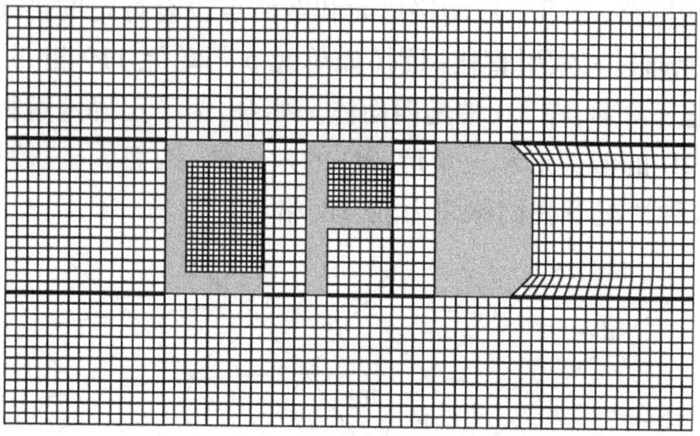

图 3.4　具有对接/非对接块边界的多块结构网格，粗线表示块边界

在相遇的地方相互重叠，以这样的方式将网格组合在一起，图 3.5 描述了一个简单外形的情况。嵌套网格技术的关键操作是在重叠区域的不同网格之间变量的精确传递，因此，需要根据所需的插值精度调整重叠的范围。与多块方法相比，嵌套网格技术的优势在于可以生成完全相互独立的特定网格，而不必考虑网格之间的边界。另外，嵌套网格技术的问题是，控制方程组的守恒特性在通过重叠区域时不能得到满足。文献[47-48]中报道了该方法在非结构网格中的一个应用。

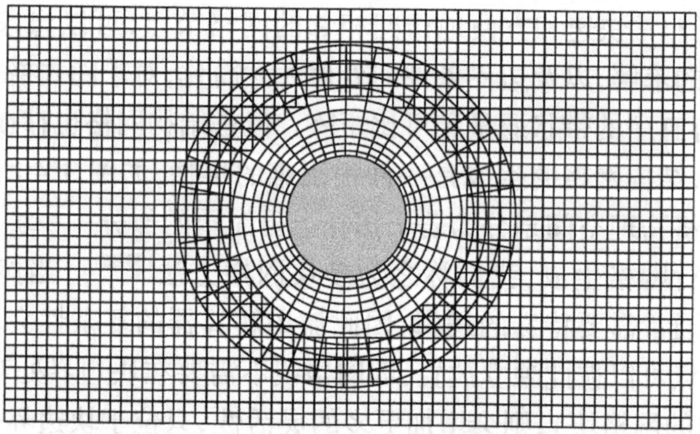

图 3.5　嵌套网格技术示意图（2D 示例）

第二类网格是非结构网格，它为复杂几何外形的处理提供了最大的灵活性。非结构网格的主要优点是，它可以自动生成三角形网格（2D 情况）或四面体网格（3D 情况），而不受流场域的复杂性影响。在实际应用中，当然仍然需要设置一些合适的参数，以获得质量良好的网格。此外，为了精确地解析边界层，最好在固壁附近采用矩形单元（2D）、三棱柱或六面体单元（3D）[50-60]，这种混合网格

的另一个好处是减少了网格单元、边、面以及网格点的数量。尽管对于几何外形复杂的情况，混合网格的生成并不容易，然而，为复杂外形构建非结构混合网格所需的时间仍然比构建多块结构网格要低得多。由于流动模拟的几何逼真度正在迅速提高，快速且用户交互少的网格生成能力变得越来越重要，在工业环境中尤其如此。非结构网格另一个优点是，依赖于网格细化和粗化的求解能以相对自然和无缝的方式进行处理。非结构网格方法的缺点之一是流场解算器必须使用复杂的数据结构，这种数据结构使用间接寻址，根据计算机硬件的不同，或多或少都会降低计算效率。另外，与结构网格相比，非结构网格的内存需求通常更高。尽管存在以上这些困难，然而，与之相比，在短时间内处理复杂几何外形问题的能力要重要得多，从这一点来看，几乎所有商业 CFD 软件供应商都转而使用非结构流场解算器也就不足为奇了。文献[61]对非结构网格空间和时间离散的各种方法进行了详细的回顾。

生成网格之后，下一个问题是如何离散控制方程组，正如已经提到的，通常存在三种选择：有限差分法、有限体积法和有限元法。我们将在下面几节中简要讨论这些格式。

3.1.1 有限差分法

有限差分法是最早应用于微分方程数值解的方法之一，它最初是由欧拉在 1768 年左右开始使用。有限差分法直接应用于控制方程组的微分形式，其原理是采用泰勒(Taylor)级数展开对流场变量的导数进行离散，让我们以下面的例子来说明。

假设想要计算标量函数 $U(x)$ 在 x_0 点的一阶导数，如果把 $U(x_0+\Delta x)$ 使用泰勒级数以 x 展开，得到

$$U(x_0+\Delta x)=U(x_0)+\Delta x\left.\frac{\partial U}{\partial x}\right|_{x_0}+\left.\frac{\Delta x^2}{2}\frac{\partial^2 U}{\partial x^2}\right|_{x_0}+\cdots \tag{3.1}$$

从而，U 的一阶导数可以近似为

$$\left.\frac{\partial U}{\partial x}\right|_{x_0}=\frac{U(x_0+\Delta x)-U(x_0)}{\Delta x}+O(\Delta x) \tag{3.2}$$

上面的近似是一阶精度的，因为截断误差（简称 $O(\Delta x)$）与其余的最大项成正比，随 Δx 的一次方趋于零（关于精度阶数的讨论见第 10 章），同样的过程可用于推导更精确的有限差分公式和求更高阶导数的近似。

有限差分法的一个重要优点是简单，另一个优点是容易获得高阶近似，从而实现空间离散的高阶精度。另外，由于有限差分法需要结构网格，其应用范围显然受到限制。此外，有限差分法不能直接应用于贴体（曲线）坐标，必须首先将

控制方程组转换到笛卡儿坐标系,或者说,从物理空间转换到计算空间(图3.2),这样,流动方程组中就出现了雅可比坐标变换(见A.1节),这个雅可比矩阵必须采用相容的方法进行离散,以避免引入额外的数值误差[62]。因此,有限差分法只适用于简单的几何外形,如今,它被用于研究湍流,并与浸入边界单元(图3.1(b))一起应用于生物学。关于有限差分法的更多细节可以在文献[63]中找到,也可在求解偏微分方程组的教科书中找到。

3.1.2 有限体积法

有限体积法直接使用守恒定律——积分形式的纳维-斯托克斯/欧拉方程组。McDonald[64]首次使用它进行了2D无黏流的模拟。有限体积法离散控制方程组时,首先把物理空间划分成多个任意多面体的控制体,其次将式(2.19)右侧的面积分近似为通过控制体各个面的通量之和,而空间离散的精度取决于通量计算的特定格式。

存在多种关于网格控制体的形状和位置的定义方法,可以分为以下两种基本类型:

(1) 格心(cell-centered)格式(图3.6(a)):流场变量存储在网格单元的中心,控制体与网格单元相同;

(2) 格点(cell-vertex)格式(图3.6(b)):流场变量存储在网格点上,控制体可以是共点单元的组合,也可以是围绕网格点的某个几何体,前一种情况称为重叠(overlapping)控制体,第二种情况称为二次(dual)控制体。

我们将在关于空间离散的两章(第4章和第5章)中讨论格心格式和格点格式的优缺点。

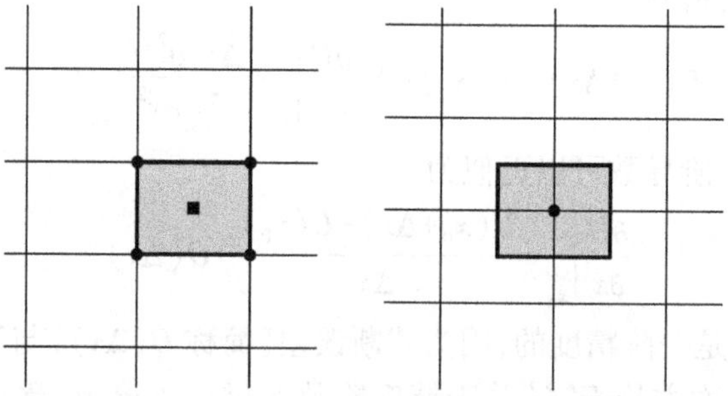

(a)　　(b)

图3.6　格心和格点(二次控制体)格式的控制体

(a) 格心格式;(b) 格点格式。

有限体积法的主要优点是直接在物理空间中进行空间离散,因此不存在有限差分法中的物理坐标系和计算坐标系之间的转换问题。与有限差分法相比,有限体积法的另一个优点是它非常灵活,可以很容易地在结构网格和非结构网格上实现,这使得有限体积法特别适合于模拟复杂外形的内流或绕流。

由于有限体积法是基于守恒定律的直接离散,所以质量、动量和能量也可以通过数值格式得到守恒,这带来了该方法的另一个重要特征,即正确计算控制方程组的弱解的能力。然而,在欧拉方程组的情况下,它必须满足一个附加条件,即熵条件。由于弱解的非唯一性,熵条件是必要条件,它可以阻止非物理特性的发生,例如膨胀激波,其违反了热力学第二定律(熵值减小)。作为守恒离散的进一步结果,在跨越间断解(如激波或接触间断)时可以直接满足兰金-许贡钮(Rankine-Hugoniot)关系。

有趣的是,可以证明,在某些条件下有限体积法与有限差分法或低阶有限元法是等价的。有限体积法由于其诱人的特性,现在非常流行,应用十分广泛,将在后面的章节中进行详细介绍。

3.1.3　有限元法

有限元法最初仅用于结构分析,它最早由 Turner 等[65]在 1956 年提出,大约 10 年之后,研究人员也开始使用有限元法求解连续介质的场方程,然而,直到 20 世纪 90 年代初,有限元法才在欧拉和纳维-斯托克斯方程组的求解中得到普及。文献[66]对经典有限元法有很好的介绍,文献[67-70]描述了有限元法在流动问题中的应用,数年前,研究焦点转移到该方法的一种变形,即间断伽辽金方法(discontinuous Galerkin scheme)[71-98],该格式的高阶形式特别适合于气动声学应用。

当有限元法应用于欧拉/纳维-斯托克斯方程组的求解时,首先需要将物理空间划分为离散单元(在 2D 中主要是三角形或四边形,在 3D 中是四面体或六面体),因此,必须生成一套非结构网格。根据单元类型和所需的精度,在边界和/或单元内部指定一定数量的点,在这些点上必须找到流动问题的解,点的总数乘以未知量的数目决定了自由度的数目。此外,必须定义型函数(shape function),它代表每个单元内解的变化,在实际实现中,通常使用线性或高阶单元(线性单元仅使用网格节点),然后型函数可以是线性或者高阶的分布,其值在相应单元之外为零,这样就使得在规则网格上得到二阶和高阶精度。

在有限元法中,需要将控制方程组由微分形式转化为等效积分形式,这可以通过两种不同的方法来实现。第一种是基于变分原理,即求一个具有极值的特定函数的物理解。第二种被称为加权余量法或弱形式法，要求残差的加权平均在物理域上等于零，而残差可以看作解的近似误差。弱形式法与守恒律的有限体积离散具有相同的优势——它允许处理如激波那样的间断，因此，弱形式法优于变分法。间断伽辽金方法的质量矩阵不同于经典有限元法，它的质量矩阵定义在局部单元上，这使得间断伽辽金方法可以通过简单的显式方法来时间推进未知解系数。

因为使用积分形式和非结构网格，有限元法适合于复杂外形内流或绕流的模拟，使其受到广泛关注。除气动声学外，该方法也特别适用于非牛顿流体的处理。有限元法具有非常严格的数学基础，特别是对于椭圆型和抛物型问题。虽然可以证明在某些情况下，有限元法在数学上与有限体积离散等价，但是其计算量却相当高，这也许可以解释为什么有限体积法在标准流动问题中更受欢迎。然而，这两种方法有时会结合使用，尤其是在非结构网格上，例如，边界的处理和黏性通量的离散通常采用从有限元法“借用”的方法。

3.1.4 其他离散方法

在某些情况下，其他一些数值格式优于上面讨论的方法，在此简要介绍三种特殊方法。

1. 谱元法

第一种方法是谱元法（spectral-element method）[99-107]，这种方法结合了有限元法的几何灵活性和高阶空间精度（如10阶）以及谱方法的快速收敛[101]。它基于解的高阶多项式表示（通常是拉格朗日插值），结合了标准伽辽金有限元法或加权残值法。谱元法既适用于保证高阶正则性的特定问题，也适用于高阶正则性不例外的问题，例如不可压缩流体力学，它特别适用于涡流问题。谱元法的优点主要在于它的对流算子的无扩散、无色散近似，以及对对流-扩散边界层的良好近似。在满足高阶正则性的条件下，该方法可以处理几何和物理上的复杂问题。除了应用范围相当窄外，谱元法的主要缺点是（如与有限体积法相比）计算量非常大。

2. 格子玻尔兹曼法

作为格子气方法的继承者，格子玻尔兹曼法（lattice Boltzmann method，LBM）[108-129]于20世纪90年代末提出，随着大规模并行计算机系统的日益普及，这种方法变得流行起来。与前面讨论的基于宏观属性（即质量、动量和能量）守恒的方法不同，LBM将流体建模为虚构的粒子，这些粒子在网格单元的离散格子上进行重复的微观传播和碰撞。该方法基于分子运动论，可以看作连续玻尔兹曼方程的一种特殊离散形式。一般来说,LBM 以松弛过程(碰撞步)的形式计算局部碰撞的影响,其结果——离散分布函数——沿特征方向发送到各自的相邻点(迁移步),然后通过取分布函数的离散矩得到宏观流场变量。此外,压力场可以直接由密度分布得到。需要指出的是,基于 Chapman-Enskog 理论,可以从 LBM 算法中恢复出连续性方程和纳维-斯托克斯方程组,最近出版的文献也显示了 LBM 和有限差分法之间有趣的相似性,并在计算效率方面对两种方法进行了比较[123]。

由于其运用了粒子动力学和局部动力学方法,LBM 具有几个相对于传统 CFD 方法的优点,特别是当涉及复杂边界或结合微观相互作用时。例如,多相/多组分流动的模拟非常具有挑战性,因为它们具有移动和变形的界面,而不同相之间的过渡边界更是如此,然而,LBM 方程通过修改碰撞算子,为解决此类问题提供了一种相对简单和相容的方法,相分离由粒子动力学自动生成,不需要特殊处理来操纵界面。因此,LBM 成功地应用于界面不稳定性模拟、气泡/液滴动力学、固体表面浸润或液滴电流体动力变形,还用在气动声学或微流体中的通过多孔介质的多相流问题。

“碰撞和流动”算法使得 LBM 非常适合使用域分解技术进行高效的并行处理,传统的 LBM 在本质上是时间推进的(无论是物理时间还是伪时间),它通常会以相当慢的速度收敛到所需的(可能是稳定的)解。此外,到目前为止,对于具有强压缩性影响的流动,还没有 LBM 的公式,这意味着跨声速或超声速流动不能用 LBM 模拟。

3. 无网格法

另一种最近引起关注的离散格式是无网格法(gridless method)[130-136],该方法仅使用点云进行空间离散,不需要像传统的结构或非结构网格方法那样将点连接起来形成网格。无网格法基于笛卡儿坐标系下的控制方程组的微分形式,流动变量的梯度由最小二乘重构确定,使用对应点周围的指定数量的相邻点进行计算。无网格法既不是有限差分法、有限体积法,也不是有限元法,因为它不需要进行坐标变换、面积或体积计算,它可以看作有限差分法和有限元法的混合。无网格法的主要优点是在解决复杂外形流场方面的灵活性(类似于非结构方法),以及在适当的地方定位或聚集点(或点云)的可能性,例如,在计算梯度时,很容易只选择特征方向上的相邻点。然而,它存在一个尚未解决的问题,虽然无网格法求解欧拉或纳维-斯托克斯方程组的守恒形式,但质量、动量和能量守恒是否真的得到保证还不是很清楚。

无论选择哪种空间离散格式,重要的是要确保格式是相容的,即当网格足够密时,它收敛到离散方程的解。因此,如果网格被细化(例如,如果将网格点的数量翻倍)，检查解的变化非常重要，当解只是略有改进时，我们称之为网格收敛解。另一个显而易见的要求是，离散格式应该具有适合于所求解的流动问题的精度阶数，为了更快地收敛，有时会放弃这一原则，特别是在工业环境中(因为有一个质量差的解总比没有解好)，这当然是一种非常危险的做法。在第 10 章，将讨论精度、稳定性和相容性问题。

3.1.5 中心格式和迎风格式

到目前为止，我们只讨论了现存的空间离散的基本选择，在上述三种主要方法(有限差分法、有限体积法和有限元法)中，存在进行空间离散的各种不同的数值格式，在这种情况下，适于将对流通量和黏性通量(式(2.19)中的$\boldsymbol{F}_c$和$\boldsymbol{F}_v$)的离散进行区别对待。由于黏性通量的物理特性，唯一合理的方法是采用中心差分(中心平均)进行离散，它在结构网格上可以直接进行，对于三角形或四面体非结构网格，可以使用伽辽金有限元法进行很好的近似[137]，即使对于有限体积法，也可以这样做。而对于非结构混合网格，情况变得非常复杂，其中修正的梯度平均法更适用[138-142]。

然而，对流通量的离散存在真正的多样性，为了对各种方法进行分类，我们将注意力限制在基于有限体积法的格式上，尽管大多数概念也可以直接应用于有限差分法和有限元法。

1. 中心格式

我们遇到的第一类格式完全基于中心差分公式或中心平均，这些格式被称为中心格式，其原理是将左右两侧的守恒变量平均，以求解控制体界面的通量。因为中心格式不能识别和抑制解的奇偶失联(即产生离散方程组的两个独立的解)，为了格式稳定，必须加入人工耗散(因为它与黏性项的相似性)，最广为人知的实现是 Jameson 等[143]的工作。在结构网格上，它基于二阶差分和四阶差分的混合，并通过对流通量雅可比矩阵的最大特征值进行调节。在非结构网格上，它使用完整的拉普拉斯(Laplacian)算子和双调和(biharmonic)算子的组合[144]。对每个方程采用不同的比例因子可以显著地改进该格式，这种方法称为矩阵耗散格式[145]。需要指出的是，在非结构混合网格上，显式龙格-库塔时间格式与传统的中心格式结合时可能会变得不稳定[146]。

2. 迎风格式

另外，存在更先进的空间离散格式，它们的构建考虑了欧拉方程组的物理性质。因为它们区分了(波传播方向的)上游和下游的影响，所以被称为迎风格式，其大致可分为四大类：

(1) 矢通量分裂格式；

(2) 通量差分裂格式;

(3) 总变差减小(total variation diminishing,TVD)格式;

(4) 波动分裂格式。

以下对每一类格式进行简要描述。

(1) 矢通量分裂格式。

一类矢通量分裂格式根据某些特征变量的符号将对流通量分解为两部分,这些特征变量与对流通量雅可比矩阵的特征值大致相似但不完全相同,然后使用迎风偏置差分对这两部分进行离散。第一个这种类型的矢通量分裂格式是在20世纪80年代初分别由Steger和Warming[147],以及van Leer[148]开发的。第二类矢通量分裂格式将对流通量分解为对流部分和压力(声学)部分。这种思想分别被Liou等的AUSM(advection upstream splitting method)格式[149-150]和Jameson的CUSP(convective upwind split pressure)格式[151-152]所利用,类似的方法还有Edwards提出的LDFSS(low-diffusion flux-splitting scheme)格式[153],以及Rossow提出的MAPS(mach number-based advection pressure splitting)格式[154-155]。第二类矢通量分裂格式最近获得了更大的普及,因为它们改进了剪切层的解析度,同时只需要中等的计算量。与通量差分裂格式或TVD格式相比,矢通量分裂格式的一个优点是,可以很容易地推广到真实气体流动,我们将在后面章节继续讨论关于真实气体的模拟。

(2) 通量差分裂格式。

第二组格式——通量差分裂格式——基于求解界面处间断的局部1D欧拉方程组,这对应于黎曼(Riemann)(激波管)问题,界面两侧的值通常被称为左状态和右状态。这种方法在两个控制体之间的界面上求解黎曼问题,它由Godunov[156]在1959年首先提出。为了降低求解精确黎曼解所需的计算量,发展了近似黎曼解算器,例如Osher和Solomon[157],以及Roe格式[158]。Roe格式由于其对边界层良好的解析度和对激波的清晰表达而经常被使用,无论在结构网格上还是在非结构网格上,它都可以很容易地实现[159]。

基于近似黎曼解的更进一步的方法被称为Harten-Lax-van Leer(HLL)解算器,Einfeldt[160]提出了一种变形(HLLE),它基于流体间断的稳定性理论,并考虑了界面上的最小和最大波速,为了恢复HLL格式中的接触间断,Toro等[161]提出了另一个版本HLLC(HLL contact)。后来其他研究者也对其进行了研究和修改[162-164],Nishikawa和Kitamura[165]进一步提出了旋转混合(rotated-hybrid)黎曼解算器,目的是解决原始Roe格式的红玉现象①,同时克服HLL解算器过度扩散

① 红玉现象是沿滞止线的弓形强激波前的扰动的增长,其出现的原因是格式无法识别声速点。

的问题。应该指出的是，红玉现象可以很容易地通过对 Roe 格式的一个简单修改来避免，这个修改被称为 Harten 熵修正，我们将在 4.3.3 节讨论 Roe 格式时给出这个修正，考虑到这一点，很难看到用 HLL 格式的变形之一取代 Roe 迎风格式有任何实际的好处。

(3) TVD 格式。

TVD 格式的思想最早由 Harten[166]在 1983 年提出，它基于这样一个概念，即防止在流动求解过程中产生新的极值。TVD 格式的主要条件是最大值不增加，最小值不减小，同时不产生新的局部极值，这样的格式称为保单调格式，因此，具有 TVD 特性的离散方法能够在不产生任何伪振荡的情况下求解激波问题。TVD 格式通常以对流通量的平均加上额外的耗散项来实现，耗散项既可以依赖于特征速度的符号，也可以不依赖，第一种情况称为迎风 TVD 格式[167]，第二种情况称为对称 TVD 格式[168]。经验表明，迎风 TVD 格式比对称 TVD 格式具有更好的激波和边界层解析度，因此，迎风 TVD 格式是首选。TVD 格式的缺点是不易于推广到高于二阶的空间精度。

(4) 波动分裂格式。

最后一组格式——波动分裂格式——提供了真正的多维迎风性质，其目的是精确求解与网格不对齐的流动特征，相对于上面提到的所有迎风格式，这是一个显著的优势，因为那些格式都只根据网格单元的方向来分裂方程组。在波动分裂方法中，流场变量与网格节点相关联，中间残差作为网格单元上的通量平衡来计算，这些网格单元在 2D 情况下由三角形组成，在 3D 情况下由四面体组成，单元的残差以迎风偏置的方式分配到节点上，最后使用节点值进行解的更新。在欧拉或纳维-斯托克斯方程组中，基于单元的残差必须分解成标量波系，在 2D 和 3D 中分解方式并不唯一，目前已经发展了多种方法，从 Roe 的波模型[169-170]到 Sidilkover 的代数格式[171]，再到最先进的特征分解方法[172-176]。尽管相对于一维分裂黎曼解算器、TVD 解算器等具有上述优势，但波动分裂方法目前仍仅在研究性的程序中使用，这可以归因于它的复杂性和巨大计算量，以及收敛问题。

3. 解的重构

以上所有格式都需要控制体边界上的流场变量的信息，如密度、速度分量或压力，可能以左、右状态的形式分别提供给界面的两侧。对于每个控制体，只有流场变量的平均值是已知的，因此某种插值——我们称之为重构——是必要的，显然，解的重构的精度决定了离散格式的空间阶数。

(1) 一阶和二阶格式。

在中心格式中，对公共面左侧和右侧的控制体的平均值进行简单的算术平均,就足以满足二阶精度(见 4.3.1 节),而对于迎风格式,将左右状态设为各自的平均值,得到的是一阶精度格式。为了在空间上达到二阶,在结构网格上需要单侧差分(见式(4.46)),在非结构网格上实现二阶迎风格式首先需要计算流场变量的梯度,其次利用梯度和平均值,用方向导数插值得到左右状态(5.3.3 节)。

(2) ENO/WENO 格式。

使用 ENO(essentially non-oscillatory)或 WENO(weighted ENO)离散格式都可以达到高于二阶的精度[177-197],它们基于如下的思想:(根据不同的方向)使用不同的离散模板,计算不同的重构(插值)多项式,以得到控制体的面上的值。然后 ENO 格式选择振荡最小的多项式,而 WENO 方法对多项式进行加权和组合,权值与振荡程度成反比,这样,在光滑区域内的解可以是空间高阶的,而真正的间断不会被过度抹平。文献[185]对 WENO 格式做了很好的介绍。

4. 中心格式与迎风格式的比较

现在你可能会问,各种空间离散方法的优点和缺点是什么？一般来说,与迎风格式相比,中心格式需要的数值计算量要少得多,因此每次求解需要的 CPU 时间也要少得多,另外,迎风格式比中心格式能更准确地捕捉间断。此外,由于迎风格式的数值扩散较小,可以用更少的网格点来解析边界层,特别是 Roe 通量差分裂格式和 CUSP 矢通量分裂格式可以非常精确地计算边界层。迎风格式的缺点出现在二阶或高阶空间精度,为了防止在强间断附近产生虚假振荡,必须使用限制器函数(或简称为限制器),因为它在光滑流动区域存在意外的切换,限制器经常会阻止迭代格式的收敛。Venkatakrishnan[198-200]对此提出了一种补救方法,该方法对大多数实际算例都有令人满意的效果,然而必须接受解中存在小的扰动。限制器函数的另一个缺点是它们需要很大的计算量,特别是在非结构网格上,我们将在后面讨论迎风离散格式时更详细地讨论限制器的问题。

5. 真实气体的迎风格式

对于真实气体模拟,特别是化学反应流动,目前提出了迎风离散格式的几个扩展。对于处于热力学平衡和化学平衡的流体,文献[201-203]描述了 van Leer 矢通量分裂[148]和 Roe 近似黎曼解算器[158]的修正。对于化学非平衡和热力学非平衡的更复杂的流动情况,两种迎风方法的公式参见文献[204-209]及其引用的文献。特别是文献[203,207-208]分别对迎风离散格式所采用的方法进行了很好的综述,文献[210]对化学反应流的控制方程组、雅可比矩阵和变换矩阵进行了总结,文献[211]对平衡真实气体相应情况进行了总结。

3.2 时间离散

正如本章开始所提到的,求解欧拉和纳维-斯托克斯方程组的数值格式普遍采用直线法,即在空间和时间上分别进行离散。这种方法提供了最大的灵活性,因为对流通量和黏性通量以及时间积分可以很容易地选择不同的近似级别——正如需要解决的问题所要求的那样,因此,我们在此将遵循这一方法。关于时间和空间耦合离散的其他方法的讨论,请参阅文献[63]。

当将直线法应用于控制方程组(2.19)时,对于每个控制体,它会生成一个关于时间的耦合的常微分方程组:

$$\frac{\mathrm{d}(\Omega \boldsymbol{M}\boldsymbol{W})}{\mathrm{d}t}=-\boldsymbol{R} \tag{3.3}$$

为了清晰起见,忽略了所有单元索引。式(3.3)中,Ω 为控制体的体积,$\boldsymbol{R}$ 表示完整的空间(有限体积)离散,包括源项——称为残差,它是关于守恒变量 $\boldsymbol{W}$ 的非线性函数。最后,$\boldsymbol{M}$ 表示质量矩阵,对于格点格式,它将控制体中 $\boldsymbol{W}$ 的平均值与相关的内部节点以及相邻节点的值联系起来[212-213],对于格心格式,可以用单位矩阵代替质量矩阵,而不影响格式的时间精度,同样的处理方法也适用于均匀网格上的格点格式,因为在此情况下节点与控制体的质心重合。质量矩阵仅是网格的函数,并与微分方程组(3.3)耦合,对于不考虑时间精度的定常状态,质量矩阵可以被"归并",即用单位矩阵替代,这样就避免了昂贵的 $\boldsymbol{M}$ 求逆运算,并且使方程组(3.3)解耦。在这一方面,重要的是要认识到,对于定常状态,解的精度完全由残差的近似阶数决定,因此,质量矩阵只对非定常流场的格点(中点-二次(median-dual))格式重要。

对于静止网格,我们可以把体积 Ω 和质量矩阵从时间导数中取出,然后用下面的非线性格式来近似时间导数[63]

$$\frac{\Omega \boldsymbol{M}}{\Delta t}\Delta \boldsymbol{W}^{n}=-\frac{\beta}{1+\omega}\boldsymbol{R}^{n+1}-\frac{1-\beta}{1+\omega}\boldsymbol{R}^{n}+\frac{\omega \Omega \boldsymbol{M}}{(1+\omega)\Delta t}\Delta \boldsymbol{W}^{n-1} \tag{3.4}$$

其中

$$\Delta \boldsymbol{W}^{n}=\boldsymbol{W}^{n+1}-\boldsymbol{W}^{n} \tag{3.5}$$

是解的修正,上标 n 和($n+1$)表示时间层(n 为当前层),Δt 表示时间步长。如果满足如下条件,则式(3.4)的格式为时间二阶精度

$$\beta=\omega+\frac{1}{2} \tag{3.6}$$

否则时间精度降为一阶。根据参数 β 和 ω 的设置,可以得到显式($\beta=0$)或隐

式时间推进格式。我们将在下面的章节简要讨论这两个主要分类,并在第 6 章进行更详细的介绍。

3.2.1　显式格式

在式(3.4)中设置 $\beta=0$ 和 $\omega=0$,得到最简单的显式时间积分格式,时间导数用前向差分近似,而残差仅在当前时间层上求解(基于已知的流场变量),即有

$$\Delta \boldsymbol{W}^n = -\frac{\Delta t}{\Omega}\boldsymbol{R}^n \tag{3.7}$$

在此,忽略了质量矩阵,它表示一个单步格式,因为一个新的解 $\boldsymbol{W}^{n+1}$ 只来自残差的一次求解。式(3.7)中的格式没有实际价值,因为它只有在与一阶迎风空间离散相结合时才是稳定的。

更流行的是多步时间推进格式(龙格-库塔法),该格式进行多步求解[143],并在中间状态计算残差,使用系数加权每一步的残差,可以对系数进行优化,以扩大稳定区间,改善格式的阻尼特性,从而提高其收敛性和鲁棒性[143,214-215]。此外,根据步系数和步数,多步法可以在时间上扩展到二阶或高阶精度,并设计了特殊的龙格-库塔法,以保持 TVD 和 ENO 空间离散方法的性质,同时使可用的时间步长最大化[216]。

显式多步时间推进格式可用于任意空间离散格式,可以很容易地在串行、矢量以及并行计算机上实现,它在数值上很廉价,而且仅需要少量的计算机内存。另外,由于稳定性的限制,最大时间步长受到严格限制,特别是对于黏性流动和高度拉伸的网格单元,收敛到定常状态的速度大大减慢。此外,对于刚性方程组(如真实气体模拟、湍流模型)或刚性源项,可能需要非常长的时间才能达到定常状态,更坏的情况下,显式格式可能变得不稳定或者导致虚假的稳态解[217]。

如果我们只对定常状态解感兴趣,则可以选择(或组合)几种收敛加速方法。第一种,也是非常常见的技术是局部时间步长方法,其思想是在每个控制体中使用当地允许的最大时间步长来进行解的推进。因此,收敛到定常状态的速度大大加快,然而,瞬态解在时间上不再准确。另一种方法是特征时间步方法,不仅使用局部变化的时间步长,而且每个方程(连续性、动量和能量方程)都使用自己的时间步长进行积分,文献[218]在 2D 欧拉方程组中展示了这一概念的潜力。与特征时间步方法相似的另一种加速技术是雅可比预处理[219-221],从根本上说它是一个点隐式的雅可比松弛,在龙格-库塔法的每一步执行。雅可比预处理可以看作一种时间步方法,其中所有的波分量(通量雅可比矩阵的特征值)被缩放到具有相同的有效速度,另外它还将隐式分量添加到基础显式格

式中。

另一种非常流行的加速方法是通过在显式格式中引入一定量的隐式因素来增加可能的最大时间步长，它被称为隐式残差光顺或残差平均[222-223]。在结构网格上，该方法要求对每个守恒变量求解一个三对角矩阵；对于非结构网格，通常采用雅可比迭代法求矩阵的逆。标准的隐式残差光顺允许时间步长增加 2~3 倍。另外还开发了其他几种隐式残差光顺技术，例如迎风型隐式残差光顺方法[224]，该方法被设计为与迎风空间离散一起使用，与标准技术相比，它允许更大的时间步长，也提高了时间推进过程的鲁棒性[225]。另一种方法是隐-显式残差光顺[226-227]，其目的是改进时间离散在较大时间步长下的阻尼特性。

这里提到的最后一种但可能是最重要的收敛加速技术是多重网格方法，它是 20 世纪 60 年代由 Fedorenko[228] 和 Bakhvalov[229] 在俄罗斯开发的，他们将多重网格应用于椭圆边值问题的求解，Brandt 进一步推动并促进了该方法的发展[230-231]。多重网格思想是基于这样的发现：迭代格式通常可以非常有效地消除求解过程中的高频误差（即控制体之间的振荡），而在减少解的低频（即全局）误差方面表现相当差。因此，在给定的网格上进行求解后，将解转移到更粗的网格上，在粗网格上，一部分低频误差变成高频误差，可以再次通过迭代解算器进行有效的消除。这个过程在一个逐渐粗化的网格序列上递归地重复，其中每个多重网格层有助于抑制一定的误差频率带宽。到达最粗网格后，依次收集解的修正值并插值回初始的细网格中，然后在细网格中进行解的更新。重复这个完整的多重网格循环，直到解的变化小于给定的阈值。为了进一步加速收敛，可以先在粗网格上启动多重网格过程，执行一定数量的循环后，将解转移到更细的网格上，再次执行多重网格循环，然后重复这个过程，直至到达最细的网格，这种方法被称为完全多重网格法（full multigrid，FMG）[231]。

如前所述，多重网格法最初是为了求解椭圆型边值问题（泊松方程）而发展起来的，在这方面它非常有效。Jameson 首先提出采用多重网格求解欧拉方程组[222,232]，基于完全近似存储（full approximation storage，FAS）格式[231]，其中多重网格直接应用于非线性控制方程组。目前，多重网格是求解纳维-斯托克斯方程组的一种标准加速技术，在文献[233-239]中可以找到结构网格下的实现范例，文献[240-250]中有用于非结构网格的范例。虽然没有在椭圆微分方程中的速度那么快，但它经常被证明可以使欧拉或纳维-斯托克斯方程组的求解加速 5~10 倍，图 3.7 是跨声速流动的一个例子。最近的研究也表明，将控制方程组分解为双曲和椭圆部分可以达到更快的收敛[251]。我们将在 9.4 节再次对多重网格方法进行讨论。

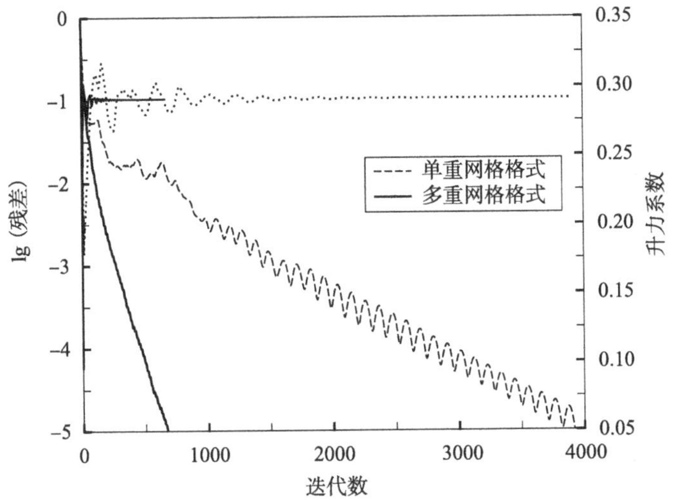

图 3.7 跨声速机翼无黏流动的收敛历程对比

3.2.2 隐式格式

通过设置 $\beta\neq0$，由式(3.4)得到一组隐式时间积分格式。在非定常流场模拟中，非常流行的是二阶时间精度的三点隐式后差格式，其中 $\beta=1,\omega=1/2$，该格式通常用于双时间步方法[212-213,252-254]，其中在每个物理时间步上用伪时间步求解定常状态问题。

对于求解定常流动问题，$\omega=0$ 的格式更合适，因为它需要的计算机内存更少，针对当前时间层线性化式(3.4)中的残差 $\boldsymbol{R}^{n+1}$，得到如下格式：

$$\left(\boldsymbol{M}\frac{\Omega}{\Delta t}+\beta\frac{\partial\boldsymbol{R}}{\partial\boldsymbol{W}}\right)\Delta\boldsymbol{W}^n=-\boldsymbol{R}^n \tag{3.8}$$

$\partial\boldsymbol{R}/\partial\boldsymbol{W}$ 表示通量雅可比矩阵，它构成一个大的稀疏矩阵。式(3.8)左侧括号内的表达式也被称为隐式算子。如前所述，质量矩阵 $\boldsymbol{M}$ 可以被单位矩阵替换，而不会影响定常状态解。式(3.8)中的参数 β 一般设为 1，得到一阶精度的时间离散。$\beta=1/2$ 时得到二阶时间精度的格式，然而，不建议取 $\beta=1/2$，因为 $\beta=1$ 的格式要鲁棒得多，而且时间精度对定常问题不起作用。

与显式格式相比，隐式格式的主要优点是可以使用相当大的时间步长，而不会影响时间积分过程的稳定性。实际上，当 $\Delta t\to\infty$ 时，式(3.8)转换成了标准牛顿法，具有二次收敛性，但是，二次收敛的条件是通量雅可比矩阵包含残差的完整线性化。隐式格式的另一个重要优点是，在刚性方程组和/或刚性源项的情况下，它具有优越的鲁棒性和收敛速度，这在真实气体模拟、湍流模拟或高拉伸网格(高 Re 数流动)的情况下经常遇到。另外,隐式格式(在时间步或迭代方面)越快和越鲁棒,通常每个时间步或迭代的计算量就越大,因而采用多重网格加速的显式格式可以具有同等甚至更高的效率。此外,隐式格式比显式格式更难以矢量化或并行化。

对每个控制体写出式(3.8)中的隐式格式,我们得到一个大型线性方程组,在每个时间步长 Δt,都需要进行求解以更新 $\Delta \boldsymbol{W}^n$,这项工作可以使用直接求解方法或迭代方法来完成。

直接方法基于使用高斯消元法或一些直接稀疏矩阵方法对式(3.8)的左端项进行精确求逆[255-256]。虽然在结构网格[257-260]和非结构网格[261]上都证明其具有二次收敛性,但对于3D问题,直接方法不是一个可用的选项,因为它需要极大的计算量和计算机内存。

因此,对于较大规模的网格或3D问题,唯一实用的方法是迭代方法,使用迭代方法时,线性系统在每一个时间步用迭代矩阵求逆方法来求解 $\Delta \boldsymbol{W}^n$。为了减少内存需求,也为了增加对角占优,通量雅可比 $\partial \boldsymbol{R}/\partial \boldsymbol{W}$ 主要基于右端项一阶精度空间离散的线性化,这种近似产生了两个主要后果:一方面是不能达到牛顿格式的二次收敛,且最大时间步长受到限制;另一方面则大大减少了每个迭代步的数值计算量,从而得到了计算效率高的格式。

对于结构网格,主要使用的迭代方法有交替方向隐式(alternating direction implicit,ADI)格式[262-265]、(线性)雅可比或高斯-赛德尔松弛格式[266-270],以及上下对称高斯-赛德尔(lower-upper symmetric Gauss-Seidel,LU-SGS)格式(也被称为LU-SSOR格式——上下对称超松弛(lower-upper symmetric successive overrelaxation))[271-275]。所有这些方法都是基于将隐式算子分裂为多个部分的和或乘积,而每个部分都可以更容易地求逆。由于相关的分解误差(相对于原始矩阵的差异)和对通量雅可比矩阵的简化,它不能很准确地求解线性系统。实际上,ADI方法和LU-SGS方法在每个时间步都只进行一次迭代。

非结构网格的隐式迭代方法大多基于高斯-赛德尔松弛格式[276-279]。为了提高收敛性,可以采用红-黑高斯-赛德尔方法,它在非结构网格上的推广见文献[280-282]。LU-SGS格式在非结构网格上的实现也提供了一种特别令人瞩目的发展前景[27,283-284],因为它对内存的需求非常低,而且计算量也非常小。

由于线隐方法在结构网格上的成功,人们在非结构网格上也进行了一些尝试[285-286]。这种方法需要构造连续的线,这样每个网格点或(在格心格式的情况下的)每个网格单元只访问一次——哈密顿循环(Hamiltonian tour)[287],这些线主要在坐标方向上,但它们需要在边界上和其他必要的地方进行折叠(因此它们被戏称为“蛇”),然后使用三对角线解算器来对式(3.8)的左端项求逆。后来，人们认识到，折叠这些线会减慢收敛速度，为了克服这个问题，每条线被分解成多段[288]，然而，这样在矢量计算机上的性能相当差。使用分段的思想，在高拉伸的黏性非结构网格上，沿垂直边界层的方向使用隐式解算器，可以改进显式格式的收敛性[248]。

更复杂的迭代技术是 Krylov 子空间方法，它以更全局的方式处理线性方程组，源于 Hestenes 和 Stiefel 提出的求解大型稀疏线性系统的高效迭代格式——共轭梯度（conjugate gradient）法[289]。原始的共轭梯度法仅适用于 Hermitian 正定矩阵，对于 $n\times n$ 矩阵，最多经过 n 次迭代就可收敛。针对 CFD 应用中出现的任意非奇异矩阵的求解，人们提出了各种 Krylov 子空间方法，例如共轭梯度平方（conjugate gradient squared，CGS）[290]、双共轭梯度稳定（bi-conjugate gradient stabilized，Bi-CGSTAB）[291]、无转置准最小残差（transpose-free quasi-minimum residual，TFQMR）[292]等方法。

然而，使用最广泛的可能是 Saad 和 Schultz 提出的广义最小残差（generalized minimal residual，GMRES）格式[293]。将式（3.8）中的隐式格式重写为

$$\boldsymbol{J}\Delta\boldsymbol{W}^n=-\boldsymbol{R}^n \tag{3.9}$$

其中，$\boldsymbol{J}$ 表示一个大型稀疏非对称矩阵（左端项）。GMRES（m）方法从一个初始的猜测 $\Delta\boldsymbol{W}_0$ 开始，求出一个形式为 $\Delta\boldsymbol{W}^n=\Delta\boldsymbol{W}_0^n+\boldsymbol{y}_m$ 的解，其中 $\boldsymbol{y}_m$ 属于 Krylov 子空间

$$\begin{aligned}&K_m\equiv\mathrm{span}\{\boldsymbol{r}_0,\boldsymbol{J}\boldsymbol{r}_0,\boldsymbol{J}^2\boldsymbol{r}_0,\cdots,\boldsymbol{J}^{m-1}\boldsymbol{r}_0\}\\&\boldsymbol{r}_0=\boldsymbol{J}\Delta\boldsymbol{W}_0^n+\boldsymbol{R}^n\end{aligned} \tag{3.10}$$

使得残差 $\|\boldsymbol{J}\Delta\boldsymbol{W}^n+\boldsymbol{R}^n\|$ 变为最小值。参数 m 指定 Krylov 子空间的维数，也就是搜索方向（$\boldsymbol{J}^i\boldsymbol{r}_0$）的个数，因为必须存储所有方向，所以 m 通常在 10~40 之间选择，对于（在湍流模拟、真实气体等情况下出现的）条件差的矩阵，需要较高的 m 值，如果在 m 个子迭代中没有实现收敛，GMRES 必须重新启动。与 Bi-CGSTAB 或 TFQMR 相比，GMRES 方法需要相当多的内存，但它更鲁棒、收敛更平稳，通常也更快。文献[294]对各种方法进行了非常详细的比较。

与其他共轭梯度方法一样，对于 CFD 问题，预处理是绝对必要的。在此，求解下式以代替式（3.9）：

$$(\boldsymbol{P}_{\mathrm{L}}\boldsymbol{J})\Delta\boldsymbol{W}^n=-\boldsymbol{P}_{\mathrm{L}}\boldsymbol{R}^n \quad \text{或} \quad \boldsymbol{J}\boldsymbol{P}_{\mathrm{R}}(\boldsymbol{P}_{\mathrm{R}}^{-1}\Delta\boldsymbol{W}^n)=-\boldsymbol{R}^n \tag{3.11}$$

矩阵 $\boldsymbol{P}_{\mathrm{L}}$ 和 $\boldsymbol{P}_{\mathrm{R}}$ 分别表示左、右预处理器。为了使特征值更接近 1，预处理器应该尽可能接近 $\boldsymbol{J}^{-1}$，当然它也应该易于求逆。一种特别有效的预处理器是具有填充零的不完全 LU 分解法（incomplete lower upper factorization method with zero fill-in，ILU（0））[295]。与 GMRES 相关的不同预处理技术的讨论，请参阅文献

[198,296-299]。

由于 GMRES 方法需要相当大的计算机内存来存储搜索方向以及预处理矩阵,因此避免显式构造和存储通量雅可比矩阵 $\partial \boldsymbol{R}/\partial \boldsymbol{W}$ 是一种好方法,这需要由无矩阵方法来提供,它是基于观察到 GMRES(和其他一些 Krylov 子空间方法)只使用如下形式的矩阵矢量积:

$$\frac{\partial \boldsymbol{R}}{\partial \boldsymbol{W}}\Delta \boldsymbol{W}^n$$

可以用有限差分来简单地近似为

$$\frac{\partial \boldsymbol{R}}{\partial \boldsymbol{W}}\Delta \boldsymbol{W}^n = \frac{\boldsymbol{R}(\boldsymbol{W}+\epsilon\Delta \boldsymbol{W}^n)-\boldsymbol{R}(\boldsymbol{W})}{\epsilon} \tag{3.12}$$

这样就只需要计算残差即可。参数 ϵ 必须小心选取,以使数值误差最小(参见文献[300-301])。无矩阵方法更重要的优点是高阶残差 $\boldsymbol{R}^n$(数值上)的精确线性化易于在隐式格式中使用,因此在适度的成本下,牛顿格式可以实现二次收敛,在这种情况下,我们称之为 Newton-Krylov 方法[212,301-305]。实践经验表明,在所有 Krylov 子空间方法中,GMRES 最适合无矩阵实现[306]。一种令人感兴趣的方法是利用 LU-SGS 格式作为无矩阵 GMRES 方法的预处理器,因为 LU-SGS 格式也不需要显式存储通量雅可比矩阵,内存需求可以进一步降低,对于 3D 非结构网格的无黏流和层流,该方法的计算效率得到了验证[307]。

采用多重网格方法也可以提高隐式格式的收敛性,通常有两种可能的方法:首先,可以在隐式格式中使用多重网格作为每个时间步产生的线性方程组(3.9)的解算器,或者作为共轭梯度法的一个预处理器[308-309];其次,隐式格式本身可以作为直接应用于控制方程组的 FAS 多重网格法中的光滑器[282,310-313]。一些研究表明,至少对于纯空气动力学问题,相当"简单"的隐式格式(如高斯-赛德尔方法)与多重网格相结合产生的解算器比 GMRES(在 CPU 时间方面)效率更高[282,302]。

3.3 湍流模拟

在无黏或层流的情况下,控制方程组(2.19)的求解并不存在任何根本性的困难,然而,湍流模拟却是一个重大问题。尽管现代超级计算机的性能很好,但是用时间相关纳维-斯托克斯方程组(式(2.19))直接模拟湍流,即 DNS,仍然只能用于低 Re 数的相当简单的流动情况。考虑到满足空间分辨率所需的网格点数为 $Re^{9/4}$ 量级,CPU 时间为 Re^3 量级,DNS 的限制就变得非常明显。这并不意味着 DNS 完全无用,它是理解湍流结构和层流向湍流转捩的重要工具,在发展和校准新的或改进的湍流模型中也发挥着至关重要的作用。然而,在工程应用中,湍流的影响只能使用不同复杂度的模型来近似考虑。

达到了第一级近似的是 LES 方法。LES 发展的基础是,湍流运动中小尺度结构比传输湍流能量的大尺度结构更具有普适性,因此,我们可以只精确地求解大涡,而通过相对简单的亚格子尺度模型来近似小尺度的影响。由于 LES 所需要的网格点数比 DNS 少得多,因此研究高 *Re* 数下的湍流变得可行。但是,由于 LES 本质上是 3D 的和非定常的,它仍然保持非常高的计算需求,距离成为工程工具还有很长的路要走。LES 非常适合复杂流动物理现象的详细研究,包括大分离非定常流动、大尺度混合(如燃料和氧化剂)、气动噪声或流动控制策略的研究。LES 在燃烧室或发动机内的流动、传热和旋转流动的更精确计算方面也很有前景,文献[314]对 LES 的研究活动进行了综述。

下一级近似用 RANS 表示,这种方法由雷诺在 1895 年提出,其基础是将流场变量根据时间平均或系综平均分解为平均部分和波动部分[315](另见文献[316-317])。在密度不恒定的情况下,建议对速度分量应用密度(质量)加权或 Favre 分解[318-319],否则,由于涉及密度波动的其他相关性,平均控制方程组将变得相当复杂。通常假定 Morkovin 假设[320]是有效的,即当马赫数小于 5 时,边界层和尾迹的湍流结构不受密度波动的明显影响。

通过将分解的变量(平均部分和波动部分)代入到纳维-斯托克斯方程组(式(2.19))中,并进行平均,得到了形式上相同的平均变量的方程组(两个附加项除外)。黏性应力张量增加了一项——雷诺应力张量[315]

$$\boldsymbol{\tau}_{ij}^{R} = -\bar{\rho}\,\widetilde{v_i'' v_j''} \tag{3.13}$$

式中:v''_i,v''_j 为速度分量 u,v,w 的密度加权波动部分;$\bar{\ }$和$\tilde{\ }$分别表示系综平均和密度加权平均。雷诺应力张量表示湍流波动引起的平均动量的输运。此外,能量方程(参见式(2.8))中的热扩散通量 $k\nabla T$ 增加了湍流热通量[63],即

$$\boldsymbol{F}_{\mathrm{D}}^{\mathrm{T}} = -\bar{\rho}\,\widetilde{h''\boldsymbol{v}''} \tag{3.14}$$

因此,求解 RANS 方程组需要对雷诺应力(式(3.13))和湍流热通量(式(3.14))进行建模。这种方法的优点是,与 LES 相比可以使用相当粗的网格,并且可以假设(至少对于附着流或适度分离的流动)存在不变的平均解,与 LES 甚至 DNS 相比,这两个特性都显著减少了计算量。因此,RANS 方法在工程应用中非常流行。当然,由于采用了平均步骤,无法得到湍流结构的详细信息。

人们设计了各种湍流模型来封闭 RANS 方程组,并且研究仍在继续。这些模型可分为一阶封闭和二阶封闭。

最复杂,但也最灵活的是二阶封闭模型。由 Rotta[321]首先提出的雷诺应力输运(Reynolds-stress transport,RST)模型求解雷诺应力张量的模型传输方程,六个应力分量的偏微分方程组必须由一个附加关系来封闭,这个附加关系通常使用湍流耗散率方程。RST 模型能够考虑强烈的非局部影响和历史影响,还可以捕捉到流线曲率或系统旋转对湍流的影响。

与 RST 方法密切相关的是代数雷诺应力(algebraic Reynolds-stress,ARS)模型,它可以视为低层次模型与 RST 方法的组合。ARS 模型只采用两个输运方程,通常是湍动能和耗散率的输运方程,雷诺应力张量的分量通过非线性代数方程组与输运量相关联[322]。ARS 方法能够预测管道中的旋转湍流和二次流动,其精度类似于 RST 模型。RST 和 ARS 模型的详细综述可见文献[323-324]。

由于 RST 模型和 ARS 模型存在的数值问题,主要是 RST 模型的刚性和 ARS 方程的非线性,使得一阶封闭在实践中得到更广泛的应用。在这些模型中,雷诺应力由一个简单的标量值,即湍流涡黏性来表示,它基于 Boussinesq 的涡黏性假设[325-326],该假设认为与层流类似,湍流剪切应力与平均应变率呈线性关系,从而黏性应力张量(式(2.15))及控制方程组(2.19)中的动力学黏性 μ 被层流部分和湍流部分的和所代替,即

$$\mu=\mu_{\mathrm{L}}+\mu_{\mathrm{T}} \tag{3.15}$$

如前所述,层流黏性可以通过萨瑟兰公式(2.30)计算。类似地,湍流热通量(式(3.14))被模拟为

$$\boldsymbol{F}_{\mathrm{D}}^{\mathrm{T}}=-k_{\mathrm{T}}\nabla T \tag{3.16}$$

式中:k_{T} 为湍流热传导系数。因此,式(2.24)中的热传导系数变为

$$k=k_{\mathrm{L}}+k_{\mathrm{T}}=c_p\left(\frac{\mu_{\mathrm{L}}}{Pr_{\mathrm{L}}}+\frac{\mu_{\mathrm{T}}}{Pr_{\mathrm{T}}}\right) \tag{3.17}$$

通常假设流场中的湍流 Pr 数为常数(空气的 $Pr_{\mathrm{T}}=0.9$),湍流涡黏性系数 μ_{T} 必须借助于湍流模型来确定。涡黏性方法的局限性在于湍流与平均应变场之间的平衡假设,以及未考虑系统旋转的影响,采用修正项[327-328],或者使用非线性涡黏性方法[329-331],都可以显著提高基于涡黏性的模型的精度。

一阶封闭根据使用的输运方程的数量可以分为零方程、一方程和多方程模型。零方程模型也称为代数模型,它根据经验关系来计算湍流涡黏性,在计算中仅使用当地平均流场变量,因此不能模拟历史效应,这妨碍了对分离流的可靠预测。最广泛使用的代数模型是 Baldwin 和 Lomax[332]发展的,目前仍在某些应用中使用。

一方程模型和两方程模型考虑到了历史效应,其中湍流的对流和扩散由输运方程模拟。最广泛使用的一方程湍流模型是由 Spalart 和 Allmaras[333]提出的,

该模型基于类涡黏性变量,它在数值上非常稳定,且易于在结构网格和非结构网格上实现。

在两方程模型中,几乎所有的方法都使用湍动能的输运方程。在大量的两方程模型中,Launder 和 Spalding[334] 的 K-ϵ 模型和 Wilcox[335] 的 K-ω 模型在工程应用中最常使用,它们在计算量和精度之间采取了一个合理的折中。文献[336]对 Spalart-Allmaras 湍流模型和各种两方程湍流模型进行了有益的比较。湍流模型的另一个有用的比较可见文献[337]。

3.4 初边值条件

无论选择哪种数值方法来求解控制方程组(2.19),都必须指定合适的初始条件和边界条件。初始条件确定流体在时间 $t=0$ 时的状态,或者迭代格式第一步的状态。显然,初始猜测越好(越接近解),得到最终解的速度就越快,数值求解过程失败的概率也相应降低,因此初始解至少满足控制方程以及附加的热力学关系很重要。在外流空气动力学中,常见做法是将整个流场指定为自由来流的压力、密度和速度分量(以马赫数、迎角和侧滑角给出)。在叶轮机械中,尽最大努力来指定整个流场的流动方向是很重要的。压力场也是如此。因此使用低阶近似(如位势方法)来生成物理上有意义的初始估计解是非常值得的。

任何数值流动模拟都只考虑了真实物理域的某一部分,计算域的截断建立了人工边界,在边界上必须指定物理量的值。例如外流空气动力学中的远场边界,内流中的入口、出口和周期边界,以及对称面边界。当构造这样的边界条件时,主要的问题当然是在截断区域的解应该尽可能地接近于考虑整个物理区域时得到的解。在远场、入口和出口边界中,通常使用特征边界条件[338-340]来抑制流场中非物理扰动的产生,但尽管如此,远场或者入口和出口边界仍然不能放置得离研究对象(机翼、叶片等)太近,否则会降低解的精度。对于外流,当研究升力体时,可以使用以翼型或机翼中心的单涡来修正远场边界的流场变量[340-342],这样可以显著减小物体与远场边界之间的距离而不至于削弱解的准确性,或者在给定外边界位置情况下提高解的精度[264,342-343]。对于内流问题,建立了基于线性化欧拉方程组和小扰动傅里叶级数展开的进出口边界公式[344-346],这些公式允许将入口和出口边界放置在与叶片相对接近的位置,而不会显著影响求解。

当物体的表面与流体接触时,会出现另一种不同类型的边界条件。对于由欧拉方程组(2.44)控制的无黏流动,恰当的边界条件是要求流动与物面相切,即

$$\text{在物面上}\quad \boldsymbol{v}\cdot\boldsymbol{n}=0$$

相反，对于纳维-斯托克斯方程组，假定物面与靠近物面的流体之间没有相对速度，即无滑移边界条件

$$\text{在物面上} \quad u=v=w=0$$

在某些情况下，物面的处理会变得更加复杂，例如，必须满足给定的壁面温度分布，或者必须考虑热辐射（见文献[347-348]）。

此外，对于不同流体（如空气和水）的交界面，必须定义边界条件[349-353]。除了物理边界条件和截断流域所产生的边界条件外，还存在由数值求解方法本身产生的边界，例如坐标切割、块或者分区边界[32-38]。

边界条件的正确实现是流场解算器的关键所在，不仅解的精度强烈依赖于对边界的正确的物理和数值处理，而且鲁棒性和收敛速度也受其很大的影响。更多关于各种重要边界条件的细节将在第 8 章中给出。

参考文献

[1] Frymier PD, Hassan HA, Salas MD. Navier-Stokes calculations using cartesian grids: I. Laminar flows. AIAA J 1988; 26: 1181-8.

[2] De Zeeuw D, Powell KG. An adaptively refined Cartesian mesh solver for the Euler equations. J Comput Phys 1993; 104: 56-68.

[3] Coirier WJ, Powell KG. An accuracy assessment of Cartesian-mesh approaches for the Euler equations. J Comput Phys 1995; 117: 121-31.

[4] Ye T, Mittal R, Udaykumar HS, Shyy W. An accurate Cartesian grid method for viscous incompressible flows with complex immersed boundaries. J Comput Phys 1999; 156: 209-40.

[5] Kirshman D, Liu F. Cartesian grid solution of the Euler equations using a gridless boundary condition treatment. AIAA Paper 2003-3974; 2003.

[6] Carolina L, Tsai HM, Liu F. An embedded Cartesian grid Euler solver with radial basis function for boundary condition implementation. AIAA Paper 2008-532; 2008.

[7] Ishida T, Kawai S, Nakahashi K. A high-resolution method for flow simulations with Cartesian mesh method. AIAA Paper 2011-1296; 2011.

[8] Uddin H, Kramer RMJ, Pantano C. A Cartesian-based embedded geometry technique with adaptive high-order finite differences for compressible flow around complex geometries. J Comput Phys 2014; 262: 379-407.

[9] Richtmyer RD, Morton KW. Difference Methods for Initial Value Problems. 2nd ed. London: Wiley-Interscience; 1967.

[10] Volpe G. Performance of compressible flow codes at low mach number. AIAA J 1993; 31: 49-56.

[11] Harlow FH, Welch JE. Numerical calculation of time-dependent viscous incompressible flow with free surface. Phys Fluids 1965; 8: 2182-9.

[12] Patankar SV. Numerical heat transfer and fluid flow. New York:McGraw-Hill;1980.

[13] Ho Y-H, Lakshminarayana B. Computation of unsteady viscous flow using a pressure-based algorithm. AIAA J 1993;31:2232-40.

[14] Javadia K, Darbandia M, Taeibi-Rahni M. Three-dimensional compressible-incompressible turbulent flow simulation using a pressure-based algorithm. Comput Fluids 2008;37:747-66.

[15] Darwish M, Sraj I, Moukalled F. A coupled finite volume solver for the solution of incompressible flows on unstructured grids. J Comput Phys 2009;228:180-201.

[16] Shterev KS, Stefanov SK. Pressure based finite volume method for calculation of compressible viscous gas flows. J Comput Phys 2010;229:461-80.

[17] Chen ZJ, Przekwas AJ. A coupled pressure-based computational method for incompressible/compressible flows. J Comput Phys 2010;229:9150-65.

[18] Chorin AJ. A numerical method for solving incompressible viscous flow problems. J Comput Phys 1967;2:12-26.

[19] Turkel E. Preconditioned methods for solving the incompressible and low speed compressible equations. J Comput Phys 1987;72:277-98.

[20] van Leer B, Lee WT, Roe PL. Characteristic time-stepping or local preconditioning of the euler equations. AIAA Paper 91-1552;1991.

[21] Choi YH, Merkle CL. The application of preconditioning in viscous flows. J Comput Phys 1993;105:207-33.

[22] Turkel E. Review of preconditioning methods for fluid dynamics. Appl Numer Math 1993;12:257-84.

[23] Weiss J, Smith WA. Preconditioning applied to variable and constant density flows. AIAA J 1995;33:2050-7.

[24] Lee D. Local preconditioning of the Euler and Navier-Stokes equations. PhD Thesis, University of Michigan;1996.

[25] Jespersen D, Pulliam T, Buning P. Recent enhancements to OVERFLOW. AIAA Paper 97-0644;1997.

[26] Merkle CL, Sullivan JY, Buelow PEO, Venkateswaran S. Computation of flows with arbitrary equations of state. AIAA J 1998;36:515-21.

[27] Sharov D, Nakahashi K. Low speed preconditioning and LU-SGS scheme for 3-D viscous flow computations on unstructured grids. AIAA Paper 98-0614;1998.

[28] Mulas M, Chibbaro S, Delussu G, Di Piazza I, Talice M. Efficient parallel computations of flows of arbitrary fluids for all regimes of Reynolds, Mach and Grashof numbers. Int J Numer Meth Heat Fluid Flow 2002;12:637-57.

[29] Puoti V. Preconditioning method for low-speed flows. AIAA J 2003;41:817-30.

[30] Briley WR, Taylor LK, Whitfield DL. High-resolution viscous flow simulations at arbitrary Mach number. J Comput Phys 2003;184:79-105.

[31] Sockol PM. Multigrid solution of the Navier-Stokes equations at low speeds with large temperature variations. J Comput Phys 2003;192:570-92.

[32] Lee KD, Rubbert PE. Transonic flow computations using grid systems with block structure.

Lecture Notes in Physics, vol. 141. Berlin: Springer Verlag; 1981. p. 266-71.
[33] Rossow C-C. Efficient computation of inviscid flow fields around complex configurations using a multiblock multigrid method. Commun Appl Numer Meth 1992; 8: 735-47.
[34] Kuerten H, Geurts B. Compressible turbulent flow simulation with a multigrid multiblock method. In: Proceedings of the 6th Copper Mountain conference on multigrid methods; 1993.
[35] Rizzi A, Eliasson P, Lindblad I, Hirsch C, Lacor C, Haeuser J. The engineering of multiblock/multigrid software for navier stokes flows on structured meshes. Comput Fluids 1993; 22: 341-67.
[36] Enander R, Sterner E. Analysis of internal boundary conditions and communication strategies for multigrid multiblock methods. Dept. of Scientific Computing, Uppsala University, Sweden, Report No. 191; 1997.
[37] Rai MM. A conservative treatment of zonal boundaries for Euler equations calculations. J Comput Phys 1986; 62: 472-503.
[38] Kassies A, Tognaccini R. Boundary conditions for euler equations at internal block faces of multi-block domains using local grid refinement. AIAA Paper 90-1590; 1990.
[39] Benek JA, Buning PG, Steger JL. A 3-D chimera grid embedding technique. AIAA Paper 85-1523; 1985.
[40] Buning PG, Chu IT, Obayashi S, Rizk YM, Steger JL. Numerical simulation of the integrated space shuttle vehicle in ascent. AIAA Paper 88-4359; 1988.
[41] Chesshire G, Henshaw WD. Composite overlapping meshes for the solution of partial differential equations. J Comput Phys 1990; 90: 1-64.
[42] Pearce DG, Stanley SA, Martin Jr FW, Gomez RJ, Le Beau GJ, Buning PG, et al. Development of a large scale chimera grid system for the space shuttle launch vehicle. AIAA Paper 93-0533; 1993.
[43] Kao H-J, Liou M-S, Chow C-Y. Grid adaptation using chimera composite overlapping meshes. AIAA J 1994; 32: 942-9.
[44] Nakahashi K, Togashi F, Sharov D. Intergrid-boundary definition method for overset unstructured grid approach. AIAA J 2000; 38: 2077-84.
[45] Henshaw WD, Schwendeman DW. An adaptive numerical scheme for high-speed reactive flow on overlapping grids. J Comput Phys 2003; 191: 420-47.
[46] Xu L, Weng P. High order accurate and low dissipation method for unsteady compressible viscous flow computation on helicopter rotor in forward flight. J Comput Phys 2014; 258: 470-88.
[47] Roget B, Sitaraman J. Robust and efficient overset grid assembly for partitioned unstructured meshes. J Comput Phys 2014; 260: 1-24.
[48] Wang G, Duchaine F, Papadogiannis D, Duran I, Moreau S, Gicquel, LYM. An overset grid method for large eddy simulation of turbomachinery stages. J Comput Phys 2014; 274: 333-55.
[49] Thompson JF, Weatherill NP. Aspects of numerical grid generation: current science and art. AIAA Paper 93-3539; 1993.
[50] Nakahashi K. FDM-FEM Zonal approach for computations of compressible viscous flows. Lecture Notes in Physics, vol. 264. Springer Verlag; 1986. p. 494-8.

[51] Nakahashi K. A finite-element method on prismatic elements for the three-dimensional Navier-Stokes equations. Lecture Notes in Physics, vol. 323, Springer Verlag; 1989. p. 434-8.

[52] Holmes DG, Connell SD. Solution of the two-dimensional Navier-Stokes equations on unstructured adaptive grids. AIAA Paper 89-1932; 1989.

[53] Peace AJ, Shaw J. The modeling of aerodynamic flows by solution of the euler equations on mixed polyhedral grids. Int J Numer Meth Eng 1992; 35: 2003-29.

[54] Kallinderis Y, Khawaja A, McMorris H. Hybrid prismatic/tetrahedral grid generation for viscous flows around complex geometries. AIAA J 1996; 34: 291-8.

[55] Sharov D, Nakahashi K. Hybrid prismatic/tetrahedral grid generation for viscous flow applications. AIAA Paper 96-2000; 1996.

[56] McMorris H, Kallinderis Y. Octree-advancing front method for generation of unstructured surface and volume meshes. AIAA J 1997; 35: 976-84.

[57] Peraire J, Morgan K. Unstructured mesh generation for 3-d viscous flow. AIAA Paper 98-3010; 1998.

[58] Marcum DL, Gaither JA. Mixed element type unstructured grid generation for viscous flow applications. AIAA Paper 99-3252; 1999.

[59] Ito Y, Nakahashi K. Unstructured hybrid grid generation based on isotropic tetrahedral grids. AIAA Paper 2002-0861; 2002.

[60] Tsoutsanis P, Antoniadis AF, Drikakis D. WENO schemes on arbitrary unstructured meshes for laminar, transitional and turbulent flows. J Comput Phys 2014; 256: 254-76.

[61] Mavriplis DJ. Unstructured grid techniques. Annu Rev Fluid Mech 1997; 29: 473-514.

[62] Abe Y, Nonomura T, Iizuka N, Fujii K. Geometric interpretations and spatial symmetry property of metrics in the conservative form for high-order finite-difference schemes on moving and deforming grids. J Comput Phys 2014; 260: 163-203.

[63] Hirsch C. Numerical computation of internal and external flows, vols. 1 and 2. Chichester (UK): John Wiley and Sons; 1988.

[64] McDonald PW. The computation of transonic flow through two-dimensional gas turbine cascades. ASME Paper 71-GT-89; 1971.

[65] Turner MJ, Clough RW, Martin HC, Topp LP. Stiffness and deflection analysis of complex structures. J Aeronaut Soc 1956; 23: 805.

[66] Zienkiewicz OC, Taylor RL. The finite element method. 4th ed. Maidenhead; McGraw-Hill; 1991.

[67] Pironneau O. Finite element methods for fluids. Chichester: John Wiley; 1989.

[68] Hassan O, Probert EJ, Morgan K, Peraire J. Adaptive finite element methods for transient compressible flow problems. In: Brebbia CA, Aliabadi MH, editors. Adaptive finite and boundary element methods. London: Elsevier Applied Science; 1993. p. 119-60.

[69] Reddy JN, Gartling DK. The finite element method in heat transfer and fluid dynamics. Boca Raton, FL: CRC Press; 1994.

[70] Shadid JN, Moffat HK, Hutchinson SA, Hennigan GL, Devine KD, Salinger AG. MPSalsa: a finite element computer program for reacting flow problems. Part 1-Theoretical development.

SAND95-2752, May 1996.
[71] Atkins HL, Shu C-W. Quadrature-free implementation of discontinuous galerkin method for hyperbolic equations. AIAA J 1998;36(5):775-82.
[72] Van der Vegt JJW, van der Ven H. Discontinuous Galerkin finite element method with anisotropic local grid refinement for inviscid compressible flows. J Comput Phys 1998;141:46-77.
[73] Lomtev I, Kirby RM, Karniadakis GE. A discontinuous Galerkin ALE method for compressible viscous flows in moving domains. J Comput Phys 1999;155:128-59.
[74] Burbeau A, Sagaut P, Bruneau, Ch-H. A problem-independent limiter for high-order Runge-Kutta discontinuous Galerkin methods. J Comput Phys 2001;169:111-50.
[75] Van der Vegt JJW, van der Ven H. Space-time discontinuous Galerkin finite element method with dynamic grid motion for inviscid compressible flows: I. General formulation. J Comput Phys 2002;182:546-85.
[76] Hartmann R, Houston P. Adaptive discontinuous Galerkin finite element methods for the compressible euler equations. J Comput Phys 2002;183:508-32.
[77] Fidkowski KJ, Darmofal DL. Development of a higher-order solver for aerodynamic applications. AIAA Paper 2004-0436;2004.
[78] Aliabadi S, Tu S-Z, Watts M. An alternative to limiter in discontinuous Galerkin finite element method for simulation of compressible flows. AIAA Paper 2004-0076;2004.
[79] Özdemir H, Blom CPA, Hagmeijer R, Hoeijmakers HWM. Development of higher-order discontinuous galerkin method on hexahedral elements. AIAA Paper 2004-2961;2004.
[80] Dolejší V, Feistauer M. A semi-implicit discontinuous Galerkin finite element method for the numerical solution of inviscid compressible flow. J Comput Phys 2004;198:727-46.
[81] Atkins HL, Helenbrook BT. Numerical evaluation of P-multigrid method for the solution of discontinuous Galerkin discretizations of diffusive equations. AIAA Paper 2005-5110;2005.
[82] Toulopoulos I, Ekaterinaris JA. Discontinuous-Galerkin discretizations for viscous flow problems in complex domains. AIAA Paper 2005-1264;2005.
[83] Bustinza R, Gatica GN. A mixed local discontinuous Galerkin method for a class of nonlinear problems in fluid mechanics. J Comput Phys 2005;207:427-56.
[84] Fidkowski KJ, Oliver TA, Lu J, Darmofal DL. p-Multigrid solution of high-order discontinuous Galerkin discretizations of the compressible Navier-Stokes equations. J Comput Phys 2005;207:92-113.
[85] Chevaugeon N, Remacle J-F, Gallez X, Ploumhans P, Caro S. Efficient discontinuous galerkin methods for solving acoustic problems. AIAA Paper 2005-2823;2005.
[86] Luo H, Baum JD, Löhner R. A p-multigrid discontinuous Galerkin method for the Euler equations on unstructured grids. J Comput Phys 2006;211:767-83.
[87] Marchandise E, Remacle J-F, Chevaugeon N. A quadrature-free discontinuous Galerkin method for the level set equation. J Comput Phys 2006;212:338-57.
[88] Qiu J, Khoo BC, Shu C-W. A numerical study for the performance of the Runge-Kutta discontinuous Galerkin method based on different numerical fluxes. J Comput Phys 2006;212:540-65.

[89] Nastase CR, Mavriplis DJ. High-order discontinuous Galerkin methods using hp-multigrid approach. J Comput Phys 2006;213:330-57.
[90] Jacobs GB, Hesthaven JS. High-order nodal discontinuous Galerkin particle-in-cell method on unstructured grids. J Comput Phys 2006;214:96-121.
[91] Lörcher F, Gassner G, Munz C-D. A discontinuous Galerkin scheme based on a space-time expansion. I. Inviscid compressible flow in one space dimension. J Sci Comput 2007; 32:175-99.
[92] Gassner G, Lörcher F, Munz C-D. A discontinuous Galerkin scheme based on a space-time expansion. II. Viscous flow equations in multi dimensions. J Sci Comput 2007;34:260-86.
[93] Gassner G, Lörcher F, Munz C-D. A contribution to the construction of diffusion fluxes for finite volume and discontinuous Galerkin schemes. J Comput Phys 2007;224:1049-63.
[94] Wang L, Mavriplis DJ. Implicit solution of the unsteady Euler equations for high-order accurate discontinuous Galerkin discretizations. J Comput Phys 2007;225:1994-2015.
[95] Zhu J, Qiu J, Shu C-W, Dumbser M. Runge-Kutta discontinuous Galerkin method using WENO limiters. II: Unstructured meshes. J Comput Phys 2008;227:4330-53.
[96] Shahbazi K, Mavriplis DJ, Burgess NK. Multigrid algorithms for high-order discontinuous Galerkin discretizations of the compressible Navier-Stokes equations. J Comput Phys 2009; 228:7917-40.
[97] Mavriplis DJ, Nastase CR. On the geometric conservation law for high-order discontinuous Galerkin discretizations on dynamically deforming meshes. J Comput Phys 2011; 230:4285-300.
[98] Zhu J, Zhong X, Shu C-W, Qiu J. Runge-Kutta discontinuous Galerkin method using a new type of WENO limiters on unstructured meshes. J Comput Phys 2013;248:200-20.
[99] Patera AT. A spectral element method for fluid dynamics; laminar flow in a channel expansion. J Comput Phys 1984;54:468-88.
[100] Maday Y, Patera AT. Spectral element methods for the incompressible Navier-Stokes equations. ASME, State of the art surveys on computational mechanics; 1987. p. 71-143.
[101] Canuto C, Hussaini MY, Quarteroni A, Zang TA. Spectral methods in fluid dynamics. Berlin: Springer Verlag; 1987.
[102] Henderson RD, Karniadakis G. Unstructured spectral element methods for simulation of turbulent flows. J Comput Phys 1995;122:191-217.
[103] Amon CH. Spectral element-fourier method for unsteady conjugate heat transfer in complex geometry flows. J Thermophys Heat Transfer 1995;9:247-53.
[104] Henderson RD, Meiron DI. Dynamic refinement algorithms for spectral element methods. AIAA Paper 97-1855; 1997.
[105] Lomtev I, Quillen CB, Karniadakis GE. Spectral/hp methods for viscous compressible flows on unstructured 2D meshes. J Comput Phys 1998;144:325-57.
[106] Bouffanais R, Deville MO, Fischer PF, Leriche E, Weill D. Large-eddy simulation of the lid-driven cubic cavity flow by the spectral element method. J Sci Comput 2006;27:151-62.
[107] Pasquetti R, Rapetti F. Spectral element methods on unstructured meshes: comparisons and

recent advances. J Sci Comput 2006;27:377-87.
[108] Mei R, Luo L-S, Shyy W. An accurate curved boundary treatment in the lattice Boltzmann method. AIAA Paper 99-3353;1999.
[109] Hsu A, Yang T, Sun C, Lopez I. A lattice Boltzmann method for turbomachinery simulations. AIAA Paper 2002-3537;2002.
[110] Yu D, Mei R, Shyy W. A unified boundary treatment in lattice Boltzmann method. AIAA Paper 2003-0953;2003.
[111] Mavriplis DJ. Multigrid solution of the Lattice-Boltzmann equation. AIAA Paper 2005-5104;2005.
[112] Imamura T, Suzuki K, Nakamura T, Yoshida M. Flow simulation around an airfoil by lattice Boltzmann method on generalized coordinates. AIAA J 2005;43:1968-73.
[113] Crouse B, Freed D, Balasubramanian G, Senthooran S, Lew P-T, Mongeau L. Fundamental aeroacoustics capabilities of the lattice-Boltzmann method. AIAA Paper 2006-2571;2006.
[114] Marié S, Ricot D, Sagaut P. Accuracy of lattice Boltzmann method for aeroacoustic simulations. AIAA Paper 2007-3515;2007.
[115] Najafiyazdi A, Mongeau L. A perfectly matched layer formulation for lattice Boltzmann method. AIAA Paper 2009-3117;2009.
[116] Aono H, Gupta A, Qi D, Shyy W. The lattice Boltzmann method for flapping wing aerodynamics. AIAA Paper 2010-4867;2010.
[117] So RMC, Fu SC, Leung RCK. Finite difference lattice Boltzmann method for compressible thermal fluids. AIAA J 2010;48:1059-71.
[118] Eitel-Amor G, Meinke M, Schröder W. Lattice Boltzmann simulations with locally refined meshes. AIAA Paper 2011-3398;2011.
[119] Satti R, Li Y, Shock R, Noelting S. Unsteady flow analysis of a multi-element airfoil using lattice Boltzmann method. AIAA J 2012;50:1805-16.
[120] Perot F, Meskine M, LeGoff V, Vidal V, Gille F, Vergne S, et al. HVAC noise predictions using a lattice Boltzmann method. AIAA Paper 2013-2228;2013.
[121] Chen L, He Y-L, Kang Q, Tao W-Q. Coupled numerical approach combining finite volume and lattice Boltzmann methods for multi-scale multi-physicochemical processes. J Comput Phys 2013;255:83-105.
[122] Li Z, Yang M, Zhang Y. Hybrid lattice Boltzmann and finite volume method for natural convection. J Thermophys Heat Transfer 2014;28:68-77.
[123] Löhner R, Corrigan AT, Wichmann K-R, Wall W. Comparison of lattice-Boltzmann and finite difference solvers. AIAA Paper 2014-1439;2014.
[124] Touil H, Ricot D, Lévêque E. Direct and large-eddy simulation of turbulent flows on composite multi-resolution grids by the lattice Boltzmann method. J Comput Phys 2014;256:220-33.
[125] Yoshida H, Nagaoka M. Lattice Boltzmann method for the convection-diffusion equation in curvilinear coordinate systems. J Comput Phys 2014;257:884-900.
[126] Dellar PJ. Lattice Boltzmann algorithms without cubic defects in Galilean invariance on standard lattices. J Comput Phys 2014;259:270-83.

[127] Guzika SM, Weisgraber TH, Colella P, Alder BJ. Interpolation methods and the accuracy of lattice-Boltzmann mesh refinement. J Comput Phys 2014;259:461-87.

[128] Faviera J, Revell A, Pinelli A. A lattice Boltzmann-immersed boundary method to simulate the fluid interaction with moving and slender flexible objects. J Comput Phys 2014;261:145-61.

[129] Patil DV, Premnath KN, Banerjee S. Multigrid lattice Boltzmann method for accelerated solution of elliptic equations. J Comput Phys 2014;265:172-94.

[130] Batina JT. A gridless Euler/Navier-Stokes solution algorithm for complex-aircraft applications. AIAA Paper 93-0333;1993.

[131] Shih S-C, Lin S-Y. A weighting least squares method for Euler and Navier-Stokes equations. AIAA Paper 94-0522;1994.

[132] Series on gridless methods. Comput Meth Appl Mech Eng 1996;139:1-440 [also 2004;193:933-1321].

[133] Luo H, Baum J, Löhner R. A hybrid building-block and gridless method for compressible flows. AIAA Paper 2006-3710;2006.

[134] Tota P, Wang Z. Meshfree Euler solver using local radial basis functions for inviscid compressible flows. AIAA Paper 2007-4581;2007.

[135] Katz A, Jameson A. A comparison of various meshless schemes within a unified algorithm. AIAA Paper 2009-0596;2009.

[136] Tang L, Yang J, Lee J. Hybrid Cartesian grid/gridless algorithm for store separation prediction. AIAA Paper 2010-0508;2010.

[137] Barth TJ. Numerical aspects of computing high-Reynolds number flows on unstructured meshes. AIAA Paper 91-0721;1991.

[138] Mavriplis DJ, Venkatakrishnan V. A unified multigrid solver for the Navier-Stokes equations on mixed element meshes. ICASE Report No. 95-53;1995.

[139] Braaten ME, Connell SD. Three dimensional unstructured adaptive multigrid scheme for the Navier-Stokes equations. AIAA J 1996;34:281-90.

[140] Haselbacher AC, McGuirk JJ, Page GJ. Finite volume discretisation aspects for viscous flows on mixed unstructured grids. AIAA Paper 97-1946;1997 [also AIAA J 1999;37:177-84].

[141] Crumpton PI, Moiner P, Giles MB. An unstructured algorithm for high Reynolds number flows on highly-stretched grids. In: 10th Int. Conf. Num. Meth. for Laminar and Turbulent Flows, Swansea, England, July 21-25;1997.

[142] Weiss JM, Maruszewski JP, Smith WA. Implicit solution of preconditioned Navier-Stokes equations using algebraic multigrid. AIAA J 1999;37:29-36.

[143] Jameson A, Schmidt W, Turkel E. Numerical solutions of the Euler equations by finite volume methods using Runge-Kutta time-stepping schemes. AIAA Paper 81-1259;1981.

[144] Jameson A, Baker TJ, Weatherill NP. Calculation of inviscid transonic flow over a complete aircraft. AIAA Paper 86-0103;1986.

[145] Swanson RC, Turkel E. On central difference and upwind schemes. J Comput Phys 1992;101:292-306.

[146] Haselbacher A, Blazek J. On the accurate and efficient discretisation of the Navier-Stokes e-

quations on mixed grids. AIAA Paper 99-3363;1999 [also AIAA J 2000;38:2094-102].
[147] Steger JL, Warming RF. Flux vector splitting of the inviscid gasdynamic equations with application to finite difference methods. J Comput Phys 1981;40:263-93.
[148] van Leer B. Flux vector splitting for the Euler equations. In: Proc. 8th int. conf. on numerical methods in fluid dynamics, Springer Verlag; 1982. p. 507-12 [also ICASE Report 82-30; 1982].
[149] Liou M-S, Steffen Jr CJ. A new flux splitting scheme. NASA TM-104404;1991 [also J Comput Phys 1993;107:23-39].
[150] Liou M-S. A sequel to AUSM: AUSM+. J Comput Phys 1996;129:364-82.
[151] Jameson A. Positive schemes and shock modelling for compressible flow. Int J Numer Meth Fluids 1995;20:743-76.
[152] Tatsumi S, Martinelli L, Jameson A. A new high resolution scheme for compressible viscous flow with shocks. AIAA Paper 95-0466;1995.
[153] Edwards JR. A low-diffusion flux-splitting scheme for Navier-Stokes calculations. Comput Fluids 1997;26:653-9.
[154] Rossow C-C. A simple flux splitting scheme for compressible flows. In: Proc. 11th DGLR-Fach-symposium, Berlin, Germany, November 10-12;1998.
[155] Rossow C-C. A flux splitting scheme for compressible and incompressible flows. AIAA Paper 99-3346;1999.
[156] Godunov SK. A difference scheme for numerical computation discontinuous solution of hydrodynamic equations. Math. Sbornik [in Russian] 1959;47:271-306 [translated US Joint Publ. Res. Service, JPRS 7226;1969].
[157] Osher S, Solomon F. Upwind difference schemes for hyperbolic systems of conservation laws. Math Comput 1982;38:339-74.
[158] Roe PL. Approximate Riemann solvers, parameter vectors, and difference schemes. J Comput Phys 1981;43:357-72.
[159] Barth TJ, Jespersen DC. The design and application of upwind schemes on unstructured meshes. AIAA Paper 89-0366;1989.
[160] Einfeldt B. On Godunov-type methods for gas dynamics. SIAM J Numer Anal 1988; 25:294-318.
[161] Toro EF, Spruce M, Speares W. Restoration of the contact surface in the HLL-Riemann solver. Shock Waves 1994;4:25-34.
[162] Quirk JJ. A contribution to the great Riemann solver debate. Int J Numer Meth Fluids 1994; 18:555-74.
[163] Kim SD, Lee BJ, Lee HJ, Jeung I-S. Robust HLLC Riemann solver with weighted average flux scheme for strong shock. J Comput Phys 2009;228:7634-42.
[164] Balsara DS, Dumbser M, Abgrall R. Multidimensional HLLC Riemann solver for unstructured meshes—with application to Euler and MHD flows. J Comput Phys 2014;261:172-208.
[165] Nishikawa H, Kitamura K. Very simple, carbuncle-free, boundary-layer-resolving, rotated-hybrid Riemann solvers. J Comput Phys 2008;227:2560-81.

[166] Harten A. High resolution schemes for hyperbolic conservation laws. J Comput Phys 1983; 49:357-93.

[167] Yee HC, Harten A. Implicit TVD schemes for hyperbolic conservation laws in curvilinear coordinates. AIAA J 1987;25:266-74.

[168] Yee HC. Construction of implicit and explicit symmetric TVD schemes and their applications. J Comput Phys 1987;68:151-79.

[169] Roe PL. Discrete models for the numerical analysis of time-dependent multidimensional gas dynamics. J Comput Phys 1986;63:458-76.

[170] Powell KG, van Leer B, Roe PL. Towards a genuinely multi-dimensional upwind scheme. Rhode-St-Genèse: VKI Lecture Series 1990-03;1990.

[171] Sidilkover D. A genuinely multidimensional upwind scheme and efficient multigrid solver for the compressible Euler equations. ICASE Report, No. 94-84;1994.

[172] Struijs R, Roe PL, Deconinck H. Fluctuation splitting schemes for the 2D Euler equations. Rhode-St-Genèse: VKI Lecture Series 1991-01;1991.

[173] Paillére H, Deconinck H, Roe PL. Conservative upwind residual-distribution schemes based on the steady characteristics of the Euler equations. AIAA Paper 95-1700;1995.

[174] Issman E, Degrez G, Deconinck H. Implicit upwind residual-distribution Euler and Navier-Stokes solver on unstructured meshes. AIAA J 1996;34:2021-28.

[175] Van der Weide E, Deconinck H. Compact residual-distribution scheme applied to viscous flow simulations. Rhode-St-Genèse: VKI Lecture Series 1998-03;1998.

[176] Abgrall R, Larat A, Ricchiuto M. Construction of very high order residual distribution schemes for steady inviscid flow problems on hybrid unstructured meshes. J Comput Phys 2011; 230:4103-36.

[177] Harten A, Engquist B, Osher S, Chakravarthy S. Uniformly high order accurate essentially non-oscillatory schemes III. J Comput Phys 1987; 71: 231-303 [also ICASE Report No. 86-22;1986].

[178] Casper J, Atkins HL. A finite-volume high-order ENO scheme for two-dimensional hyperbolic systems. J Comput Phys 1993;106:62-76.

[179] Godfrey AG, Mitchell CR, Walters RW. Practical aspects of spatially high-order accurate methods. AIAA J 1993;31:1634-42.

[180] Abgrall R, Lafon FC. ENO schemes on unstructured meshes. Rhode-St-Genèse: VKI Lecture Series 1993-04;1993.

[181] Abgrall R. On essentially non-oscillatory schemes on unstructured meshes: analysis and implementation. J Comput Phys 1994;114:45-58.

[182] Jiang G-S, Shu C-W. Efficient implementation of weighted ENO schemes. J Comput Phys 1996;126:202-28.

[183] Ollivier-Gooch CF. High-Order ENO schemes for unstructured meshes based on least-squares reconstruction. AIAA Paper 97-0540;1997.

[184] Stanescu D, Habashi WG. Essentially nonoscillatory euler solutions on unstructured meshes using extrapolation. AIAA J 1998;36:1413-16.

[185] Friedrich O. Weighted essentially non-oscillatory schemes for the interpolation of mean values on unstructured grids. J Comput Phys 1998;144:194-212.

[186] Shi J, Hu C, Shu C-W. A technique of treating negative weights in WENO schemes. J Comput Phys 2002;175:108-27.

[187] Martín MP, Taylor EM, Wu M, Weirs VG. A bandwidth-optimized WENO scheme for the effective direct numerical simulation of compressible turbulence. J Comput Phys 2006;220:270-89.

[188] Dumbser M, Käser M, Titarev VA, Toro EF. Quadrature-free non-oscillatory finite volume schemes on unstructured meshes for nonlinear hyperbolic systems. J Comput Phys 2007;226:204-43.

[189] Borges R, Carmona M, Costa B, Don WS. An improved weighted essentially non-oscillatory scheme for hyperbolic conservation laws. J Comput Phys 2008;227:3191-211.

[190] Dumbser M, Enaux C, Toro EF. Finite volume schemes of very high order of accuracy for stiff hyperbolic balance laws. J Comput Phys 2008;227:3971-4001.

[191] Castro M, Costa B, Don WS. High order weighted essentially non-oscillatory WENO-Z schemes for hyperbolic conservation laws. J Comput Phys 2011;230:1766-92.

[192] Tsoutsanis P, Antoniadis AF, Drikakis D. WENO schemes on arbitrary unstructured meshes for laminar, transitional and turbulent flows. J Comput Phys 2014;256:254-76.

[193] Ivan L, Groth, CPT. High-order solution-adaptive central essentially non-oscillatory (CENO) method for viscous flows. J Comput Phys 2014;257:830-62.

[194] Xu L, Weng P. High order accurate and low dissipation method for unsteady compressible viscous flow computation on helicopter rotor in forward flight. J Comput Phys 2014;258:470-88.

[195] Huang C-S, Arbogast T, Hung C-H. A re-averaged WENO reconstruction and a third order CWENO scheme for hyperbolic conservation laws. J Comput Phys 2014;262:291-312.

[196] Boscheri W, Balsara DS, Dumbser M. Lagrangian ADER-WENO finite volume schemes on unstructured triangular meshes based on genuinely multidimensional HLL Riemann solvers. J Comput Phys 2014;267:112-38.

[197] Fan P, Shen Y, Tian B, Yang C. A new smoothness indicator for improving the weighted essentially non-oscillatory scheme. J Comput Phys 2014;269:329-54.

[198] Venkatakrishnan V. Preconditioned conjugate gradient methods for the compressible Navier-Stokes equations. AIAA J 1991;29:1092-110.

[199] Venkatakrishnan V. On the accuracy of limiters and convergence to steady state solutions. AIAA Paper 93-0880;1993.

[200] Venkatakrishnan V. Convergence to steady-state solutions of the Euler equations on unstructured grids with limiters. J Comput Phys 1995;118:120-30.

[201] Vinokur M. Flux Jacobian matrices and generalized roe average for an equilibrium real gas. NASA CR-177512;1988.

[202] Vinokur M, Liu Y. Equilibrium gas flow computations: II. An analysis of numerical formulations of conservation laws. AIAA Paper 88-0127;1988.

[203] Vinokur M, Montagné J-L. Generalized flux-vector splitting and Roe average for an equilibrium real gas. J Comput Phys 1990;89:276-300.

[204] Vinokur M, Liu Y. Nonequilibrium flow computations: I. An analysis of numerical formulations of conservation laws. NASA CR-177489;1988.

[205] Molvik GA, Merkle CL. A set of strongly coupled, upwind algorithms for computing flows in chemical nonequilibrium. AIAA Paper 89-0199;1989.

[206] Liu Y, Vinokur M. Upwind algorithms for general thermo-chemical nonequilibrium flows. AIAA Paper 89-0201;1989.

[207] Grossman B, Cinnella P. Flux-split algorithms for flows with non-equilibrium chemistry and vibrational relaxation. J Comput Phys 1990;88:131-68.

[208] Shuen J-S, Liou M-S, van Leer B. Inviscid flux-splitting algorithms for real gases with non-equilibrium chemistry. J Comput Phys 1990;90:371-95.

[209] Slomski JF, Anderson JD, Gorski JJ. Effectiveness of multigrid in accelerating convergence of multidimensional flows in chemical nonequilibrium. AIAA Paper 90-1575;1990.

[210] Yu S-T, Chang S-C, Jorgenson PCE, Park S-J, Lai M-C. Basic equations of chemically reactive flow for computational fluid dynamics. AIAA Paper 98-1051;1998.

[211] Rinaldi E, Pecnik R, Colonna P. Exact Jacobians for implicit Navier-Stokes simulations of equilibrium real gas flows. J Comput Phys 2014;270:459-77.

[212] Venkatakrishnan V. Implicit schemes and parallel computing in unstructured grid CFD. ICASE Report No. 95-28;1995.

[213] Venkatakrishnan V, Mavriplis DJ. Implicit method for the computation of unsteady flows on unstructured grids. J Comput Phys 1996;127:380-97.

[214] van Leer B, Tai CH, Powell KG. Design of optimally smoothing multi-stage schemes for the Euler equations. AIAA Paper 89-1933;1989.

[215] Chang HT, Jiann HS, van Leer B. Optimal multistage schemes for Euler equations with residual smoothing. AIAA J 1995;33:1008-16.

[216] Shu CW, Osher S. Efficient implementation of essentially non-oscillatory shock capturing schemes. J Comput Phys 1988;77:439-71.

[217] Lafon A, Yee HC. On the numerical treatment of nonlinear source terms in reaction-convection equations. AIAA Paper 92-0419;1992.

[218] van Leer B, Lee WT, Roe P. Characteristic time-stepping or local preconditioning of the Euler equations. AIAA Paper 91-1552;1991.

[219] Riemslagh K, Dick E. A multigrid method for steady Euler equations on unstructured adaptive grids. In: Proc. 6th copper mountain conf. on multigrid methods, NASA Conf. Publ. 1993; 3224:527-42.

[220] Morano E, Dervieux A. Looking for O(N) Navier-Stokes solutions on non-structured meshes. In: Proc. 6th Copper Mountain conf. on multigrid methods, NASA Conf. Publ. 1993; 3224:449-64.

[221] Ollivier-Gooch CF. Towards problem-independent multigrid convergence rates for unstructured mesh methods. In: Proc. 6th Int. Symp. CFD, Lake Tahoe, NV;1995.

[222] Jameson A. Solution of the Euler equations by a multigrid method. Appl Math Comput 1983;13:327-56.
[223] Jameson A. Computational transonics. Commun Pure Appl Math 1988;41:507-49.
[224] Blazek J, Kroll N, Radespiel R, Rossow C-C. Upwind implicit residual smoothing method for multi-stage schemes. AIAA Paper 91-1533;1991.
[225] Blazek J, Kroll N, Rossow C-C. A comparison of several implicit smoothing methods. In: Proc. ICFD conf. on numerical meth. for fluid dynamics, Reading;1992. p. 451-60.
[226] Enander E. Improved implicit residual smoothing for steady state computations of first-order hyperbolic systems. J Comput Phys 1993;107:291-6.
[227] Enander E, Sjögreen B. Implicit explicit residual smoothing for upwind schemes. Internal Report 96-179, Department of Scientific Computing, Uppsala University, Sweden;1996.
[228] Fedorenko RP. A relaxation method for solving elliptic difference equations. USSR Comput Math Math Phys 1962;1(5):1092-6.
[229] Bakhvalov NS. On the convergence of a relaxation method with natural constraints on the elliptic operator. USSR Comput Math Math Phys 1966;6(5):101-35.
[230] Brandt A. Multi-level adaptive solutions to boundary-value problems. Math Comput 1977;31:333-90.
[231] Brandt A. Guide to multigrid development. Multigrid Methods I, Lecture Notes in Mathematics, vol. 960. Berlin: Springer Verlag;1981.
[232] Jameson A. Multigrid algorithms for compressible flow calculations. Multigrid Methods II, Lecture Notes in Mathematics, vol. 1228. Berlin: Springer Verlag;1985. p. 166-201.
[233] Martinelli L. Calculation of viscous flows with a multigrid method. Ph. D. Thesis, Dept. of Mech. and Aerospace Eng., Princeton University;1987.
[234] Mulder WA. A new multigrid approach to convection problems. J Comput Phys 1989;83:303-23.
[235] Koren B. Multigrid and defect correction for the steady Navier-Stokes equations. J Comput Phys 1990;87:25-46.
[236] Turkel E, Swanson RC, Vatsa VN, White JA. Multigrid for hypersonic viscous two-and three-dimensional flows. AIAA Paper 91-1572;1991.
[237] Radespiel R, Swanson RC. Progress with multigrid schemes for hypersonic flow problems. ICASE Report, No. 91-89;1991 [also J Comput Phys 1995;116:103-22].
[238] Arnone A, Pacciani R. Rotor-stator interaction analysis using the Navier-Stokes equations and a multigrid method. Trans ASME J Turbomach 1996;118:679-89.
[239] Sockol PM. Multigrid solution of the Navier-Stokes equations at low speeds with large temperature variations. J Comput Phys 2003;192:570-92.
[240] Mavriplis DJ. Three-dimensional multigrid for the Euler equations. AIAA J 1992;30:1753-61.
[241] Peraire J, Peiró J, Morgan K. A 3D finite-element multigrid solver for the Euler equations. AIAA Paper 92-0449;1992.
[242] Lallemand M, Steve H, Dervieux A. Unstructured multigridding by volume agglomeration: cur-

rent status. Comput Fluids 1992;21:397-433.

[243] Crumpton PI, Giles MB. Aircraft computations using multigrid and an unstructured parallel library. AIAA Paper 95-0210;1995.

[244] Ollivier-Gooch CF. Multigrid acceleration of an upwind Euler solver on unstructured meshes. AIAA J 1995;33:1822-27.

[245] Mavriplis DJ. Multigrid techniques for unstructured meshes. ICASE Report No. 95-27;1995.

[246] Mavriplis DJ, Venkatakrishnan V. A 3D agglomeration multigrid solver for the Reynolds-averaged Navier-Stokes equations on unstructured meshes. Int J Numer Meth Fluids 1996; 23:527-44.

[247] Mavriplis DJ. Directional agglomeration multigrid techniques for high Reynolds number viscous flow solvers. AIAA Paper 98-0612;1998.

[248] Mavriplis DJ. Multigrid strategies for viscous flow solvers on anisotropic unstructured meshes. J Comput Phys 1998;145:141-65.

[249] Okamoto N, Nakahashi K, Obayashi S. A coarse grid generation algorithm for agglomeration multigrid method on unstructured grids. AIAA Paper 98-0615;1998.

[250] Carré G, Lanteri S. Parallel linear multigrid by agglomeration for the acceleration of 3d compressible flow calculations on unstructured meshes. Numer Algorit 2000;24:309-32.

[251] Roberts TW, Sidilkover D, Swanson RC. Textbook multigrid efficiency for the steady Euler equations. AIAA Paper 97-1949;1997.

[252] Pulliam TH. Time accuracy and the use of implicit methods. AIAA Paper 93-3360;1993.

[253] Arnone A, Liou M-S, Povinelli LA. Multigrid time-accurate integration of Navier-Stokes equations. AIAA Paper 93-3361;1993.

[254] Alonso J, Martinelli L, Jameson A. Multigrid unsteady Navier-Stokes calculations with aeroelastic applications. AIAA Paper 95-0048;1995.

[255] George A, Liu JW. Computer solution of large sparse positive definite systems. Prentice Hall Series in Comput. Math. Englewood Cliffs, NJ: Prentice Hall;1981.

[256] Pothen A, Simon HD, Liou KP. Partitioning sparse matrices with eigenvectors of graphs. SIAM J Matrix Anal Appl 1990;11:430-52.

[257] Bender EE, Khosla PK. Solution of the two-dimensional Navier-Stokes equations using sparse matrix solvers. AIAA Paper 87-0603;1987.

[258] Venkatakrishnan V. Newton solution of inviscid and viscous problems. AIAA J 1989; 27:885-91.

[259] Beam RM, Bailey HS. Viscous computations using a direct solver. Comput Fluids 1990; 18:191-204.

[260] Vanden KJ, Whitfield DL. Direct and iterative algorithms for the three-dimensional Euler equations. AIAA J 1995;33:851-8.

[261] Venkatakrishnan V, Barth TJ. Application of direct solvers to unstructured meshes for the Euler and Navier-Stokes equations using upwind schemes. AIAA Paper 89-0364;1989.

[262] Briley WR, McDonald H. Solution of the multi-dimensional compressible Navier-Stokes equations by a generalized implicit method. J Comput Phys 1977;24:372-97.

[263] Beam R, Warming RF. An implicit factored scheme for the compressible Navier-Stokes equations. AIAA J 1978;16:393-402.

[264] Pulliam TH, Steger JL. Recent improvements in efficiency, accuracy and convergence for implicit approximate factorization scheme. AIAA Paper 85-0360;1985.

[265] Rosenfeld M, Yassour Y. The alternating direction multi-zone implicit method. J Comput Phys 1994;110:212-20.

[266] Golub GH, Van Loan CF. Matrix computations. The Johns Hopkins University Press, Baltimore, Maryland;1983.

[267] Chakravarthy SR. Relaxation methods for unfactored implicit schemes. AIAA Paper 84-0165; 1984.

[268] Napolitano M, Walters RW. An incremental line Gauss-Seidel method for the incompressible and compressible Navier-Stokes equations. AIAA Paper 85-0033;1985.

[269] Thomas JL, Walters RW. Upwind relaxation algorithms for the Navier-Stokes equations. AIAA J 1987;25:527-34.

[270] Jenssen CB. Implicit multiblock Euler and Navier-Stokes calculations. AIAA J 1994; 32:1808-14.

[271] Yoon S, Jameson A. Lower-upper implicit schemes with multiple grids for the Euler equations. AIAA Paper 86-0105 [also AIAA J 1987;7:929-35].

[272] Yoon S, Jameson A. An LU-SSOR scheme for the Euler and Navier-Stokes equations. AIAA Paper 87-0600 [also AIAA J 1988;26:1025-26].

[273] Rieger H, Jameson A. Solution of steady three-dimensional compressible Euler and Navier-Stokes equations by an implicit LU scheme. AIAA Paper 88-0619;1988.

[274] Shuen J-S. Upwind differencing and LU factorization for chemical non-equilibrium Navier-Stokes equations. J Comput Phys 1992;99:233-50.

[275] Blazek J. Investigations of the implicit LU-SSOR scheme. DLR Research Report, No. 93-51;1993.

[276] Fezoui L, Stoufflet B. A class of implicit upwind schemes for Euler simulations with unstructured meshes. J Comput Phys 1989;84:174-206.

[277] Karman SL. Development of a 3D unstructured CFD method. PhD thesis, The University of Texas at Arlington;1991.

[278] Batina JT. Implicit upwind solution algorithms for three-dimensional unstructured meshes. AIAA J 1993;31:801-5.

[279] Slack DC, Whitaker DL, Walters RW. Time integration algorithms for the two-dimensional Euler equations on unstructured meshes. AIAA J 1994;32:1158-66.

[280] Anderson WK. A grid generation and flow solution method for the Euler equations on unstructured grids. J Comput Phys 1994;110:23-38.

[281] Anderson WK, Bonhaus DL. An implicit upwind algorithm for computing turbulent flows on unstructured grids. Comput Fluids 1994;23:1-21.

[282] Anderson WK, Rausch RD, Bonhaus DL. Implicit/multigrid algorithms for incompressible turbulent flows on unstructured grids. J Comput Phys 1996;128:391-408 [also AIAA Paper

95-1740;1995].

[283] Tomaro RF, Strang WZ, Sankar LN. An implicit algorithm for solving time dependent flows on unstructured grids. AIAA Paper 97-0333;1997.

[284] Kano S, Nakahashi K. Navier-Stokes computations of HSCT off-design aerodynamics using unstructured hybrid grids. AIAA Paper 98-0232;1998.

[285] Hassan O, Morgan K, Peraire J. An adaptive implicit/explicit finite element scheme for compressible high speed flows. AIAA Paper 89-0363;1989.

[286] Hassan O, Morgan K, Peraire J. An implicit finite element method for high speed flows. AIAA Paper 90-0402;1990.

[287] Gibbons A. Algorithmic graph theory. New York: Cambridge University Press;1985.

[288] Löhner R, Martin D. An implicit linelet-based solver for incompressible flows. AIAA Paper 92-0668;1992.

[289] Hestenes MR, Stiefel EL. Methods of conjugate gradients for solving linear systems. J Res Nat Bur Stand 1952;49:409.

[290] Sonneveld P. CGS, a fast Lanczos-type solver for nonsymmetric linear systems. SIAM J Sci Stat Comput 1989;10:36-52.

[291] Van der Vorst HA. BiCGSTAB: a fast and smoothly converging variant of Bi-CG for the solution of nonsymmetric linear systems. SIAM J Sci Stat Comput 1992;13:631-44.

[292] Freund RW. A transpose-free quasi-minimal residual algorithm for non-Hermitian linear systems. SIAM J Sci Comput 1993;14:470-82.

[293] Saad Y, Schultz MH. GMRES: a generalized minimum residual algorithm for solving nonsymmetric linear systems. SIAM J Sci Stat Comput 1986;7:856-69.

[294] Meister A. Comparison of different Krylov subspace methods embedded in an implicit finite volume scheme for the computation of viscous and inviscid flow fields on unstructured grids. J Comput Phys 1998;140:311-45.

[295] Meijerink JA, Van der Vorst HA. Guidelines for usage of incomplete decompositions in solving sets of linear equations as they occur in practical problems. J Comput Phys 1981;44:134-55.

[296] Wigton LB, Yu NJ, Young DP. GMRES acceleration of computational fluid dynamics codes. AIAA Paper 85-1494;1985.

[297] Kadioğlu M, Mudrick S. On the implementation of the GMRES(m) method to elliptic equations in meteorology. J Comput Phys 1992;102:348-59.

[298] Venkatakrishnan V, Mavriplis DJ. Implicit solvers for unstructured meshes. J Comput Phys 1993;105:83-91.

[299] Ajmani K, Ng W-F, Liou M-S. Preconditioned conjugate gradient methods for low speed flow calculations. AIAA Paper;93-0881;1993.

[300] Dennis JE, Schnabel RB. Numerical methods for unconstrained optimization and nonlinear equations. Englewood Cliffs, NJ: Prentice-Hall;1983.

[301] Brown PN, Saad Y. Hybrid Krylov methods for nonlinear systems of equations. SIAM J Sci Stat Comput 1990;11:450-81.

[302] Nielsen EJ, Anderson WK, Walters RW, Keyes DE. Application of Newton-Krylov methodology to a three-dimensional unstructured Euler code. AIAA Paper 95-1733;1995.

[303] Cai X, Keyes DE, Venkatakrishnan V. Newton-Krylov-Schwarz: an implicit solver for CFD. ICASE Report 95-87;1995.

[304] Tidriri MD. Schwarz-based algorithms for compressible flows. ICASE Report No. 96-4;1996.

[305] Zingg D, Pueyo A. An efficient Newton-GMRES solver for aerodynamic computations. AIAA Paper 97-1955;1997.

[306] McHugh PR, Knoll DA. Inexact Newton' s method solutions to the incompressible Navier-Stokes and energy equations using standard and matrix-free implementations. AIAA Paper 93-3332;1993.

[307] Luo H, Baum JD, Löhner R. A fast, matrix-free implicit method for compressible flows on unstructured grids. AIAA Paper 99-0936;1999.

[308] Raw M. Robustness of coupled algebraic multigrid for the Navier-Stokes equations. AIAA Paper 96-0297;1996.

[309] Brieger L, Lecca G. Parallel multigrid preconditioning of the conjugate gradient method for systems of subsurface hydrology. J Comput Phys 1998;142:148-62.

[310] Yoon S, Kwak D. Multigrid convergence of an implicit symmetric relaxation scheme. AIAA Paper 93-3357;1993.

[311] Blazek J. A multigrid LU-SSOR scheme for the solution of hypersonic flow problems. AIAA Paper 94-0062;1994.

[312] Oosterlee CW. A GMRES-based plane smoother in multigrid to solve 3D anisotropic fluid flow problems. J Comput Phys 1997;130:41-53.

[313] Gerlinger P, Stoll P, Brüggemann D. An implicit multigrid method for the simulation of chemically reacting flows. J Comput Phys 1998;146:322-45.

[314] Piomelli U. Large-eddy simulation: present state and future perspectives. AIAA Paper 98-0534;1998.

[315] Reynolds O. On the dynamical theory of incompressible viscous fluids and the determination of the criterion. Phil Trans R Soc Lond A 1895;186:123-64.

[316] Schlichting H. Boundary layer theory. New York: McGraw Hill;1968.

[317] Young AD. Boundary layers. BSP Professional Books. Oxford: Blackwell Scientific Publication Ltd.;1989.

[318] Favre A. Equations des gaz turbulents compressibles, part 1: formes générales. J Mécan 1965; 4:361-90.

[319] Favre A. Equations des gaz turbulents compressibles, part 2: méthode des vitesses moyennes; méthode des vitesses moyennes pondérées par la masse volumique. J Mécan 1965; 4: 391-421.

[320] Morkovin MV. Effects of compressibility on turbulent flow. In: Favre A, editor. The mechanics of turbulence. New York: Gordon & Breach;1964.

[321] Rotta J. Statistische theorie nichthomogener turbulenz I. Z Phys 1951;129:547-72.

[322] Rodi W. A new algebraic relation for calculating the Reynolds stresses. ZAMM 1976;

56:219-21.
[323] Speziale CG. A review of Reynolds stress models for turbulent shear flows. ICASE Report No. 95-15;1995.
[324] Hallbäck M, Henningson DS, Johansson AV, Alfredsson PH, editors. Turbulence and transition modelling. ERCOFTAC Series, vol. 2, Kluwer Academic Publishers, Dordrecht, The Netherlands;1996.
[325] Boussinesq J. Essai sur la théorie des eaux courantes. Mem Pres Acad Sci 1877;XXIII:46.
[326] Boussinesq J. Theorie de l' écoulement tourbillonant et tumulteur des liquides dans les lits rectiligues. Comptes Rendus Acad Sci 1896;CXXII:1293.
[327] Spalart P, Shur M. On the sensitization of turbulence models to rotation and curvature. Aerospace Sci Tech 1997;5:297-302.
[328] Shur M, Strelets M, Travin A, Spalart P. Turbulence modeling in rotating and curved channels: assessment of the Spalart-Shur correction term. AIAA Paper 98-0325;1998.
[329] Shih TH, Zhu J, Liou WW, Chen K-H, Liu N-S, Lumley J. Modeling of turbulent swirling flows. In: Proc. 11th symposium on turbulent shear flows, Grenoble, France;1997 [also NASA TM-113112;1997].
[330] Chen K-H, Liu N-S. Evaluation of a non-linear turbulence model using mixed volume unstructured grids. AIAA Paper 98-0233;1998.
[331] Abdel Gawad AF, Abdel Latif OE, Ragab SA, Shabaka IM. Turbulent flow and heat transfer in rotating non-circular ducts with nonlinear k-ϵ model. AIAA Paper 98-0326;1998.
[332] Baldwin BS, Lomax HJ. Thin Layer Approximation and Algebraic Model for Separated Turbulent Flow. AIAA Paper 78-257;1978.
[333] Spalart P, Allmaras S. A one-equation turbulence model for aerodynamic flows. AIAA Paper 92-0439;1992.
[334] Launder BE, Spalding B. The numerical computation of turbulent flows. Comput Meth Appl Mech Eng 1974;3:269-89.
[335] Wilcox DC. Reassessment of the scale determining equation for advanced turbulence models. AIAA J 1988;26:1299-310.
[336] Bardina JE, Huang PG, Coakley TJ. Turbulence modeling validation, testing, and development. NASA TM-110446;1997.
[337] McParlin S, Adamczak D. Comparison of turbulence models for the RAE Model 2155 test cases. AIAA Paper 2003-598;2003.
[338] Jameson A. A non-oscillatory shock capturing scheme using flux limited dissipation. MAE Report No. 1653, Dept. of Mechanical and Aerospace Engineering, Princeton University;1984.
[339] Whitfield DL, Janus JM. Three-dimensional unsteady Euler equations solution using flux vector splitting. AIAA Paper 84-1552;1984.
[340] Thomas JL, Salas MD. Far field boundary conditions for transonic lifting solutions to the Euler equations. AIAA J 1986;24:1074-80.
[341] Usab WJ, Murman EM. Embedded mesh solution of the Euler equations using a multiple-grid method. AIAA Paper 83-1946;1983.

[342] Radespiel R. A cell-vertex multigrid method for the Navier-Stokes equations. NASA TM-101557;1989.

[343] Kroll N, Jain RK. Solution of two-dimensional Euler equations-experience with a finite volume code. DFVLR-FB 87-41;1987.

[344] Verhoff A. Modeling of computational and solid surface boundary conditions for fluid dynamics calculations. AIAA Paper 85-1496;1985.

[345] Hirsch C, Verhoff A. Far field numerical boundary conditions for internal and cascade flow computations. AIAA Paper 89-1943;1989.

[346] Giles MB. Non-reflecting boundary conditions for Euler equation calculations. AIAA Paper 89-1942;1989.

[347] Tan Z, Wang DM, Srininvasan K, Pzekwas A, Sun R. Numerical simulation of coupled radiation and convection from complex geometries. AIAA Paper 98-2677;1998.

[348] Agarwal RK, Schulte P. A new computational algorithm for the solution of radiation heat transfer problems. AIAA Paper 98-2836;1998.

[349] Hino T. An unstructured grid method for incompressible viscous flows with a free surface. AIAA Paper 97-0862;1997.

[350] Löhner R, Yang C, Onate E, Idelsson S. An unstructured grid-based, parallel free surface solver. AIAA Paper 97-1830;1997.

[351] Cowles G, Martinelli L. A cell-centered parallel multiblock method for viscous incompressible flows with a free surface. AIAA Paper 97-1865;1997.

[352] Wagner CA, Davis DW, Slimon SA, Holligsworth TA. Computation of free surface flows using a hybrid multiblock/chimera approach. AIAA Paper 98-0228;1998.

[353] Calderer A, Kang S, Sotiropoulos F. Level set immersed boundary method for coupled simulation of air/water interaction with complex floating structures. J Comput Phys 2014; 277:201-27.

第4章 结构网格有限体积法

正如在第3章的引言中提到的，绝大多数欧拉和纳维-斯托克斯方程组的数值格式都采用直线法，也就是说，在空间和时间上分别进行离散。因此，它允许我们根据所求解问题的需求，对空间和时间导数使用不同精度的数值近似，从而得到极大的灵活性，在此我们还将采用直线法。有关基于空间和时间耦合离散的数值方法的详细讨论，如 Lax-Wendroff 格式族（如显式 MacCormack 预估-校正格式、隐式 Lerat 格式等），参见文献[1]。

一般的结构有限体积法都基于由纳维-斯托克斯方程组(2.19)或欧拉方程组(2.44)表示的守恒形式。首先，将物理空间细分为若干网格单元——2D 情况下为四边形，3D 情况下为六面体，生成的网格应具有如下特征：

(1) 物理域完全被网格覆盖；

(2) 网格单元之间没有剩余的自由空间；

(3) 网格单元互不重叠。

得到的结构网格由网格点（网格单元的角）的坐标 x,y,z 和计算空间（见图3.2）中的索引 i,j,k 唯一表述。为了计算对流通量、黏性通量以及源项的积分，基于网格定义了控制体。为了简单起见，假设控制体不随时间变化（否则参见 A.5 节），守恒变量 $\boldsymbol{W}$ 的时间导数可以写为

$$\frac{\partial}{\partial t}\int_{\Omega}\boldsymbol{W}\mathrm{d}\Omega=\Omega\frac{\partial \boldsymbol{W}}{\partial t}$$

因此，式(2.19)变为

$$\frac{\partial \boldsymbol{W}}{\partial t}=-\frac{1}{\Omega}\left[\oint_{\partial\Omega}(\boldsymbol{F}_{\mathrm{c}}-\boldsymbol{F}_{\mathrm{v}})\mathrm{d}S-\int_{\Omega}\boldsymbol{Q}\mathrm{d}\Omega\right] \tag{4.1}$$

式(4.1)右侧的面积分可以近似为通过控制体的面通量之和，这种近似称为空间离散，通常假定通量沿特定面为常量，并在面的中心处进行计算。一般假定源项在控制体内为常量，然而，在源项占主导地位的情况下，建议使用相邻控制体源

项值的加权和来计算(见文献[2]及其引用的文献)。如果考虑如图4.1(b)所示的特定控制体$\Omega_{I,J,K}$,由式(4.1)可得

$$\frac{\mathrm{d}\boldsymbol{W}_{I,J,K}}{\mathrm{d}t}=-\frac{1}{\Omega_{I,J,K}}\left[\sum_{m=1}^{N_F}(\boldsymbol{F}_c-\boldsymbol{F}_v)_m\Delta S_m-(\boldsymbol{Q}\Omega)_{I,J,K}\right] \tag{4.2}$$

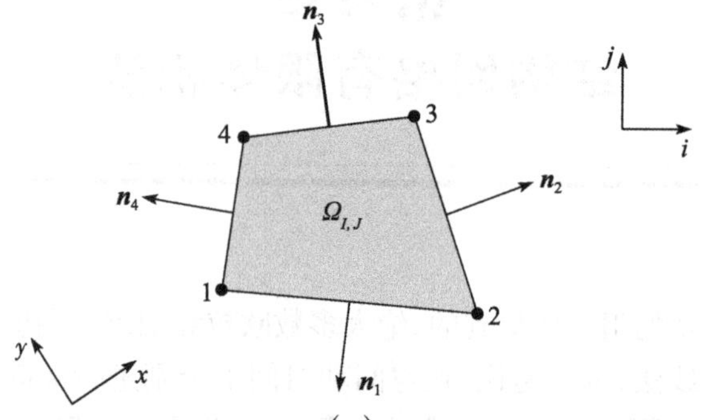

(a)

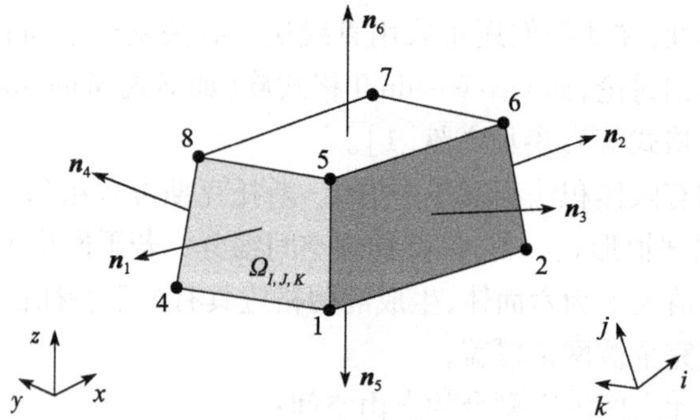

(b)

图4.1 结构网格的控制体(Ω)和相关面的单位法向矢量($\boldsymbol{n}_m$)

(a) 2D单元;(b) 3D单元。在(a)中,单位法向矢量$\boldsymbol{n}_2$和$\boldsymbol{n}_4$与计算空间的i坐标(方向)相关。$\boldsymbol{n}_1$和$\boldsymbol{n}_3$与j坐标相关。在(b)中,单位法向矢量$\boldsymbol{n}_1$和$\boldsymbol{n}_2$与i坐标相关,$\boldsymbol{n}_5$和$\boldsymbol{n}_6$与j坐标相关,$\boldsymbol{n}_3$和$\boldsymbol{n}_4$与k坐标有关。

在上面的表达式中,大写字母I,J,K表示计算空间中控制体的索引(见3.1节),正如稍后将看到的,控制体不一定与网格相同。此外,N_F表示控制体的面数(2D情况下$N_F=4$,3D情况下$N_F=6$),变量ΔS_m表示第m个面的面积。式(4.2)右侧方括号中的项一般称为残差,在此用$\boldsymbol{R}_{I,J,K}$表示,因此式(4.2)可以简写为

$$\frac{\mathrm{d}\boldsymbol{W}_{I,J,K}}{\mathrm{d}t}=-\frac{1}{\Omega_{I,J,K}}\boldsymbol{R}_{I,J,K} \tag{4.3}$$

将所有控制体 $\Omega_{I,J,K}$ 的关系式写成式(4.3)，我们得到一个一阶常微分方程组，该方程组在时间上是双曲型的，因此必须从已知的初始解开始，在时间上进行推进求解。此外，如第 8 章所述，我们还必须为黏性通量和无黏通量提供适当的边界条件。

在数值求解离散后的控制方程组(4.3)时，第一个问题是相对于计算网格，我们如何定义控制体，以及将流场变量定位于何处。在结构有限体积法的框架下，有三种基本策略可用：

(1) 格心格式。控制体与网格单元相同，流场变量与它们的质心相关联。

(2) 具有重叠控制体的格点格式。流场变量指定到网格点(顶点、节点)上，控制体定义为具有共同顶点的所有网格单元的组合(2D 情况下为 4 个单元，3D 情况下为 8 个单元)，此时与两个相邻网格点相关联的控制体是相互重叠的。

(3) 具有二次控制体的格点格式。流场变量存储在网格顶点，但控制体通过连接具有相同顶点的单元的中点来创建，这样，任意两个网格点所对应的控制体是不重叠的。

所有这三种方法将在 4.2 节概述，该节将专门讨论离散格式的一般概念。需要指出的是，目前具有重叠控制体的格点格式已经很少使用，尽管如此，为完整性起见，本书还是将其列出进行介绍。

需要注意的是，在我们所讨论的情况下，所有的流场变量，即守恒变量(ρ，ρu，ρv，ρw 和 ρE)和因变量(p，T，c，…)均关联于相同的位置，如单元中心或网格节点，这种方法称为同位网格格式。相比之下，许多基于压力的老方法(参见 3.1 节)使用交错网格格式，其中压力和速度分量存储在不同的位置，以抑制由中心差分引起的解的振荡。

关于对流通量的求解，可选的范围很广，最基础的问题是，我们需要知道控制体所有 N_{F} 个面上的值。事实上我们并不能直接获得面上的流场变量值，而必须通过插值的方式将通量或者流场变量插值到控制体的面上，大体上可以通过以下两种方式来实现：

(1) 算术平均，类似于中心离散格式；

(2) 偏置插值，类似于迎风离散格式，考虑了流动方程组的特征。

除此之外，我们还将在 4.3 节中讨论应用最广泛的对流通量离散格式的精度、适用性和计算量等方面的问题。

控制体黏性通量的计算方法最常用的是基于流场变量的算术平均方法。相对于流场变量，式(2.15)和式(2.24)中速度梯度和温度梯度的计算更复杂，我们将在 4.4 节提出适当的方法。

4.1 控制体的几何量

在介绍离散方法之前,我们首先讨论一下控制体的几何量的计算问题,这部分内容对后续的离散方法具有指导意义。在此,几何量包括体积 $\Omega_{I,J,K}$、面 m 的单位法向矢量 $\boldsymbol{n}_m$(方向定义为向外)和面积 ΔS_m,法向矢量和面积也可以表示为控制体的度量值。在接下来的章节中,我们将分别考虑 2D 情况和 3D 情况下的一般四边形或六面体控制体。

4.1.1 2D 情况

一般来说,我们把平面内的流动看作 3D 问题的一种特殊情况,在这种情况下,解关于一个坐标方向对称(例如,关于 z 方向)。由于对称性,为了获得正确的体积、压力等物理量的单位,我们将所有网格单元和控制体的厚度设为常数值 b,从而控制体的体积等于它的面积与厚度 b 的乘积。四边形的面积可以使用高斯公式精确计算。对于如图 4.1(a)所示的控制体,经过一些代数运算后得到

$$\Omega_{I,J}=\frac{b}{2}[(x_1-x_3)(y_2-y_4)+(x_4-x_2)(y_1-y_3)] \tag{4.4}$$

其中,假设控制体位于 x-y 平面,z 坐标是对称轴。由于厚度 b 是任意的,为了方便起见,可以将 b 设为 1。在 2D 情况下,控制体的面由直线给出,因此单位法向矢量沿直线为常值,当我们根据式(4.2)的近似对通量进行积分时,必须计算面的面积 ΔS 与对应的单位法向矢量 $\boldsymbol{n}$ 的乘积,即

$$\boldsymbol{S}_m=\begin{bmatrix} S_{x,m} \\ S_{y,m} \end{bmatrix}=\boldsymbol{n}_m\Delta S_m \tag{4.5}$$

因为对称,面矢量(以及单位法向矢量)的 z 方向分量必为零,故从表达式中去掉了此项。图 4.1(a)中控制体的面矢量为

$$\begin{aligned}
\boldsymbol{S}_1&=b\begin{bmatrix} y_2-y_1 \\ x_1-x_2 \end{bmatrix}\\
\boldsymbol{S}_2&=b\begin{bmatrix} y_3-y_2 \\ x_2-x_3 \end{bmatrix}\\
\boldsymbol{S}_3&=b\begin{bmatrix} y_4-y_3 \\ x_3-x_4 \end{bmatrix}\\
\boldsymbol{S}_4&=b\begin{bmatrix} y_1-y_4 \\ x_4-x_1 \end{bmatrix}
\end{aligned} \tag{4.6}$$

由式(4.5)得到面 m 的单位法向矢量为

$$\boldsymbol{n}_m=\frac{\boldsymbol{S}_m}{\Delta S_m} \tag{4.7}$$

其中

$$\Delta S_m=|\boldsymbol{S}_m|=\sqrt{S_{x,m}^2+S_{y,m}^2}$$

实际计算中，对于每个控制体 $\Omega_{I,J}$，只需计算并存储面矢量 $\boldsymbol{S}_1$ 和 $\boldsymbol{S}_4$ 即可。为了节省内存和减少点操作数，面矢量 $\boldsymbol{S}_2$ 和 $\boldsymbol{S}_3$ 可以从相邻控制体中获取（为了使面法向朝外，需要取反号）。

4.1.2　3D 情况

与 2D 情况不同，3D 情况下，面矢量和体积的计算存在一些问题，主要原因是，组成控制体的每一个面所包含的四个顶点可能不在一个平面上，此时法向矢量在面上不再是常数（图 4.2）。为了克服这个困难，可以将控制体的所有六个面都分解成两个或多个三角形，其体积可以由分解后的四面体构成，并以适当的方式进行基于这种细分的离散，在任意网格上至少是一阶精度[3]。因为必须在每个剖分的三角形上分别进行通量的积分，点操作数将至少增加一倍，计算量将会大大增加。然而，文献[3-4]表明，对于足够光滑的网格，当控制体的面趋近于平行四边形时，分解成三角形并不能显著提高求解精度，因此，在下面的讨论中，将采用基于平均法向矢量的四边形面简化处理方法。

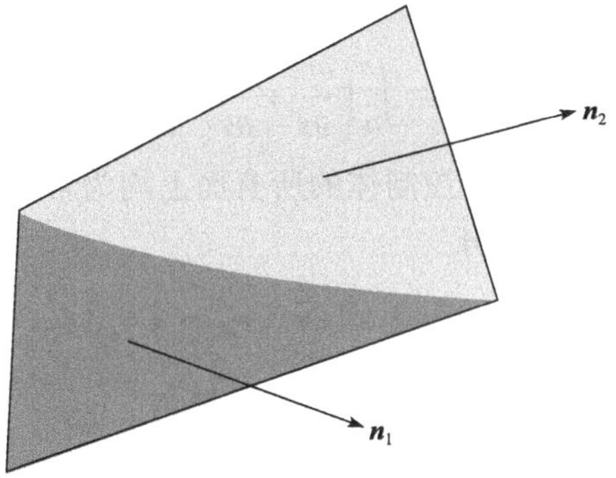

图 4.2　3D 情况下具有变化的法向矢量的控制体的面

如图 4.1(b)所示，六面体控制体的面矢量 $\boldsymbol{S}$ 最简便的计算方法是使用与 2D 情况下四边形面积计算相同的高斯公式。例如，对于面 $m=1$（图 4.1(b)中的点 1、5、8 和 4），我们首先定义坐标差为

$$\begin{aligned}
\Delta x_A &= x_8 - x_1 \quad \Delta x_B = x_5 - x_4 \\
\Delta y_A &= y_8 - y_1 \quad \Delta y_B = y_5 - y_4 \\
\Delta z_A &= z_8 - z_1 \quad \Delta z_B = z_5 - z_4
\end{aligned} \tag{4.8}$$

则面矢量 $\boldsymbol{S}_1 = \boldsymbol{n}_1 \Delta S_1$ 为

$$\boldsymbol{S}_1 = \frac{1}{2}\begin{bmatrix} \Delta z_A \Delta y_B - \Delta y_A \Delta z_B \\ \Delta x_A \Delta z_B - \Delta z_A \Delta x_B \\ \Delta y_A \Delta x_B - \Delta x_A \Delta y_B \end{bmatrix} \tag{4.9}$$

其余五个面矢量以类似的方式计算。对于每个控制体 $\Omega_{I,J,K}$，非常实用的做法是只存储6个面矢量中的3个（例如，$\boldsymbol{S}_1$，$\boldsymbol{S}_3$ 和 $\boldsymbol{S}_5$）。其余的面矢量 $\boldsymbol{S}_2$，$\boldsymbol{S}_4$ 和 $\boldsymbol{S}_6$ 从相邻控制体中获取（为了使面法向朝外，需要取反号）。式(4.8)和式(4.9)提供一个平均的面矢量，当面趋近于平行四边形时，即当面的顶点都位于一个平面时，这种近似就变为精确。由式(4.7)和式(4.10)可得到单位法向矢量。

$$\Delta S_m = \sqrt{S_{x,m}^2 + S_{y,m}^2 + S_{z,m}^2} \tag{4.10}$$

计算一般六面体的体积存在多种精度相当的计算公式（如文献[1]），其中基于散度定理[5]的方法在各种应用中均有较好的表现，它将某个矢量的散度的体积分与它的面积分关联起来，其关键思想是利用控制体 Ω 的某个点在空间中的位置矢量，我们将之设为 $\boldsymbol{r} = [r_x, r_y, r_z]^{\mathrm{T}}$，根据散度定理，有

$$\int_\Omega \operatorname{div}(\boldsymbol{r})\,\mathrm{d}\Omega = \oint_{\partial\Omega} (\boldsymbol{r} \cdot \boldsymbol{n})\,\mathrm{d}S \tag{4.11}$$

可以很容易计算出式(4.11)的左端项，它给出了我们需要的体积 Ω

$$\int_\Omega \operatorname{div}(\boldsymbol{r})\,\mathrm{d}\Omega = \int_\Omega \left(\frac{\partial r_x}{\partial x} + \frac{\partial r_y}{\partial y} + \frac{\partial r_z}{\partial z} \right) \mathrm{d}\Omega = 3\Omega \tag{4.12}$$

如果假设单位法向矢量在控制体的所有面上均为常值，则可以采用下式求解式(4.11)中右端项的面积分

$$\oint_{\partial\Omega} (\boldsymbol{r} \cdot \boldsymbol{n})\,\mathrm{d}S \approx \sum_{m=1}^{6} (\boldsymbol{r}_{\mathrm{mid}} \cdot \boldsymbol{n})_m \Delta S_m \tag{4.13}$$

式中：$\boldsymbol{r}_{\mathrm{mid},m}$ 为控制体的第 m 个面的中心。例如：

$$\boldsymbol{r}_{\mathrm{mid},1} = \frac{1}{4}(\boldsymbol{r}_1 + \boldsymbol{r}_5 + \boldsymbol{r}_8 + \boldsymbol{r}_4)$$

其中，矢量 $\boldsymbol{r}_1$，$\boldsymbol{r}_5$，$\boldsymbol{r}_8$ 和 $\boldsymbol{r}_4$ 对应图4.1(b)中面 $m=1$ 的顶点1，5，8和4。其余面的中心有类似的关系。式(4.13)中面 m 的面积 ΔS_m 可从式(4.10)得到。结合式(4.12)和式(4.13)，并用面矢量 $\boldsymbol{S}$ 代替乘积 $\boldsymbol{n}_m \Delta S_m$，最后得到控制体 $\Omega_{I,J,K}$ 体积的关系式为

$$\Omega_{I,J,K}=\frac{1}{3}\sum_{m=1}^{6}(\boldsymbol{r}_{\text{mid}}\cdot\boldsymbol{S})_m \tag{4.14}$$

文献[6]提供了面三角化情况下的类似公式。

坐标系的原点理论上可以移动到任意位置，而不影响式(4.14)中体积的计算。因此建议将原点定位于控制体的一个顶点(例如，图 4.1(b)中的点 1)，以便于达到更好的数值标度。用转换后的矢量 $\boldsymbol{r}^*$ 替换上面表达式(4.11)~式(4.14)中的 $\boldsymbol{r}$，该矢量定义为

$$\boldsymbol{r}^*=\boldsymbol{r}-\boldsymbol{r}_{\text{origin}}$$

特别的，当控制体的面为平面时，式(4.14)计算得到的体积是精确的。

4.2　一般离散方法

在本章的引言中，已经提到了定义控制体和流场变量位置的三种方法。本节将对这三种方法进行更详细的介绍，并对它们的优点和缺点进行讨论。

4.2.1　格心格式

如图 4.3 所示，如果控制体与网格单元相同，且流场变量位于网格单元的质心，则称为格心格式。在求解离散的流动方程组(4.2)时，还必须提供单元面的对流和黏性通量[7]，它们可以用以下三种方法来近似：

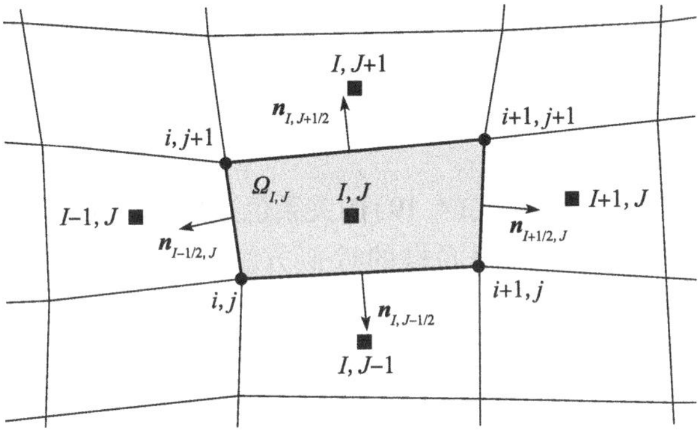

图 4.3　格心格式的控制体(2D 情况)

(1) 先求出面左右两侧网格单元体心处的通量，再将该通量值平均到面上，计算中需要使用相同的面矢量(通常仅应用于对流通量)；

(2) 先将面左右两侧网格单元体心处的变量平均到面上，再通过面上的变量计算通量；

(3) 先将流场变量分别插值到面左右两侧,再用来计算通量(仅用于对流通量)。

以图 4.3 中单元面 $\boldsymbol{n}_{I+1/2,J}$ 为例,第一种方法——通量平均——在 2D 情况下为

$$(\boldsymbol{F}_c\Delta S)_{I+1/2,J} \approx \frac{1}{2}[\boldsymbol{F}_c(\boldsymbol{W}_{I,J})+\boldsymbol{F}_c(\boldsymbol{W}_{I+1,J})]\Delta S_{I+1/2,J} \tag{4.15}$$

其中,$\Delta S_{I+1/2,J}$ 由式(4.6)和式(4.7)计算得到。

第二种方法——变量平均——可以表述为

$$(\boldsymbol{F}\Delta S)_{I+1/2,J} \approx \boldsymbol{F}(\boldsymbol{W}_{I+1/2,J})\Delta S_{I+1/2,J} \tag{4.16}$$

其中,控制体的面 $\boldsymbol{n}_{I+1/2,J}$ 处的守恒变量/因变量定义为两个相邻单元值的算术平均,即

$$\boldsymbol{W}_{I+1/2,J} = \frac{1}{2}(\boldsymbol{W}_{I,J}+\boldsymbol{W}_{I+1,J}) \tag{4.17}$$

式(4.16)中矢通量 $\boldsymbol{F}$ 既可以表示对流通量,也可以表示黏性通量。

第三种方法首先将流场变量(主要是速度分量、压力、密度和总焓)分别插值到单元面的两侧,插值后的量——称为左侧状态和右侧状态(见 4.3 节开始部分)——在两侧通常不相同,然后使用某些非线性函数,从左右状态之差计算得到通过单元面的通量,即

$$(\boldsymbol{F}_c\Delta S)_{I+1/2,J} \approx f_{\text{Flux}}(\boldsymbol{U}_L, \boldsymbol{U}_R, \Delta S_{I+1/2,J}) \tag{4.18}$$

其中

$$\begin{aligned} \boldsymbol{U}_L &= f_{\text{Interp}}(\cdots, \boldsymbol{U}_{I-1,J}, \boldsymbol{U}_{I,J}, \cdots) \\ \boldsymbol{U}_R &= f_{\text{Interp}}(\cdots, \boldsymbol{U}_{I,J}, \boldsymbol{U}_{I+1,J}, \cdots) \end{aligned} \tag{4.19}$$

表示插值的状态。

当然,类似于式(4.15)~式(4.19)的关系也适用于其他单元面。

在 3D 情况下也可以采用同样的近似方法,例如,在单元面 $\boldsymbol{n}_{I+1/2,J,K}$(如图 4.1(b),等于 $\boldsymbol{n}_2$)上,式(4.15)中的通量平均变为

$$(\boldsymbol{F}_c\Delta S)_{I+1/2,J,K} \approx \frac{1}{2}[\boldsymbol{F}_c(\boldsymbol{W}_{I,J,K})+\boldsymbol{F}_c(\boldsymbol{W}_{I+1,J,K})]\Delta S_{I+1/2,J,K} \tag{4.20}$$

其中,$\Delta S_{I+1/2,J,K}$ 的定义如式(4.8)和式(4.9),变量的平均类似于式(4.16),即

$$(\boldsymbol{F}\Delta S)_{I+1/2,J,K} \approx \boldsymbol{F}(\boldsymbol{W}_{I+1/2,J,K})\Delta S_{I+1/2,J,K} \tag{4.21}$$

其中

$$\boldsymbol{W}_{I+1/2,J,K} = \frac{1}{2}(\boldsymbol{W}_{I,J,K}+\boldsymbol{W}_{I+1,J,K}) \tag{4.22}$$

流场变量的插值方法与式(4.18)相似,即

$$(\boldsymbol{F}_c\Delta S)_{I+1/2,J,K}\approx f_{\mathrm{Flux}}(\boldsymbol{U}_L,\boldsymbol{U}_R,\Delta S_{I+1/2,J,K})\tag{4.23}$$

其中,$\boldsymbol{U}_L$ 和 $\boldsymbol{U}_R$ 是单元面上的插值量。

离散流动方程组(4.2)中最后一项尚未计算的是源项 $\boldsymbol{Q}$。正如我们在引言中讲过的,通常可以假定源项在控制体内为常量,使用对应单元格心的流场变量计算源项即可。因此我们定义

$$(\boldsymbol{Q}\Omega)_{I,J,K}=\boldsymbol{Q}(\boldsymbol{W}_{I,J,K})\Omega_{I,J,K}\tag{4.24}$$

利用上述关系式,可以计算出流过面的通量,进一步根据式(4.2)可以对 $\Omega_{I,J,K}$ 的边界进行数值积分,也就是说,得到完整的残差 $\boldsymbol{R}_{I,J,K}$。在 4.3 节和 4.4 节中,将更详细地介绍关于对流通量和黏性通量的求解。

4.2.2　格点格式：重叠控制体

在格点格式中,所有的流场变量都与计算网格的节点相关联。在基于重叠控制体的方法中,就像格心格式一样,网格单元仍然代表控制体,不同之处在于,控制体的残差必须分配到网格点上[4,8-9],如图 4.4 所示。

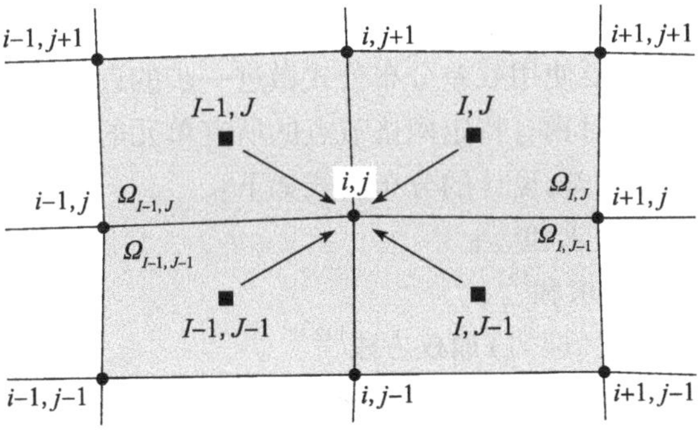

图 4.4　格点格式的重叠控制体(2D),箭头表示从单元质心到公共节点 i,j 的残差分配

考虑图 4.4 中由下列节点定义的控制体 $\Omega_{I,J}$:

$$(i,j)\quad(i+1,j)\quad(i+1,j+1)\quad(i,j+1)$$

注意到点(i,j)位于 $\Omega_{I,J}$ 的左下角,对于由点(i,j)和$(i+1,j)$给出的面 $\Delta S_{I,J-1/2}$ 的对流通量,可以近似为

$$(\boldsymbol{F}_c\Delta S)_{I,J-1/2}\approx\boldsymbol{F}_c(\boldsymbol{W}_{I,J-1/2})\Delta S_{I,J-1/2}\tag{4.25}$$

面心值通过对组成面的节点处的变量进行算术平均得到,即

$$\boldsymbol{W}_{I,J-1/2}=\frac{1}{2}(\boldsymbol{W}_{i,j}+\boldsymbol{W}_{i+1,j})\tag{4.26}$$

面积 $\Delta S_{I,J-1/2}$ 由关系式(4.6)和式(4.7)计算得到。

这种方法在 3D 情况下同样适用,如图 4.1(b)中与法向矢量 $\boldsymbol{n}_3$ 相关的面,其平均变量为

$$\boldsymbol{W}_{I,J,K-1/2}=\frac{1}{4}(\boldsymbol{W}_1+\boldsymbol{W}_2+\boldsymbol{W}_5+\boldsymbol{W}_6) \tag{4.27}$$

假设边 1-2 在 i 方向上,边 1-5 在 j 方向上,边 1-4 在 k 方向上,如果我们最终将点 (i,j,k) 与图 4.1(b)中的角 1 相关联,则式(4.27)中的平均值也可表示为

$$\boldsymbol{W}_{I,J,K-1/2}=\frac{1}{4}(\boldsymbol{W}_{i,j,k}+\boldsymbol{W}_{i+1,j,k}+\boldsymbol{W}_{i,j+1,k}+\boldsymbol{W}_{i+1,j+1,k}) \tag{4.28}$$

对流通量由下式得到:

$$(\boldsymbol{F}_c\Delta S)_{I,J,K-1/2}\approx\boldsymbol{F}_c(\boldsymbol{W}_{I,J,K-1/2})\Delta S_{I,J,K-1/2} \tag{4.29}$$

其中,面积 $\Delta S_{I,J,K-1/2}$ 由式(4.8)和式(4.9)计算得到。

黏性通量通常采用与二次控制体格式相同的方法计算[10-11],形成更为紧致的格式(即,涉及更少的节点)。

对由关系式(4.25)和式(4.26)、式(4.28)和式(4.29)求出的所有面的贡献求和,得到所有网格单元的中间残差 $\boldsymbol{R}_{I,J,K}$。为了将基于单元的残差与基于节点的残差联系起来,需要使用残差分布公式做进一步的近似,残差分布公式从根本上说是一个函数,它对具有特定网格节点的所有单元的残差进行加权求和,以得到基于节点的未知残差,设计的分布公式如下:

(1) Ni 的体积加权求和[8];

(2) Hall 的无加权求和[9];

(3) Rossow 的特征(迎风)加权方法[12-13]。

截断误差[4]的理论研究表明,Ni 的格式比 Hall 的方法更精确。然而在实际应用中,Ni 的分布公式在网格强烈扭曲和拉伸的地方会产生问题,例如对于 O 型网格,翼型后缘附近的压力场会出现强烈的振荡[4,14],只有大幅增加数值黏性时才能实现收敛。迎风加权方法[12-13]的基本思想与波动分裂格式[15-18](见 3.1.5 节)非常相似,但是数值实现要简单得多,从根本上说,残差只向特征方向的上游发送。

在上述三种方法中,Hall 的分布格式被证明是最鲁棒的,其特定节点的残差由包含该节点的所有单元的中间残差 $\boldsymbol{R}_{I,J,K}$ 通过简单求和得到,因此在图 4.4 所示的 2D 情况下,可以得到

$$\boldsymbol{R}_{i,j}=\boldsymbol{R}_{I,J}+\boldsymbol{R}_{I-1,J}+\boldsymbol{R}_{I-1,J-1}+\boldsymbol{R}_{I,J-1} \tag{4.30}$$

在 3D 情况下,总共有 8 个基于单元的残差以同样的方式进行求和,对式(4.30)仔细考察可以发现,因为通过内部面的通量会相互抵消,$\boldsymbol{R}_{i,j}$ 就是通过

超单元边界的净通量。该超单元由下列单元组成

$$\Omega_{i,j}=\Omega_{I,J}+\Omega_{I-1,J}+\Omega_{I-1,J-1}+\Omega_{I,J-1} \tag{4.31}$$

该超单元也代表以点 i,j 为中心的“总”的控制体。在 3D 情况下，总体积由下列单元组成：

$$\Omega_{i,j,k}=\Omega_{I,J,K}+\Omega_{I-1,J,K}+\Omega_{I-1,J-1,K}+\Omega_{I,J-1,K}+ \\ \Omega_{I,J,K-1}+\Omega_{I-1,J,K-1}+\Omega_{I-1,J-1,K-1}+\Omega_{I,J-1,K-1} \tag{4.32}$$

从图 4.4 可以看出，控制体之间至少有一个单元重叠，重叠控制体格式即得名于此。

源项采用相应网格节点的流场变量计算，即

$$(\boldsymbol{Q}\Omega)_{i,j,k}=\boldsymbol{Q}(\boldsymbol{W}_{i,j,k})\Omega_{i,j,k} \tag{4.33}$$

根据以上定义，式(4.2)中的时间步格式变为

$$\frac{\mathrm{d}\boldsymbol{W}_{i,j,k}}{\mathrm{d}t}=-\frac{1}{\Omega_{i,j,k}}\boldsymbol{R}_{i,j,k} \tag{4.34}$$

这是一个常微分方程组，它需要在每个网格点 i,j,k 上用与格心格式相同的方法进行求解。需要注意的是，在式(4.34)中使用的是总体积，如式(4.31)或式(4.32)。

4.2.3　格点格式：二次控制体

在这种格式中，控制体以特定的网格节点(顶点)为中心，所有流场变量都存储在网格节点(顶点)上[19-20]。如图 4.5 所示，在 2D 情况下，通过连接共享节点的四个单元的质心来构建二次控制体；在 3D 情况下，为了形成控制体的面，需要将八个单元的质心连接起来。此外，也可以将单元的质心与边的中点连接起来(在 2D 情况下)，然后再连接到相邻单元的质心[21]，因此控制体的面由具有不同法向矢量的两部分组成，这在非结构网格中是很常见的(见第 5 章)。然而，如图 4.6 所示，在结构网格中，这种控制体的定义仅在边界处成立，否则将改变面的离散，在内部相对光滑的网格上使用第二种方法并不会带来精度方面明显的改进，因此下面将采用二次控制体的简单定义，而边界的处理将在第 8 章中讨论。

在求解离散的流动方程组(4.2)时，需要计算控制体面的对流通量和黏性通量，这可以通过以下三种方法来实现：

(1) 由控制体面的左侧和右侧节点的值计算出的通量进行平均，但是需要使用相同的面矢量(通常仅应用于对流通量)；

(2) 由面的左侧和右侧节点的变量进行平均，使用得到的值计算通量；

(3) 由分别插值到面的左侧和右侧的流场变量来计算通量(仅用于对流通量)。

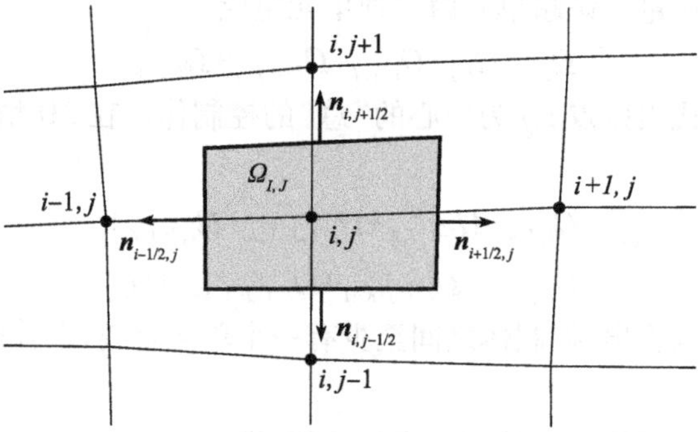

图 4.5　2D 格点格式的二次控制体

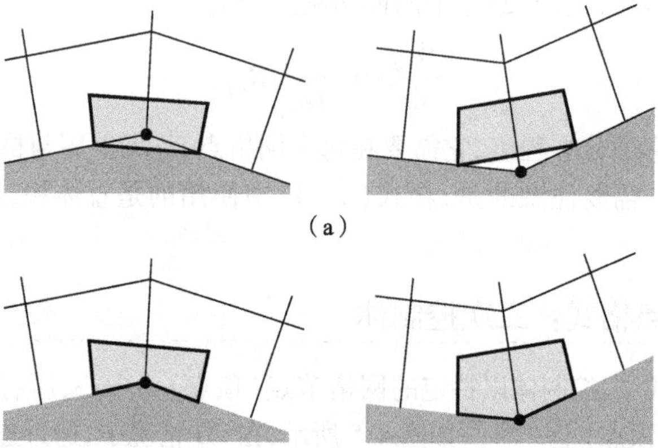

（a）

（b）

图 4.6　边界处二次控制体的定义(2D 情况)

(a) 由边的中点连接而成；(b) 由边的中点和边界的中心节点连接而成。

如图 4.5 中的单元面 $\boldsymbol{n}_{i+1/2,j}$，2D 情况下，第一种方法——通量平均为

$$(\boldsymbol{F}_c\Delta S)_{i+1/2,j}\approx\frac{1}{2}[\boldsymbol{F}_c(\boldsymbol{W}_{i,j})+\boldsymbol{F}_c(\boldsymbol{W}_{i+1,j})]\Delta S_{i+1/2,j} \tag{4.35}$$

其中，$\Delta S_{i+1/2,j}$ 由式(4.6)和式(4.7)根据已知的单元质心坐标计算。

第二种方法——变量平均——可以表述为

$$(\boldsymbol{F}\Delta S)_{i+1/2,j}\approx\boldsymbol{F}(\boldsymbol{W}_{i+1/2,j})\Delta S_{i+1/2,j} \tag{4.36}$$

其中，控制体的面 $\boldsymbol{n}_{i+1/2,j}$ 的守恒(或依赖)变量由两个相邻节点值的算术平均得到。因此

$$\boldsymbol{W}_{i+1/2,j}=\frac{1}{2}(\boldsymbol{W}_{i,j}+\boldsymbol{W}_{i+1,j}) \tag{4.37}$$

式(4.36)中的矢通量 $\boldsymbol{F}$ 既可以表示对流通量，也可以表示黏性通量。

第三种方法将流场变量（主要是速度分量、压力、密度和总焓）分别插值到面两侧，插值后的量——称为左侧状态和右侧状态（见 4.3 节的开始部分）——通常在两侧是不同的，然后利用某个非线性函数从左右状态的差值计算面的通量。因此

$$(\boldsymbol{F}_{c}\Delta S)_{i+1/2,j} \approx f_{\text{Flux}}(\boldsymbol{U}_{L}, \boldsymbol{U}_{R}, \Delta S_{i+1/2,j}) \tag{4.38}$$

其中

$$\begin{aligned} \boldsymbol{U}_{L} &= f_{\text{Interp}}(\cdots, \boldsymbol{U}_{i-1,j}, \boldsymbol{U}_{i,j}, \cdots) \\ \boldsymbol{U}_{R} &= f_{\text{Interp}}(\cdots, \boldsymbol{U}_{i,j}, \boldsymbol{U}_{i+1,j}, \cdots) \end{aligned} \tag{4.39}$$

表示插值状态。类似式(4.35)~式(4.39)的关系同样适用于控制体的其他面。

3D 情况下也可以采用相同的近似方法。例如，在单元面 $\boldsymbol{n}_{i+1/2,j,k}$（在图 4.1(b) 中，对应于 $\boldsymbol{n}_2$）上，式(4.35)中的通量平均变为

$$(\boldsymbol{F}_{c}\Delta S)_{i+1/2,j,k} \approx \frac{1}{2}[\boldsymbol{F}_{c}(\boldsymbol{W}_{i,j,k}) + \boldsymbol{F}_{c}(\boldsymbol{W}_{i+1,j,k})]\Delta S_{i+1/2,j,k} \tag{4.40}$$

其中，$\Delta S_{i+1/2,j,k}$ 由式(4.8)和式(4.9)得到。流场变量平均的结果类似于式(4.36)

$$(\boldsymbol{F}\Delta S)_{i+1/2,j,k} \approx \boldsymbol{F}(\boldsymbol{W}_{i+1/2,j,k})\Delta S_{i+1/2,j,k} \tag{4.41}$$

其中

$$\boldsymbol{W}_{i+1/2,j,k} = \frac{1}{2}(\boldsymbol{W}_{i,j,k} + \boldsymbol{W}_{i+1,j,k}) \tag{4.42}$$

最后，对应于式(4.38)的流场变量的插值方法为

$$(\boldsymbol{F}_{c}\Delta S)_{i+1/2,j,k} \approx f_{\text{Flux}}(\boldsymbol{U}_{L}, \boldsymbol{U}_{R}, \Delta S_{i+1/2,j,k}) \tag{4.43}$$

其中，$\boldsymbol{U}_{L}$ 和 $\boldsymbol{U}_{R}$ 是单元面上的插值量。关于对流通量和黏性通量求解的几种方法将分别在 4.3 节和 4.4 节中详细介绍。

离散流动方程组(4.2)的最后一项是源项 $\boldsymbol{Q}$。正如在引言中所述，通常认为源项在控制体内是恒定的，为此可以使用相应网格点的流场变量计算源项，定义为

$$(\boldsymbol{Q}\boldsymbol{\Omega})_{i,j,k} = \boldsymbol{Q}(\boldsymbol{W}_{i,j,k})\boldsymbol{\Omega}_{i,j,k} \tag{4.44}$$

利用上述关系，可以计算出通过面的通量，并且根据式(4.2)可以对 $\Omega_{i,j,k}$ 的边界进行数值积分。这样，就得到了包含源项的完整的残差 $\boldsymbol{R}_{i,j,k}$。对于每个网格点，守恒变量随时间的变化为

$$\frac{\mathrm{d}\boldsymbol{W}_{i,j,k}}{\mathrm{d}t} = -\frac{1}{\Omega_{i,j,k}}\boldsymbol{R}_{i,j,k} \tag{4.45}$$

在第 6 章中，我们将给出一些合适的求解方法。

4.2.4　格心格式与格点格式的比较

我们在前 3 节中概述了格心和格点离散方法，在此比较这三种格式，并讨论

它们的优缺点,其中的个别观点目前尚存争议。

首先是离散精度。由文献[4,22]的讨论可知,在扭曲的网格上,格点格式(重叠或二次控制体)可以达到一阶精度;在笛卡儿或光滑网格上(即相邻单元之间的体积变化不大且仅略有畸变),根据通量计算格式的不同,格点格式可以达到二阶或更高阶精度[23]。与之相对,格心格式的离散误差很大程度上依赖于网格的光滑性,例如,对于图 4.7 中的单元排列,即使对于线性变化的函数,平均方法也不能为面的中点提供正确的值。研究表明,在斜率不连续的网格上,即使网格被无限细化,离散误差也不会减小,如文献[4]所示,这种零阶误差表现为等值线的振荡或扭结,而格点格式在相同的情况下却不会出现问题。然而,在笛卡儿网格或者足够光滑的网格上,格心格式也可以达到二阶或更高的精度。文献[24-27]对离散误差作了进一步的分析。

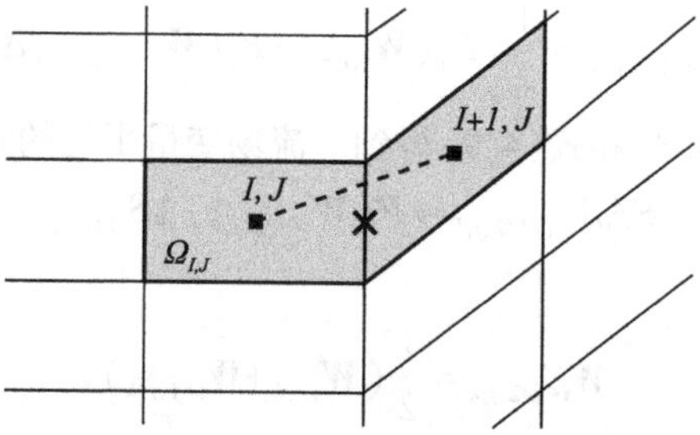

图 4.7 畸形网格上的格心通量平衡,叉号表示单元面的中点

其次,比较三种方法以及它们在边界上的特点。具有二次控制体的格点格式在固壁边界处会遇到困难,让我们再次回到图 4.6,很明显在边界处只有大约一半的控制体,沿着面的通量积分得到位于控制体内部的残差——理想情况下位于质心,但是此处的残差与直接固定在固壁上的节点相关联,与格心格式相比,这种不匹配导致离散误差的增加。对于二次控制体而言,定义在尖角(如后缘)处也会产生压力或密度的非物理极值的问题,更复杂的情况出现在坐标切割或者周期边界处(见第 8 章),控制体的两个部分的通量必须正确地求和,所有的格点格式还需要额外的逻辑,以确保在多块网格的边界点上得到相容的解,而格心格式没有这样的问题。

重叠控制体的格点格式在处理固壁边界方面比二次控制体格式具有优势,但是它不能与流行的 TVD、AUSM 或 CUSP 等迎风离散方法相结合,与格心格式和二次控制体格式相比,离散需要的点更多(3D 情况下为 27 个而非 7 个),在隐式时间离散的情况下,会抹平间断和增加内存开销。

格心格式和格点格式的最后一个主要区别出现在非定常流问题中。如 3.2

节所述,格点格式至少需要对质量矩阵进行近似处理[28-29]。与之相反,在格心格式中,因为残差天然地与控制体的质心相关联,完全可以抛弃质量矩阵。

综上所述,二次控制体的格点格式和格心格式对于定常流场的数值模拟非常相似,主要的区别在于畸变网格、边界处理和非定常流等方面,在后两种情况下,格心格式比格点格式具有优势,可以在流场解算器中更直接地实现。

4.3　对流通量的离散

在前面的章节中,我们从总体上讨论了空间离散方法。本节将更详细地学习如何近似计算对流通量。

正如在3.1.5节中看到的,在有限体积法的框架下,有以下几种格式选择:

(1) 中心格式;

(2) 矢通量分裂格式;

(3) 通量差分裂格式;

(4) TVD格式;

(5) 波动分裂格式。

在此,我们聚焦于最重要和最流行的方法,并忽略对基础格式修改的详细介绍,代之以给出相关的参考文献。

在开始详细介绍各种离散格式之前,我们首先应该解释左侧状态和右侧状态以及模板或计算分子的定义。

某些格心格式和二次控制体格式需要将流场变量插值到控制体的各个面上,图4.8描述了i方向网格的情况,中心格式(4.3.1节)的方式是使用面左右两侧相同数量的模板值进行线性插值,即它以面为中心进行插值。基于欧拉方程组特征的离散——迎风格式——使用非对称方式分别从面的左侧和右侧插值流场变量,这两个值被命名为左侧状态和右侧状态,然后利用它们来计算通过该面的对流通量(见式(4.18)、式(4.23)、式(4.38)或式(4.43))。几乎所有插值公式(TVD格式除外)都基于van Leer的MUSCL(monotone upstream-centered schemes for conservation laws)方法[30]。对于一般流场变量U,有

$$
\begin{aligned}
U_{\mathrm{R}} &= U_{I+1}-\frac{\epsilon}{4}\left[(1+\hat{k})\Delta_{-}+(1-\hat{k})\Delta_{+}\right]U_{I+1} \\
U_{\mathrm{L}} &= U_{I}+\frac{\epsilon}{4}\left[(1+\hat{k})\Delta_{+}+(1-\hat{k})\Delta_{-}\right]U_{I}
\end{aligned}
\qquad (4.46)
$$

前向(Δ_+)和后向(Δ_-)差分算子定义为

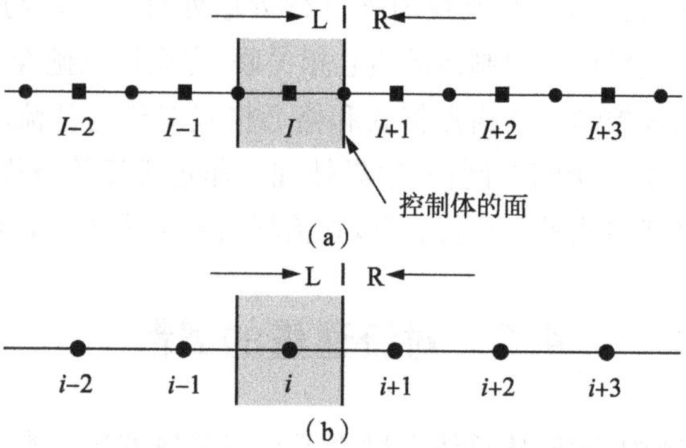

图 4.8　单元面 $I+1/2$ 和 $i+1/2$ 的左右状态

(a) 格心格式；(b) 二次控制体格点格式。圆圈表示节点，矩形表示单元质心。

$$\begin{aligned}\Delta_+ U_I &= U_{I+1} - U_I \\ \Delta_- U_I &= U_I - U_{I-1}\end{aligned} \tag{4.47}$$

相应地需要将索引移位，对于二次控制体的格点格式，只需将 I 替换为节点索引 i，上述关系仍然有效。将参数 ϵ 设为零可以得到一阶精度的迎风离散。参数 $\hat{k}$ 决定插值的空间精度，对于 $\epsilon=1$ 和 $\hat{k}=-1$，上面的插值公式(4.46)给出流场变量的完全单边插值，在等间距网格上得到二阶精度的迎风近似；$\hat{k}=0$ 对应于二阶精度迎风偏置的线性插值；$\hat{k}=1/3$ 得到一个三点插值公式，它构成(在有限体积架构下——参见文献[31,49])一种二阶迎风偏置格式，其截断误差比 $\hat{k}=-1$ 和 $\hat{k}=0$ 格式小；最后，如果指定 $\hat{k}=1$，那么 MUSCL 方法就会简化为一个纯粹的中心格式——变量的平均。在实际应用中，$\hat{k}=0$ 和 $\hat{k}=1/3$ 是最常用的。

如果流动区域存在强梯度，则 MUSCL 插值(式(4.46))必须通过限制器函数或限制器来改进，限制器的目的是抑制解的非物理振荡，关于限制器的讨论将在 4.3.5 节进一步展开。

模板或计算分子是指残差、梯度等计算所包含的单元质心或网格点的组合。例如，如果根据式(4.15)、式(4.20)或者式(4.35)、式(4.40)在控制体的面上做平均通量，2D 情况下我们得到一个 5 点模板，它由如下的单元或点组成：

$$(I,J)\quad (I+1,J)\quad (I,J+1)\quad (I-1,J)\quad (I,J-1)$$

在 3D 情况下，会产生一个 7 点模板，包括的单元或点为

$$\begin{array}{llll}(I,J,K) & (I+1,J,K) & (I,J+1,K) & (I-1,J,K)\\ (I,J-1,K) & (I,J,K-1) & (I,J,K+1) & \end{array}$$

图 4.9 显示了这两种模板。注意，在笛卡儿网格中，这对应于 i,j 和 k 方向上一阶导数的二阶精度中心差分近似。因此，对于微分形式的控制方程，采用中

心格式的有限差分,将得到相同的结果。

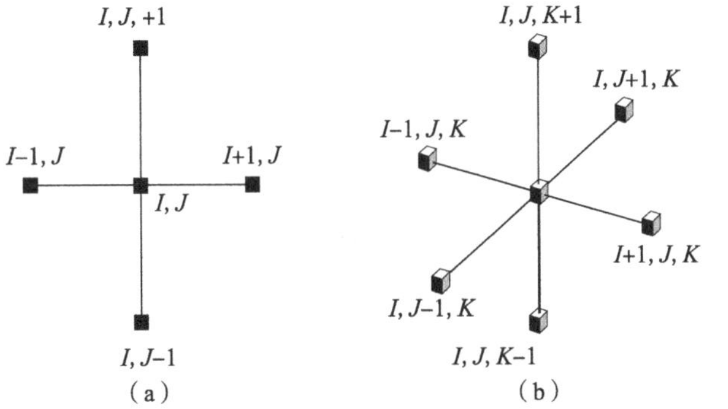

图 4.9　中心离散格式的模板(计算分子)

(a) 2D;(b) 3D。

4.3.1　具有人工耗散的中心格式

与其他离散方法相比,具有人工耗散的中心格式非常简单,在格心格式和格点格式中都很容易实现,因此,该格式获得了非常广泛的应用。

中心格式的基本思想是,首先求出控制体面两侧守恒变量的算术平均值,再基于此计算控制体面上的对流通量。这种做法使得解的奇偶解耦(生成离散方程的两个独立解),并在激波处产生过冲,因此,为了格式稳定,必须添加人工耗散(类似于黏性通量)。中心格式由 Jameson 等[7]首次在欧拉方程组中实现,根据作者的名字,它也被缩写为 JST 格式。

中心格式在间断面和边界层的解析度上,通常比迎风格式的精度要低。不过,鉴于它的计算量相当小,人们仍热衷于在保持低计算量的前提下试图提高中心格式的精度,诸如减少边界层中的人工耗散[32-33],或提高激波分辨率[34-36]等改进措施。遵循文献[36]的另一种思想是利用对流通量的雅可比矩阵独立地衡量每个守恒方程的耗散,这种成功的方法被称为矩阵耗散格式,它可以看作原始标量格式和迎风格式的折中;其基本格式也是采用了一个单一的、基于压力的探测器,在间断处从二阶精度切换到一阶精度,以防止流场变量的非物理振荡。在文献[37]中,Jameson 发展了一个称为 SLIP(Symmetric Limited Positive)格式的概念,对每个守恒方程分别应用一个限制器,文献[38]简要介绍了以上方法,并在无黏和黏性 2D 流动中进行了比较。

1. 标量耗散格式

根据式(4.15)、式(4.21)、式(4.36)或式(4.41),通过控制体面的对流通量（式(2.21)）可以使用变量的平均值来近似求解，然后将人工耗散加到中心通量中以保持格式稳定[7,39]。因此，面$(I+1/2,J,K)$的对流通量为

$$(\boldsymbol{F}_{\mathrm{c}}\Delta S)_{I+1/2,J,K}\approx\boldsymbol{F}_{\mathrm{c}}(\boldsymbol{W}_{I+1/2,J,K})\Delta S_{I+1/2,J,K}-\boldsymbol{D}_{I+1/2,J,K} \tag{4.48}$$

其中，流场变量的平均值为（图4.8）

$$\boldsymbol{W}_{I+1/2,J,K}=\frac{1}{2}(\boldsymbol{W}_{I,J,K}+\boldsymbol{W}_{I+1,J,K}) \tag{4.49}$$

在二次控制体的格点格式中，下标使用节点索引(i,j,k)来替换，为简单起见，$(I+1/2,J,K)$在下文中简称为$(I+1/2)$。人工耗散通量由自适应的二阶和四阶差分混合组成，它由一阶和三阶差分算子之和产生，即

$$\begin{aligned}\boldsymbol{D}_{I+1/2}=\hat{\Lambda}^{\mathrm{S}}_{I+1/2}[&\epsilon^{(2)}_{I+1/2}(\boldsymbol{W}_{I+1}-\boldsymbol{W}_{I})-\\&\epsilon^{(4)}_{I+1/2}(\boldsymbol{W}_{I+2}-3\boldsymbol{W}_{I+1}+3\boldsymbol{W}_{I}-\boldsymbol{W}_{I-1})]\end{aligned} \tag{4.50}$$

由式(4.50)可知，该格式具有紧致的9点模板(2D)和13点模板(3D)。

耗散由所有坐标方向上对流通量雅可比矩阵的谱半径之和进行标定，即

$$\hat{\Lambda}^{\mathrm{S}}_{I+1/2}=(\hat{\Lambda}^{I}_{\mathrm{c}})_{I+1/2}+(\hat{\Lambda}^{J}_{\mathrm{c}})_{I+1/2}+(\hat{\Lambda}^{K}_{\mathrm{c}})_{I+1/2} \tag{4.51}$$

例如，在I方向上（由上标I表示）的单元面$(I+1/2)$的谱半径，由平均计算，即

$$(\hat{\Lambda}^{I}_{\mathrm{c}})_{I+1/2}=\frac{1}{2}[(\hat{\Lambda}^{I}_{\mathrm{c}})_{I}+(\hat{\Lambda}^{I}_{\mathrm{c}})_{I+1}] \tag{4.52}$$

谱半径为

$$\hat{\Lambda}_{\mathrm{c}}=(|V|+c)\Delta S \tag{4.53}$$

式中：V为逆变速度（式(2.22)）；c为声速。

四阶差分项在激波处会导致解的强烈振荡，可以使用压力探测器将其关闭，此外，探测器还可以在流场光滑区域关闭二阶差分，以尽可能减小耗散。因此，式(4.50)中的系数$\epsilon^{(2)}$和$\epsilon^{(4)}$定义为

$$\begin{aligned}\epsilon^{(2)}_{I+1/2}&=k^{(2)}\max(\Upsilon_{I},\Upsilon_{I+1})\\\epsilon^{(4)}_{I+1/2}&=\max[0(k^{(4)}-\epsilon^{(2)}_{I+1/2})]\end{aligned} \tag{4.54}$$

压力探测器为

$$\Upsilon_{I}=\frac{|p_{I+1}-2p_{I}+p_{I-1}|}{p_{I+1}+2p_{I}+p_{I-1}} \tag{4.55}$$

典型参数值为$k^{(2)}=1/2$和$1/128\leqslant k^{(4)}\leqslant 1/64$。

为了减少跨越黏性剪切层的人工耗散的量，可以将式(4.50)中的比例因子重新定义为[32-33]

$$\hat{\Lambda}^{\mathrm{S}}_{I+1/2}=\frac{1}{2}[(\phi^{I}\hat{\Lambda}^{I}_{\mathrm{c}})_{I}+(\phi^{I}\hat{\Lambda}^{I}_{\mathrm{c}})_{I+1}]$$

$$\hat{\Lambda}_{J+1/2}^{S}=\frac{1}{2}[(\phi^J\hat{\Lambda}_c^J)_J+(\phi^J\hat{\Lambda}_c^J)_{J+1}] \tag{4.56}$$

$$\hat{\Lambda}_{K+1/2}^{S}=\frac{1}{2}[(\phi^K\hat{\Lambda}_c^K)_K+(\phi^K\hat{\Lambda}_c^K)_{K+1}]$$

然后使用上式代替式(4.51)。方向相关系数 ϕ 为

$$\begin{aligned}\phi^I&=1+\max\left[\left(\frac{\hat{\Lambda}_c^J}{\hat{\Lambda}_c^I}\right)^{\sigma},\left(\frac{\hat{\Lambda}_c^K}{\hat{\Lambda}_c^I}\right)^{\sigma}\right]\\ \phi^J&=1+\max\left[\left(\frac{\hat{\Lambda}_c^I}{\hat{\Lambda}_c^J}\right)^{\sigma},\left(\frac{\hat{\Lambda}_c^K}{\hat{\Lambda}_c^J}\right)^{\sigma}\right]\\ \phi^K&=1+\max\left[\left(\frac{\hat{\Lambda}_c^I}{\hat{\Lambda}_c^K}\right)^{\sigma},\left(\frac{\hat{\Lambda}_c^J}{\hat{\Lambda}_c^K}\right)^{\sigma}\right]\end{aligned} \tag{4.57}$$

参数 σ 通常设置为 1/2 或 2/3，这个公式减小了沿控制体较短边方向耗散项的比例，而控制体较长边与流动方向一致。

2. 矩阵耗散格式

为了通过减少数值耗散来提高精度，可以对前面的 JST 格式进行修改，使其更接近于迎风格式，这种思想使用一个矩阵——对流通量雅可比矩阵——代替标量值 $\hat{\Lambda}^S$ 来标定耗散项[36]，使每个方程都能用相应的特征值进行适当的标定。因此，式(4.50)变为

$$\boldsymbol{D}_{I+1/2}=|\boldsymbol{A}_c|_{I+1/2}[\epsilon_{I+1/2}^{(2)}(\boldsymbol{W}_{I+1}-\boldsymbol{W}_I)-\epsilon_{I+1/2}^{(4)}(\boldsymbol{W}_{I+2}-3\boldsymbol{W}_{I+1}+3\boldsymbol{W}_I-\boldsymbol{W}_{I-1})] \tag{4.58}$$

缩放矩阵对应于用特征值的绝对值对角化的对流通量雅可比矩阵($\boldsymbol{A}_c=\partial\boldsymbol{F}_c/\partial\boldsymbol{W}$)。特征值的绝对值为

$$|\boldsymbol{A}_c|=\boldsymbol{T}|\Lambda_c\Delta S|\boldsymbol{T}^{-1} \tag{4.59}$$

右($\boldsymbol{T}$)和左($\boldsymbol{T}^{-1}$)特征矢量矩阵以及特征值 Λ_c 的对角矩阵见 A.11 节。需要注意的是，必须限制驻点和声速线处的特征值，以防止耗散为零。计算 $|\boldsymbol{\Lambda}_c|$ 和 $\boldsymbol{W}$ 乘积的有效方法参见文献[36]，需要注意的是，通过设置 $\epsilon^{(2)}=1/2$ 和 $\epsilon^{(4)}=0$，可以得到一个一阶精度的完全迎风的格式。

正如前面所讲，此格式的基本出发点是发展一个精度接近于迎风格式，但在计算量上只比标量耗散方法略大(15%～20%)的格式，与矢通量分裂格式(CUSP 和 AUSM)的比较结果见文献[38]。

4.3.2　矢通量分裂格式

矢通量分裂方法可以看作迎风格式的第一层次，因为它只考虑了波的传播

方向。矢通量分裂格式将对流通量分解为两个部分——可以根据某些特征变量的符号进行分解，也可以按对流和压力部分进行分解。著名的 van Leer 矢通量分裂格式[40]属于基于特征分解的第一类格式，第二类是较新的方法，如 Liou 等的 AUSM 格式[41-42]和 Jameson 的 CUSP 格式[43-44]，类似的方法还有 Edwards 提出的 LDFSS 格式[45]，以及 Rossow 提出的 MAPS 格式[46-47]。

矢通量分裂格式只能在格心格式（4.2.1 节）或二次控制体的格点格式（4.2.3 节）框架下实现，与采用标量人工耗散的中心格式相比，它的优点是只增加极少的计算量，但对激波和边界层的分辨率要高得多，当然矩阵耗散格式（式（4.58））也可以产生精度相当的结果[38]。当前，研究人员仍然在对基本格式（特别是 AUSM）进行各种改良，以解决此类格式在某些数值上的困难。接下来，将介绍 van Leer 格式、AUSM 格式和 CUSP 格式的基础版本，对最重要的修改给出一些提示，并提供相关文献作为参考。

1. van Leer 格式

van Leer 矢通量分裂格式[40]基于对流通量的特征分解，文献[48-49]将其推广到贴体网格上。

根据垂直于控制体面（例如，图 4.8 中的$(I+1/2)$）的马赫数，对流通量被分裂为正负两部分，即

$$\boldsymbol{F}_{\mathrm{c}}=\boldsymbol{F}_{\mathrm{c}}^{+}+\boldsymbol{F}_{\mathrm{c}}^{-} \tag{4.60}$$

其中，面上的马赫数定义为

$$(M_n)_{I+1/2}=\left(\frac{V}{c}\right)_{I+1/2} \tag{4.61}$$

式中：V 为逆变速度（式（2.22））；c 为声速。在二次控制体的格点格式中，需用节点索引代替单元索引。

首先需要将变量 ρ, u, v, w 和 p 的值按式（4.19）或式（4.39）插值到控制体的面上，然后使用左状态计算正通量，右状态计算负通量，对流马赫数$(M_n)_{I+1/2}$由下式得到[40]

$$(M_n)_{I+1/2}=M_{\mathrm{L}}^{+}+M_{\mathrm{R}}^{-} \tag{4.62}$$

其中，分裂马赫数定义为

$$M_{\mathrm{L}}^{+}=\begin{cases} M_{\mathrm{L}}, & M_{\mathrm{L}}\geqslant +1 \\ \dfrac{1}{4}(M_{\mathrm{L}}+1)^{2}, & |M_{\mathrm{L}}|<1 \\ 0, & M_{\mathrm{L}}\leqslant -1 \end{cases} \tag{4.63}$$

和

$$M_{\mathrm{R}}^{-}=\begin{cases}0, & M_{\mathrm{R}}\geqslant +1\\ -\dfrac{1}{4}(M_{\mathrm{R}}-1)^2, & |M_{\mathrm{R}}|<1\\ M_{\mathrm{R}}, & M_{\mathrm{R}}\leqslant -1\end{cases}\tag{4.64}$$

马赫数 M_{L} 和 M_{R} 分别用左状态和右状态计算，即

$$M_{\mathrm{L}}=\frac{V_{\mathrm{L}}}{c_{\mathrm{L}}},\quad M_{\mathrm{R}}=\frac{V_{\mathrm{R}}}{c_{\mathrm{R}}}\tag{4.65}$$

在 $|M_n|<1$（亚声速流动）的情况下，正通量和负通量部分为

$$\boldsymbol{F}_{\mathrm{c}}^{\pm}=\begin{bmatrix} f_{\mathrm{mass}}^{\pm}\\ f_{\mathrm{mass}}^{\pm}[n_x(-V\pm 2c)/\gamma+u]\\ f_{\mathrm{mass}}^{\pm}[n_y(-V\pm 2c)/\gamma+v]\\ f_{\mathrm{mass}}^{\pm}[n_z(-V\pm 2c)/\gamma+w]\\ f_{\mathrm{energy}}^{\pm}\end{bmatrix}\tag{4.66}$$

质量和能量的通量定义为

$$\begin{aligned} f_{\mathrm{mass}}^{+}&=+\rho_{\mathrm{L}}c_{\mathrm{L}}\frac{(M_{\mathrm{L}}+1)^2}{4}\\ f_{\mathrm{mass}}^{-}&=-\rho_{\mathrm{R}}c_{\mathrm{R}}\frac{(M_{\mathrm{R}}-1)^2}{4}\\ f_{\mathrm{energy}}^{\pm}&=f_{\mathrm{mass}}^{\pm}\left\{\frac{[(\gamma-1)V\pm 2c]^2}{2(\gamma^2-1)}+\frac{u^2+v^2+w^2-V^2}{2}\right\}_{\mathrm{L/R}}\end{aligned}\tag{4.67}$$

对于超声速流动，即 $|M_n|\geqslant 1$ 时，通量为

$$\begin{aligned} &\boldsymbol{F}_{\mathrm{c}}^{+}=\boldsymbol{F}_{\mathrm{c}}\quad \boldsymbol{F}_{\mathrm{c}}^{-}=0,\quad M_n\geqslant +1\\ &\boldsymbol{F}_{\mathrm{c}}^{+}=0\quad \boldsymbol{F}_{\mathrm{c}}^{-}=\boldsymbol{F}_{\mathrm{c}},\quad M_n\leqslant -1\end{aligned}\tag{4.68}$$

左右状态的求解一般采用 MUSCL 方法[30]，该方法由式（4.46）给出，如果流场包含像激波这样的间断，则对于 $\hat{k}=-1$，$\hat{k}=0$ 和 $\hat{k}=1/3$ 的高阶格式需要使用限制器，更多细节参见 4.3.5 节。

van Leer 矢通量分裂格式在欧拉方程组中表现很好，但是，使用纳维-斯托克斯方程组进行的几项研究[50-51]显示，动量和能量方程中的分裂误差对边界层造成的污染，导致驻点和壁面温度的计算不准确。因此，文献[52]建议对垂直于边界层方向的动量通量进行修正，文献[53]中提出了类似的能量通量补救方法，这两种修正都消除了分裂误差，从而大大提高了解的精度[54]。

2. AUSM 格式

Liou 和 Steffen[41]、Liou[55] 提出了 AUSM，随后 Wada 和 Liou[56] 对其进行了修改，并更名为 AUSMD/V，最后，Liou 提出了一个改进的版本 AUSM+[42,57]。

该方法的基本思想是基于如下发现：对流通量（式(2.21)）由对流部分和压力部分这两个物理上截然不同的部分组成，即

$$\boldsymbol{F}_{\mathrm{c}}=V\begin{bmatrix}\rho\\ \rho u\\ \rho v\\ \rho w\\ \rho H\end{bmatrix}+\begin{bmatrix}0\\ n_x p\\ n_y p\\ n_z p\\ 0\end{bmatrix} \tag{4.69}$$

式(4.69)中的第一项为由逆变速度 V 对流传送的标量，与之相对，压力项是由声速决定的。这样一来，对流部分使用纯迎风方式离散，（即使对于亚声速流动）根据 V 的符号，确定使用左状态还是右状态，另外，压力项在亚声速情况下包含两种状态，只有在超声速流动中才会变为完全迎风状态。

根据文献[41]的基础 AUSM 公式，我们从式(4.61)中引入对流马赫数 $(M_n)_{I+1/2}$，这样，可以将控制体的面 $(I+1/2)$ 或 $(i+1/2)$ 的对流通量重写为

$$(\boldsymbol{F}_{\mathrm{c}})_{I+1/2}=(M_n)_{I+1/2}\begin{bmatrix}\rho c\\ \rho cu\\ \rho cv\\ \rho cw\\ \rho cH\end{bmatrix}_{\mathrm{L/R}}+\begin{bmatrix}0\\ n_x p\\ n_y p\\ n_z p\\ 0\end{bmatrix}_{I+1/2} \tag{4.70}$$

其中

$$(\cdot)_{\mathrm{L/R}}=\begin{cases}(\cdot)_{\mathrm{L}}, & M_{I+1/2}\geqslant 0\\ (\cdot)_{\mathrm{R}}, & \text{其他}\end{cases} \tag{4.71}$$

与 van Leer 矢通量分裂格式相似，根据式(4.62)和式(4.63)～式(4.65)的关系，对流马赫数取为左、右分裂马赫数之和，左右流场变量(ρ,u,v,w 和 p)遵循 MUSCL 方法[30]，按式(4.19)或式(4.39)分别插值到控制体的面上，如式(4.46)所示。如果流场包含激波等大的梯度，所有高阶 MUSCL 格式($\hat{k}=-1$，$\hat{k}=0$ 和 $\hat{k}=1/3$)都需要使用限制器，更多细节请参阅 4.3.5 节。

控制体的面 $(I+1/2)$ 上的压力由如下的分裂得到[41]

$$p_{I+1/2}=p_{\mathrm{L}}^{+}+p_{\mathrm{R}}^{-} \tag{4.72}$$

根据文献[40]，分裂压力为

$$p_{\mathrm{L}}^{+}=\begin{cases}p_{\mathrm{L}}, & M_{\mathrm{L}}\geqslant +1\\ \dfrac{p_{\mathrm{L}}}{4}(M_{\mathrm{L}}+1)^{2}(2-M_{\mathrm{L}}), & |M_{\mathrm{L}}|<1\\ 0, & M_{\mathrm{L}}\leqslant -1\end{cases}\tag{4.73}$$

和

$$p_{\mathrm{R}}^{-}=\begin{cases}0, & M_{\mathrm{R}}\geqslant +1\\ \dfrac{p_{\mathrm{R}}}{4}(M_{\mathrm{R}}-1)^{2}(2+M_{\mathrm{R}}), & |M_{\mathrm{R}}|<1\\ p_{\mathrm{R}}, & M_{\mathrm{R}}\leqslant -1\end{cases}\tag{4.74}$$

对于 $|M_{\mathrm{L/R}}|<1$，也可以使用下面的低阶展开：

$$p_{\mathrm{L/R}}^{\pm}=\frac{p_{\mathrm{L/R}}}{2}(1\pm M_{\mathrm{L/R}})\tag{4.75}$$

也可以将 AUSM 写成

$$\begin{aligned}(\boldsymbol{F}_{\mathrm{c}})_{I+1/2}=&\frac{1}{2}(M_n)_{I+1/2}\left\{\begin{bmatrix}\rho c\\ \rho cu\\ \rho cv\\ \rho cw\\ \rho cH\end{bmatrix}_{\mathrm{L}}+\begin{bmatrix}\rho c\\ \rho cu\\ \rho cv\\ \rho cw\\ \rho cH\end{bmatrix}_{\mathrm{R}}\right\}-\\ &\frac{1}{2}|(M_n)_{I+1/2}|\left\{\begin{bmatrix}\rho c\\ \rho cu\\ \rho cv\\ \rho cw\\ \rho cH\end{bmatrix}_{\mathrm{R}}-\begin{bmatrix}\rho c\\ \rho cu\\ \rho cv\\ \rho cw\\ \rho cH\end{bmatrix}_{\mathrm{L}}\right\}+\begin{bmatrix}0\\ n_x(p_{\mathrm{L}}^{+}+p_{\mathrm{R}}^{-})\\ n_y(p_{\mathrm{L}}^{+}+p_{\mathrm{R}}^{-})\\ n_z(p_{\mathrm{L}}^{+}+p_{\mathrm{R}}^{-})\\ 0\end{bmatrix}\end{aligned}\tag{4.76}$$

式(4.76)右侧第一项表示左右状态的马赫数加权平均——类似于式(4.15)、式(4.20)或式(4.35)、式(4.40)的通量方程的平均。第二项具有耗散特性，它可以用标量值 $|(M_n)_{I+1/2}|$ 来标定。

AUSM 可以清晰分辨强激波，并能准确模拟边界层，然而，最初的 AUSM[41,55] 在激波位置以及流动与网格对齐的情况下会产生局部压力振荡[58]，因此，文献[58-59]建议在激波位置改用 van Leer 格式。当对流马赫数 $(M_n)_{I+1/2}$ 趋于零时，式(4.76)中的耗散项也趋于零，此时的格式不能消除扰动，为解决流动对齐问题，文献[58]对耗散项的标度进行了修正，具体如下：

$$|(M_n)_{I+1/2}| = \begin{cases} |(M_n)_{I+1/2}|, & |(M_n)_{I+1/2}| > \delta \\ \dfrac{(M_n)^2_{I+1/2} + \delta^2}{2\delta}, & |(M_n)_{I+1/2}| \leqslant \delta \end{cases} \tag{4.77}$$

式中：δ 是一个小量（$0<\delta\leqslant 0.5$）。因此将保持足够的数值耗散，另外，为了不影响 AUSM 在边界层的精度，可以采用与式（4.57）中所述的类似中心格式的思路，在壁面法向方向减小 δ 的值。

文献[42,57]进一步改进了基础 AUSM 格式，包括采用新的马赫分裂和压力分裂，分别取代式（4.63）和式（4.64），以及式（4.73）和式（4.74），改进的格式称为 AUSM+格式，它在激波附近具有更好的性能。

3. CUSP 格式

对流迎风分裂压力（CUSP）格式的概念与 AUSM 格式非常相似，然而，CUSP 方法的优点在于可以用通量的平均值（但不像 AUSM 中那样加权）减去耗散项来表示，这种特性对于显式混合多步格式的实现至关重要。此外，由于与 AUSM 相比，CUSP 格式具有不同的标度因子，因此 CUSP 格式在流动对齐的情况下表现更佳。CUSP 格式由 Jameson 提出[43,60-61]，并经 Tatsumi 等修改[44,62]，它可以在格心格式或（格点）二次控制体类型的空间离散中实现。

通过控制体面的对流通量（式（2.21））可以由式（4.15）、式（4.20）、式（4.35）或式（4.40）使用通量的算术平均值来近似得到，然后从中心通量中减去耗散项以达到稳定的目的。因此，我们得到面（$I+1/2$）的对流通量为

$$(\boldsymbol{F}_c)_{I+1/2} = \frac{1}{2}[\boldsymbol{F}_c(\boldsymbol{W}_R) + \boldsymbol{F}_c(\boldsymbol{W}_L)] - \boldsymbol{D}_{I+1/2} \tag{4.78}$$

在二次控制体离散的情况下，相应的下标为 $i+1/2$。由状态差和矢通量差线性组合而成的耗散项可以表示为

$$(\boldsymbol{D})_{I+1/2} = \frac{1}{2}(\alpha^* c)_{I+1/2}\left\{\begin{bmatrix}\rho \\ \rho u \\ \rho v \\ \rho w \\ \rho\phi\end{bmatrix}_R - \begin{bmatrix}\rho \\ \rho u \\ \rho v \\ \rho w \\ \rho\phi\end{bmatrix}_L\right\} + \frac{1}{2}\beta_{I+1/2}\left\{\begin{bmatrix}\rho V \\ \rho uV + n_x p \\ \rho vV + n_y p \\ \rho wV + n_z p \\ \rho HV\end{bmatrix}_R - \begin{bmatrix}\rho V \\ \rho uV + n_x p \\ \rho vV + n_y p \\ \rho wV + n_z p \\ \rho HV\end{bmatrix}_L\right\} \tag{4.79}$$

式(4.79)中的 ϕ 有两种选择,可以是总能 E,也可以是总焓 H,第一种情况称为 E-CUSP 格式[63],第二种选择称为 H-CUSP 格式。$\phi=H$ 的形式保持了总焓[44],因此适用于无黏流动。左(L)状态和右(R)状态的计算类似于 MUSCL 方法[30],使用限制的插值(式(4.118)~式(4.121))。此外,式(4.79)中的两个因子 $\alpha^* c$ 和 β 定义为

$$\alpha^* c=\begin{cases}|V|, & \beta=0 \\ -(1+\beta)\Lambda^-, & \beta>0 \text{ 且 } 0<M_n<1 \\ +(1-\beta)\Lambda^+, & \beta<0 \text{ 且} -1<M_n<0 \\ 0, & |M_n|\geqslant 1\end{cases} \tag{4.80}$$

和

$$\beta=\begin{cases}+\max\left(0,\dfrac{V+\Lambda^-}{V-\Lambda^-}\right), & 0\leqslant M_n<1 \\ -\max\left(0,\dfrac{V+\Lambda^+}{V-\Lambda^+}\right), & -1\leqslant M_n<0 \\ \mathrm{sign}(M_n), & |M_n|\geqslant 1\end{cases} \tag{4.81}$$

其中,$M_n=V/c$。式(4.80)和式(4.81)中的逆变速度 V 由速度分量的算术平均值计算,即

$$V_{I+1/2}=\frac{1}{2}[(u_I+u_{I+1})n_x+(v_I+v_{I+1})n_y+(w_I+w_{I+1})n_z] \tag{4.82}$$

计算 M_n 时的声速 c 也由算术平均量得到。

式(4.80)和式(4.81)中的正负特征值 Λ^+ 和 Λ^- 为 Roe 矩阵的特征值[64],将在 4.3.3 节中展开讨论,文献[61]给出了特征值,即

$$\Lambda^{\pm}=\frac{\gamma+1}{2\gamma}\tilde{V}\pm\sqrt{\left(\frac{\gamma-1}{2\gamma}\tilde{V}\right)^2+\frac{\tilde{c}^2}{\gamma}} \tag{4.83}$$

其中,$\tilde{V}$ 表示逆变速度,γ 为比热比,$\tilde{c}$ 为声速。式(4.83)中的流场变量,必须在控制体的面上使用 Roe 平均进行求解[64],Roe 平均公式为

$$\begin{aligned}\tilde{u}_{I+1/2}&=\frac{u_L\sqrt{\rho_L}+u_R\sqrt{\rho_R}}{\sqrt{\rho_L}+\sqrt{\rho_R}}\\ \tilde{v}_{I+1/2}&=\frac{v_L\sqrt{\rho_L}+v_R\sqrt{\rho_R}}{\sqrt{\rho_L}+\sqrt{\rho_R}}\\ \tilde{w}_{I+1/2}&=\frac{w_L\sqrt{\rho_L}+w_R\sqrt{\rho_R}}{\sqrt{\rho_L}+\sqrt{\rho_R}}\end{aligned} \tag{4.84}$$

$$\widetilde{H}_{I+1/2}=\frac{H_{\mathrm{L}}\sqrt{\rho_{\mathrm{L}}}+H_{\mathrm{R}}\sqrt{\rho_{\mathrm{R}}}}{\sqrt{\rho_{\mathrm{L}}}+\sqrt{\rho_{\mathrm{R}}}}$$

$$\widetilde{c}_{I+1/2}=\sqrt{(\gamma-1)\left(\widetilde{H}-\frac{\widetilde{u}^2+\widetilde{v}^2+\widetilde{w}^2}{2}\right)_{I+1/2}}$$

$$\widetilde{V}_{I+1/2}=\widetilde{u}_{I+1/2}n_x+\widetilde{v}_{I+1/2}n_y+\widetilde{w}_{I+1/2}n_z$$

对因子 $\alpha^* c$ 和 β 的定义使得对流通量在超声速流中为完全迎风，即 $\alpha^* c=0$ 和 $\beta=\mathrm{sign}(M_n)$。另外，在亚声速流动中（$\beta=0$），通过 $|V|$ 进行耗散的标定，这是计算黏性层所需要的性质。对于大长高比单元，显式时间格式通常需要在单元较长边的方向上增加数值耗散以保持鲁棒性，这可以通过采用与式（4.57）相似的谱半径方式来实现，更多的细节可参考文献[38]，其中还包含了 CUSP 格式与标量以及矩阵人工耗散（4.3.1 节）格式的比较。

另外，将式（4.84）中的 Roe 平均替换为算术平均，并将式（4.83）中的特征值替换为 $\Lambda^{\pm}=V\pm c$，可以大大简化 CUSP 格式的实现，引入如下定义：

$$\alpha^* c=\alpha c-\beta V \tag{4.85}$$

修改的格式的因子为[44,61]

$$\alpha=\begin{cases}|M_n|, & |M_n|\geqslant\delta\\ \dfrac{M_n^2+\delta^2}{2\delta}, & |M_n|<\delta\end{cases} \tag{4.86}$$

和

$$\beta=\begin{cases}\max(0,2M_n-1), & 0\leqslant M_n<1\\ \min(0,2M_n+1), & -1<M_n<0\\ \mathrm{sign}(M_n), & |M_n|\geqslant 1\end{cases} \tag{4.87}$$

在此，逆变速度 $M_n=V/c$ 按式（4.82）计算，式（4.86）中的参数 δ 旨在防止驻点的耗散消失，但这似乎并不总是必要的。通过上述简化得到一个非常有效率的格式，但是与 Roe 平均的原始公式相比，其激波分辨率略有降低。

4.3.3 通量差分裂格式

通量差分裂格式通过求解黎曼（激波管）问题，从（通常不连续的）左右状态计算控制体的面的对流通量，这个思想最初由 Godunov 提出[65]。与矢通量分裂格式相比，通量差分裂格式不仅考虑波（信息）传播的方向，而且还考虑了波本身。为了减少 Godunov 格式求解黎曼问题精确解的计算量，Osher 等[66]和 Roe[64]提出了近似黎曼解算器，特别是 Roe 方法，由于其在边界层流动中的精度高，以及对激波的分辨率好，获得了广泛的应用。因此，我们将更详细地介绍 Roe 解算器。

Roe 近似黎曼解算器既可以在格心格式中实现，也可以在二次控制体格式中实现，它将控制体面的通量差分解为波的贡献之和，同时保证了欧拉方程组的守恒性质。在面$(I+1/2)$或$(i+1/2)$上，通量差表示为[64]

$$(\boldsymbol{F}_c)_R-(\boldsymbol{F}_c)_L=(\boldsymbol{A}_{Roe})_{I+1/2}(\boldsymbol{W}_R-\boldsymbol{W}_L) \tag{4.88}$$

其中，$\boldsymbol{A}_{Roe}$ 表示 Roe 矩阵，L 和 R 分别表示左右状态（见图 4.8）。Roe 矩阵是将流场变量替换为 Roe 平均变量的对流通量雅可比矩阵 $\boldsymbol{A}_c$（见 A.9 节），如果 Roe 平均由左右状态根据以下公式计算[64,67]，则式(4.88)中的通量差是精确的。

$$\begin{aligned}
&\tilde{\rho}=\sqrt{\rho_L\rho_R}\\
&\tilde{u}=\frac{u_L\sqrt{\rho_L}+u_R\sqrt{\rho_R}}{\sqrt{\rho_L}+\sqrt{\rho_R}}\\
&\tilde{v}=\frac{v_L\sqrt{\rho_L}+v_R\sqrt{\rho_R}}{\sqrt{\rho_L}+\sqrt{\rho_R}}\\
&\tilde{w}=\frac{w_L\sqrt{\rho_L}+w_R\sqrt{\rho_R}}{\sqrt{\rho_L}+\sqrt{\rho_R}}\\
&\tilde{H}=\frac{H_L\sqrt{\rho_L}+H_R\sqrt{\rho_R}}{\sqrt{\rho_L}+\sqrt{\rho_R}}\\
&\tilde{c}=\sqrt{(\gamma-1)\left(\tilde{H}-\frac{1}{2}\tilde{q}^2\right)}\\
&\tilde{V}=\tilde{u}n_x+\tilde{v}n_y+\tilde{w}n_z\\
&\tilde{q}^2=\tilde{u}^2+\tilde{v}^2+\tilde{w}^2
\end{aligned} \tag{4.89}$$

当我们将 Roe 矩阵的对角化，即 $\boldsymbol{A}_{Roe}=\boldsymbol{T}\boldsymbol{\Lambda}_c\boldsymbol{T}^{-1}$，插入式(4.88)，可以使 Roe 格式中的波分解更加清晰，即

$$(\boldsymbol{F}_c)_R-(\boldsymbol{F}_c)_L=\boldsymbol{T}\boldsymbol{\Lambda}_c(\boldsymbol{C}_R-\boldsymbol{C}_L) \tag{4.90}$$

左($\boldsymbol{T}^{-1}$)和右($\boldsymbol{T}$)特征矢量矩阵，以及特征值($\boldsymbol{\Lambda}_c$)对角矩阵使用 Roe 平均(式(4.89))求解。在式(4.90)中，特征变量 $\boldsymbol{C}$ 表示波幅，特征值 Λ_c 为近似黎曼问题的相关波速，最后右特征矢量是波本身。

根据前面的讨论，控制体面的对流通量根据如下公式计算[64]：

$$(\boldsymbol{F}_c)_{I+1/2}=\frac{1}{2}\left[\boldsymbol{F}_c(\boldsymbol{W}_R)+\boldsymbol{F}_c(\boldsymbol{W}_L)-|\boldsymbol{A}_{Roe}|_{I+1/2}(\boldsymbol{W}_R-\boldsymbol{W}_L)\right] \tag{4.91}$$

$|\boldsymbol{A}_{Roe}|$ 与左右状态差的乘积可以采用如下形式有效地计算：

$$|\boldsymbol{A}_{Roe}|(\boldsymbol{W}_R-\boldsymbol{W}_L)=|\Delta\boldsymbol{F}_1|+|\Delta\boldsymbol{F}_{2,3,4}|+|\Delta\boldsymbol{F}_5| \tag{4.92}$$

其中

$$|\Delta \boldsymbol{F}_1| = |\widetilde{V}-\widetilde{c}|\left(\frac{\Delta p-\widetilde{\rho}\,\widetilde{c}\,\Delta V}{2\widetilde{c}^2}\right)\begin{bmatrix}1\\ \widetilde{u}-\widetilde{c}n_x\\ \widetilde{v}-\widetilde{c}n_y\\ \widetilde{w}-\widetilde{c}n_z\\ \widetilde{H}-\widetilde{c}\,\widetilde{V}\end{bmatrix} \tag{4.93}$$

$$|\Delta \boldsymbol{F}_{2,3,4}| = |\widetilde{V}|\left\{\left(\Delta\rho-\frac{\Delta p}{\widetilde{c}^2}\right)\begin{bmatrix}1\\ \widetilde{u}\\ \widetilde{v}\\ \widetilde{w}\\ \frac{1}{2}\widetilde{q}^2\end{bmatrix}+\widetilde{\rho}\begin{bmatrix}0\\ \Delta u-\Delta V n_x\\ \Delta v-\Delta V n_y\\ \Delta w-\Delta V n_z\\ \widetilde{u}\Delta u+\widetilde{v}\Delta v+\widetilde{w}\Delta w-\widetilde{V}\Delta V\end{bmatrix}\right\} \tag{4.94}$$

$$|\Delta \boldsymbol{F}_5| = |\widetilde{V}+\widetilde{c}|\left(\frac{\Delta p+\widetilde{\rho}\,\widetilde{c}\,\Delta V}{2\widetilde{c}^2}\right)\begin{bmatrix}1\\ \widetilde{u}+\widetilde{c}n_x\\ \widetilde{v}+\widetilde{c}n_y\\ \widetilde{w}+\widetilde{c}n_z\\ \widetilde{H}+\widetilde{c}\,\widetilde{V}\end{bmatrix} \tag{4.95}$$

跳跃条件定义为 $\Delta(\cdot)=(\cdot)_R-(\cdot)_L$，Roe 平均变量由式(4.89)给出。

左右状态使用如式(4.46)的 MUSCL 格式[30]确定。如果流场包含任何间断，则所有高阶格式($\hat{k}=-1,\hat{k}=0$ 和 $\hat{k}=1/3$)都必须增加限制器(4.3.5 节)。

根据式(4.88)，在固定膨胀的情况下，$(\boldsymbol{F}_c)_L=(\boldsymbol{F}_c)_R$，但是 $\boldsymbol{W}_L\neq\boldsymbol{W}_R$，因此 Roe 近似黎曼解算器会产生非物理的膨胀激波。此外，还可能会出现“红玉现象”，即扰动在弓形强激波前沿滞止线增长[68-69]，可参见文献[70]中的讨论。产生这种现象的根源在于原始格式不能识别声速点，为了解决这个问题，可使用 Harten 熵修正[71-72]对特征值的模 $|\Lambda_c|=|\widetilde{V}\pm\widetilde{c}|$ 进行修改：

$$|\Lambda_c| = \begin{cases}|\Lambda_c|, & |\Lambda_c|>\delta\\ \dfrac{\Lambda_c^2+\delta^2}{2\delta}, & |\Lambda_c|\leqslant\delta\end{cases} \tag{4.96}$$

其中，δ 是一个小量，可以设置为当地声速的某个分数(如 1/10 或 1/20)。为了防止线性波 $|\Delta \boldsymbol{F}_{2,3,4}|$ 在 $\widetilde{V}\to 0$(例如，在驻点处或者对于网格对齐流动)时消失，也可将上述方法应用于 $|\widetilde{V}|$。

与中心格式或矢通量分裂格式相比，Roe 解算器的一个明显缺点出现在真实气体模拟中，即 Roe 矩阵和 Roe 平均需要相应的改变，它们可能会变得相当复杂。读者可以在文献[73-76]以及其引用的文献中找到平衡和非平衡真实气体流动公式的例子，在文献[77]中描述了用于任意可压缩和不可压缩流体的 Roe 格式的实现，此外，9.5 节中包含了一个 Roe 格式的低马赫数预处理公式（见式(9.82)）。

4.3.4 总变差减小格式

为了防止在流动解中产生新的极值，Harten 首先提出了 TVD 格式的概念[78]，TVD 格式的主要条件是对于标量守恒方程，解的总变差随时间减小，其中总变差定义为

$$TV \equiv \sum_{I} |U_{I+1} - U_I| \tag{4.97}$$

这意味着解的最大值必须是非递增的，最小值必须是非递减的，从而，在时间演化过程中不会产生新的局部极值。因此，具有 TVD 特性的离散方法可以精确地解析强激波，而不会像具有标量或矩阵人工耗散的中心格式（4.3.1 节）那样，产生解的任何虚假振荡。

TVD 格式一般写成对流通量的平均加上符合 TVD 条件的额外的耗散项的形式（有限的通量耗散）[78-79]。如果耗散项取决于特征速度的符号，我们称之为迎风向 TVD 格式[82-86]，否则称为对称 TVD 格式[80-81]。经验表明，迎风 TVD 格式比对称 TVD 格式具有更高的精度[87]，迎风 TVD 格式特别适合于超声速和高超声速流场的模拟[88]，如果采用文献[89]中介绍的改进，它也能够在边界层中具有精确的分辨率[54]。

在该框架下，通过控制体（图 4.8）的面 $(I+1/2)$ 的对流通量可以表示为

$$(\boldsymbol{F}_c)_{I+1/2} = \frac{1}{2}[(\boldsymbol{F}_c)_{I+1} + (\boldsymbol{F}_c)_I] + \frac{1}{2}\boldsymbol{T}_{I+1/2}\boldsymbol{\Theta}_{I+1/2} \tag{4.98}$$

对于具有二次控制体的格点格式（4.2.3 节），索引变为 $(i+1/2)$、$(i+1)$。矩阵 $\boldsymbol{T}$ 为雅可比矩阵 $\boldsymbol{A}_c = \partial \boldsymbol{F}_c / \partial \boldsymbol{W}$ 的右特征矢量，矩阵的各项可在 A.11 节中找到。在式(4.98)中，$\boldsymbol{\Theta}$ 项考虑了特征速度的方向，它控制差分算子的迎风方向，矢量 $\boldsymbol{\Theta}$ 的第 l 个分量定义为[85]

$$\Theta^l_{I+1/2} = \frac{1}{2}\psi(\Lambda^l_{I+1/2})(\Psi^l_{I+1} + \Psi^l_I) - \psi(\Lambda^l_{I+1/2} + \chi^l_{I+1/2})\Delta C^l_{I+1/2} \tag{4.99}$$

其中，Λ^l 表示对角矩阵 $\boldsymbol{\Lambda}_c$（见 A.11 节）的单个特征值，Ψ 为限制器函数（式(4.122)）。此外

$$\chi^l_{I+1/2} = \frac{1}{2}\psi(\Lambda^l_{I+1/2}) \cdot \begin{cases} \dfrac{(\Psi^l_{I+1/2} - \Psi^l_I)}{\Delta C^l_{I+1/2}}, & \Delta C^l_{I+1/2} \neq 0 \\ 0, & \Delta C^l_{I+1/2} = 0 \end{cases} \tag{4.100}$$

最后 ΔC^{l} 为特征变量之差的元素，即

$$\Delta \boldsymbol{C}_{I+1/2}=\boldsymbol{T}_{I+1/2}^{-1}(\boldsymbol{W}_{I+1}-\boldsymbol{W}_{I}) \tag{4.101}$$

其中，$\boldsymbol{T}^{-1}$ 是左特征矢量矩阵。使用 Harten 熵修正[71-72]来防止当 $|z|\to 0$ 时，$\psi(z)$ 为零，即

$$\psi(z)=\begin{cases}|z|, & |z|>\delta_1\\ \dfrac{z^2+\delta_1^2}{2\delta_1}, & |z|\leqslant\delta_1\end{cases} \tag{4.102}$$

参数 δ_1 表示为速度分量和声速的函数[85]

$$(\delta_1)_{I+1/2}=\delta(|u_{I+1/2}|+|v_{I+1/2}|+|w_{I+1/2}|+c_{I+1/2}) \tag{4.103}$$

其中，$0.05\leqslant\delta\leqslant 0.5$。面$(I+1/2)$上原始变量的值可以由 I 和 $I+1$ 处的状态采用 Roe 平均（式(4.89)）或简单算术平均得到。限制器函数 $\boldsymbol{\Psi}$ 可以防止在强梯度附近产生虚假解，该内容将在 4.3.5 节介绍。需要强调的是，上述迎风 TVD 格式并没有使用 MUSCL 方法来实现高阶精度。

在间断点处，式(4.99)和式(4.100)中的限制器函数会被设为零，此时迎风 TVD 方法在空间上是精确的一阶精度[90]，反之，如上所述的迎风 TVD 格式在光滑流动区域为二阶精度。

4.3.5 限制器函数

二阶和高阶迎风空间离散需要使用限制器或限制器函数，以防止在具有大梯度的区域（如在激波处）产生振荡和伪解。因此，我们要寻找的至少是一个保持单调性的格式，这意味着流场中的极大值必须是不递增的，极小值必须是不递减的，并且在时间演化过程中不能产生新的局部极值。换句话说，如果初始数据是单调的，那么解也必须保持单调。保单调格式的严格条件（或者 TVD 格式的严格条件）通常被放弃，取而代之的是局部极值减小（local extremum diminishing，LED）条件[61]，在此，只有包含在模板中的局部极值必须减小。

然而，根据 Godunov 理论，高阶线性格式（如 MUSCL 方法）不可能保持单调性[91]，因此有必要使用非线性限制器函数来构造一个保持单调性或 TVD 的离散。如图 4.10 所示，其中使用了式(4.98)的迎风 TVD 格式带限制器以及不带限制器计算经过 NACA0012 翼型的 2D 跨声速流动，可以清楚地看到，如果没有限制器，在翼型上下侧激波附近会出现较大的振荡，另外，在远离激波的位置，有限制器的和无限制器的解则变得几乎相同。

限制器的目的是减小流场变量插值到控制体面的斜率（即 $(U_{I+1}-U_I)/\Delta x$），以限制解的变化。在强间断处，限制器必须将斜率减小到零，以防止产生新的极

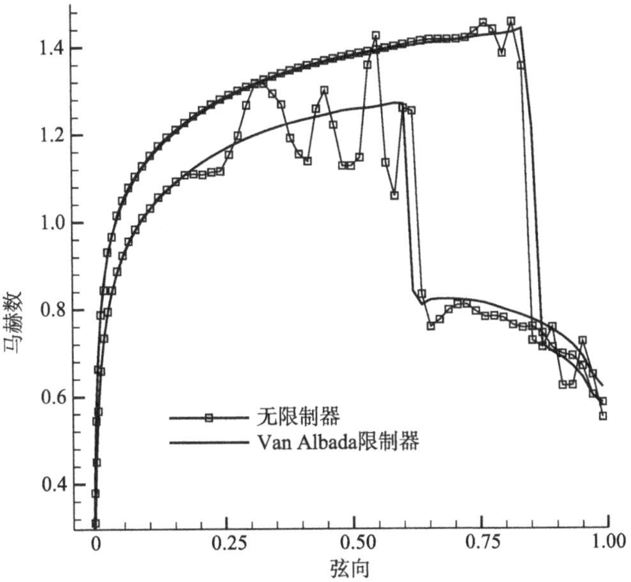

图 4.10　有无限制器时无黏跨声速流动计算比较。NACA0012 翼型，$M_\infty=0.85$，$\alpha=1°$

值，这意味着对于 MUSCL 方法以及 TVD 格式，在紧邻大梯度的区域，恢复为（单调的）一阶迎风格式（式（4.46）中 $\epsilon=0$）。对限制器的最后一个要求是很明显的——在光滑流动区域必须保持原始的无限制离散，以使数值耗散尽可能地低。限制器对左右状态插值的影响如图 4.11 所示，这个例子显示了在局部最小值 I 处斜率的减小以及单元 $I+1$、$I+2$ 处斜率的变化，以确保解的单调。需要记住的是，面左右状态的差异可能（而且通常会）仍然存在。

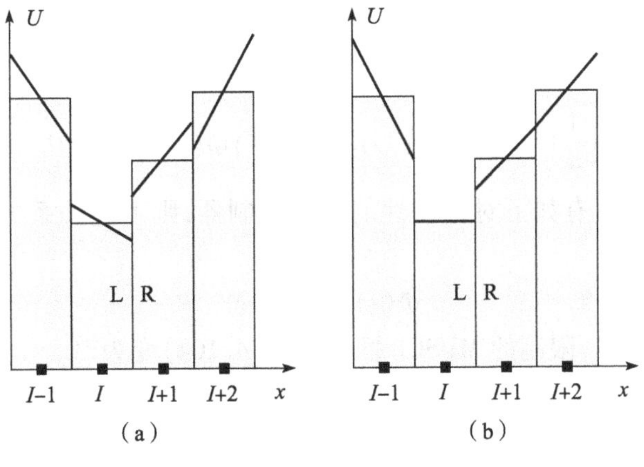

图 4.11　面的插值方式的比较

(a) 直接插值；(b) 有限制插值。粗线表示斜率 $\Delta U/\Delta x$，横线表示格心值。

下面介绍四种不同的限制器函数，它们在实践中得到了充分的确认和证明，将考虑用于二阶 MUSCL、CUSP 和迎风 TVD 格式的限制器。

1. 用于 MUSCL 插值的限制器函数

van Leer 的 MUSCL 方法[30]在必要时通过使用限制器函数来减小式(4.47)中 $\Delta_+ U_I$ 和 $\Delta_- U_I$ 的差异，从而转化为一个保持单调性的格式。通过引入斜率限制器 $\Phi^{\pm}$，式(4.46)的 MUSCL 插值公式被修改为(见图 4.8)

$$
\begin{aligned}
U_{\mathrm{R}} &= U_{I+1} - \frac{1}{4}\left[(1+\hat{k})\Phi^{+}_{I+1/2}\Delta_- + (1-\hat{k})\Phi^{+}_{I+3/2}\Delta_+\right]U_{I+1} \\
U_{\mathrm{L}} &= U_I + \frac{1}{4}\left[(1+\hat{k})\Phi^{-}_{I+1/2}\Delta_+ + (1-\hat{k})\Phi^{+}_{I-1/2}\Delta_-\right]U_I
\end{aligned}
\tag{4.104}
$$

其中，式(4.46)中的参数 ϵ 被设为 1。斜率限制器是相邻解变化的比值的函数，即 $\Phi^{\pm}_{I+1/2} = \Phi(r^{\pm}_{I+1/2})$，其中[1]

$$
\begin{aligned}
r^{+}_{I+1/2} &= \frac{U_{I+2} - U_{I+1}}{U_{I+1} - U_I} \\
r^{-}_{I+1/2} &= \frac{U_I - U_{I-1}}{U_{I+1} - U_I}
\end{aligned}
\tag{4.105}
$$

如果用 r_{L} 替代 $r^{+}_{I-1/2}$，用 r_{R} 替代 $r^{-}_{I+3/2}$，则

$$
\begin{aligned}
r_{\mathrm{R}} &= \frac{U_{I+1} - U_I}{U_{I+2} - U_{I+1}} = \frac{\Delta_-}{\Delta_+}U_{I+1} \\
r_{\mathrm{L}} &= \frac{U_{I+1} - U_I}{U_I - U_{I-1}} = \frac{\Delta_+}{\Delta_-}U_I
\end{aligned}
\tag{4.106}
$$

可以将式(4.104)写成如下形式：

$$
\begin{aligned}
U_{\mathrm{R}} &= U_{I+1} - \frac{1}{4}\left[(1+\hat{k})r_{\mathrm{R}}\Phi(1/r_{\mathrm{R}}) + (1-\hat{k})\Phi(r_{\mathrm{R}})\right](U_{I+2} - U_{I+1}) \\
U_{\mathrm{L}} &= U_I + \frac{1}{4}\left[(1+\hat{k})r_{\mathrm{L}}\Phi(1/r_{\mathrm{L}}) + (1-\hat{k})\Phi(r_{\mathrm{L}})\right](U_I - U_{I-1})
\end{aligned}
\tag{4.107}
$$

如果只考虑具有如下对称特性的斜率限制器，则上述关系式(4.107)可以得到简化，即

$$
\Phi(r) = \Phi(1/r) \tag{4.108}
$$

根据这个定义，限制的 MUSCL 插值公式(4.104)变为[92]

$$
\begin{aligned}
U_{\mathrm{R}} &= U_{I+1} - \frac{1}{2}\Psi_{\mathrm{R}}(U_{I+2} - U_{I+1}) \\
U_{\mathrm{L}} &= U_I + \frac{1}{2}\Psi_{\mathrm{L}}(U_I - U_{I-1})
\end{aligned}
\tag{4.109}
$$

限制器函数定义为

$$\Psi_{L/R}=\frac{1}{2}[(1+\hat{k})r_{L/R}+(1-\hat{k})]\Phi_{L/R} \tag{4.110}$$

可以对式(4.110)中的斜率限制器 Φ 进行不同的表述，根据特定的 $\hat{k}$ 值进行调整，从而得到最准确、最稳定，并且保持单调性的 MUSCL 格式。

2. $\hat{k}=0$ 的 MUSCL 格式

$\hat{k}=0$，特别适合于二阶迎风向偏置格式的组合是[93]

$$\Phi(r)=\frac{2r}{r^2+1} \tag{4.111}$$

在这种情况下，函数 $\Psi(r)$ 对应于 van Albada 限制器[94]，即

$$\Psi(r)=\frac{r^2+r}{1+r^2} \tag{4.112}$$

由式(4.109)可得左右状态的表达式为

$$\begin{aligned}U_R&=U_{I+1}-\frac{1}{2}\delta_R\\U_L&=U_I+\frac{1}{2}\delta_L\end{aligned} \tag{4.113}$$

函数 δ 在两种状态下的形式是相同的，为

$$\delta=\frac{a(b^2+\epsilon)+b(a^2+\epsilon)}{a^2+b^2+2\epsilon} \tag{4.114}$$

左右状态的系数 a 和 b 定义为

$$\begin{aligned}a_R&=\Delta_+U_{I+1},\quad b_R=\Delta_-U_{I+1}\\a_L&=\Delta_+U_I,\quad b_L=\Delta_-U_I\end{aligned} \tag{4.115}$$

差分算子 $\Delta_\pm$ 由式(4.47)给出。式(4.114)中增加的参数 ϵ 使限制器在光滑流动区域保持激活状态，以防止小尺度的振荡[93]，确保达到完全收敛的定常状态解。参数 ϵ 通常设置为与当地网格尺度(例如在 3D 情况下为 $\Omega^{1/3}$)成正比[93,95]。如果状态变量 U 是以物理单位给出的，则需要对参数 ϵ 进行额外的缩放。可以看出，对于光滑流动，式(4.113)中的关系式与 $\hat{k}=0$ 的原始(无限制的)MUSCL 格式(式(4.46))相同，因此不影响解的精度，另外，函数 δ 在局部极值处为零，则将精度降低到所需的一阶。

3. $\hat{k}=1/3$ 的 MUSCL 格式

对于 $\hat{k}=1/3$ 的三点、二阶精度，迎风偏置 MUSCL 格式，可推导相应的限制器函数，其斜率限制器为

$$\Phi(r)=\frac{3r}{2r^2-r+2} \tag{4.116}$$

在这种情况下，函数 $\Psi(r)$ 对应于 Hemker 和 Koren 限制器[96]，按照与上一种情况相同的方式处理，对于面 $(I+1/2)$ 的左右状态而言，可得到与式(4.113)相同的公式，其中 δ 为[93]

$$\delta=\frac{(2a^2+\epsilon)b+(b^2+2\epsilon)a}{2a^2+2b^2-ab+3\epsilon} \tag{4.117}$$

系数 a、b 和参数 ϵ 的定义保持不变。

4. CUSP 格式的限制器

在 CUSP 格式的框架(4.3.2 节)中，可根据文献[44]的方法计算二阶精度的左(L)和右(R)状态值，即

$$\begin{aligned} U_{\mathrm{R}} &= U_{I+1}-\frac{1}{2}L(\Delta U_{I+3/2},\Delta U_{I-1/2}) \\ U_{\mathrm{L}} &= U_{I}+\frac{1}{2}L(\Delta U_{I+3/2},\Delta U_{I-1/2}) \end{aligned} \tag{4.118}$$

其中

$$\begin{aligned} \Delta U_{I-1/2} &= U_I-U_{I-1} \\ \Delta U_{I+3/2} &= U_{I+2}-U_{I+1} \end{aligned} \tag{4.119}$$

在式(4.118)和式(4.119)中，U 表示一个因变量，$L(\,)$ 为有限制的平均，即

$$L(\Delta_1,\Delta_2)=\frac{1}{2}\Psi(\Delta_1,\Delta_2)(\Delta_1+\Delta_2) \tag{4.120}$$

限制器定义为

$$\Psi(\Delta_1,\Delta_2)=1-\left|\frac{\Delta_1-\Delta_2}{|\Delta_1|+|\Delta_2|+\epsilon}\right|^{\sigma} \tag{4.121}$$

其中，σ 是一个正的系数，通常设为 2，常数 ϵ 用于防止被零除（例如 $\epsilon=10^{-20}$）。如果 Δ_1 和 Δ_2 大小相同但正负相反，则限制器为 $\Psi=0$，这意味着此时得到左、右状态的一阶精度近似。

此外，也可以使用式(4.113)～式(4.115)所示的 $\hat{k}=0$ 的 MUSCL 格式和 van Albada 限制器来代替上述关系。

5. TVD 格式的限制器

与前面的方法相比，在此处限制器不是作用于守恒变量或原始变量，而是作用于特征变量 $\boldsymbol{C}$，文献[85]给出了一个特别合适的限制器函数为

$$\Psi_I^l=\frac{\Delta C_{I-1/2}^l\Delta C_{I+1/2}^l+|\Delta C_{I-1/2}^l\Delta C_{I+1/2}^l|}{\Delta C_{I-1/2}^l+\Delta C_{I+1/2}^l+\epsilon} \tag{4.122}$$

其中，$\Delta C_{I+1/2}^{l}$ 表示控制体面（$I+1/2$）上特征变量的差值（式（4.101）），分母上的正常数 $\epsilon \approx 10^{-20}$ 用于防止被零除。在大梯度区域，限制器函数变为零，使式（4.99）和式（4.98）变为一阶精度的迎风格式。在流场变量光滑变化的区域，式（4.98）中的迎风 TVD 格式保持二阶精度，其中 $\Psi_{I}^{l}=C_{I}^{l}-C_{I-1}^{l}$。

4.4　黏性通量的离散

为了获得相容的空间离散，黏性通量通常选择与对流通量相同的控制体，对于基于重叠控制体的格点格式（4.2.2 节），出于稳定性原因，需要用二次控制体（4.2.3 节）来替换[97-99]。对于离散控制方程组（4.2）中的黏性通量 $\boldsymbol{F}_{\mathrm{v}}$，可利用控制体面上的变量的平均值，通过类似于式（4.17）、式（4.22）、式（4.37）或式（4.42）计算得到，这与黏性通量的椭圆性质相符。黏性项计算（式（2.23）~式（2.24））和应力（式（2.15））所需的面的速度分量（u,v,w）、动力学黏性系数 μ 以及热传导系数 k 的值，只需采用简单平均即可。在格心格式中（图 4.3 和图 4.8），控制体的面（$I+1/2$）处的值为

$$U_{I+1/2}=\frac{1}{2}(U_{I}+U_{I+1}) \tag{4.123}$$

其中，U 是上述流场变量，对于两种格点格式的面（$i+1/2$）也是如此，分别见图 4.5 和图 4.8。

剩下的任务是计算式（2.15）中的速度和式（2.24）中的温度的一阶导数（梯度），可以通过下述两种方式之一来完成：

（1）有限差分；

（2）格林公式。

第一种方法用到从笛卡儿坐标（x,y,z）到曲线坐标（ξ,η,ζ）的局部变换，例如：

$$\frac{\partial U}{\partial x}=\frac{\partial U}{\partial \xi}\frac{\partial \xi}{\partial x}+\frac{\partial U}{\partial \eta}\frac{\partial \eta}{\partial x}+\frac{\partial U}{\partial \zeta}\frac{\partial \zeta}{\partial x}\cdots \tag{4.124}$$

导数 U_{ξ}，U_{η} 和 U_{ζ} 由有限差分近似得到，更多细节见文献[97-99]，坐标导数和雅可比矩阵变换见 A.1 节。

对于第二种方法，则更符合本书中讨论的有限体积法，它需要构建一个额外的控制体来计算导数，下面将分别讨论其在格心格式和格点格式中的形式。

只要得到了控制体面上的流场变量值和一阶导数值，就可以根据式（4.2）将黏性通量的贡献加起来，再加上无黏通量的贡献，就完成了空间离散，从而可以对近似控制方程组进行时间积分。

4.4.1 格心格式

格林公式将 U 的一阶导数的体积分与 U 的面积分联系起来，为了应用格林公式，我们必须首先定义一个合适的控制体。由于式(4.2)中的求和需要面心处的导数，我们把定义相邻网格单元的边的中点连接起来，构建一个以面为中心的辅助控制体[20,32,99]，如图4.12(a)所示。

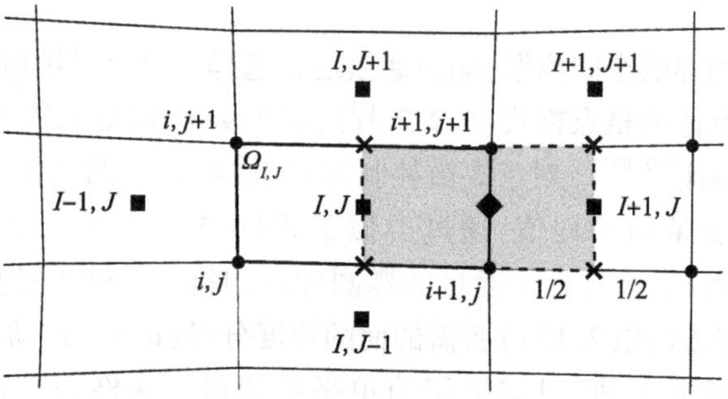

(a)

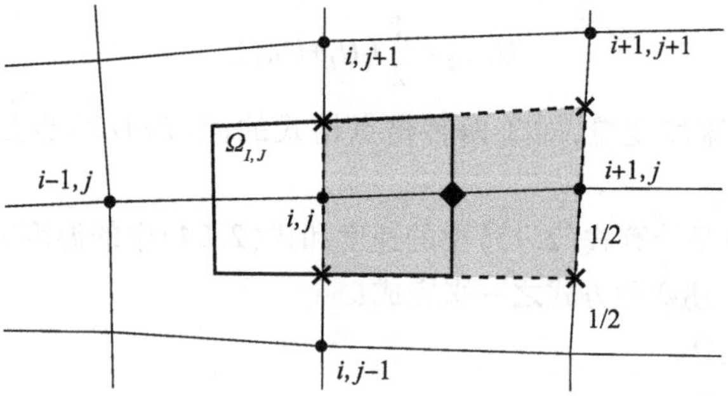

(b)

图4.12 2D情况下，求解一阶导数的辅助控制体 Ω'（灰色区域）

(a) 格心格式；(b) 格点格式。菱形符号表示求解的一阶导数所在的位置。

为了计算面$(I+1/2)$上的一阶导数——图4.12(a)中由菱形符号标记的面——必须在辅助控制体（以下用上标′表示）的边界上对相应的流场变量 U 进行积分，如对于 x 方向的导数，有

$$\frac{\partial U}{\partial x}=\frac{1}{\Omega'}\int_{\partial\Omega'} U\mathrm{d}S'_x \approx \frac{1}{\Omega'}\sum_{m=1}^{N_F} U_m S'_{x,m} \tag{4.125}$$

其中，N_F 表示面数（2D情况下 $N_F=4$，3D情况下 $N_F=6$）。体积 Ω'，面矢量 $S'_m=[S'_{x,m},S'_{y,m},S'_{z,m}]$ 的分量的计算方法已在4.1节中介绍。面上 U_m 的值可以

直接采用格心值(例如,面左右的 $U_{i,j}$ 和 $U_{i+1,j}$),或者通过对上下面进行平均,例如,在 $J+1/2$ 处有

$$U_{mI,J+1/2}=\frac{1}{4}(U_{I,J}+U_{I+1,J}+U_{I,J+1}+U_{I+1,J+1}) \tag{4.126}$$

在 3D 情况下可以使用相同的方法,同样使用四个格心值进行平均。因此

$$\begin{aligned}U_{mI,J+1/2,K}&=\frac{1}{4}(U_{I,J,K}+U_{I+1,J,K}+U_{I,J+1,K}+U_{I+1,J+1,K})\\U_{mI,J+1/2,K+1/2}&=\frac{1}{4}(U_{I,J,K}+U_{I,J,K+1}+U_{I,J+1,K}+U_{I,J+1,K+1})\end{aligned} \tag{4.127}$$

上面的格式非常紧凑,在 2D 情况下计算模板仅扩展到 9 个单元,3D 情况下是 15 个单元。需要注意的是,这种计算一阶导数的方法不能抑制两类伪模态(相邻单元体心解的解耦)的产生[20,100]:棋盘模态——产生于控制体的积分形式;一对波纹或搓板模态——产生于相邻单元值的平均,然而,在实践中通常不存在这样的问题。如果首先计算每个单元梯度(类似于对流通量),然后在单元面上取平均值,则会出现更严重的问题,尽管这种方法看起来比上述方法更有吸引力,但不推荐使用,因为它会导致强烈的奇偶失联。

上述格式有一个缺点,即对于非均匀网格,会导致精度损失[99-100]。对于任意拉伸的网格,导数的近似变得不相容(辅助控制体的质心不再对应于面中心),因此,只有在适度且光滑的拉伸网格中,黏性通量才能以二阶精度离散。

最后,需要注意的是,在计算梯度时,通过恰当地略去部分贡献,可以很容易地实现纳维-斯托克斯方程组的 TSL 近似(2.4.3 节),例如,在图 4.12(a)中,如果边界层沿 I 方向,则放弃辅助控制体的左侧(I,J)和右侧$(I+1,J)$的贡献。

4.4.2 格点格式

如前所述,两种格点格式都采用二次控制体(4.2.3 节)来离散黏性通量,因此,问题是如何计算这种控制体的面上的一阶导数。考虑图 4.12(b),一种做法是首先在网格单元上积分来计算体心的梯度,这在任意拉伸的网格上都是一阶精度。其次,如文献[32,101]那样,在控制体的面上将基于单元的梯度进行平均,但是这种方法不能防止解的奇偶失联。

另一种做法类似于格心格式,通过连接相邻网格单元的边的中点来构建一个围绕面的辅助控制体[102-103],如图 4.12(b)所示。一阶差分的求解过程与格心格式相同,必要时使用平均量。这种方法在形式上与有限差分近似相同[97-99],它求得的黏性通量在任意拉伸网格上具有一阶精度,在光滑网格上具有二阶精度[97,99],它的另一个优点是计算模板所用节点少,在 2D 情况下最多需要 9 个节

点，在 3D 情况下最多需要 15 个节点。

最后，还应该提到另一种方法，它选择一个更复杂的积分路径，将所有相邻节点进行平均[21]。这种方法具有所需节点庞大的缺点，即使在 2D 情况下，也包含 25 个节点的模板，与紧致模板相比，它通常会增加更多的数值耗散。此外，如果时间积分采用隐式格式，则通量雅可比矩阵的带宽将变得非常大。关于梯度求解的各种方法的详细讨论也可以参见文献[104]。

参考文献

[1] Hirsch C. Numerical computation of internal and external flows, vols. 1 and 2. Chichester, UK: John Wiley and Sons; 1988.

[2] Mohanraj R, Neumeier Y, Zinn BT. Characteristic-based treatment of source terms in Euler equations for Roe scheme. AIAA J 1999;37:417-24.

[3] Jameson A, Baker TJ. Solution of the Euler equations for complex configurations. AIAA Paper 83-1929; 1983.

[4] Rossow C-C. Berechnung von Strömungsfeldern durch Lösung der Euler-Gleichungen mit einer erweiterten Finite-Volumen Diskretisierungsmethode [Calculation of flow fields by the solution of Euler equations using an extended finite volume discretization scheme]. DLR Research Report No. 89-38; 1989.

[5] Bruner, CWS. Geometric properties of arbitrary polyhedra in terms of face geometry. AIAA J 1995;33:1350.

[6] www. ma. ic. ac. uk/~rn/centroid. pdf

[7] Jameson A, Schmidt W, Turkel E. Numerical solutions of the euler equations by finite volume methods using Runge-Kutta time-stepping schemes. AIAA Paper 81-1259; 1981.

[8] Ni RH. Multiple grid scheme for solving the Euler equations. AIAA Paper 81-1025; 1981.

[9] Hall MG. Cell-vertex multigrid scheme for solution of the Euler equations. In: Proc. int. conf. on numerical methods for fluid dynamics. Reading, UK: Springer Verlag; 1985.

[10] Radespiel R. A cell-vertex multigrid method for the Navier-Stokes equations. NASA TM-101557; 1989.

[11] Radespiel R, Rossow C-C, Swanson RC. An efficient cell-vertex multigrid scheme for the three-dimensional Navier-Stokes equations. AIAA Paper 89-1953; 1989 [also in AIAA J 1990;28:1464-72].

[12] Rossow C-C. Flux balance splitting—a new approach for a cell-vertex upwind scheme. In: Proc. 12th int. conf. on numerical methods in fluid dynamics. Oxford, UK: Springer Verlag; 1990.

[13] Rossow C-C. Accurate solution of the 2D Euler equations with an efficient cell-vertex upwind scheme. AIAA Paper 93-0071; 1993.

[14] Usab WJ. Embedded mesh solutions of the Euler equations using a multiple-grid method.

Ph. D. Thesis, MIT, Cambridge, MA, USA; 1983.
[15] Sidilkover D. A genuinely multidimensional upwind scheme and efficient multigrid solver for the compressible Euler equations. ICASE Report No. 94-84; 1994.
[16] Paillére H, Deconinck H, Roe PL. Conservative upwind residual-distribution schemes based on the steady characteristics of the Euler equations. AIAA Paper 95-1700; 1995.
[17] Issman E, Degrez G, Deconinck H. Implicit upwind residual-distribution Euler and Navier-Stokes solver on unstructured meshes. AIAA J 1996; 34: 2021-8.
[18] Van der Weide E, Deconinck H. Compact residual-distribution scheme applied to viscous flow simulations. VKI Lecture Series 1998-03; 1998.
[19] Dick E. A flux-difference splitting method for steady Euler equations. J Comput Phys 1988; 76: 19-32.
[20] Hall M. A vertex-centroid scheme for improved finite-volume solution of the Navier-Stokes equations. AIAA Paper 91-1540; 1991.
[21] Crumpton PI, Shaw GJ. A vertex-centred finite volume method with shock detection. Int J Numer Meth Fluids 1994; 18: 605-25.
[22] Roe PL. Error estimates for cell-vertex solutions of the compressible Euler equation. ICASE Report No. 87-6; 1987.
[23] Hoffman JD. Relationship between the truncation errors of centered finite-difference approximations on uniform and nonuniform meshes. J Comput Phys 1982; 46: 464-74.
[24] Arts T. On the consistency of four different control surfaces used for finite area blade-to-blade calculations. Int J Numer Meth Fluids 1984; 4: 1083-96.
[25] Turkel E. Accuracy of schemes with non-uniform meshes for compressible fluid flows. ICASE Report No. 85-43; 1985.
[26] Turkel E, Yaniv S, Landau U. Accuracy of schemes for the Euler equations with non-uniform meshes. ICASE Report No. 85-59; 1985.
[27] Rossow C-C. Comparison of cell-centered and cell-vertex finite volume schemes. In: Proc. 7th GAMM conference, notes on numerical fluid mechanics. Wiesbaden: Vieweg Publishing; 1987.
[28] Venkatakrishnan V. Implicit schemes and parallel computing in unstructured grid CFD. ICASE Report No. 95-28; 1995.
[29] Venkatakrishnan V, Mavriplis DJ. Implicit method for the computation of unsteady flows on unstructured grids. J. Comput Phys 1996; 127: 380-97.
[30] van Leer B. Towards the ultimate conservative difference scheme. V. A second order sequel to Godunov' s method. J Comput Phys 1979; 32: 101-36.
[31] Leonard BP. Comparison of truncation error of finite-difference and finite-volume formulations of convection terms. NASA TM-105861; 1992.
[32] Martinelli L. Calculation of viscous flows with a multigrid method. Ph. D. Thesis, Dept. of Mech. and Aerospace Eng., Princeton University; 1987.
[33] Martinelli L, Jameson A. Validation of a multigrid method for Reynolds averaged equations. AIAA Paper 88-0414; 1988.
[34] Yoon S, Kwak D. Artificial dissipation models for hypersonic external flow. AIAA Paper

88-3708;1988.
[35] Turkel E, Swanson RC, Vatsa VN, White JA. Multigrid for hypersonic viscous two- and three-dimensional flows. AIAA Paper 91-1572;1991.
[36] Swanson RC, Turkel E. On central difference and upwind schemes. J Comput Phys 1992;101:292-306.
[37] Jameson A. Artificial diffusion, upwind biasing, limiters and their effect on accuracy and multigrid convergence in transonic and hypersonic flow. AIAA Paper 93-3559;1993.
[38] Swanson RC, Radespiel R, Turkel E. Comparison of several dissipation algorithms for central difference schemes. ICASE Report No. 97-40;1997 [also AIAA Paper 97-1945, 1997].
[39] Pulliam TH. Artificial dissipation models for the Euler equations. AIAA J 1986;24:1931-40.
[40] van Leer B. Flux-vector splitting for the Euler equations. In: Proc. 8th Int. conf. on numerical methods in fluid dynamics, Springer Verlag;1982. p. 507-12 [also ICASE Report 82-30;1982].
[41] Liou M-S, Steffen Jr CJ. A new flux splitting scheme. NASA TM-104404, 1991; also J Comput Phys 1993;107:23-39.
[42] Liou M-S. A sequel to AUSM: AUSM+. J Comput Phys 1996;129:364-82.
[43] Jameson A. Artificial diffusion, upwind biasing, limiters and their effect on accuracy and multigrid convergence in transonic and hypersonic flow. AIAA Paper 93-3559;1993.
[44] Tatsumi S, Martinelli L, Jameson A. A new high resolution scheme for compressible viscous flow with shocks. AIAA Paper 95-0466;1995.
[45] Edwards JR. A low-diffusion flux-splitting scheme for Navier-Stokes calculations. Comput Fluids 1997;26:653-9.
[46] Rossow C-C. A simple flux splitting scheme for compressible flows. In: Proc. 11th DGLR-Fach-symposium, Berlin, Germany; November 10-12, 1998.
[47] Rossow C-C. A flux splitting scheme for compressible and incompressible flows. AIAA Paper 99-3346;1999.
[48] Thomas JL, van Leer B, Walters RW. Implicit flux-split schemes for the Euler equations. AIAA Paper 85-1680;1985.
[49] Anderson WK, Thomas JL, van Leer B. A comparison of finite volume flux vector splittings for the Euler equations. AIAA J 1986;24:1453-60.
[50] Hänel D, Schwane R, Seider G. On the accuracy of upwind schemes for the solution of the Navier-Stokes equations. AIAA Paper 87-1105;1987.
[51] van Leer B, Thomas JL, Roe PL, Newsome RW. A comparison of numerical flux formulas for the Euler and Navier-Stokes equations. AIAA Paper 87-1104;1987.
[52] Hänel D, Schwane R. An implicit flux-vector scheme for the computation of viscous hypersonic flows. AIAA Paper 89-0274;1989.
[53] van Leer B. Flux vector splitting for the 1990's. In: Invited Lecture for the CFD Symposium on Aeropropulsion, Cleveland, OH;1990.
[54] Seider G, Hänel D. Numerical influence of upwind TVD schemes on transonic airfoil drag prediction. AIAA Paper 91-0184;1991.
[55] Liou MS. On a new class of flux splittings. In: Proc. 13th int. conf. on numerical methods in

fluid dynamics, Rome, Italy; 1992.

[56] Wada Y, Liou M-S. A flux splitting scheme with high-resolution and robustness for discontinuities. AIAA Paper 94-0083; 1994.

[57] Liou M-S. Progress towards an improved CFD method-AUSM+. AIAA Paper 95-1701; 1995.

[58] Radespiel R, Kroll N. Accurate flux vector splitting for shocks and shear layers. J Comput Phys 1995; 121: 66-78.

[59] Edwards JR. Numerical implementation of a modified Liou-Steffen upwind scheme. AIAA J 1994; 32: 2120-22.

[60] Jameson A. Positive schemes and shock modelling for compressible flow. Int J Numer Meth Fluids 1995; 20: 743-76.

[61] Jameson A. Analysis and design of numerical schemes for gas dynamics. II: Artificial diffusion and discrete shock structure. Int J Comput Fluid Dynam 1995; 5: 1-38.

[62] Tatsumi S, Martinelli L, Jameson A. A design, implementation and validation of flux limited schemes for the solution of the compressible Navier-Stokes equations. AIAA Paper 94-0647; 1994.

[63] Nemec M, Zingg DW. Aerodynamic computations using the convective upstream split pressure scheme with local preconditioning. AIAA Paper 98-2444; 1998.

[64] Roe PL. Approximate Riemann solvers, parameter vectors, and difference schemes. J Comput Phys 1981; 43: 357-72.

[65] Godunov SK. A difference scheme for numerical computation discontinuous solution of hydrodynamic equations. Math Sbornik [in Russian] 1959; 47: 271-306 [translated US Joint Publ. Res. Service, JPRS 7226, 1969].

[66] Osher S, Solomon F. Upwind difference schemes for hyperbolic systems of conservation laws. Math Comput 1982; 38: 339-74.

[67] Roe PL, Pike J. Efficient construction and utilisation of approximate Riemann solutions. In: Glowinski R, Lions JL, editors. Computing methods in applied sciences and engineering. The Netherlands: North Holland Publishing; 1984.

[68] Peery KM, Imlay ST. Blunt-body flow simulations. AIAA Paper 88-2904; 1988.

[69] Lin HC. Dissipation additions to flux-difference splitting. AIAA Paper 91-1544; 1991.

[70] Quirk JJ. A contribution to the great Riemann solver debate. ICASE Report No. 92-64; 1992.

[71] Harten A, Lax PD, van Leer B. On upstream differencing and Godunov-type schemes for hyperbolic conservation laws. Soc Indust Appl Math Rev 1983; 25(1): 35-61.

[72] Harten A, Hyman JM. Self adjusting grid methods for one-dimensional hyperbolic conservation laws. J Comput Phys 1983; 50: 235-69.

[73] Vinokur M. Flux Jacobian matrices and generalized Roe average for an equilibrium real gas. NASA CR-177512; 1988.

[74] Vinokur M, Montagné J-L. Generalized flux-vector splitting and Roe average for an equilibrium real gas. J Comput Phys 1990; 89: 276-300.

[75] Grossman B, Cinnella P. Flux-split algorithms for flows with non-equilibrium chemistry and vibrational relaxation. J Comput Phys 1990; 88: 131-68.

[76] Shuen J-S, Liou M-S, van Leer B. Inviscid flux-splitting algorithms for real gases with non-equilibrium chemistry. J Comput Phys 1990;90:371-95.

[77] Mulas M, Chibbaro S, Delussu G, Di Piazza I, Talice M. Efficient parallel computations of flows of arbitrary fluids for all regimes of Reynolds, Mach and Grashof Numbers. Int J Numer Meth Heat Fluid Flow 2002;12:637-57.

[78] Harten A. High resolution schemes for hyperbolic conservation laws. J Comput Phys 1983;49:357-93.

[79] Jameson A, Lax PD. Conditions for the construction of multi-point total variation diminishing difference schemes. MAE Report 1650, Dept. of Mechanical and Aerospace Engineering, Princeton University;1984.

[80] Davis SF. TVD finite difference schemes and artificial viscosity. ICASE Report No. 84-20;1984.

[81] Yee HC. Construction of implicit and explicit symmetric TVD schemes and their applications. J Comput Phys 1987;68:151-79.

[82] Yee HC, Kutler P. Application of second-order accurate total variation diminishing (TVD) schemes to the Euler equations in general geometries. NASA TM-85845;1983.

[83] Yee HC. Upwind and symmetric shock-capturing schemes. NASA TM-89464;1987.

[84] Yee HC, Harten A. Implicit TVD schemes for hyperbolic conservation laws in curvilinear coordinates. AIAA J 1987;25:266-74.

[85] Yee HC, Klopfer GH, Montagné J-L. High-resolution shock-capturing schemes for inviscid and viscous hypersonic flows. NASA TM-100097;1988.

[86] Yee HC. A class of high-resolution explicit and implicit shock-capturing methods. VKI Lecture Series 1989-04;1989 [also NASA TM-101088;1989].

[87] Kroll N, Gaitonde D, Aftosmis M. A systematic comparative study of several high resolution schemes for complex problems in high speed flows. In:29th AIAA Aerospace Sci. Meeting and Exhibit, Reno, USA;1991.

[88] Kroll N, Rossow C-C. A high resolution cell vertex tvd scheme for the solution of the twoand three-dimensional Euler equations. 12th International Conf. on Numerical Methods in Fluid Dynamics, Oxford, UK;1990.

[89] Müller B. Simple improvements of an upwind TVD scheme for hypersonic flow. AIAA Paper 89-1977;1989.

[90] Blazek J. Methods to accelerate the solutions of the Euler- and Navier-Stokes equations for steady-state super- and hypersonic flows. Translation of DLR Research Report 94-35, ESA-TT-1331;1995.

[91] LeVeque RJ. Numerical methods for conservation laws. Basel, Switzerland: Birkhäuser Verlag;1992.

[92] Spekreijse SP. Multigrid solution of the steady Euler equations. Ph. D. Dissertation, Centrum voor Wiskunde en Informatica, Amsterdam, The Netherlands;1987.

[93] Venkatakrishnan V. Preconditioned conjugate gradient methods for the compressible Navier-Stokes equations. AIAA J 1991;29:1092-110.

[94] van Albada GD, van Leer B, Roberts WW. A comparative study of computational methods in cosmic gas dynamics. Astron Astrophys 1982;108:76-84.

[95] Venkatakrishnan V. On the accuracy of limiters and convergence to steady state solutions. AIAA Paper 93-0880;1993.

[96] Hemker PW, Koren B. Multigrid, defect correction and upwind schemes for the steady Navier-Stokes equations. In: Synopsis, HERMES Hypersonic Research Program Meeting, Stuttgart, Germany; 1987.

[97] Radespiel R, Swanson RC. An investigation of cell centred and cell vertex multigrid schemes for the Navier-Stokes equations. AIAA Paper 89-0548;1989.

[98] Radespiel R. A cell-vertex multigrid method for the Navier-Stokes equations. NASA TM-101557;1989.

[99] Swanson RC, Radespiel R. Cell centered and cell vertex multigrid schemes for the Navier-Stokes equations. AIAA J 1991;29:697-703.

[100] Morton KW, Paisley MF. On the cell-centre and cell-vertex approaches to the steady Euler equations and the use of shock fitting. In: Proc. int. conf. num. meth. fluid dynamics 10, Beijing, China; 1986.

[101] Dimitriadis KP, Leschziner MA. Multilevel convergence acceleration for viscous and turbulent transonic flows computed with cell-vertex method. In: Proc. 4th Copper Mountain conference on multigrid methods. Colorado: SIAM; 1989. p. 130-48.

[102] Dick E. A flux-vector splitting method for steady Navier-Stokes equations. Int J Numer Meth Fluids 1988;8:317-26.

[103] Dick E. A multigrid method for steady incompressible Navier-Stokes equations based on partial flux splitting. Int J Numer Meth Fluids 1989;9:113-20.

[104] Crumpton PI, Mackenzie JA, Morton KW. Cell vertex algorithms for the compressible Navier-Stokes equations. J Comput Phys 1993;109:1-15.

第5章 非结构有限体积法

正如我们在第3章引言中提到的,绝大多数求解欧拉和纳维-斯托克斯方程组的数值格式都采用直线法,也就是说,对空间和时间进行分开离散。这种方法的主要优点是可以对空间和时间导数选择不同精度的数值近似。与像 Lax-Wendroff 格式族(如显式 MacCormack 预估-校正格式、隐式 Lerat 格式等——详细信息见文献[1])那样的基于空间和时间耦合离散的方法相比,直线法提供了更大的灵活性,由于它应用广泛,在此我们也将采用这种方法。

本章讨论的有限体积格式基于守恒形式的纳维-斯托克斯方程组(2.19)或欧拉方程组(2.44)。在预处理步骤中,首先将物理域划分成许多个网格单元。2D 情况下,单元类型主要是三角形,有时也与四边形结合。3D 情况下四面体是最常用的单元类型[2-7],然而,越来越多的流场解算器使用混合类型单元来模拟高 *Re* 数黏性流动[8-16],其中包括四面体、三棱柱、金字塔,在某些情况下还使用六面体(图 5.1)。由各种单元类型组成的非结构网格称为混合网格(mixed grids),如图 3.3 和图 5.2 所示。英文"mixed grids"不应与"hybrid grids"混淆,"hybrid grids"指的是结构网格和非结构网格组成的混合网格(参见文献[17-19])。

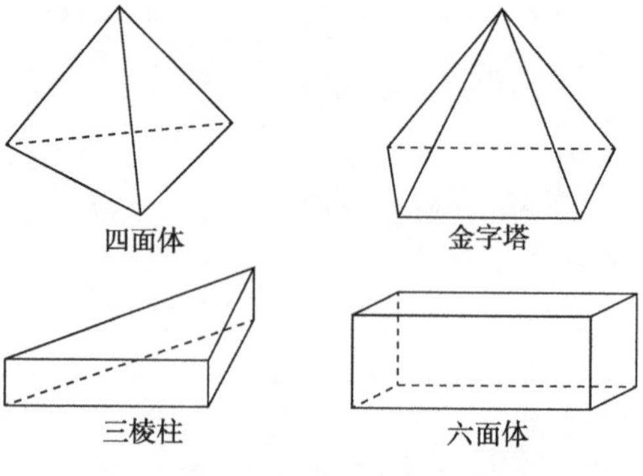

图 5.1 用于 3D 非结构网格生成的单元

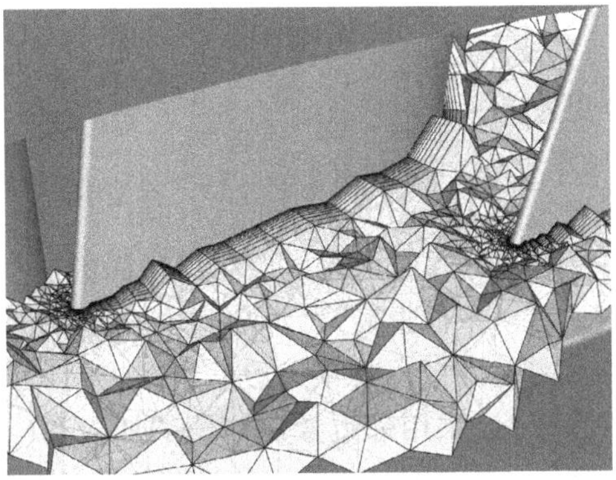

(a)

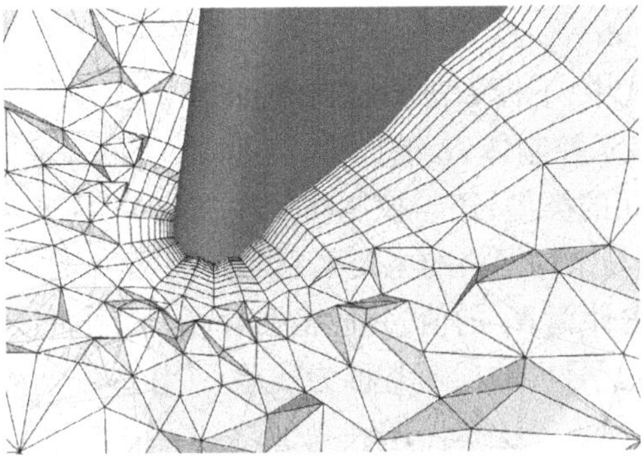

(b)

图5.2 涡轮导向器周围的3D混合网格的粗略切割。网格使用CENTAUR™[20-21]生成。注意物面附近的多层四边形面是三棱柱单元的侧面

网格生成必须以一种能保持控制方程组的守恒性质的方式进行,即:

(1) 物理域必须完全被网格覆盖;

(2) 单元之间没有剩余的自由空间;

(3) 单元互不重叠。

除满足上述要求外,网格还应该是光滑的,即相邻网格单元的体积或拉伸比不应有太大的变化。此外,单元应尽可能的规则,尤其重要的是避免产生退化的网格单元,比如变为薄片(图11.24),否则,数值误差会完全破坏解的精度[22-23],这样的网格单元还会导致严重的收敛问题。

为了计算对流和黏性通量以及源项的积分,需要基于网格定义合适的控制体。为简单起见,我们假设控制体不随时间变化(否则请参见附录A.5),守恒变量 $\boldsymbol{W}$ 的时间导数可转换为

$$\frac{\partial}{\partial t}\int_{\Omega}\boldsymbol{W}\mathrm{d}\Omega=\Omega\frac{\partial \boldsymbol{W}}{\partial t}$$

因此,式(2.19)变为

$$\frac{\partial \boldsymbol{W}}{\partial t}=-\frac{1}{\Omega}\left[\oint_{\partial\Omega}(\boldsymbol{F}_{\mathrm{c}}-\boldsymbol{F}_{\mathrm{v}})\mathrm{d}S-\int_{\Omega}\boldsymbol{Q}\mathrm{d}\Omega\right] \tag{5.1}$$

式(5.1)右侧的面积分可近似为通过控制体各面的通量之和,这种近似称为空间离散。通常假定在一个面上通量为常量,并在面心处进行求解,这种近似处理对于二阶精度格式是足够的。源项一般假定在控制体内为常量,但是在源项占主导地位的情况下,建议采用邻近控制体值的加权和(参见文献[24]及其引用的文献)来求解 $\boldsymbol{Q}$。如果考虑一个特定的控制体 Ω_I,从式(5.1)可以得到

$$\frac{\mathrm{d}\boldsymbol{W}_I}{\mathrm{d}t}=-\frac{1}{\Omega_I}\left[\sum_{m=1}^{N_{\mathrm{F}}}(\boldsymbol{F}_{\mathrm{c}}-\boldsymbol{F}_{\mathrm{v}})_m\Delta S_m-(\boldsymbol{Q}\Omega)_I\right] \tag{5.2}$$

在上面的表达式中,下标 I 表示控制体,我们在后面将看到,它不一定与网格重合。此外,N_{F} 表示控制体 Ω_I 的面数,变量 ΔS_m 表示面 m 的面积。面数 N_{F} 的值不但取决于单元的类型,还与控制体的类型相关,一般来说,不同控制体的面数也会不同,这是与结构网格的主要区别之一。然而,数值处理过程和数据结构的发展可避免预先计算 N_{F} 的值,我们将在后面几节中重提这一点。

式(5.2)右侧方括号内的项通常称为残差。因此,可以将式(5.2)简写为

$$\frac{\mathrm{d}\boldsymbol{W}_I}{\mathrm{d}t}=-\frac{1}{\Omega_I}\boldsymbol{R}_I \tag{5.3}$$

将所有控制体 Ω_I 的关系式写成式(5.3),得到一个一阶常微分方程组。方程组在时间上是双曲型的,这意味着我们必须从已知初值开始,在时间上进行推进。另外,还必须为黏性和无黏性通量提供适当的边界条件,详细内容见第8章。

在数值求解离散控制方程组(5.3)时,第一个问题是如何定义控制体以及流场变量相对于网格点的位置。在有限体积格式框架下,可以采取以下三种基本策略:

(1) 格心格式[2,16,25-26]——控制体与网格单元相同,流场变量位于网格单元的质心(图5.6);

(2) 控制体重叠的格点格式[27-28]——流场变量位于网格节点,并将控制体定义为具有相同节点的所有网格单元的并集,这意味着两个相邻节点的控制体会相互重叠;

(3) 中点-二次网格的格点格式[13,15,29-33]——流场变量也位于网格节点,但此时控制体是通过连接周围单元的体心、面心和边中点来创建的(图5.7和图5.8),用这种方法,网格节点被它们相应的控制体包围起来——组成了一个

不重叠的二次网格。

由于控制体重叠的格点格式已不再使用，在此我们将集中讨论格心格式和中点-二次网格的格点格式，这两种方法将在 5.2 节中详细讨论。

需要强调的是，我们讨论的所有流场变量，包括守恒变量($\rho,\rho u,\rho v,\rho w,\rho E$)和原始变量($p,T,c$ 等)，都位于相同的位置——格心或者格点处，这种方法被称为同位网格格式。相比之下，许多比较老的(结构的)基于压力的方法(参见 3.1 节)使用交错网格格式，其中压力和速度分量存储在不同的位置，以抑制由空间中心差分所引起的解的振荡。

对流通量的求解存在多种选择，最基本的问题是，我们必须知道一个控制体的所有 N_F 个面上的对流通量值。因为流场变量并不位于面上，不能直接使用，因此必须将通量或流场变量插值到控制体的面上，这种插值被称为用控制体内的值进行解的重构(见 5.3.3 节)，大体上，可以通过以下两种方式进行插值：

(1) 算术平均，类似于中心离散格式；

(2) 偏置插值，如针对流动方程组特性的迎风离散格式。

除上述方法外，我们还将在 5.3 节中讨论使用最广泛的对流通量离散格式的精度、适用范围和数值计算量等内容。

求解控制体面上黏性通量的常用方法是基于流场变量的算术平均。式(2.15)和式(2.24)中速度梯度和温度梯度的计算比较困难，特别是在混合网格的情况下，其完整的计算过程将在 5.4 节中介绍。

5.1　控制体的几何量

在开始讨论对流通量和黏性通量的离散方法之前，一项重要的工作是计算控制体 Ω_I 的几何量——它的体积、面 m 的单位法向矢量 $\boldsymbol{n}_m$(定义方向为指向控制体外)和面积 ΔS_m，以及单元的质心。法向矢量和面积也称为控制体的度量值。在下面几节中，将分别讨论 2D 和 3D 情况。

5.1.1　2D 情况

通常我们认为平面流动是 3D 问题的特殊情况，它的解关于一个坐标方向对称(例如，z 方向)。为了使体积、压力等量的物理单位正确，根据对称性，所有网格单元的厚度可设定为一个常数值 b。因而，2D 情况下控制体的体积等于它的面积与厚度 b 的乘积。因为 b 是任意的，为了方便，可令 $b=1$。下面仅讨论三角形和四边形单元，尽管中点-二次格式的控制体的形状可能相当复杂，但它总是可以被剖分为若干个三角形和/或四边形。

1. 三角形单元

一般三角形的面积可以方便、精确地采用高斯公式来计算。因此，使用图 5.3(a)所示的节点编号，体积为

$$\Omega=\frac{b}{2}[(x_1-x_2)(y_1+y_2)+(x_2-x_3)(y_2+y_3)+(x_3-x_1)(y_3+y_1)] \qquad (5.4)$$

为了得到正的体积值，节点必须按逆时针方向编号。

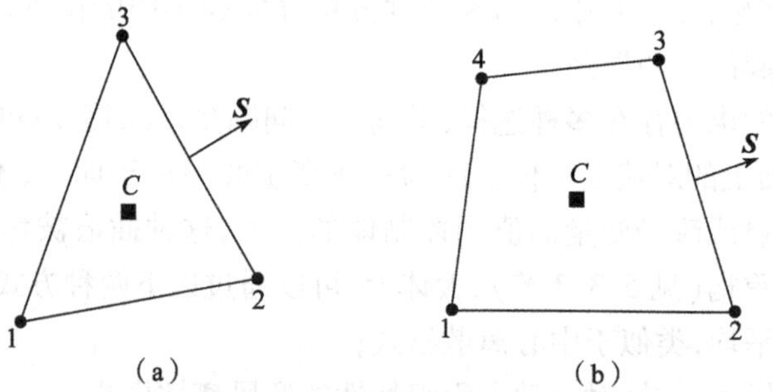

图 5.3　节点的编号和面矢量

(a) 三角形单元；(b) 四边形单元。字母 C 表示单元的中心。

2. 四边形单元

一般四边形的面积可以用高斯公式精确地计算出来，经过代数运算可得

$$\Omega=\frac{b}{2}[(x_1-x_3)(y_2-y_4)+(x_4-x_2)(y_1-y_3)] \qquad (5.5)$$

其中，节点按图 5.3(b) 逆时针方向编号。在上面的讨论中，假设控制体位于 x-y 平面，z 轴为对称轴。

在 2D 情况下，控制体的边为直线，因此它的单位法向矢量为常量。当我们根据式(5.2)对通量进行近似积分时，必须计算面的面积 ΔS 与对应的单位法向矢量 $\boldsymbol{n}$ 的乘积，将其记为面矢量 $\boldsymbol{S}$。以图 5.3 的边 2-3 为例，指向外的面矢量可按下式计算：

$$\boldsymbol{S}_{23}=\boldsymbol{n}_{23}\Delta S_{23}=b\begin{bmatrix}y_3-y_2\\x_2-x_3\end{bmatrix} \qquad (5.6)$$

由于对称性，面矢量（以及单位法向矢量）在 z 方向的分量为零，因此在式(5.6)中被省略。由式(5.6)可得单位法向矢量，有

$$\Delta S=|\boldsymbol{S}|=\sqrt{S_x^2+S_y^2} \qquad (5.7)$$

其中，S_x 和 S_y 表示面矢量的笛卡儿坐标分量。

3. 单元中心

图 5.3(a) 中三角形的中心定义为

$$\boldsymbol{r}_{\mathrm{c}}=\frac{1}{3}(\boldsymbol{r}_1+\boldsymbol{r}_2+\boldsymbol{r}_3) \tag{5.8}$$

$\boldsymbol{r}_{1/2/3}$ 表示节点的笛卡儿坐标。四边形单元的中心根据文献[34]给出的公式计算,四边形被分解成共用两点的两个三角形。节点编号如图 5.3(b)所示,1 和 3 为共用节点,关系式为

$$\boldsymbol{r}_{\mathrm{c}}=\frac{\Omega_{123}\boldsymbol{r}_{\mathrm{c},123}+\Omega_{134}\boldsymbol{r}_{\mathrm{c},134}}{\Omega_{123}+\Omega_{134}} \tag{5.9}$$

式(5.4)计算两个三角形 Ω_{123} 和 Ω_{134} 的体积,用式(5.8)计算它们的质心 $\boldsymbol{r}_{\mathrm{c}}$。一般多边形的求解公式可参考文献[35]。

5.1.2　3D 情况

与 2D 情况不同,3D 情况下,含有四边形面的单元或控制体的面矢量和体积的计算可能会出现一些问题。主要原因是,在一般情况下,控制体的四边形面的四个顶点可能不在同一个平面上,此时该面的法向矢量不再是常量(见图 4.2)。为了克服这个困难,我们可以把每个四边形剖分为两个甚至更多的三角形,然而,对于光滑网格上的二阶格式,剖分带来的精度收益可以忽略不计,只有在三阶和更高阶的空间离散中才有效果——并且实际上也是必要的。因此在下面的讨论中,四边形面的处理采用基于平均法向矢量的简化方法。

1. 三角形面

利用高斯公式可以精确计算出三角形的面矢量 $\boldsymbol{S}$。根据图 5.4(a)定义的节点顺序,我们得到三角形 1-2-3 的“边差”值为

$$\begin{aligned}
\Delta xy_A&=(x_1-x_2)(y_1+y_2)\\
\Delta yz_A&=(y_1-y_2)(z_1+z_2)\\
\Delta xy_B&=(x_2-x_3)(y_2+y_3)\\
\Delta yz_B&=(y_2-y_3)(z_2+z_3)\\
\Delta xy_C&=(x_3-x_1)(y_3+y_1)\\
\Delta yz_C&=(y_3-y_1)(z_3+z_1)\\
\Delta zx_A&=(z_1-z_2)(x_1+x_2)\\
\Delta zx_B&=(z_2-z_3)(x_2+x_3)\\
\Delta zx_C&=(z_3-z_1)(x_3+x_1)
\end{aligned} \tag{5.10}$$

然后可以从下式计算指向外的面矢量 $\boldsymbol{S}=\boldsymbol{n}\Delta S$ 为

$$\boldsymbol{S}=\frac{1}{2}\begin{bmatrix}\Delta yz_A+\Delta yz_B+\Delta yz_C\\ \Delta zx_A+\Delta zx_B+\Delta zx_C\\ \Delta xy_A+\Delta xy_B+\Delta xy_C\end{bmatrix} \tag{5.11}$$

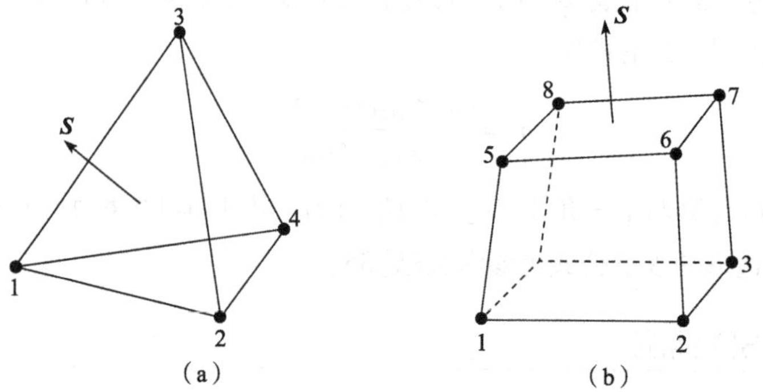

图 5.4　节点的编号和面矢量

(a) 四面体单元；(b) 六面体单元。

2. 四边形面

如图 5.4(b)所示，计算四边形面的平均面矢量 $\boldsymbol{S}$ 最简便的方法是使用 2D 四边形面积的高斯公式。对于图 5.4(b)中节点 5、6、7 和 8 构成的面，首先定义坐标差：

$$\begin{aligned}\Delta x_A&=x_8-x_6,\quad \Delta x_B=x_7-x_5\\ \Delta y_A&=y_8-y_6,\quad \Delta y_B=y_7-y_5\\ \Delta z_A&=z_8-z_6,\quad \Delta z_B=z_7-z_5\end{aligned} \tag{5.12}$$

然后，由下面的关系计算向外的面矢量 $\boldsymbol{S}=\boldsymbol{n}\Delta S$：

$$\boldsymbol{S}=\frac{1}{2}\begin{bmatrix}\Delta z_A\Delta y_B-\Delta y_A\Delta z_B\\ \Delta x_A\Delta z_B-\Delta z_A\Delta x_B\\ \Delta y_A\Delta x_B-\Delta x_A\Delta y_B\end{bmatrix} \tag{5.13}$$

当面趋近于平面四边形时，即当面的顶点都位于同一平面时，近似值变为精确值。

三角形和四边形的单位法向矢量均由 $\boldsymbol{n}=\boldsymbol{S}/\Delta S$ 和下式得到：

$$\Delta S=\sqrt{S_x^2+S_y^2+S_z^2} \tag{5.14}$$

其中，S_x, S_y, S_z 分别为由式(5.11)或式(5.13)给出的面矢量的笛卡儿坐标分量。

3. 体积

正如在介绍 3D 结构有限体积格式时所述，一种非常方便的体积计算方法是基于散度定理[36]的方法。4.1.2 节的讨论最终得到了如下计算体积的表达式：

$$\Omega=\frac{1}{3}\sum_{m=1}^{N_{\mathrm{F}}}(\boldsymbol{r}_{\mathrm{c}}\cdot\boldsymbol{S})_{m} \tag{5.15}$$

其中，N_{F} 为控制体的面数，$(\boldsymbol{r}_{\mathrm{c}})_m$ 为控制体的第 m 个面的中心，$\boldsymbol{S}_m$ 是第 m 个面的面矢量(方向向外)。式(5.15)直接适用于非结构网格，对于由三角形面组成的体，或者由平面四边形面组成的体，该式精确成立。

4. 单元质心

前面提到的中点-二次格式的控制体的构造需要网格单元格质心的信息。一般控制体的质心定义为

$$\boldsymbol{r}_{\mathrm{c}}=\frac{1}{\Omega}\int_{\Omega}\boldsymbol{r}\mathrm{d}\Omega \tag{5.16}$$

根据文献[34]的推导，式(5.16)可离散为

$$\boldsymbol{r}_{\mathrm{c}}=\frac{3\sum_{m=1}^{N_{\mathrm{F}}}(\boldsymbol{r}_{\mathrm{c}}\cdot\boldsymbol{n})_{m}(\boldsymbol{r}_{\mathrm{c}})_{m}\Delta S_{m}}{4\sum_{m=1}^{N_{\mathrm{F}}}(\boldsymbol{r}_{\mathrm{c}}\cdot\boldsymbol{n})_{m}\Delta S_{m}} \tag{5.17}$$

其中，对于三角形面，面 m 的面心 $(\boldsymbol{r}_{\mathrm{c}})_m$ 由式(5.8)求得；对于四边形面，由式(5.9)求得。可以发现，式(5.17)中的分母等于式(5.15)中的 Ω 的 12 倍。文献[35]给出了由三角形面组成的多面体的类似公式。

5.2　一般离散方法

本章开始已经提到，存在两种最流行的定义控制体和流场变量位置的方法，分别是格心格式和中点-二次格式，本节将详细介绍这两种方法。

在开始之前，先简要讨论一下非结构流场解算器所需的基本数据结构。事实上，存取灵活且操作高效的数据结构对任何非结构格式都至关重要，网格中缺失的结构信息必须在流场解算器中提供，而网格信息至少需要以下数据：

(1) 网格节点(顶点)的坐标；

(2) 从单元指向网格节点的指针；

(3) 从边界面指向网格节点的指针。

离散格式需要的额外的数据结构可以从这些信息中生成。我们以图 5.4(a)中的四面体为例来说明上述数据是如何存储的，如果假设面 1-2-4 为边界面(壁面、入口、远场等)，则可以采用以下存储方式：

```
#nodes(x,y,z):
P1.x  P1.y  P1.z
P2.x  P2.y  P2.z
P3.x  P3.y  P3.z
P4.x  P4.y  P4.z
…
#tetrahedra:
…
P1  P2  P3  P4
…
#boundaries:
…
type  P1  P4  P2
…
```

本书附带的源程序中的2D非结构流场解算器就使用了类似的格式。

关于计算域边界的数据结构，需要强调两点。首先，存储边界面比存储节点更方便。这点可以通过分析图5.5所示的情况来理解，节点P_1由三个不同物理类型的边界共享，而节点P_2和P_3由两个边界共享，因此，应用确切的边界条件会变得非常麻烦，而一个面只能属于一个边界，如面P_1-P_2-P_4只属于边界1。

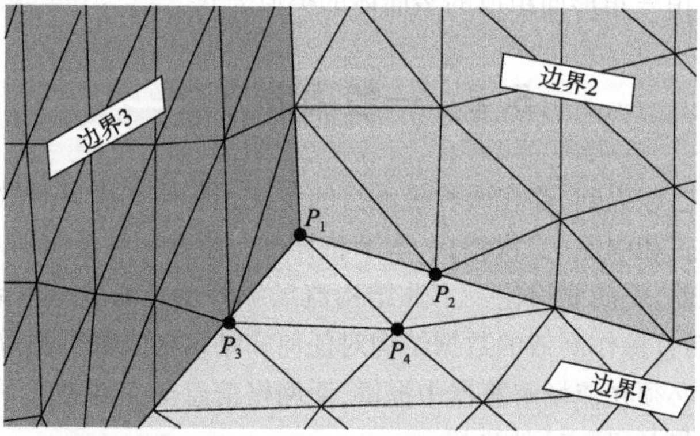

图5.5 角点周围的三个边界——网格点关于边界类型的不确定性

第二点涉及边界面节点的编号顺序。编号顺序需要采用一致的方式，例如，从流场外看时，均为逆时针方向——以使得所有的面矢量（式(5.6)和式(5.11)或式(5.13)）都指向流场外或流场内，这不仅对通量积分很重要，对网格生成也很重要。

5.2.1　格心格式

格心格式的控制体与网格单元相同，并且流场变量位于质心，如图 5.6 所示。当求解离散的流动方程组(5.2)时，必须计算控制体面的面心处的对流和黏性通量，以满足在光滑网格(四边形面采用平均法向矢量)上进行二阶精度的离散。通量求解有以下三种近似方法：

(1) 由面左侧和右侧网格单元质心处的值计算通量，然后进行平均，使用相同的单位法向矢量(一般只适用于对流通量)进行计算；

(2) 使用与面左侧和右侧相邻网格单元质心变量的平均值；

(3) 用周围单元的值分别重构面两侧的值，然后使用重构的值计算通量(仅用于对流通量)。

因此，以图 5.6 中单位法向矢量 $\boldsymbol{n}_{01}$ 所在的面为例，第一种方法(通量平均)在 2D 情况下为

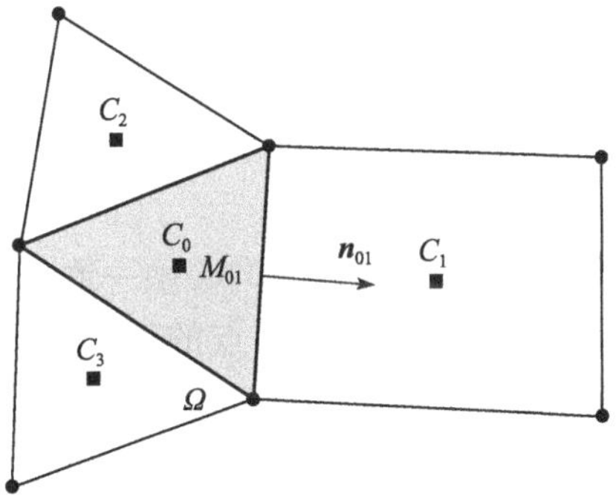

图 5.6　格心格式的控制体(2D)。网格节点用圆表示，单元中心用矩形表示

$$(\boldsymbol{F}_c \Delta S)_{01} \approx \frac{1}{2}[\boldsymbol{F}_c(\boldsymbol{W}_0, \boldsymbol{n}_{01}) + \boldsymbol{F}_c(\boldsymbol{W}_1, \boldsymbol{n}_{01})]\Delta S_{01} \tag{5.18}$$

面积 ΔS_{01} 使用式(5.6)和式(5.7)计算。

第二种方法(变量平均)可以表述为

$$(\boldsymbol{F}\Delta S)_{01} \approx \boldsymbol{F}(\boldsymbol{W}_{01}, \boldsymbol{n}_{01})\Delta S_{01} \tag{5.19}$$

其中，单位法向矢量 $\boldsymbol{n}_{01}$ 所在的面的守恒/因变量定义为两个相邻单元体心值的算术平均，即

$$\boldsymbol{W}_{01} = \frac{1}{2}(\boldsymbol{W}_0 + \boldsymbol{W}_1) \tag{5.20}$$

式(5.19)中的通量矢量 $\boldsymbol{F}$ 既代表对流通量，也可以表示黏性通量。

第三种方法首先将流场变量(通常是速度分量、压力、密度和总焓)分别插

值到单元面两侧。重构量——称为左侧状态量和右侧状态量（见 5.3.3 节）——通常是不相同的值。然后使用适当的非线性近似函数从左右两侧状态量的差值计算通过面的通量。因此

$$(\boldsymbol{F}_c\Delta S)_{01} \approx f_{\text{Flux}}(\boldsymbol{U}_L, \boldsymbol{U}_R, \Delta S_{01}) \tag{5.21}$$

其中

$$\begin{aligned} \boldsymbol{U}_L &= f_{\text{Rec}}(\cdots, \boldsymbol{U}_2, \boldsymbol{U}_0, \cdots) \\ \boldsymbol{U}_R &= f_{\text{Rec}}(\cdots, \boldsymbol{U}_1, \boldsymbol{U}_0, \cdots) \end{aligned} \tag{5.22}$$

表示重构状态量。

类似的关系也适用于控制体的其他面，上述近似方法在 3D 中也可以使用。面矢量 $\boldsymbol{S}$ 由式（5.11）或式（5.13）计算。

正如我们在本节序言中所述，为了支持离散方法，必须以适当的方式扩展描述单元的基本数据结构。从前面的讨论可以明显看出，数值运算的开展主要使用（控制体的）面及其相邻单元质心的值，因此，很自然地可以采用基于面的数据结构进行空间离散。针对每个特定网格面（图 5.6），这种数据结构存储了以下内容：

（1）指向共用该网格面的两个单元的指针——用于访问与这两个单元（C_0，C_1）相关的流场变量；

（2）面矢量（$\boldsymbol{S}_{01}=\boldsymbol{n}_{01}\Delta S_{01}$）——它的方向必须一致向外或向内；

（3）从每个单元的质心到面 M_{01} 的中点的两个矢量，即（C_0-M_{01}），（C_1-M_{01}）——将流场变量精确地插值到面上需要这两个矢量。对于纯四面体网格，没有必要存储这两个矢量，因为流场插值可使用一个简单的外插公式[2,26]（参见式（5.44））来完成。

因此，通量的积分（例如，根据式（5.19））可以实现为一个遍历所有网格面（包括内部和边界）的循环：

```
DO face =1,nfaces
  I =pointer_to_left_cell(face)
  J =pointer_to_right_cell(face)
  (FΔS)_IJ ≈ F(W_IJ,n_IJ) ΔS_IJ
  R_I =R_I + (FΔS)_IJ
  R_J =R_J - (FΔS)_IJ
ENDDO
```

循环完成后，加上源项 $\boldsymbol{Q}_I\Omega_I$，就得到了所有单元的最终残差（$\boldsymbol{R}$）。而关于单元的循环是一种效率较低的方法，因为面矢量必须存储两次，通量被计算了两次

(边界除外)。此外,因为使用完全相同的面矢量 $\boldsymbol{S}_{IJ}$ 计算进入控制体 Ω_I 和 Ω_J 的部分通量,确保了控制方程组的守恒特性。

5.2.2　中点-二次格点格式

在格点格式中,流场变量位于网格节点(顶点)。通过连接共享特定节点的所有单元的质心、面心和边的中点,形成中点-二次控制体,四面体如图 5.7(a)所示,六面体如图 5.7(b)所示。中点-二次控制体的定义使得在每个网格节点周围形成一个多面体外壳,如图 5.8 中 2D 混合网格所示。这种多面体可以看作一个二次网格——格式的名字即从此处来。值得注意的是,中点-二次有限体积离散等效于使用线性单元的伽辽金有限元格式(见文献[37])。

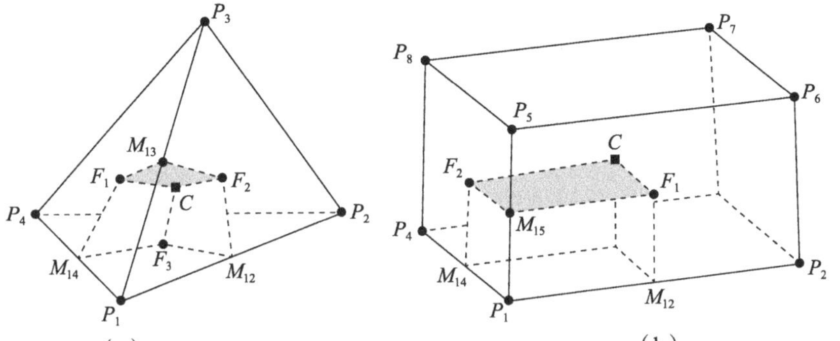

(a)　　　　　　　　　　　　　　　(b)

图 5.7　中点-二次格式的部分控制体和面(阴影部分)

(a)四面体;(b) 六面体。P 表示网格节点,C 表示网格质心(式(5.17)),F 表示面心(式(5.8)或式(5.9)),M 表示边的中点。阴影区域代表控制体面的一部分,分别分配给边 P_1-P_3 或 P_1-P_5。

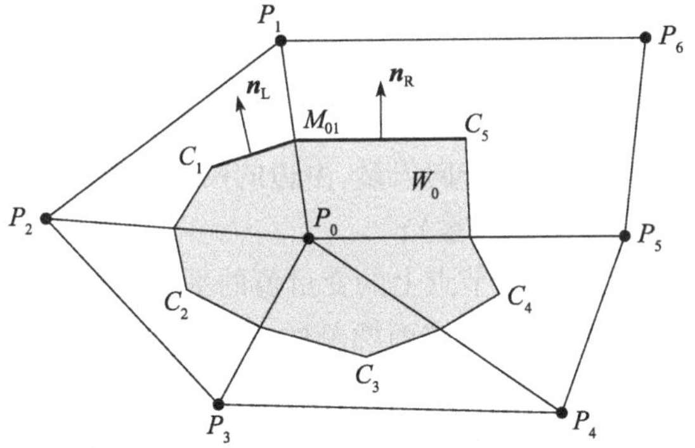

图 5.8　中点-二次格式的控制体(2D)。C_1、C_2 等表示单元中心;P_1、P_2 等表示网格节点。与边 P_0-P_1 关联的面区域用粗线表示。

为了对离散的流动方程组(5.2)求解,我们必须对控制体面的对流和黏性通量进行积分。因此必须分别计算每个子面(例如图5.7(a)中的 $F_1-M_{13}-F_2-C$)的通量,然而,只有三阶或更高精度的离散[38-39]才需要这么做。对于二阶格式,通常可以假定流场变量在围绕一个特定边的所有面上是常数,然后使用来自两个节点的变量和梯度在边的中点处计算通量。这种方法允许我们为与每条边相关联的所有面定义一个平均单位法向矢量和总面积。参照图5.8,以边 P_0-P_1 为例,其平均法向矢量为

$$\boldsymbol{n}_{01}\Delta S_{01}=\boldsymbol{n}_{\mathrm{L}}\Delta S_{\mathrm{L}}+\boldsymbol{n}_{\mathrm{R}}\Delta S_{\mathrm{R}} \tag{5.23}$$

总面的面积由 $\Delta S_{01}=\|\boldsymbol{n}_{\mathrm{L}}\Delta S_{\mathrm{L}}+\boldsymbol{n}_{\mathrm{R}}\Delta S_{\mathrm{R}}\|_2$ 给出。这同样适用于3D情况,在3D中,平均法向矢量由所有共用特定边中点的子面的法向矢量求和得到,如图5.9所示。在2D情况下面矢量($\boldsymbol{S}=\boldsymbol{n}\Delta S$)由式(5.6)计算。3D情况下,子面总是四边形,我们可以把它们剖分成三角形并使用式(5.11)计算,也可以根据式(5.13)来简化,该简化处理对于光滑网格来说是足够的。单元质心和面心分别由式(5.17)、式(5.8)或式(5.9)得到。

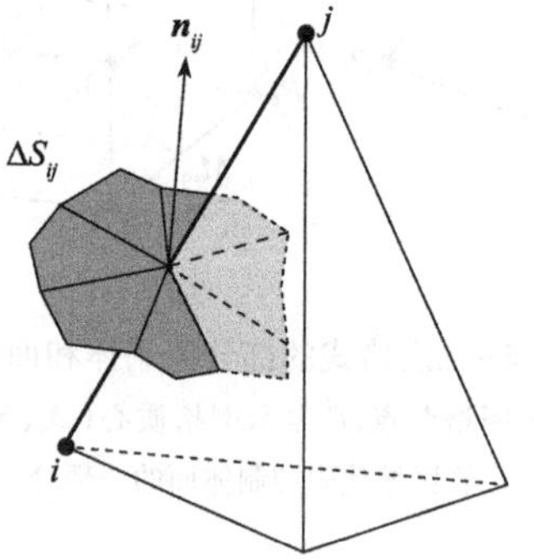

图5.9 3D中点-二次格点格式中与边 ij 相关的总面面积和平均单位法向矢量

通量求解有以下三种方法:

(1) 使用相同的平均单位法向矢量,由边的两个节点上的值计算出通量,然后取平均(通常只适用于对流通量);

(2) 使用存储在边的两个节点上的变量值的平均来计算通量;

(3) 从周围节点的值分别重构面两侧的流场变量,然后使用两侧的变量来计算通量(只适用于对流通量)。

通量的计算方法形式上与格心格式相同,因此,式(5.18)~式(5.22)也适用于中点-二次格式。如果利用上述方法将每条边与一个平均单位法向矢量相关联,采用基于边的数据结构进行空间离散是最有效的方法。针对每一条特定

的边，基于边的数据结构存储(图5.9)：

(1) 指向定义边的两个节点的指针——用于访问与两个控制体 Ω_i 和 Ω_j 相关的流场变量；

(2) 面矢量($\boldsymbol{S}_{ij}=\boldsymbol{n}_{ij}\Delta S_{ij}$)，方向必须一致向外或向内；

(3) 从节点 i 到节点 j 的边矢量——用于面上流场变量的插值(解的重构)，边矢量也可以临时从节点坐标计算。另外，增加人工耗散的中心格式(5.3.1节)不需要边矢量。

基于这样的数据结构，通量的积分(例如根据式(5.19))可以实现为一个遍历所有网格边的循环：

```
DO edge =1,nedges
  i =pointer_to_left_node(edge)
  j =pointer_to_right_node(edge)
  (FΔS)ij ≈F(Wij,nij)ΔSij
  Ri =Ri + (FΔS)ij
  Rj =Rj - (FΔS)ij
ENDDO
```

在循环完成后，加上源项 $\boldsymbol{Q}_i\Omega_i$，就得到了所有节点的最终残差($\boldsymbol{R}$)。这种方法比分别对每个控制体上的通量求和效率高得多，因为每个平均面矢量只存储一次，每条边只需要访问一次。此外，由于使用完全相同的平均面矢量 $\boldsymbol{S}_{ij}$ 计算进入控制体 Ω_i 和 Ω_j 的通量，因而质量、动量和能量可以保持精确守恒。

5.2.3 格心格式与中点-二次格式的比较

格心格式和中点-二次格式的优缺点一直是争论的焦点，主要原因是这两种方法在精度、计算时间和实际外形的内存需求方面缺乏公平的比较。在此我们收集了支持和反对这两种格式的最重要的指标：

(1) 精度；

(2) 计算量；

(3) 内存需求；

(4) 灵活性。

这有助于更好地理解每种格式所固有的问题，并可能有助于为预期的应用范围选择最合适的格式。

1. 精度

三角形/四面体网格的格心格式与中点-二次格式相比，控制体数量是后者的两倍/六倍，从而自由度也是后者的两倍/六倍[37]。由四面体和三棱柱组成的典型混合网格，格心格式比中点-二次格式的自由度多大约三倍。这表明在相同的网格上，格心格式比格点格式更精确。但与中点-二次格式相比，计算残差时格心格式的通量数要少得多（四面体网格大约为 3 : 7），这可能会影响精度。在精度方面，没有明确的证据表明哪种格式更好。

中点-二次格式在拉伸三角形和四面体网格上会遇到一种特殊的问题。图 5.10 显示了由直角三角形组成的一种三角化网格，常用于模拟壁面附近的黏性流动。从图 5.10(a)可以看出，与边 ij 相关的面 ΔS_{ij} 变得高度倾斜，空间离散格式（特别是黎曼解算器）大多假定通量正交于面，因此这样就引入了一个对于一阶格式特别重要的误差[39]。使用包含-二次（containment-dual）控制体可以改善这种情况[40]，如图 5.11 所示，包含-二次方法使用最小包含圆/球体的中心而不是单元的质心来定义控制面，这使得控制体与四边形网格相同（图 5.10(b)）。需要注意的是，像 ij' 这样的斜边将没有相关的面，同时还需要额外的预处理工作，但是可以显著提高解的精度[41]。当然，另一种可能是在边界层中直接使用四边形或六面体单元。文献[22-23]对网格引起的误差有进一步的讨论。

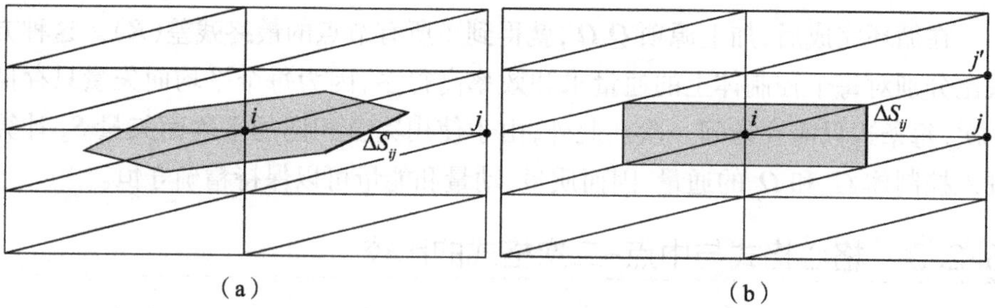

图 5.10　拉伸直角三角形网格的控制体比较

(a) 中点-二次控制体；(b) 包含-二次控制体。

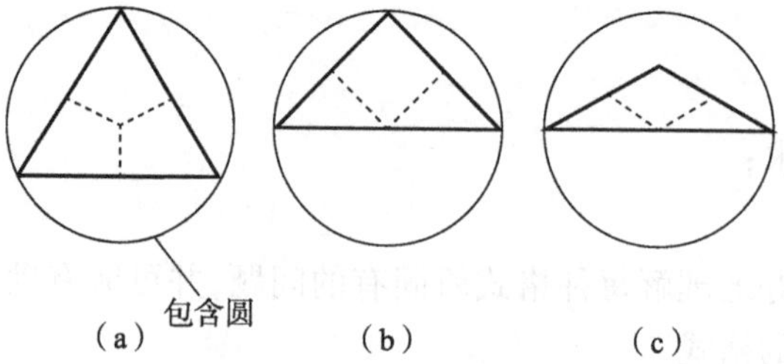

图 5.11　三角形中的包含-二次网格[41]（虚线）

(a) 锐角；(b) 直角；(c) 钝角。包含圆是包含三角形的最小圆，对于钝角三角形，包含圆的圆心在最长边上。

中点-二次格式的另一个固有问题是在物理域边界上的离散。在边界处它只有大约一半的控制体(参见图4.6),通量的积分产生一个位于内部的残差,理想情况下它位于控制体的中心,然而中点-二次格式的残差却是与直接位于边界上的节点相关联的。与格心格式相比,这种不匹配导致离散误差的增加,在固壁边界上尤其严重。二次控制体的定义在尖角处(如后缘)也会出现问题,表现为压力或密度出现非物理极值。更复杂的情况出现在周期边界(见8.8节),在这种边界上,控制体的两个部分的通量必须正确地加起来。

控制体的质心与残差存储节点之间的不匹配也会影响中点-二次格式。在非定常流动情况下,它以质量矩阵的形式出现,我们在3.2节开始时已经讨论过这一点。格心格式的优点是质量矩阵可以从方程中消除,而且不会影响解的精度,相比之下,中点-二次格式需要对质量矩阵进行特殊处理[42-43]。

2. 计算量

为了判断两种格式所需的计算量,首先我们必须考虑通量的积分。从前面的讨论知道,格心格式使用基于面的循环,而中点-二次格式基于边的循环,由于两种格式在界面处的通量计算非常相似,因此面数与边数之比给就出了计算量之比。在四面体网格上,面数(如果每两个单元只计算一次)大约是边数的两倍,对于相同的网格,格心格式的计算量是中点-二次格式的两倍[37]。然而,在包含三棱柱单元的混合网格上,格心方法更具竞争力。除了边界外,两种方法在六面体网格上计算量相近,因为六面体网格的面数等于其边数。

3. 内存需求

与中点-二次格式相比,格心格式在四面体网格上存储约6倍的流场变量,在标准混合网格上存储约3倍的流场变量。正如我们所看到的,在每个面或者边上,这两种格式都需要存储两个整数和三个实数(指针和面矢量)。此外,对于每个面,格心格式必须在内存中保存两个到面心的矢量——6个实数,与之相反,中点-二次格式只需处理节点坐标,从而只需相当少的内存。总而言之,格心格式需要的计算机内存大约是中点-二次格式的两倍。

4. 网格生成/自适应

格心格式在非对接单元界面下具有显著的优势,如图3.4中字母“F”处,与中点-二次方法相比,计算这种界面处的通量不需要特殊和昂贵的处理,这样就增加了网格生成、网格自适应以及网格运动的灵活性,格心格式的这个优点也被应用于嵌套笛卡儿网格的流场解算器中。

5.3 对流通量离散

在前几节中,我们考虑了空间离散方法的一般问题以及必要的数据结构,在接下来的内容中,将讨论对流通量求解的更多细节。

正如在 3.1.5 节中已经看到的,在有限体积方法的框架内,基本上有以下格式可供选择:

(1) 中心格式;

(2) 矢通量分裂格式;

(3) 通量差分裂格式;

(4) 总变差减少(TVD)格式;

(5) 波动分裂格式。

首先,我们以较长篇幅讨论非结构网格上的中心离散格式,因为它与结构网格上的中心离散有很大的不同。与之相反,在结构网格和非结构网格上,迎风格式的基本原理是相同的,因此,相关细节可在 4.3.2 节~4.3.4 节中找到。非结构网格的不同之处在于解的重构,即计算控制体的面上的流场变量值,因此我们将在 5.3.3 节中详细讨论常用的重构方法。由于篇幅的限制,在此不讨论波动分裂方法,关于波动分裂格式的参考书目,请参阅 3.1.5 节。

5.3.1 带人工耗散的中心格式

中心格式的基本思想是根据式(5.20)对控制体的面两侧单元的守恒变量进行算术平均,然后使用算术平均值计算面的对流通量。但是这将导致解的奇偶解耦(生成离散方程的两个独立解),并在激波处振荡,为了稳定性考虑,必须添加人工耗散,人工耗散是基于二阶和四阶差分的混合。Jameson 等[44]首次将该格式应用于结构网格上的欧拉方程组,根据作者的名字,它也被缩写为 JST 格式。

JST 格式在非结构网格上的实现中,应用拉普拉斯算子计算二阶差分,使用拉普拉斯算子的拉普拉斯算子计算四阶差分[27,45]。为了减少计算量,采用伪拉普拉斯算子代替真拉普拉斯算子。利用这种简化,文献[46]首次提出了一个 2D 公式,然后文献[47]进行了改进,随后文献[2]将该格式推广到 3D。它使用了一个距离加权步骤,对于任意网格上的线性变化函数,可消去伪拉普拉斯算子。对于单元 I 中的一般标量 U,伪拉普拉斯算子假定为

$$L(\boldsymbol{U}_I)=\sum_{J=1}^{N_A}\theta_{IJ}(\boldsymbol{U}_J-\boldsymbol{U}_I) \tag{5.24}$$

其中,N_A 表示相邻控制体的数量。在中点-二次格式中,单元索引用节点索引

(i,j)代替。式(5.24)中的求和最好使用类似于通量计算的基于面的循环(格心格式)或基于边的循环(中点-二次格式)。几何权值 θ 定义为

$$\theta_{IJ}=1+\Delta\theta_{IJ} \tag{5.25}$$

它的值是一个最优化问题的解[2],使用拉格朗日乘子求解最优化问题,因此得到如下式所示的几何权值:

$$\Delta\theta_{IJ}=\lambda_{x,I}(x_J-x_I)+\lambda_{y,I}(y_J-y_I)+\lambda_{z,I}(z_J-z_I) \tag{5.26}$$

其中,x,y,z 是单元(在中点-二次格式中是节点)质心的笛卡儿坐标。根据文献[2]计算每个单元(节点)的拉格朗日乘子 λ,得到

$$\begin{aligned}
\lambda_x&=\frac{R_x a_{11}+R_y a_{12}+R_z a_{13}}{d}\\
\lambda_y&=\frac{R_x a_{21}+R_y a_{22}+R_z a_{23}}{d}\\
\lambda_z&=\frac{R_x a_{31}+R_y a_{32}+R_z a_{33}}{d}
\end{aligned} \tag{5.27}$$

系数为

$$\begin{aligned}
a_{11}&=I_{yy}I_{zz}-I_{yz}^2\\
a_{12}&=I_{xz}I_{yz}-I_{xy}I_{zz}\\
a_{13}&=I_{xy}I_{yz}-I_{xz}I_{yy}\\
a_{21}&=I_{xz}I_{yz}-I_{xy}I_{zz}\\
a_{22}&=I_{xx}I_{zz}-I_{xz}^2\\
a_{23}&=I_{xy}I_{xz}-I_{xx}I_{yz}\\
a_{31}&=I_{xy}I_{yz}-I_{xz}I_{yy}\\
a_{32}&=I_{xz}I_{xy}-I_{xx}I_{yz}\\
a_{33}&=I_{xx}I_{yy}-I_{xy}^2\\
d&=I_{xx}I_{yy}I_{zz}-I_{xx}I_{yz}^2-I_{yy}I_{xz}^2-I_{zz}I_{xy}^2+2I_{xy}I_{xz}I_{yz}
\end{aligned} \tag{5.28}$$

对于单元 I,一阶矩为

$$\begin{aligned}
R_{x,I}&=\sum_{J=1}^{N_A}(x_J-x_I)\\
R_{y,I}&=\sum_{J=1}^{N_A}(y_J-y_I)\\
R_{z,I}&=\sum_{J=1}^{N_A}(z_J-z_I)
\end{aligned} \tag{5.29}$$

此外,二阶矩为

$$
\begin{aligned}
I_{xx,I} &= \sum_{J=1}^{N_A} (x_J - x_I)^2 \\
I_{yy,I} &= \sum_{J=1}^{N_A} (y_J - y_I)^2 \\
I_{zz,I} &= \sum_{J=1}^{N_A} (z_J - z_I)^2 \\
I_{xy,I} &= \sum_{J=1}^{N_A} (x_J - x_I)(y_J - y_I) \\
I_{xz,I} &= \sum_{J=1}^{N_A} (x_J - x_I)(z_J - z_I) \\
I_{yz,I} &= \sum_{J=1}^{N_A} (y_J - y_I)(z_J - z_I)
\end{aligned}
\tag{5.30}
$$

对于严重扭曲的网格，几何权（式（5.25））可能导致拉普拉斯近似的非正逼近，从而造成稳定性丧失，因此文献[46]建议将权值限制在[0,2]的范围内，然而，这一措施有损离散精度，更多相关细节请参见文献[12]。

使用拉普拉斯算子的拉普拉斯算子计算四阶差分，即用 $L(U)$ 代替式（5.24）中的 U。因此，单元 I 的人工耗散项的最终形式为

$$
\begin{aligned}
\boldsymbol{D}_I = &\sum_{J=1}^{N_A} (\hat{\Lambda}_c)_{IJ} \epsilon_{IJ}^{(2)} \theta_{IJ} (\boldsymbol{W}_J - \boldsymbol{W}_I) - \\
&\sum_{J=1}^{N_A} (\hat{\Lambda}_c)_{IJ} \epsilon_{IJ}^{(4)} \theta_{IJ} [L(\boldsymbol{W}_J) - L(\boldsymbol{W}_I)]
\end{aligned}
\tag{5.31}
$$

加上人工耗散项，则式（5.2）中的方程组变为

$$
\Omega_I \frac{\mathrm{d}\boldsymbol{W}_I}{\mathrm{d}t} = -\left[\sum_{m=1}^{N_F} (\boldsymbol{F}_c - \boldsymbol{F}_v)_m \Delta S_m\right] + \boldsymbol{D}_I + \boldsymbol{Q}_I \Omega_I \tag{5.32}
$$

其中，N_F 表示控制体的面数（它可能与相邻控制体的数量不同，例如当一个四边形面被分成了两个三角形时）。

式（5.31）中的二阶和四阶项用对流通量雅可比矩阵的谱半径进行缩放。根据文献[28]，单元 I 的谱半径为

$$
(\hat{\Lambda}_c)_I = \sum_{m=1}^{N_F} (|V_m| + c_m) \Delta S_m \tag{5.33}
$$

式中：V_m 为逆变速度（式（2.22））；c_m 为声速，由流场变量计算的两个量在面上取平均值，得到控制体面的谱半径为

$$(\hat{\Lambda}_c)_{IJ}=\frac{1}{2}[(\hat{\Lambda}_c)_I+(\hat{\Lambda}_c)_J] \tag{5.34}$$

在流场中使用基于压力的探测器关闭激波位置的四阶差分和光滑部分的二阶差分。因此,式(5.31)中系数 $\epsilon_{IJ}^{(2)}$ 和 $\epsilon_{IJ}^{(4)}$ 定义为

$$\begin{aligned}\epsilon_{IJ}^{(2)}&=k^{(2)}\max(Y_I,Y_J)\\ \epsilon_{IJ}^{(4)}&=\max[0,(k^{(4)}-\epsilon_{IJ}^{(2)})]\end{aligned} \tag{5.35}$$

压力探测器为

$$Y_I=\frac{\left|\sum_{J=1}^{N_A}\theta_{IJ}(p_J-p_I)\right|}{\sum_{J=1}^{N_A}(p_J+p_I)} \tag{5.36}$$

典型参数值为 $k^{(2)}=1/2$ 和 $1/128\leqslant k^{(4)}\leqslant 1/64$。

正如在4.3.1节已经讨论过的,当我们将式(5.31)中的谱半径 $(\hat{\Lambda}_c)_{IJ}$ 替换为矩阵时,可以提高上述格式的精度,这种矩阵耗散格式[48]在非结构网格上的实现与结构网格相同,缩放矩阵的定义如式(4.59)所示。文献[49]讨论了矩阵耗散格式在3D混合网格中的应用。

对于三角形/四面体以外的单元类型,流行的显式龙格-库塔类型的时间离散与中心格式耦合求解时,会遇到严重的稳定性问题[50],其原因是用拉普拉斯算子的拉普拉斯算子来表示四阶差分。一种解决方法是使用左右状态差(参见4.3节)来近似四阶差分[50],即

$$\boldsymbol{D}_I=\sum_{J=1}^{N_A}(\hat{\Lambda}_c)_{IJ}\epsilon_{IJ}^{(2)}\theta_{IJ}(\boldsymbol{W}_J-\boldsymbol{W}_I)-\sum_{J=1}^{N_A}4(\hat{\Lambda}_c)_{IJ}\epsilon_{IJ}^{(4)}[\boldsymbol{W}_L-\boldsymbol{W}_R] \tag{5.37}$$

这种方法使得四边形/六面体网格的模板与对应结构网格格式相同,左右状态的计算可以使用如5.3.3节中介绍的线性重构。

5.3.2 迎风格式

在非结构网格中,迎风格式似乎比上述中心格式更受欢迎,至少目前如此。事实上,Roe通量差分裂格式[51]是非结构网格中应用最广泛的方法,与中心格式相比,它的边界层分辨率更高,对网格畸变的敏感性更低,这也是Roe格式更具吸引力的原因。然而,为这些更好的性能所付出的代价是需要更高的计算量,如果必须使用限制器来抑制解的振荡(5.3.5节),计算量将更大。

4.3节中介绍的任意结构网格的迎风格式均适用于非结构网格,而无须对基本方法进行修改。只有左右状态量(式(5.22))的计算,即解的重构,以及限

制器函数的计算需要新的公式。因此,在此只讨论解的重构和限制器。关于各种迎风方法的细节,请参阅4.3.2节~4.3.4节。文献[33]给出了在非结构网格中点-二次格式下Roe格式的实现。

5.3.3 解的重构

正如我们在4.3.2节~4.3.4节所看到的,迎风格式需要求解控制体面左侧和右侧的流动状态,式(5.37)中修正的人工耗散格式也是如此。

第一种方法,假设每个控制体内的解是恒定的,因此面左侧和右侧的状态量简单地取左侧和右侧控制体的流动变量。例如,在中点-二次格式(图5.9)的情况下,有

$$
\begin{aligned}
U_{\mathrm{L}} &= U_i \\
U_{\mathrm{R}} &= U_j
\end{aligned} \tag{5.38}
$$

U 表示某个标量流场变量。这使得空间离散只是一阶精度,对于黏性流动,一阶精度的解扩散性太强,导致剪切层的过度增长,因此黏性流场计算必须采用具有更高阶空间精度的方法。

如果假设解在控制体内变化,我们可以实现二阶和更高阶的空间精度。对于高阶方法中最常用的二阶精度方法,假定解在控制体内以线性方式变化。为了计算左右状态,需要对假设解的变化进行重构。接下来,我们将讨论最流行的线性和二次变化重构方法,感兴趣的读者可以参考文献[39]来比较各种线性重构技术。

1. 基于MUSCL方法的重构

实现二阶精度的一种方法是将MUSCL方法[52]推广到非结构网格。当应用于中点-二次格式时,该方法为每条边 ij 生成两个"虚构"节点 i' 和 j'[53-57],如图5.12所示,这两个虚构节点是边 ij 在其两个方向上各延长它的长度后所得到的线的端点。将解从周围单元(图5.12中灰色单元)插值到虚构节点后,就可以使用式(4.46)中的MUSCL公式计算左右状态:

$$
\begin{aligned}
U_{\mathrm{R}} &= U_j - \frac{1}{4}\left[(1+\hat{\kappa})\Delta_- + (1-\hat{\kappa})\Delta_+\right]U_j \\
U_{\mathrm{L}} &= U_i + \frac{1}{4}\left[(1+\hat{\kappa})\Delta_+ + (1-\hat{\kappa})\Delta_-\right]U_i
\end{aligned} \tag{5.39}
$$

前向(Δ_+)和后向(Δ_-)差分算子定义为

$$
\begin{aligned}
\Delta_+ U_i &= U_j - U_i, \quad \Delta_- U_i = U_i - U_{i'} \\
\Delta_+ U_j &= U_{j'} - U_j, \quad \Delta_- U_j = U_j - U_i
\end{aligned} \tag{5.40}
$$

在存在强间断的情况下,MUSCL插值式(式(5.39))必须使用限制器函数

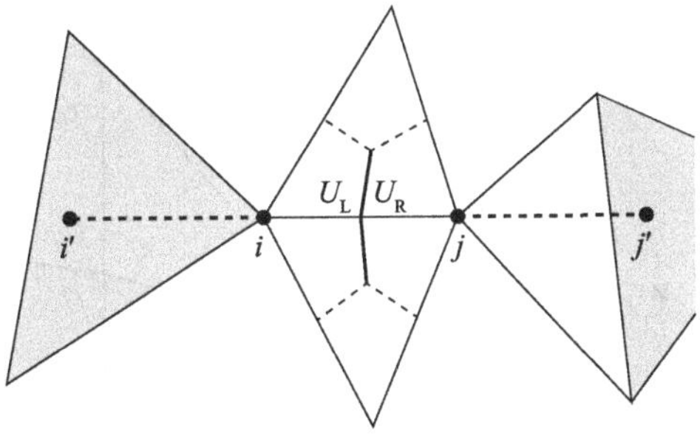

图 5.12　在边 ij 方向上的基于单元插值的左右状态求解(2D 中点-二次格式)

(根据 4.3.5 节)来改善。

这种方法的一个缺点是必须为每条边存储包含虚构节点的单元。另一个概念上的缺点是,对于控制体没有统一的梯度重建。此外,它在边界上可能会出现困难,因为其中的一个虚构点位于物理域之外。

2. 分段线性重构

Barth 和 Jespersen 在文献[30]中提出了一种与有限元格式密切相关的重构方法。假设解是在控制体上分段线性分布,利用下面的关系找到格心格式的左右状态:

$$U_L = U_I + \Psi_I(\nabla U_I \cdot \boldsymbol{r}_L) \tag{5.41}$$
$$U_R = U_J + \Psi_J(\nabla U_J \cdot \boldsymbol{r}_R)$$

其中,∇U_I 是 U 在单元 I 中心的梯度($=[\partial U/\partial x, \partial U/\partial y, \partial U/\partial z]^T$),$\Psi$ 表示限制器函数(参见 5.3.5 节)。矢量 $\boldsymbol{r}_L$ 和 $\boldsymbol{r}_R$ 从单元质心指向面心,如图 5.13(a)所示。

同样的方法也适用于中点-二次格式[30],即

$$U_L = U_i + \frac{1}{2}\Psi_i(\nabla U_i \cdot \boldsymbol{r}_{ij}) \tag{5.42}$$
$$U_R = U_j - \frac{1}{2}\Psi_j(\nabla U_j \cdot \boldsymbol{r}_{ij})$$

由图 5.9 或图 5.13(b)可知

$$\boldsymbol{r}_{ij} = \boldsymbol{r}_j - \boldsymbol{r}_i \tag{5.43}$$

$\boldsymbol{r}_{ij}$ 表示从节点 i 到节点 j 的矢量。

容易看到,Barth 和 Jespersen 方法对应于围绕面的相邻体心/节点展开的泰勒级数,其中仅保留线性项。在规则网格上,线性重构在形式上是二阶精度[39]。如果梯度 ∇U 的计算没有误差,则该格式在任意网格上可以精确地重构线性函数。在各种重构方法中,线性重构可能是最常用的一种。

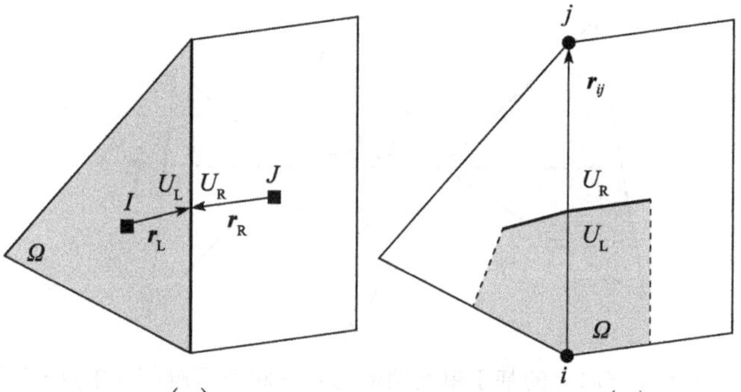

(a)　　　　　　　　　　(b)

图 5.13　2D 情况下的线性重构

(a) 格心格式;(b) 中点-二次格式。

上述格式需要计算体心或者节点的梯度,这可以通过 Green-Gauss 法或最小二乘法来实现,具体方法见 5.3.4 节。此外,限制器函数在非结构网格上的实现将在 5.3.5 节详细介绍。

3. 基于节点加权的线性重构

Frink[26] 证明,对于格心格式,线性重构(式(5.41))不需要在纯三角形或四面体网格上显式计算出梯度,原因是这些单元具有两个不变的几何特征:①从节点通过体心的线总是与相对的面的中点相交;②从体心到面心的距离是面心到对面节点距离的 1/4(对于三角形,为 1/3)。因此,体心的梯度可以用简单的有限差分来近似[26],例如,如果我们重构图 5.7(a)中面心 F_3 处的解,则式(5.41)将变成

$$U_{L/R}=U_C+\frac{\Psi_C}{4}\left[\frac{1}{3}(U_1+U_2+U_4)-U_3\right] \tag{5.44}$$

式中:U_C 为体心处的值;U_1、U_2 等为节点值;Ψ 为限制器函数。

Frink 设计了两种不同的方法来确定节点值。第一种方法是基于距离倒数加权。其中,周围单元对节点的贡献与节点到体心的距离成反比[26,58],即

$$U_i=\left(\sum_{J=1}^{N_A}\theta_{iJ}U_J\right)\Big/\left(\sum_{J=1}^{N_A}\theta_{iJ}\right) \tag{5.45}$$

权 $\theta_{iJ}=1/r_{iJ}$。距离为

$$r_{iJ}=\sqrt{(x_J-x_i)^2+(y_J-y_i)^2+(z_J-z_i)^2} \tag{5.46}$$

下标 J 和 i 分别表示体心和节点。上述重构方法的精度低于二阶,然而 Frink 指出,至少对于无黏流计算,这种重构不需要限制器[26],这样就大大减少了计算量。

第二种方法是基于 Holmes 和 Connell[46]，以及 Rausch 等[47]在 2D 情况下的工作，它被 Frink[2]推广到 3D。如果变量是线性变化的，则定义式（5.45）中的权 θ_{iJ} 的取值需确保可以精确计算节点的值，这产生了与伪拉普拉斯算子（式（5.24））计算相同的约束条件，同时权也相同，可由式（5.25）~式（5.30）获得，只是式中体心坐标 x_I,y_I,z_I 被节点坐标 x_i,y_i,z_i 所替换。该格式在形式上是二阶精度，对于线性函数，节点值可以精确计算。为了保证在畸形网格上的正定性，权必须限制在[0,2]的范围内[46]，但是这降低了重构的精度。Frink 和 Pirzadeh[4]也研究了在纳维-斯托克斯方程组重构的一些异常情况。

4. 分段二次重构

为了在多项式重构中达到高于二阶的精度，我们必须在对控制体的面相邻的体心/节点的泰勒级数展开中保留更多的项。基于 Barth 和 Frederickson 的工作[59]，Barth 发展了一种 k-exact 重构格式的概念[60]，即对于 k 个自由度的多项式的精确重构。Barth 方法中多项式的定义在某种程度上保证了均值的守恒，即重构多项式的平均值等于控制体内的平均解，这个性质保证了在重构过程中质量、动量和能量的守恒。该方法在中点-二次格式下实现了 $k=3$ 的重构，采用最小二乘法计算多项式的系数。Mitchell 和 Walters[61]，以及 Mitchell[62]的格心格式也遵循类似的思想，但是这些方法需要极高的数值计算量和复杂的数据结构，因而阻碍了它们的广泛应用。

Delanaye 和 Essers[63]，以及 Delanaye[64]发展了格心格式下二次重构的一种特殊形式，其计算效率高于 Barth 方法，左右状态用二次项之后截断的泰勒级数近似[63-64]为

$$
\begin{aligned}
U_{\mathrm{L}} &= U_I+\boldsymbol{\Psi}_{I,1}(\nabla U_I\cdot\boldsymbol{r}_{\mathrm{L}})+\frac{1}{2}\boldsymbol{\Psi}_{I,2}(\boldsymbol{r}_{\mathrm{L}}^{\mathrm{T}}H_I\boldsymbol{r}_{\mathrm{L}})\\
U_{\mathrm{R}} &= U_J+\boldsymbol{\Psi}_{J,1}(\nabla U_J\cdot\boldsymbol{r}_{\mathrm{R}})+\frac{1}{2}\boldsymbol{\Psi}_{J,2}(\boldsymbol{r}_{\mathrm{R}}^{\mathrm{T}}H_J\boldsymbol{r}_{\mathrm{R}})
\end{aligned}
\tag{5.47}
$$

其中，$\boldsymbol{H}_I$ 表示在体心 I 上的 Hessian 矩阵，即

$$
\boldsymbol{H}_I=\begin{bmatrix}\partial^2_{xx}U & \partial^2_{xy}U & \partial^2_{xz}U\\ \partial^2_{xy}U & \partial^2_{yy}U & \partial^2_{yz}U\\ \partial^2_{xz}U & \partial^2_{yz}U & \partial^2_{zz}U\end{bmatrix}_I
\tag{5.48}
$$

变量 $\boldsymbol{\Psi}_{I,1}$ 和 $\boldsymbol{\Psi}_{I,2}$ 分别表示线性和二次项的两个不同的限制器函数[63]。二次重构方法在规则网格上是三阶精度，在任意网格上由于误差项的抵消至少是二阶精度[64]。然而，达到这些性质的必要条件是，式（5.47）中的梯度 ∇U 至少需要用二阶精度计算，同时 Hessian 矩阵具有一阶精度计算，这可以通过

Green-Gauss 梯度求解与基于最小二乘的二阶导数近似相结合来实现[63-64]，从而产生了一个数值上高效的格式。但与线性重构相比，内存和 CPU 时间开销仍然相当大。该方法使用与面和节点相邻的单元组成的固定模板，图 5.14 给出了模板示例，以及用于 Green-Gauss 梯度计算的积分路径。为了确定二次多项式的所有系数，模板必须提供至少 6（3D 为 10）个值。为了保证由二次重构得到的精度，有必要考虑在面上使用解的线性变化代替常值，这意味着必须在 2D 面的两个点（三角形面上的三个点）——高斯积分点（图 5.14）——上重构解，并且通量必须在控制体面上以分段方式积分[38]。

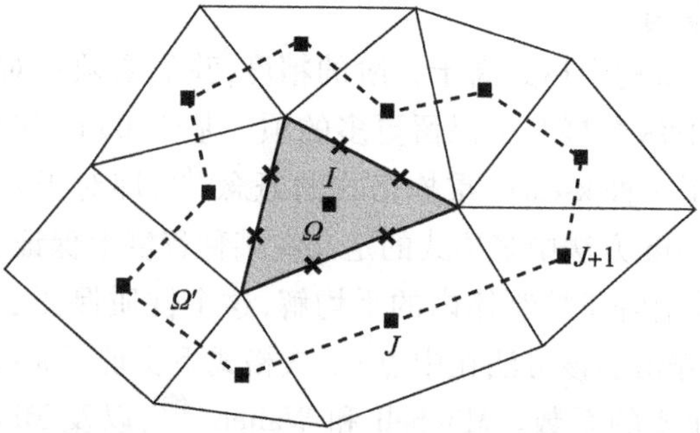

图 5.14　2D 情况下由 Delanaye[63-64] 提出的二次重构方法的模板图（实心方块）。虚线表示 Green-Gauss 梯度计算的积分路径（控制体 Ω′）。叉号表示通量积分的求积点

如今，研究的焦点转向了 ENO（essentially non-oscillatory）格式，特别是二次高阶重构的 WENO（加权 ENO）格式[65-85]。这种格式的思想是使用不同的（与它们的方向相关的）离散模板，为每个控制体计算多个不同的重构多项式。ENO 格式选择多项式的最小振荡，而 WENO 方法加权并混合这些多项式，权重与振荡程度成反比。这样，解在光滑区域实现了空间上的高阶，同时真正的间断不被抹平。文献[73]对 WENO 格式有很好的介绍。

5.3.4　梯度求解

在分段线性重构和二次重构的讨论中留下的一个待解决的问题，即梯度的求解，在黏性通量的计算（5.4 节）中还需要速度分量的梯度和温度的梯度。下面将介绍两种方法：第一种基于 Green-Gauss 定理，第二种使用最小二乘方法。

1. Green-Gauss 方法

此方法将标量函数 U 的梯度近似为 U 与单位外法向矢量的乘积在控制体 Ω' 上的面积分，即

$$\nabla U \approx \frac{1}{\Omega'}\int_{\partial\Omega'} U\boldsymbol{n}\mathrm{d}S \tag{5.49}$$

1）中点-二次格式

Barth 和 Jespersen[30]从伽辽金有限元法推导出了 Green-Gauss 方法的一种特殊离散形式，随后，Barth 将这种离散推广到 3D[86]。Barth 和 Jespersen 将式(5.49)应用于共点单元组成的区域，证明了该方法可以与基于边的数据结构兼容，然而，这仅适用于三角形/四面体网格上的中点-二次格式。公式如下：

$$\nabla U_i \approx \frac{1}{\Omega}\sum_{j=1}^{N_F}\frac{1}{2}(U_i+U_j)\boldsymbol{n}_{ij}\Delta S_{ij} \tag{5.50}$$

其中，式(5.49)中的 Ω' 等于中点-二次控制体 Ω 的体积，求和包括与节点 i 相关的所有 N_F 个边，此外，$\boldsymbol{n}_{ij}$ 根据式(5.23)表示平均单位法向矢量，ΔS_{ij} 为总面积。同样的公式(5.50)适用于 2D 或 3D。需要重视的是，为了获得相容的近似，必须对边界处的求和进行变化[87]（另见 8.10 节）。

2）格心格式

在格心格式中也使用 Green-Gauss 方法。单元质心 I 处的梯度为

$$\nabla U_I \approx \frac{1}{\Omega}\sum_{J=1}^{N_F}\frac{1}{2}(U_I+U_J)\boldsymbol{n}_{IJ}\Delta S_{IJ} \tag{5.51}$$

对体积为 Ω 的单元的所有面进行求和。式(5.51)中，$\boldsymbol{n}_{IJ}$ 为单位法向矢量，ΔS_{IJ} 为面积。

3）混合网格

式(5.50)或式(5.51)的 Green-Gauss 梯度求解方法的主要吸引力在于它与通量计算（如式(5.19)）相似，这样重构梯度时不需要额外的数据结构。而主要缺点是式(5.50)和式(5.51)的近似在混合网格上不成立，文献[50]证明，在不同类型的网格单元交界处，梯度可能变得非常不准确。在中点-二次格式中，只要保证式(5.49)中的控制体 Ω' 等于与节点 i 共点的所有单元之和时，就可以解决这个问题。如图 5.8 所示，在 2D 中梯度为

$$\nabla U_i \approx \frac{1}{\Omega'}\sum_{j=1}^{N_o}\frac{1}{2}(U_j+U_{j+1})\boldsymbol{n}_j\Delta S_j \tag{5.52}$$

$i=0$ 时，外部面的数量为 $N_o=6$（由点 P_1-P_6 给出），当 $j=6$ 时，$j+1$ 取 1。另外，$\boldsymbol{n}_j$ 和 ΔS_j 代表单位法向矢量和外部面的面积。在 3D 中，当假设所有的面都是三角形时——无论是自然的还是剖分的，我们可以使用下式求解梯度：

$$\nabla U_i \approx \frac{1}{\Omega'}\sum_{j=1}^{N_o}\frac{1}{3}(U_{j,1}+U_{j,2}++U_{j,3})\boldsymbol{n}_j\Delta S_j \tag{5.53}$$

同样的补救方法也适用于格心格式，如图 5.14 所示，控制体 Ω' 的面由单元 Ω 的相邻单元的质心以及它们的相邻单元的质心定义[63-64]，梯度计算时相应地将式(5.52)或式(5.53)中的节点索引替换为单元索引。

这种基于单元的方法的最明显的缺点是需要额外的数据结构，用来提供中心节点/质心和控制体 Ω' 的外部面之间的联系，因此该方法不再与网格信息无关，也不可能构造一个有效的收集-分发循环，这使得在混合网格中后面所讲的最小二乘法更具有吸引力。

在三角形或四面体网格上采用基于边/面实现式(5.50)或式(5.51)，在混合网格上采用基于单元实现式(5.52)或式(5.53)，Green-Gauss 方法的精度至少为一阶[64]。这种方法是相容的，即对于线性函数的梯度计算可以准确到舍入误差的水平。对于线性重构来说一阶精度是足够的，对于二次重构，需要在任意网格上达到二阶精度，这可以通过从梯度的一阶近似中减去截断误差的估计值来实现[64]。

2. 最小二乘法

采用最小二乘法计算梯度最早是由 Barth 提出的[37,86]。为了说明这种方法，让我们基于中点-二次格式进行讨论，从而最小二乘法是对与中心节点 i 相连的每条边使用一阶泰勒级数逼近。解沿边 ij 的变化可由下式计算：

$$(\nabla U_i)\cdot \boldsymbol{r}_{ij}=U_j-U_i \tag{5.54}$$

其中，$\boldsymbol{r}_{ij}$ 由式(5.43)给出，表示从节点 i 到节点 j 的矢量(如图 5.9 或图 5.13(b))。当我们将关系式(5.54)应用到与节点 i 相连的所有边时，得到下面的过约束线性方程组：

$$\begin{bmatrix} \Delta x_{i1} & \Delta y_{i1} & \Delta z_{i1} \\ \Delta x_{i2} & \Delta y_{i2} & \Delta z_{i2} \\ \vdots & \vdots & \vdots \\ \Delta x_{ij} & \Delta y_{ij} & \Delta z_{ij} \\ \vdots & \vdots & \vdots \\ \Delta x_{iN_\mathrm{A}} & \Delta y_{iN_\mathrm{A}} & \Delta z_{iN_\mathrm{A}} \end{bmatrix} \begin{bmatrix} \partial_x U \\ \partial_y U \\ \partial_z U \end{bmatrix}_i = \begin{bmatrix} \theta_1(U_1-U_i) \\ \theta_2(U_2-U_i) \\ \vdots \\ \theta_j(U_j-U_i) \\ \\ \theta_{N_\mathrm{A}}(U_{N_\mathrm{A}}-U_i) \end{bmatrix} \tag{5.55}$$

其中，$\Delta(\cdot)_{ij}=(\cdot)_j-(\cdot)_i$，$\partial_m(\cdot)=\partial(\cdot)/\partial m$。此外，$N_\mathrm{A}$ 表示与节点 i 相连的节点 j 的数目，θ_j 表示特定的权系数，权重取决于几何形状和/或取决于解(见文献[41])，然而，在实际中 θ_j 通常设为 1。为方便起见，将式(5.55)简写为

$$\boldsymbol{Ax}=\boldsymbol{b} \tag{5.56}$$

求解式(5.56)的梯度矢量 $\boldsymbol{x}$ 需要对矩阵 $\boldsymbol{A}$ 求逆，为了防止病态条件问题(特别是在拉伸网格中)，Anderson 和 Bonhaus 建议采用 Gram-Schmidt 方法将 $\boldsymbol{A}$ 分解成正交矩阵 $\boldsymbol{Q}$ 与上三角矩阵 $\boldsymbol{R}$ 的乘积[88]，该方法最近被推广到 3D 情况[50]。因此，式(5.56)的解可以直接从下式得到：

$$\boldsymbol{x}=\boldsymbol{R}^{-1}\boldsymbol{Q}^{\mathrm{T}}\boldsymbol{b} \tag{5.57}$$

使用带有双下标的小写字母来表示矩阵元素，我们可以将矩阵 $\boldsymbol{A}=[\boldsymbol{a}_1,\boldsymbol{a}_2,\boldsymbol{a}_3]$ 的 Gram-Schmidt 正交化矩阵写成 $\boldsymbol{Q}=[\boldsymbol{q}_1,\boldsymbol{q}_2,\boldsymbol{q}_3]$，其中

$$\begin{aligned}
\boldsymbol{q}_1&=\frac{1}{r_{11}}\boldsymbol{a}_1\\
\boldsymbol{q}_2&=\frac{1}{r_{22}}\left(\boldsymbol{a}_2-\frac{r_{12}}{r_{11}}\boldsymbol{a}_1\right)\\
\boldsymbol{q}_3&=\frac{1}{r_{33}}\left[\boldsymbol{a}_3-\frac{r_{23}}{r_{22}}\boldsymbol{a}_2-\left(\frac{r_{13}}{r_{11}}-\frac{r_{12}}{r_{11}}\frac{r_{23}}{r_{22}}\right)\boldsymbol{a}_1\right]
\end{aligned} \tag{5.58}$$

上三角矩阵 $\boldsymbol{R}$ 的元素如下：

$$\begin{aligned}
r_{11}&=\sqrt{\sum_{j=1}^{N_{\mathrm{A}}}(\Delta x_{ij})^2}\\
r_{12}&=\frac{1}{r_{11}}\sum_{j=1}^{N_{\mathrm{A}}}\Delta x_{ij}\Delta y_{ij}\\
r_{22}&=\sqrt{\sum_{j=1}^{N_{\mathrm{A}}}(\Delta y_{ij})^2-r_{12}^2}\\
r_{13}&=\frac{1}{r_{11}}\sum_{j=1}^{N_{\mathrm{A}}}\Delta x_{ij}\Delta z_{ij}\\
r_{23}&=\frac{1}{r_{22}}\left(\sum_{j=1}^{N_{\mathrm{A}}}\Delta y_{ij}\Delta z_{ij}-\frac{r_{12}}{r_{11}}\sum_{j=1}^{N_{\mathrm{A}}}\Delta x_{ij}\Delta z_{ij}\right)\\
r_{33}&=\sqrt{\sum_{j=1}^{N_{\mathrm{A}}}(\Delta z_{ij})^2-(r_{13}^2+r_{23}^2)}
\end{aligned} \tag{5.59}$$

使用式(5.57)~式(5.59)，节点 i 的梯度由边差值的加权和得到：

$$\nabla U_i\equiv\boldsymbol{x}=\sum_{j=1}^{N_{\mathrm{A}}}\boldsymbol{w}_{ij}\theta_j(U_j-U_i) \tag{5.60}$$

权矢量 $\boldsymbol{w}_{ij}$ 定义为

$$
\boldsymbol{w}_{ij}=\begin{bmatrix} \alpha_{ij,1}-\dfrac{r_{12}}{r_{11}}\alpha_{ij,2}+\beta\alpha_{ij,3} \\ \alpha_{ij,2}-\dfrac{r_{23}}{r_{22}}\alpha_{ij,3} \\ \alpha_{ij,3} \end{bmatrix} \tag{5.61}
$$

式(5.61)中的各项如下：

$$
\begin{aligned}
\alpha_{ij,1} &= \frac{\Delta x_{ij}}{r_{11}^2} \\
\alpha_{ij,2} &= \frac{1}{r_{22}^2}\left(\Delta y_{ij}-\frac{r_{12}}{r_{11}}\Delta x_{ij}\right) \\
\alpha_{ij,3} &= \frac{1}{r_{33}^2}\left(\Delta z_{ij}-\frac{r_{23}}{r_{22}}\Delta y_{ij}+\beta\Delta x_{ij}\right)
\end{aligned} \tag{5.62}
$$

其中

$$
\beta=\frac{r_{12}r_{23}-r_{13}r_{22}}{r_{11}r_{22}} \tag{5.63}
$$

对于格心格式，最小二乘法的公式在形式上不变，只是节点用体心代替。文献[16]给出了一个例子。

最小二乘法在一般网格上具有一阶精度[64]。它也是相容的，即无论什么单元类型，线性函数的梯度计算都可以准确到舍入误差水平，因此特别适合于混合网格计算。它的计算成本与 Green-Gauss 法相当，只需要在基于面/边的单个循环中执行一次矢量-标量的乘法运算（式(5.60)），但是它必须在每个节点上预先计算并存储上三角矩阵 $\boldsymbol{R}$ 的 6 个元素（式(5.59)）。文献[89]的详细研究显示，为了在高度拉伸且弯曲的网格上获得非线性函数的梯度的准确的近似，式(5.55)～式(5.60)中的权系数 θ_j 必须设为节点 i 与节点 j 之间距离的倒数（类似于式(5.45)中的 θ_{iJ}），然而，它对三角形（四面体）网格上的格心格式没有帮助[89]。

经验还表明，如果在黏性壁面使用三棱柱或六面体单元，则对于中点-二次格式使用最小二乘法时需要特别注意。考虑图 5.15，假设我们想要计算节点 i 处的速度分量的梯度，很明显，只有边 ij 的贡献是有用的，因为在其他与 i 相连的节点上，$u=v=w=0$。为了扩大模板数量，可以插入虚拟边[50]，如图 5.15 中的虚线所示，虚拟边的引入大大提高了离散格式的精度和鲁棒性，需要强调的是，它们仅用于梯度重构，而不用于通量计算。

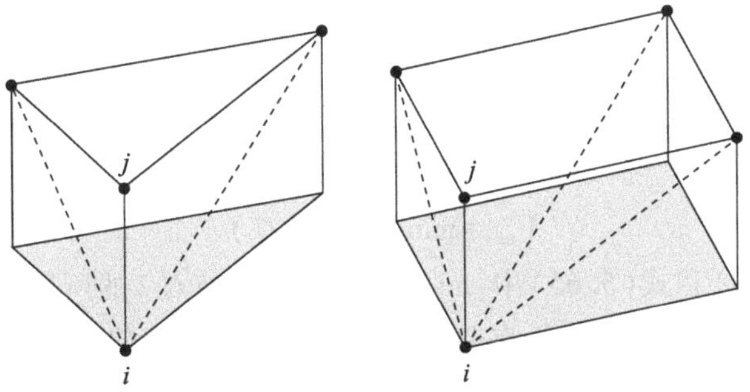

(a)　　(b)

图 5.15　用于计算节点 i 处梯度的虚拟边[50]（虚线）

(a) 三棱柱单元；(b) 六面体单元。阴影表示边界。

5.3.5 限制器函数

二阶和高阶迎风空间离散需要使用限制器或限制器函数，以防止在具有较大梯度的区域（例如在激波处）产生振荡和伪解。我们希望得到的至少是一个保持单调性的格式，这意味着在时间推进过程中，流场中的极大值必须不增大，极小值必须不减小，并且不产生新的局部极值。在 4.3.5 节我们已经讨论了结构迎风离散格式的这一点。

在非结构网格中，限制器的目的是减少用于重构控制体面左右状态的梯度值。为了得到保持单调性的一阶迎风格式，限制器函数在强间断处必须为零，使式(5.38)变为常数重构。为了使数值耗散尽可能低，必须在光滑流动区域保持原始的无限制重构。下面将介绍两个广泛使用的限制器函数——Barth 和 Jespersen 限制器[30]，以及 Venkatakrishnan 限制器[90-91]，关于多维限制器的概念，读者可参阅文献[92-93]。

1. Barth 和 Jespersen 限制器

文献[30]第一次在非结构网格上实现了限制器函数。对于中点-二次格式，在节点 i 处定义为

$$\Psi_i = \min_j \begin{cases} \min\left(1, \dfrac{U_{\max}-U_i}{\Delta_2}\right), & \Delta_2>0 \\ \min\left(1, \dfrac{U_{\min}-U_i}{\Delta_2}\right), & \Delta_2<0 \\ 1, & \Delta_2=0 \end{cases} \tag{5.64}$$

式中的简写表示

$$
\begin{aligned}
&\Delta_2=\frac{1}{2}(\nabla U_i\cdot \boldsymbol{r}_{ij})\\
&U_{\max}=\max(U_i,\max_j U_j)\\
&U_{\min}=\min(U_i,\min_j U_j)
\end{aligned}
\tag{5.65}
$$

在式(5.64)和式(5.65)中，$\min_j$ 或$\max_j$ 表示节点 i 的所有直接相连节点 j（即与 i 通过一条边连接的所有节点）的最小值或最大值。此外，如图 5.9 或图 5.13(b)所示，边矢量 $\boldsymbol{r}_{ij}$ 由式(5.43)定义，最后，U_j 表示某个相邻节点 j 处的标量值。对于格心格式也有类似的公式，只需以单元索引代替节点索引，并且：

$$
\Delta_2=\nabla U_I\cdot \boldsymbol{r}_{\mathrm{L}}
\tag{5.66}
$$

其中，$\boldsymbol{r}_{\mathrm{L}}$ 为从体心到对应单元面心的矢量。为了避免在式(5.64)中除以一个非常小的值 Δ_2，最好将 Δ_2 修改为 $\mathrm{Sign}(\Delta_2)(|\Delta_2|+\omega)$，其中 ω 近似于机器精度[90]。

Barth 限制器强制得到一个单调的解。然而它的耗散相当大，并且它往往会抹平间断，更严重的问题是，在光滑流动区域，也会因数值噪声而激活限制器，这通常会阻止完全收敛到定常状态[39,90]。因此，Venkatakrishnan 限制器变得更流行。

2. Venkatakrishnan 限制器

Venkatakrishnan 限制器[90-91]因其优越的收敛性而得到广泛应用。它使用下面的因子来减小顶点 i 处的重构梯度∇U：

$$
\Psi_i=\min_j\begin{cases}
\dfrac{1}{\Delta_2}\left[\dfrac{(\Delta_{1,\max}^2+\epsilon^2)\Delta_2+2\Delta_2^2\Delta_{1,\max}}{\Delta_{1,\max}^2+2\Delta_2^2+\Delta_{1,\max}\Delta_2+\epsilon^2}\right], & \Delta_2>0\\[2ex]
\dfrac{1}{\Delta_2}\left[\dfrac{(\Delta_{1,\min}^2+\epsilon^2)\Delta_2+2\Delta_2^2\Delta_{1,\min}}{\Delta_{1,\min}^2+2\Delta_2^2+\Delta_{1,\min}\Delta_2+\epsilon^2}\right], & \Delta_2<0\\[2ex]
1, & \Delta_2=0
\end{cases}
\tag{5.67}
$$

其中

$$
\begin{aligned}
&\Delta_{1,\max}=U_{\max}-U_i\\
&\Delta_{1,\min}=U_{\min}-U_i
\end{aligned}
\tag{5.68}
$$

式中：$U_{\max}$ 和 $U_{\min}$ 表示所有相邻节点 j，包括节点 i 本身的最大和最小值，式(5.65)给出了 $U_{\max}$、$U_{\min}$ 和 Δ_2 的定义。参数 ϵ^2 用于控制限制的量，将 ϵ^2 设置为 0 会导致完全限制，但这可能会阻碍收敛，与此相反，如果将 ϵ^2 设置为一个大值,限制器将返回一个约为 1 的值,从而完全没有限制,可能会在求解时发生波动。在实践中,我们发现 ϵ^2 应该与当地长度尺度成正比,即

$$\epsilon^2 = (K\Delta h)^3 \tag{5.69}$$

其中,K 是 $O(1)$ 的常数,Δh 是控制体体积的立方根(2D 情况是面积的平方根)。需要注意,限制器函数(式(5.67))必须用无量纲量定义。式(5.69)中系数 K 对激波分辨率的影响如图 5.16 所示。

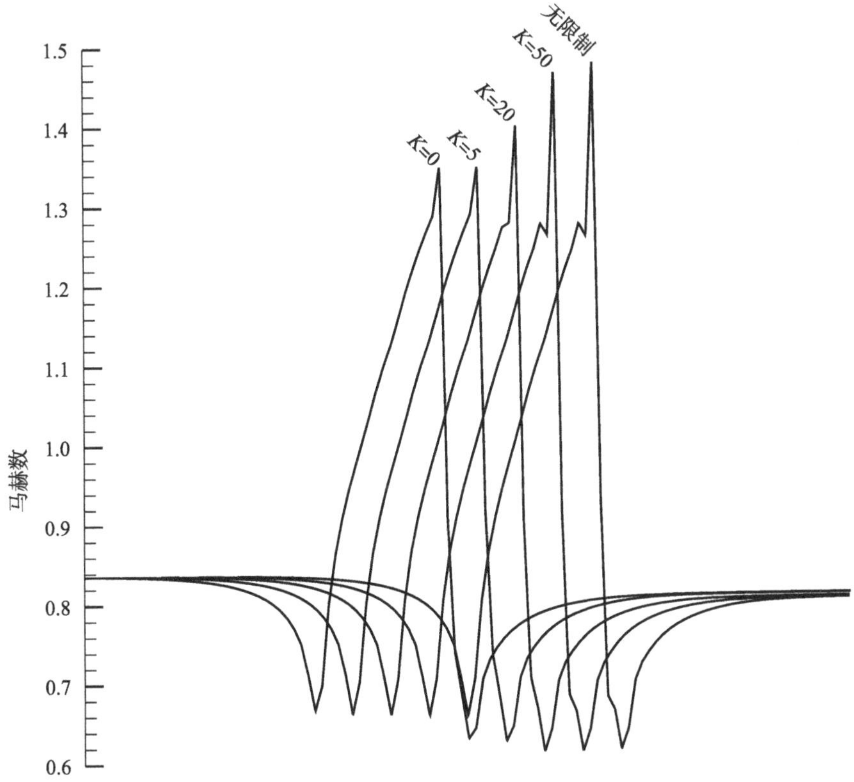

图 5.16　Venkatakrishnan 限制器的常数 K 对圆弧的无黏绕流求解的影响

可以看出,完全限制($K=0$)与 $K=5$ 的解相同。然而,当 $K=0$ 时,显式时间格式仅收敛了大约三个数量级,而对于 $K=5$,它收敛到机器零(图 5.17)。图 5.16 还显示,随着 K 值的增加,解逐渐变得无限制,使得在激波处发生过冲现象的可能性增加。

上述限制器函数的计算量相当高。为了计算 U_{max},U_{min} 以及限制器函数 Ψ 本身,需要两个基于边(格心格式时为面)的循环和一个基于节点(单元)的循环,此外,对于每个流场变量,U_{max}、U_{min} 和 Ψ 都必须存储于节点(单元)上。

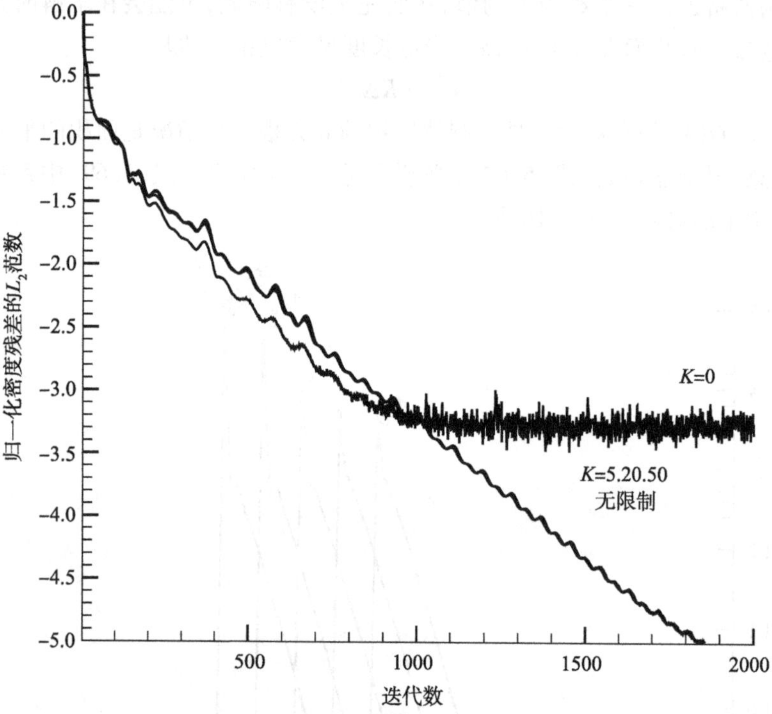

图 5.17　Venkatakrishnan 限制器中的常数 K 对圆弧的无黏绕流收敛性的影响

5.4　黏性通量离散

为了计算式(5.2)中的扩散通量 $\boldsymbol{F}_v$,必须知道控制体面上的流场变量及其一阶导数的值。为了获得相容的空间离散和简化数据结构,黏性通量通常选择与对流通量相同的控制体。由于黏性通量的椭圆型性质,计算黏性项(式(2.23)和式(2.24))和应力(式(2.15))所需的速度分量(u,v,w)、动力学黏性系数μ和热传导系数 k 在面上的值只需使用简单平均计算,在格心格式中(图 5.13(a)),控制体的面 IJ 处的值由下式计算:

$$U_{IJ}=\frac{1}{2}(U_I+U_J) \tag{5.70}$$

其中,U 是上述流场变量之一。类似的表达式也适用于中点-二次格式中的面 ij,见图 5.13(b)。

剩下的任务是计算式(2.15)中的速度分量和式(2.24)中的温度的一阶导数(梯度)。可以采用以下两种方法:

(1) 基于单元的梯度；

(2) 梯度的平均。

下面将详细介绍这两种方法。

5.4.1 基于单元的梯度

这类梯度计算的一个共同特点是必须存储网格单元的信息或与网格单元几何相关的一些系数，因此，除了针对对流通量提出的基于面/边的数据结构外，必须扩展数据结构。下面将介绍三种得到确认的用于格心和中点-二次格式的离散方法。

1. 面心控制体

计算控制体的面的梯度的一种方法是定义一个辅助的控制体，它的体心位于这个面上，然后使用 Green-Gauss 定理，在 4.4 节中，我们已经在结构有限体积离散的框架下讨论过这种方法。例如，在中点-二次格式中，可以把共享一条边的所有单元的梯度的体积加权平均作为该边中点的梯度[94]，基于单元的梯度根据式(5.49)进行计算，方法是在所有网格单元上循环，并在边上累加梯度，单元面上 U 的值通过对节点值进行平均得到，方法类似于式(5.53)。这种方法在内存和操作数方面代价相当高，然而，它可以在任意混合网格上实现。

2. 近似伽辽金有限元法

在伽辽金有限元法的基础上，推导出了适用于中点-二次格式的另一种方法[31]，从根本上说，这种方法将控制体表面的梯度积分转化为中心节点的 Hessian 矩阵(二阶导数)的计算。然后，黏性项采用笛卡儿坐标系下的纳维-斯托克斯方程组的微分形式(式(A.4)，其中 $\xi=x,\eta=y,\zeta=z,J^{-1}=1$)，它包含如 $\partial_x(\mu\partial_x u)$ 等项，其中 $\partial_m(\cdot)=\partial(\cdot)/\partial m$。因此不再需要对控制体各面的黏性通量进行进一步的积分。

最初的格式是基于纯三角形/四面体网格，它使用包含特定节点的所有单元的合集。与伽辽金方法不同的是，为了简化实现，动力学黏性系数由节点值平均获得。节点 i 处的二阶导数为[95]

$$\begin{bmatrix} \partial_x(\mu\partial_x U) & \partial_y(\mu\partial_x U) & \partial_z(\mu\partial_x U) \\ \partial_x(\mu\partial_y U) & \partial_y(\mu\partial_y U) & \partial_z(\mu\partial_y U) \\ \partial_x(\mu\partial_z U) & \partial_y(\mu\partial_z U) & \partial_z(\mu\partial_z U) \end{bmatrix}_i = \frac{1}{\Omega'}\sum_{j=1}^{N_A}\left\{\begin{bmatrix} \alpha_{xx} & \alpha_{xy} & \alpha_{xz} \\ \alpha_{yx} & \alpha_{yy} & \alpha_{yz} \\ \alpha_{zx} & \alpha_{zy} & \alpha_{zz} \end{bmatrix}_{ij} \frac{\mu_i+\mu_j}{2}(U_i-U_j)\right\} \tag{5.71}$$

体积Ω'包含所有共享节点 i 的四面体。式(5.71)中的系数矩阵 $\boldsymbol{\alpha}$ 关于对角线对称[95]，即 $\alpha_{xy}=\alpha_{yx}$，$\alpha_{xz}=\alpha_{zx}$，$\alpha_{yz}=\alpha_{zy}$。因此，每条边只需要存储 6 个系数。文献[95]给出的系数为

$$\alpha_{nk}=\sum_{e}\frac{(S_e^j)_n(S_e^i)_k}{\Omega_e} \tag{5.72}$$

其中，n,k 表示下标 x,y,z。如图 5.18 所示，$(S_e^i)_k$，$(S_e^j)_n$ 表示向外的面矢量 $\boldsymbol{S}_e^i$，$\boldsymbol{S}_e^j$。对全部共享边 ij 的四面体(具有特定体积 Ω_e)进行求和操作。如果网格是静止的，这些系数可以在预处理步骤中计算出来。它的非常理想的特性是，可以采用与对流通量相同的基于边的数据结构。

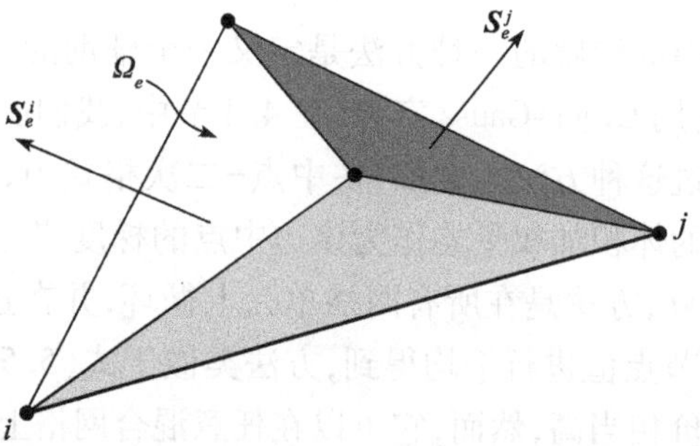

图 5.18　节点 i 处的黏性项：计算与边 ij 相关的系数的体积为 Ω_e 的四面体和三角形面[95]

这种方法的缺点是，完全黏性项只有在三角形或四面体网格中能够得到保留，对于像三棱柱或六面体这样的单元，这种技术在所有三个坐标方向上简化为纳维-斯托克斯方程组的 TSL 类近似[8]。文献[96]中提出了对非单纯形单元保留完全黏性项的推广方法，但是这种基于边的高效的数据结构将不再适用，因为推广的模板包含不与节点 i 通过边相连的节点。

3. 节点值的平均

这种格式适用于格心型控制体和纯四面体网格。它使用来自文献[62]中介绍的模板的一种改进版本[4]计算单元面上的梯度，利用定义单元面的三个节点的值的平均值，并结合了单元质心的值。面的一阶导数由下列线性方程组的解得到[4]：

$$\begin{bmatrix} x_J-x_I & y_J-y_I & z_J-z_I \\ \frac{1}{2}(x_2+x_3)-x_1 & \frac{1}{2}(y_2+y_3)-y_1 & \frac{1}{2}(z_2+z_3)-z_1 \\ \frac{1}{2}(x_1+x_3)-x_2 & \frac{1}{2}(y_1+y_3)-y_2 & \frac{1}{2}(z_1+z_3)-z_2 \end{bmatrix}\begin{bmatrix}\partial_x U\\ \partial_y U\\ \partial_z U\end{bmatrix}$$

$$
=\begin{bmatrix} U_J-U_I \\ \frac{1}{2}(U_2+U_3)-U_1 \\ \frac{1}{2}(U_1+U_3)-U_2 \end{bmatrix} \tag{5.73}
$$

根据图 5.19,下标 I、J 表示单元质心,下标 1、2、3 分别表示节点 P_1、P_2、P_3。节点上的流场变量可以通过距离倒数加权(式(5.45))或伪拉普拉斯加权[2](类似于式(5.24))来确定。

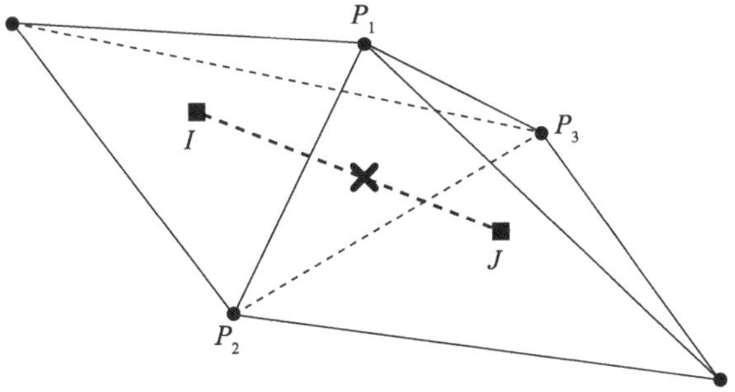

图 5.19　格心格式:四面体网格的梯度计算模板[4]。叉号表示为计算黏性通量而需要计算梯度的位置(面心 $P_1P_2P_3$)

5.4.2　梯度的平均

由于我们已经计算了每个控制体内部的梯度(例如,使用分段线性重构,式(5.41)或式(5.42)),很容易让人想起通过简单的平均来计算面心的梯度[97],即

$$
\overline{\nabla U_{IJ}}=\frac{1}{2}[\nabla U_I+\nabla U_J] \tag{5.74}
$$

这种方法特别有吸引力,因为它只需要基本的基于面或边的数据结构,而且不需要额外的存储。然而,如文献[95]所指出的,它导致模板过宽,权重分布不合理[50],此外文献[50]指出,这类模板会造成解在四边形或六面体网格上的解耦。

利用沿体心(格心格式下)连线的方向导数(在以体心为中心的情况下),可以改善该方法的性质,特别是可以阻止解耦的发生,即

$$
\left(\frac{\partial U}{\partial l}\right)_{IJ}\approx\frac{U_J-U_I}{l_{IJ}} \tag{5.75}
$$

其中,l_{IJ} 表示体心 I 和 J 之间的距离(图 5.19 中的虚线)。使用式(5.43)的 $\boldsymbol{r}_{ij}$,

中点-二次格式也有类似的表达式。在连接 I 和 J 的直线上定义单位矢量：

$$\boldsymbol{t}_{IJ}=\frac{\boldsymbol{r}_{IJ}}{l_{IJ}} \tag{5.76}$$

修正后的平均可以写为[98-99]

$$\nabla U_{IJ}=\overline{\nabla U_{IJ}}-\left[\overline{\nabla U_{IJ}}\cdot\boldsymbol{t}_{IJ}-\left(\frac{\partial U}{\partial l}\right)_{IJ}\right]\boldsymbol{t}_{IJ} \tag{5.77}$$

其中，$\overline{\nabla U_{IJ}}$ 由式(5.74)给出。这种改进在四面体以及三棱柱或六面体网格上产生了强耦合模板[50]。这种方法仍然与基于面/边的数据结构兼容，并且不需要额外的存储空间，因此只要控制体内的梯度同时用于对流通量，这种方法就比基于单元的方法更具有吸引力。

参考文献

[1] Hirsch C. Numerical computation of internal and external flows, vols. 1 and 2. Chichester, UK: John Wiley and Sons; 1988.

[2] Frink NT. Recent progress toward a three-dimensional Navier-Stokes solver. AIAA Paper 94-0061; 1994.

[3] Hassan O, Probert EJ, Weatherill NP, Marchant MJ, Morgan K, Marcum DL. The numerical simulation of viscous transonic flows using unstructured grids. AIAA Paper 94-2346; 1994.

[4] Frink NT, Pirzadeh SZ. Tetrahedral finite-volume solutions to the Navier-Stokes equations on complex configurations. In: 10th int. conf. on finite elements in fluids. Tucson, USA; 1998.

[5] Luo H, Baum JD, Löhner R. Computation of compressible flows using a two-equation turbulence model on unstructured grids. AIAA Paper 97-0430; 1997.

[6] Wang Q, Massey SJ, Abdol-Hamid KS, Frink NT. Solving Navier-Stokes equations with advanced turbulence models on three-dimensional unstructured grids. AIAA Paper 99-0156; 1999.

[7] Luo H, Baum JD, Löhner R. A fast, matrix-free implicit method for compressible flows on unstructured grids. AIAA Paper 99-0936; 1999.

[8] Mavriplis DJ, Venkatakrishnan V. A unified multigrid solver for the Navier-Stokes equations on mixed element meshes. ICASE Report No. 95-53; 1995.

[9] Khawaja A, Kallinderis Y, Parthasarathy V. Implementation of adaptive hybrid grids for 3-D turbulent flows. AIAA Paper 96-0026; 1996.

[10] Coirier WJ, Jorgenson PCE. A mixed-volume approach for the Euler and Navier-Stokes equations, AIAA Paper 96-0762; 1996.

[11] Mavriplis DJ. Adaptive meshing technique for viscous flow calculations on mixed-element unstructured meshes. AIAA Paper 97-0857; 1997.

[12] Haselbacher AC, McGuirk JJ, Page GJ. Finite volume discretization aspects for viscous flows on mixed unstructured grids. AIAA Paper 97-1946; 1997 [also AIAA J 1999; 37: 177-84].

[13] Kano S, Nakahashi K. Navier-Stokes computations of HSCT off-design aerodynamics using unstructured hybrid meshes. AIAA Paper 98-0232;1998.

[14] Sharov D, Nakahashi K. Hybrid prismatic/tetrahedral grid generation for viscous flow applications. AIAA J 1998;36:157-62.

[15] Blazek J, Irmisch S, Haselbacher A. Unstructured mixed-grid Navier-Stokes solver for turbomachinery applications. AIAA Paper 99-0664;1999.

[16] Strang WZ, Tomaro RF, Grismer MJ. The defining methods of Cobalt60: a parallel, implicit, unstructured Euler/Navier-Stokes flow solver. AIAA Paper 99-0786;1999.

[17] Nakahashi K. FDM-FEM zonal approach for computations of compressible viscous flows. Lecture Notes in Physics, vol. 264. Springer Verlag;1986. p. 494-8.

[18] Nakahashi K. A finite-element method on prismatic elements for the three-dimensional Navier-Stokes equations. Lecture Notes in Physics, vol. 323. New York: Springer Verlag;1989. p. 434-8.

[19] Soetrisno M, Imlay ST, Roberts DW, Taflin DE. Computations of viscous flows for multi-element wings using hybrid structured-unstructured grids. AIAA Paper 97-0623;1997.

[20] Kallinderis Y, Khawaja A, McMorris H. Hybrid prismatic/tetrahedral grid generation for viscous flows around complex geometries. AIAA J 1996;34:291-8.

[21] Kallinderis Y, Khawaja A, McMorris H, Irmisch S, Walker D. Hybrid prismatic/tetrahedral grids for turbomachinery applications. In: Proc. 6th int. meshing roundtable. Park City, Utah, USA;1997. p. 21-31.

[22] Baker TJ. Discretization of the Navier-Stokes equations and mesh induced errors. In: Proc. 5th int. conf. on numerical grid generation in CFD. Mississipi State University, Mississipi; April 1996.

[23] Baker TJ. Irregular meshes and the propagation of solution errors. In: Proc. 15th int. conf. on numerical methods in fluid dynamics. Monterey, CA; June 1996.

[24] Mohanraj R, Neumeier Y, Zinn BT. Characteristic-based treatment of source terms in Euler equations for Roe scheme. AIAA J 1999;37:417-24.

[25] Jameson A, Mavriplis D. Finite volume solution of the two-dimensional Euler equations on a regular triangular mesh. AIAA J 1986;24:611-18.

[26] Frink NT, Parikh P, Pirzadeh S. A fast upwind solver for the Euler equations on three-dimensional unstructured meshes. AIAA Paper 91-0102;1991.

[27] Mavriplis DJ. Multigrid solution of the two-dimensional Euler equations on unstructured triangular meshes. AIAA J 1988;26:824-31.

[28] Mavriplis DJ, Jameson A, Martinelli L. Multigrid solution of the Navier-Stokes equations on triangular meshes. ICASE Report No. 89-11;1989.

[29] Whitaker DL, Grossman B, Löhner R. Two-dimensional euler computations on a triangular mesh using an upwind, finite-volume scheme. AIAA Paper 89-0470;1989.

[30] Barth TJ, Jespersen DC. The design and application of upwind schemes on unstructured meshes. AIAA Paper 89-0366;1989.

[31] Barth TJ. Numerical aspects of computing high-Reynolds number flows on unstructured meshes. AIAA Paper 91-0721;1991.

[32] Hwang CJ, Wu SJ. Adaptive finite volume upwind approach on mixed quadrilateral-triangular meshes. AIAA J 1993;31:61-7.

[33] Whitaker DL. Three-dimensional unstructured grid Euler computations using a fully-implicit, upwind method. AIAA Paper 93-3337;1993.

[34] Wang ZJ. Improved formulation for geometric properties of arbitrary polyhedra. AIAA J 1999; 37:1326-7.

[35] www. ma. ic. ac. uk/~rn/centroid. pdf

[36] Bruner CWS. Geometric properties of arbitrary polyhedra in terms of face geometry. AIAA J 1995;33:1350.

[37] Barth TJ. Aspects of unstructured grids and finite-volume solvers for the Euler and Navier-Stokes equations. AGARD R-787, Special course on unstructured grid methods for advection dominated flows. Brussels, Belgium; 18-22 May, 1992. p. 6. 1-61.

[38] Essers JA, Delanaye M, Rogiest P. An upwind-biased finite-volume technique for solving compressible Navier-Stokes equations on irregular meshes. Applications to supersonic blunt-body flows and shock-boundary layer interactions. AIAA Paper 93-3377;1993.

[39] Aftosmis M, Gaitonde D, Tavares TS. Behavior of linear reconstruction techniques on unstructured meshes. AIAA J 1995;33:2038-49.

[40] Xia X, Nicolaides R. Covolume techniques for anisotropic media. Numer Math 1992; 61:215-34.

[41] Barth TJ, Linton SW. An unstructured mesh Newton solver for compressible fluid flow and its parallel implementation. AIAA Paper 95-0221;1995.

[42] Venkatakrishnan V. Implicit schemes and parallel computing in unstructured grid CFD. ICASE Report No. 95-28;1995.

[43] Venkatakrishnan V, Mavriplis DJ. Implicit method for the computation of unsteady flows on unstructured grids. J Comput Phys 1996;127:380-97.

[44] Jameson A, Schmidt W, Turkel E. Numerical solutions of the Euler equations by finite volume methods using Runge-Kutta time-stepping schemes. AIAA Paper 81-1259;1981.

[45] Jameson A, Baker TJ, Weatherill NP. Calculation of inviscid transonic flow over a complete aircraft. AIAA Paper 86-0103;1986.

[46] Holmes, D. G. and Connell SD. Solution of the 2D Navier-Stokes equations on unstructured adaptive meshes. AIAA Paper, No. 89-1932;1989.

[47] Rausch RD, Batina JT, Yang, HTY. Spatial adaption procedures on unstructured meshes for accurate unsteady aerodynamic flow computations. AIAA Paper 91-1106;1991.

[48] Swanson RC, Turkel E. On central difference and upwind schemes. J Comput Phys 1992;101: 292-306.

[49] Mavriplis DJ. Directional agglomeration multigrid techniques for high Reynolds number viscous flow solvers. AIAA Paper 98-0612;1998.

[50] Haselbacher A, Blazek J. On the accurate and efficient discretization of the Navier-Stokes equations on mixed grids. AIAA Paper 99-3363, 1999 [also AIAA J 2000;38:2094-102].

[51] Roe PL. Approximate Riemann solvers, parameter vectors, and difference schemes. J Comput

Phys 1981;43:357-72.
[52] van Leer B. Towards the ultimate conservative difference scheme. V. A second order sequel to Godunov's method. J Comput Phys 1979;32:101-36.
[53] Desideri JA, Dervieux A. Compressible flow solvers using unstructured grids. VKI Lecture Series 1988-05. Rhode-St-Genèse, Belgium: Von Karman Institute; 1988. p. 1-115.
[54] Fezoui L, Stoufflet B. A class of implicit upwind schemes for Euler simulations with unstructured meshes. J Comput Phys 1989;84:174-206.
[55] Whitaker DL, Slack DC, Walters RW. Solution algorithms for the two-dimensional Euler equations on unstructured meshes. AIAA Paper 90-0697;1990.
[56] Arminjon P, Dervieux A. Construction of TVD-like artificial viscosities on two-dimensional arbitrary FEM grids. J Comput Phys 1993;106:176-98.
[57] Jameson A. Positive schemes and shock modelling for compressible flows. Int J Numer Meth Fluids 1995;20:743-76.
[58] Frink NT. Upwind scheme for solving the Euler equations on unstructured tetrahedral meshes. AIAA J 1992;30:70-77.
[59] Barth TJ, Frederickson PO. Higher order solution of the Euler equations on unstructured grids using quadratic reconstruction. AIAA Paper 90-0013;1990.
[60] Barth TJ. Recent developments in high order k-exact reconstruction on unstructured meshes. AIAA Paper 93-0668;1993.
[61] Mitchell CR, Walters RW. k-Exact reconstruction for the Navier-Stokes equations on arbitrary grids. AIAA Paper 93-0536;1993.
[62] Mitchell CR. Improved reconstruction schemes for the Navier-Stokes equations on unstructured meshes. AIAA Paper 94-0642;1994.
[63] Delanaye M, Essers JA. Finite volume scheme with quadratic reconstruction on unstructured adaptive meshes applied to turbomachinery flows. ASME IGTI gas turbine conference. Houston, USA;1995.
[64] Delanaye M. Polynomial reconstruction finite volume schemes for the compressible Euler and Navier-Stokes equations on unstructured adaptive grids. PhD Thesis, The University of Liège, Belgium;1996.
[65] Harten A, Engquist B, Osher S, Chakravarthy S. Uniformly high order accurate essentially non-oscillatory schemes III. J Comput Phys 1987;71:231-303 [also ICASE Report No. 86-22;1986].
[66] Casper J, Atkins HL. A finite-volume high-order ENO scheme for two-dimensional hyperbolic systems. J Comput Phys 1993;106:62-76.
[67] Godfrey AG, Mitchell CR, Walters RW. Practical aspects of spatially high-order accurate methods. AIAA J 1993;31:1634-42.
[68] Abgrall R, Lafon FC. ENO schemes on unstructured meshes. VKI Lecture Series 1993-04. Rhode-St-Genèse, Belgium: Von Karman Institute; 1993.
[69] Abgrall R. On essentially non-oscillatory schemes on unstructured meshes: analysis and implementation. J Comput Phys 1994;114:45-58.
[70] Jiang G-S, Shu C-W. Efficient implementation of weighted ENO schemes. J Comput Phys

1996;126:202-28.
[71] Ollivier-Gooch CF. High-Order ENO schemes for unstructured meshes based on least-squares reconstruction. AIAA Paper 97-0540;1997.
[72] Stanescu D, Habashi WG. Essentially nonoscillatory Euler solutions on unstructured meshes using extrapolation. AIAA J 1998;36:1413-16.
[73] Friedrich O. Weighted essentially non-oscillatory schemes for the interpolation of mean values on unstructured grids. J Comput Phys 1998;144:194-212.
[74] Shi J, Hu C, Shu C-W. A technique of treating negative weights in WENO schemes. J Comput Phys 2002;175:108-27.
[75] Martín MP, Taylor EM, Wu M, Weirs VG. A bandwidth-optimized WENO scheme for the effective direct numerical simulation of compressible turbulence. J Comput Phys 2006;220:270-89.
[76] Dumbser M, Käser M, Titarev VA, Toro EF. Quadrature-free non-oscillatory finite volume schemes on unstructured meshes for nonlinear hyperbolic systems. J Comput Phys 2007;226:204-43.
[77] Borges R, Carmona M, Costa B, Don WS. An improved weighted essentially non-oscillatory scheme for hyperbolic conservation laws. J Comput Phys 2008;227:3191-211.
[78] Dumbser M, Enaux C, Toro EF. Finite volume schemes of very high order of accuracy for stiff hyperbolic balance laws. J Comput Phys 2008;227:3971-4001.
[79] Castro M, Costa B, Don WS. High order weighted essentially non-oscillatory WENO-Z schemes for hyperbolic conservation laws. J Comput Phys 2011;230:1766-92.
[80] Tsoutsanis P, Antoniadis AF, Drikakis D. WENO schemes on arbitrary unstructured meshes for laminar, transitional and turbulent flows. J Comput Phys 2014;256:254-76.
[81] Ivan L, Groth CPT. High-order solution-adaptive central essentially non-oscillatory (CENO) method for viscous flows. J Comput Phys 2014;257:830-62.
[82] Xu L, Weng P. High order accurate and low dissipation method for unsteady compressible viscous flow computation on helicopter rotor in forward flight. J Comput Phys 2014;258:470-88.
[83] Huang C-S, Arbogast T, Hung C-H. A re-averaged WENO reconstruction and a third order CWENO scheme for hyperbolic conservation laws. J Comput Phys 2014;262:291-312.
[84] Boscheri W, Balsara DS, Dumbser M. Lagrangian ADER-WENO finite volume schemes on unstructured triangular meshes based on genuinely multidimensional HLL Riemann solvers. J Comput Phys 2014;267:112-38.
[85] Fan P, Shen Y, Tian B, Yang C. A new smoothness indicator for improving the weighted essentially non-oscillatory scheme. J Comput Phys 2014;269:329-54.
[86] Barth TJ. A 3-D upwind euler solver for unstructured meshes. AIAA Paper 91-1548;1991.
[87] Luo H, Baum JD, Löhner R. An improved finite volume scheme for compressible flows on unstructured grids. AIAA Paper 95-0348;1995.
[88] Anderson WK, Bonhaus DL. An implicit upwind algorithm for computing turbulent flows on unstructured grids. Comput Fluids 1994;23:1-21.
[89] Mavriplis DJ. Revisiting the least-squares procedure for gradient reconstruction on unstructured

meshes. AIAA Paper 2003-3986;2003.

[90] Venkatakrishnan V. On the accuracy of limiters and convergence to steady state solutions. AIAA Paper 93-0880;1993.

[91] Venkatakrishnan V. Convergence to steady-state solutions of the Euler equations on unstructured grids with limiters. J Comput Phys 1995;118:120-30.

[92] Li W, Ren Y-X, Lei G, Luo H. The multi-dimensional limiters for solving hyperbolic conservation laws on unstructured grids. J Comput Phys 2011;230:7775-95.

[93] Li W, Ren Y-X. The multi-dimensional limiters for solving hyperbolic conservation laws on unstructured grids II: Extension to high order finite volume schemes. J Comput Phys 2012;231:4053-77.

[94] Sharov D, Nakahashi K. Low speed preconditioning and LU-SGS scheme for 3-D viscous flow computations on unstructured grids. AIAA Paper 98-0614;1998.

[95] Mavriplis DJ. Three-dimensional multigrid Reynolds-averaged Navier-Stokes solver for unstructured meshes. AIAA Paper 94-1878, 1994; also AIAA J 1995;33:445-53.

[96] Braaten ME, Connell SD. Three dimensional unstructured adaptive multigrid scheme for the Navier-Stokes equations. AIAA J 1995;34:281-90.

[97] Luo H, Baum JD, Löhner R, Cabello J. Adaptive edge-based finite element schemes for the Euler and Navier-Stokes equations on unstructured grids. AIAA Paper 93-0336;1993.

[98] Crumpton PI, Moiner P, Giles MB. An unstructured algorithm for high Reynolds number flows on highly-stretched grids. In: 10th int. conf. num. meth. for laminar and turbulent flows. Swansea, England; July 21-25, 1997.

[99] Weiss JM, Maruszewski JP, Smith WA. Implicit solution of preconditioned Navier-Stokes equations using algebraic multigrid. AIAA J 1999;37:29-36.

第6章 时间离散方法

应用直线法，即对控制方程组(2.19)分别进行空间和时间离散，得到一个关于时间的耦合常微分方程组，对于每个控制体，有

$$\frac{\mathrm{d}(\Omega \boldsymbol{M}\boldsymbol{W})_I}{\mathrm{d}t} = -\boldsymbol{R}_I \tag{6.1}$$

式中：Ω 为体积；$\boldsymbol{R}$ 为残差；$\boldsymbol{M}$ 为质量矩阵；下标 I 表示特定控制体。对式(6.1)进行时间积分，可以得到定常流动的定常解($\boldsymbol{R}_I=0$)，或复现非定常流动的时间历程。

在3.2节中，简要讨论了求解式(6.1)的各个方面，我们发现，可由基本的非线性格式推导出各种显式和隐式方法。对于静止网格，非线性格式为

$$\frac{(\Omega \boldsymbol{M})_I}{\Delta t_I}\Delta \boldsymbol{W}_I^n = -\frac{\beta}{1+\omega}\boldsymbol{R}_I^{n+1} - \frac{1-\beta}{1+\omega}\boldsymbol{R}_I^n + \frac{\omega}{1+\omega}\frac{(\Omega \boldsymbol{M})_I}{\Delta t_I}\Delta \boldsymbol{W}_I^{n-1} \tag{6.2}$$

其中

$$\Delta \boldsymbol{W}_I^n = \boldsymbol{W}_I^{n+1} - \boldsymbol{W}_I^n \tag{6.3}$$

式(6.3)表示对解的更新(修正量)。上标 n 和 $n+1$ 表示时间层，因此 $\boldsymbol{W}^n$ 表示当前时刻 t 的流场解，而 $\boldsymbol{W}^{n+1}$ 表示 $t+\Delta t$ 时刻的流场解。参数 β 和 ω 决定了离散类型(显式或隐式离散)和时间精度，例如必须满足式(3.6)所表示的条件，才能达到二阶时间精度。

下面各节将详细探讨应用最广泛的显式和隐式时间推进方法，介绍如何计算特定格式允许的最大时间步长，此外还探讨在结构和非结构网格上如何恰当地实现计算方法的问题，最后一节将讨论非定常流动问题的时间精确求解。

6.1 显式时间推进方法

显式格式从已知的解 $\boldsymbol{W}^n$ 开始，利用相应的残差 $\boldsymbol{R}^n$，获得 $t+\Delta t$ 时刻的新解，

即新解 $\boldsymbol{W}^{n+1}$ 仅取决于已知的值。这一事实使得显式格式非常简单,易于实现。

正如在 3.2.1 节所讨论的,设 $\beta=0$ 和 $\omega=0$,可由式(6.2)推导出基本的显式格式,即

$$\boldsymbol{M}_I \Delta \boldsymbol{W}_I^n = -\frac{\Delta t_I}{\Omega_I}\boldsymbol{R}_I^n \tag{6.4}$$

式(6.4)称为前向欧拉近似法。对于定常问题或格心离散,质量矩阵 $\boldsymbol{M}$ 可以被“归并”(即由单位矩阵代替原始矩阵)。

最流行且应用最广泛的显式方法是多步(龙格-库塔)时间推进法及其变体混合多步法。因此,下面将对这两种方法进行介绍。

6.1.1　多步法(龙格-库塔法)

显式多步法的概念最早由 Jameson 等[1]提出,它通过若干步(或级)推进求解,这些步可视为根据式(6.4)对解的一系列更新。应用于离散控制方程组(6.1)(式中对质量矩阵 $\boldsymbol{M}$ 进行了归并)的 m 步法为

$$\begin{aligned}
\boldsymbol{W}_I^{(0)} &= \boldsymbol{W}_I^n,\\
\boldsymbol{W}_I^{(1)} &= \boldsymbol{W}_I^{(0)} - \alpha_1 \frac{\Delta t_I}{\Omega_I}\boldsymbol{R}_I^{(0)}\\
\boldsymbol{W}_I^{(2)} &= \boldsymbol{W}_I^{(0)} - \alpha_2 \frac{\Delta t_I}{\Omega_I}\boldsymbol{R}_I^{(1)}\\
&\vdots\\
\boldsymbol{W}_I^{(n+1)} &= \boldsymbol{W}_I^{(m)} = \boldsymbol{W}_I^{(0)} - \alpha_m \frac{\Delta t_I}{\Omega_I}\boldsymbol{R}_I^{(m-1)}
\end{aligned} \tag{6.5}$$

其中,$\alpha_1,\alpha_2,\cdots,\alpha_m$ 为各步的系数,而 $\boldsymbol{R}_I^{(k)}$ 表示由第 k 步的解 $\boldsymbol{W}_I^{(k)}$ 计算的残差。

与经典的龙格-库塔法不同,多步法仅存储第零步的解和最新步的残差,以减少内存需求。对于特定的空间离散,可调节各步的系数,以增加最大时间步长和提高稳定性[2-4],只需 $\alpha_m=1$ 即可保持相容性。根据多步法对龙格-库塔法的修正,只有当 $\alpha_m=1/2$ 时才能实现二阶时间精度,否则,多步法为一阶时间精度。

上述多步法(式(6.5))特别适用于结构和非结构网格的迎风空间离散,而中心离散采用混合多步法的效率更高,6.1.2 节将介绍混合多步法。表 6.1 给出了 3~5 步方法的一阶和二阶迎风格式优化的步系数[2]。实践经验表明,在流场包含强激波的情况下,无论空间离散为多少阶,应优先考虑一阶格式的系数,这是基于以下事实,即每种高阶格式在激波处都转换为一阶格式,以防止解的振荡,而强激波处的残差对收敛到稳态解的影响最为显著。

显式格式的主要缺点是时间步长(Δt)受到控制方程组特性和网格几何结构的严格限制,我们将在6.1.4节讨论允许的最大时间步长的计算,将在10.3节稳定性分析中,探讨如何确定时间步长和CFL数的理论方法。

表6.1 多步法:一阶和二阶迎风空间离散的优化步系数(α)和CFL数(σ)

	一阶格式			二阶格式		
步数	3	4	5	3	4	5
σ	1.5	2.0	2.5	0.69	0.92	1.15
α_1	0.1481	0.0833	0.0533	0.1918	0.1084	0.0695
α_2	0.4000	0.2069	0.1263	0.4929	0.2602	0.1602
α_3	1.0000	0.4265	0.2375	1.0000	0.5052	0.2898
α_4		1.0000	0.4414		1.0000	0.5060
α_5			1.0000			1.0000

6.1.2 混合多步法

如果不在每一步都重新计算黏性通量和耗散,则可以大大减少应用于式(6.1)的显式多步法(式(6.5))的计算量,此外,还可以混合不同步的耗散项,以提高计算方法的稳定性,这类方法由Martinelli[5]及Mavriplis和Jameson[6]设计,称为混合多步法。如果对各步系数进行仔细优化,则混合法可以如基本多步法一样鲁棒。

下面以流行的五步混合法为例进行说明,该方法在奇数步计算耗散项,通常表示为(5,3)格式。首先,将空间离散分为两部分,即

$$\boldsymbol{R}_I=(\boldsymbol{R}_c)_I-(\boldsymbol{R}_d)_I \tag{6.6}$$

第一部分$\boldsymbol{R}_c$包含对流通量的中心离散,可以是变量或通量的平均,它还包括了源项。第二部分$\boldsymbol{R}_d$由黏性通量和数值耗散组成,例如对于带有人工耗散的中心格式(4.3.1节或5.3.1节),设$\boldsymbol{R}_c$和$\boldsymbol{R}_d$为

$$(\boldsymbol{R}_c)_I=\sum_{k=1}^{N_F}[\boldsymbol{F}_c(\boldsymbol{W}_{av})\Delta S]_k-(\boldsymbol{Q}\Omega)_I$$

$$(\boldsymbol{R}_d)_I=\sum_{k=1}^{N_F}[\boldsymbol{F}_v\Delta S+\boldsymbol{D}]_k$$

式中:$\boldsymbol{W}_{av}$为面k左右两侧流动变量的算术平均值。

根据式(6.6)的残差分裂,(5,3)格式可表示为

$$
\begin{aligned}
\boldsymbol{W}_I^{(0)} &= \boldsymbol{W}_I^n \\
\boldsymbol{W}_I^{(1)} &= \boldsymbol{W}_I^{(0)} - \alpha_1 \frac{\Delta t_I}{\Omega_I}[\boldsymbol{R}_c^{(0)} - \boldsymbol{R}_d^{(0)}]_I \\
\boldsymbol{W}_I^{(2)} &= \boldsymbol{W}_I^{(0)} - \alpha_2 \frac{\Delta t_I}{\Omega_I}[\boldsymbol{R}_c^{(1)} - \boldsymbol{R}_d^{(0)}]_I \\
\boldsymbol{W}_I^{(3)} &= \boldsymbol{W}_I^{(0)} - \alpha_3 \frac{\Delta t_I}{\Omega_I}[\boldsymbol{R}_c^{(2)} - \boldsymbol{R}_d^{(2,0)}]_I \\
\boldsymbol{W}_I^{(4)} &= \boldsymbol{W}_I^{(0)} - \alpha_4 \frac{\Delta t_I}{\Omega_I}[\boldsymbol{R}_c^{(3)} - \boldsymbol{R}_d^{(2,0)}]_I \\
\boldsymbol{W}_I^{n+1} &= \boldsymbol{W}_I^{(0)} - \alpha_5 \frac{\Delta t_I}{\Omega_I}[\boldsymbol{R}_c^{(4)} - \boldsymbol{R}_d^{(4,2)}]_I
\end{aligned}
\tag{6.7}
$$

其中

$$
\begin{aligned}
\boldsymbol{R}_d^{(2,0)} &= \beta_3 \boldsymbol{R}_d^{(2)} + (1-\beta_3)\boldsymbol{R}_d^{(0)} \\
\boldsymbol{R}_d^{(4,2)} &= \beta_5 \boldsymbol{R}_d^{(4)} + (1-\beta_5)\boldsymbol{R}_d^{(2,0)}
\end{aligned}
\tag{6.8}
$$

表 6.2 给出了中心格式和迎风格式在上述关系式(6.7)和式(6.8)中的各步系数 α_m 和混合系数 β_m，这两组系数针对多重网格方法(9.4 节)专门进行了优化，我们将在 10.3 节中讨论上述混合多步法的特性。

需要指出的是，仅在前两步计算耗散项 $\boldsymbol{R}_d$，而不进行任何混合也是很流行的做法。通常与中心离散一起使用的是著名的(5,2)格式，它使用表 6.2 的各步系数。但是与(5,3)格式相比，(5,2)格式不太适用于黏性流场计算和多重网格。

表 6.2 混合 5 步法：中心离散和迎风离散的优化步系数(α)和混合系数(β)以及 CFL 数(σ)

	中心格式 $\sigma=3.6$		一阶迎风格式 $\sigma=2.0$		二阶迎风格式 $\sigma=1.0$	
步数	α	β	α	β	α	β
1	0.2500	1.00	0.2742	1.00	0.2742	1.00
2	0.1667	0.00	0.2067	0.00	0.2067	0.00
3	0.3750	0.56	0.5020	0.56	0.5020	0.56
4	0.5000	0.00	0.5142	0.00	0.5142	0.00
5	1.0000	0.44	1.0000	0.44	1.0000	0.44

注意：一阶和二阶迎风格式的各步系数和混合系数相同，但 CFL 数不同。

6.1.3 源项的处理

在某些情况下,式(4.2)或式(5.2)中的源项 $\boldsymbol{Q}$ 占主导地位,比如当采用化学模型或湍流模型时。大的源项引起流场变量在空间和时间上迅速变化,它发生在比流动方程组小得多的时间尺度上,这大大增加了控制方程组的刚性。刚性定义为雅可比矩阵 $\partial\boldsymbol{R}/\partial\boldsymbol{W}$ 的最大特征值与最小特征值之比,也可以看作最大时间尺度与最小时间尺度的比值。

将上述显式多步法(或其他纯显式方法)应用于刚性方程组时,为了使时间积分稳定,必须大幅减小时间步长,因此收敛到稳态解将变得非常缓慢,更严重的是,显式方法甚至可能无法得到正确解[7]。Curtiss 和 Hirschfelder[8] 提出的解决方法是用隐式方式处理源项,为了演示该方法,将式(6.4)中的基本显式格式重写为如下形式(参见式(4.2)或式(5.2)):

$$\frac{\Omega_I}{\Delta t_I}\Delta\boldsymbol{W}_I^n = -\left[\sum_{k=1}^{N_F}(\boldsymbol{F}_c^n - \boldsymbol{F}_v^n)_k\Delta S_k - \Omega_I\boldsymbol{Q}_I^{n+1}\right] \tag{6.9}$$

$$= -(\boldsymbol{R}_Q)_I^n + \Omega_I\boldsymbol{Q}_I^{n+1}$$

其中,在新的时间层 $n+1$ 计算源项。为简便起见,式(6.9)中忽略了质量矩阵。由于 $n+1$ 层的源项值是未知的,必须对源项进行近似计算,为此,用当前时间层 n 的值对源项进行线性化,得

$$\boldsymbol{Q}^{n+1} \approx \boldsymbol{Q}^n + \frac{\partial\boldsymbol{Q}}{\partial\boldsymbol{W}}\Delta\boldsymbol{W}^n \tag{6.10}$$

将式(6.10)代入式(6.9)并整理,得[9-10]

$$\left[\frac{1}{\Delta t_I}\boldsymbol{I} - \left(\frac{\partial\boldsymbol{Q}}{\partial\boldsymbol{W}}\right)_I\right]\Delta\boldsymbol{W}_I^n = -\frac{1}{\Omega_I}\left[(\boldsymbol{R}_Q)_I^n - \Omega_I\boldsymbol{Q}_I^n\right] \tag{6.11}$$

式中:$\boldsymbol{I}$ 表示单位矩阵。因为左边方括号中的项(隐式算子)仅与控制体 Ω_I 本身的值有关,所以式(6.11)称为点隐式方法。与式(6.9)相比较,现在标量时间步长 Δt 变为一个矩阵,因此每个流动方程都由一个单独的参数进行缩放,对应于相关的特征值。通过这种方式,消除了时间尺度之间的差异,缓解了由于源项引起的对时间步长的限制。

将上述点隐式方法应用于式(6.5)中的多步法,得到第 k 步的解为

$$\boldsymbol{W}_I^{(k)} = \boldsymbol{W}_I^{(0)} - \left[(\boldsymbol{R}_Q)_I^{(k-1)} - \Omega_I\boldsymbol{Q}_I^{(0)}\right]\left[\frac{\Omega_I}{\alpha_k\Delta t_I}\boldsymbol{I} - \left(\frac{\partial\boldsymbol{Q}}{\partial\boldsymbol{W}}\right)_I\right]^{-1} \tag{6.12}$$

式中:$\boldsymbol{R}_Q$ 由式(6.9)定义。类似的表达式也适用于式(6.7)中的混合多步法。感兴趣的读者可参见文献[11-13]中关于源项对稳定性影响的详细研究。

一种更为巧妙的方法是用隐式龙格-库塔法处理刚性源项,该方法最早由

Rosenbrock[14]和 Butcher[15]提出。与式(6.12)中的方法相比,这类方法在数值上更稳定,但是它需要在每步对一个大型线性方程组求逆。隐式龙格-库塔法的详细介绍及其应用见6.2.6节。

6.1.4 最大时间步长的确定

每种显式时间推进方法仅在时间步长 Δt 不超过某个值时才能保持数值稳定,它必须满足 CFL 条件[16]。CFL 条件是指数值方法的依赖域必须包含偏微分方程的依赖域。对于基本显式格式(式(6.4)),CFL 条件意味着时间步长应等于或小于空间离散格式模板上传输信息所需的时间。因此,在1D 情况下,线性对流方程时间步长的 CFL 条件为

$$\Delta t=\sigma\frac{\Delta x}{|\Lambda_c|} \tag{6.13}$$

其中,$\Delta x/|\Lambda_c|$ 表示以速度 Λ_c 在大小为 Δx 的单元上传播信息所需的时间,速度 Λ_c 对应对流通量雅可比矩阵的最大特征值,正系数 σ 表示 CFL 数。CFL 数的大小取决于时间推进方法的类型和参数,以及空间离散方法的形式。我们将在10.3节中针对对流模型方程和对流扩散模型方程,讨论 CFL 数 σ 对数值稳定性的影响。表6.1和表6.2列出了各种多步法和离散格式的 CFL 数。

通过冯·诺依曼稳定性分析(见10.3节),可确定线性模型方程的最大时间步长。然而,在多维和非线性控制方程组中最大时间步长只能近似计算。下面将介绍在结构网格和非结构网格上,无黏流和黏性流的时间步长计算公式。

1. 结构网格的时间步长

1) 欧拉方程组

对于结构网格,由近似关系式[17-19]可确定控制体 Ω_I 的时间步长 Δt 为

$$\Delta t_I=\sigma\frac{\Omega_I}{(\hat{\Lambda}_c^I+\hat{\Lambda}_c^J+\hat{\Lambda}_c^K)_I} \tag{6.14}$$

表6.1给出了多步法的 CFL 数 σ,表6.2给出了混合多步法的 CFL 数 σ。三个网格方向的对流通量雅可比矩阵(A.9节)的谱半径分别为

$$\begin{aligned}\hat{\Lambda}_c^I&=(|\boldsymbol{v}\cdot\boldsymbol{n}^I|+c)\Delta S^I\\ \hat{\Lambda}_c^J&=(|\boldsymbol{v}\cdot\boldsymbol{n}^J|+c)\Delta S^J\\ \hat{\Lambda}_c^K&=(|\boldsymbol{v}\cdot\boldsymbol{n}^K|+c)\Delta S^K\end{aligned} \tag{6.15}$$

由控制体在各自方向上相对的两侧的对应值取平均,得到式(6.15)中的法向矢量和面积。例如,如果一个二次控制体(4.2.3节)的方向如图4.1(b)所示,则 I 方向的法向矢量和面积为

$$\boldsymbol{n}_{i,j,k}^{I}=\frac{1}{2}(\boldsymbol{n}_1-\boldsymbol{n}_2),\quad \Delta S_{i,j,k}^{I}=\frac{1}{2}(\Delta S_1+\Delta S_2) \tag{6.16}$$

J 方向和 K 方向的法向矢量和面积公式类似。

利用式(6.14),得到只对一个控制体有效的局部时间步长。如果关注定常求解,可使用局部时间步长技术来加速收敛(见 9.1 节)。然而,如果时间精度很重要,则必须对所有控制体采用全局时间步长,即

$$\Delta t=\min_I(\Delta t_I) \tag{6.17}$$

其中,最小值为所有控制体的最小值。

2) 纳维-斯托克斯方程组

对于黏性流动,在计算 Δt 时必须包含黏性通量雅可比矩阵(A.10 节)的谱半径,它们严重限制了边界层内的最大时间步长。时间步长可由下式计算[5,18-19]:

$$\Delta t_I=\sigma\frac{\Omega_I}{(\hat{\Lambda}_c^I+\hat{\Lambda}_c^J+\hat{\Lambda}_c^K)_I+C(\hat{\Lambda}_v^I+\hat{\Lambda}_v^J+\hat{\Lambda}_v^K)_I} \tag{6.18}$$

对于中心离散,黏性谱半径的系数通常设为 $C=4$,对于一阶迎风离散通常设为 $C=2$,二阶迎风离散通常设为 $C=1$。如果采用涡黏性湍流模型,则 I 方向的黏性谱半径为[18-19]

$$\hat{\Lambda}_v^I=\max\left(\frac{4}{3\rho},\frac{\gamma}{\rho}\right)\left(\frac{\mu_L}{Pr_L}+\frac{\mu_T}{Pr_T}\right)\frac{(\Delta S^I)^2}{\Omega} \tag{6.19}$$

其他方向的黏性谱半径与之类似。式(6.19)中,μ_L 和 μ_T 分别为层流和湍流动力学黏性系数,Pr_L 和 Pr_T 分别为层流和湍流的 Pr 数。表 6.1 和表 6.2 中的 CFL 数同样适用于黏性流动。对黏性流动特别有效的是由式(6.7)得到的(5,3)混合格式。

2. 非结构网格的时间步长

针对非结构网格提出了几种计算最大时间步长的方法,下面介绍两种不同的方法。

1) 方法 1

与结构网格的计算方法非常相近,这种方法已经得到了验证[6],即

$$\Delta t_I=\sigma\frac{\Omega_I}{(\hat{\Lambda}_c+C\hat{\Lambda}_v)_I} \tag{6.20}$$

式中:$\hat{\Lambda}_c$ 和 $\hat{\Lambda}_v$ 表示控制体所有面上的对流和黏性谱半径之和。与在结构网格一样,常数 C 通常为 $1\leqslant C\leqslant 4$。对于格心格式,谱半径定义为[6]

$$(\hat{\Lambda}_c)_I = \sum_{J=1}^{N_F} (\,|\boldsymbol{v}_{IJ} \cdot \boldsymbol{n}_{IJ}| + c_{IJ})\Delta S_{IJ}$$

$$(\hat{\Lambda}_v)_I = \frac{1}{\Omega_I}\sum_{J=1}^{N_F}\left[\max\left(\frac{4}{3\rho_{IJ}},\frac{\gamma_{IJ}}{\rho_{IJ}}\right)\left(\frac{\mu_L}{Pr_L}+\frac{\mu_T}{Pr_T}\right)_{IJ}(\Delta S_{IJ})^2\right] \tag{6.21}$$

使用算术平均计算控制体各面处的流动变量值。

2）方法2

由式(6.21)计算的谱半径太大，特别是在混合网格上，这使得时间步长比数值稳定所必需的时间步长要小。文献[20]中的应用提供了一种更精确的时间步长计算方法，即

$$\Delta t_I = \sigma\frac{\Omega_I}{(\hat{\Lambda}_c^x+\hat{\Lambda}_c^y+\hat{\Lambda}_c^z)_I + C(\hat{\Lambda}_v^x+\hat{\Lambda}_v^y+\hat{\Lambda}_v^z)_I} \tag{6.22}$$

其中，对流谱半径为

$$\hat{\Lambda}_c^x = (\,|u|+c)\Delta\hat{S}^x$$

$$\hat{\Lambda}_c^y = (\,|v|+c)\Delta\hat{S}^y \tag{6.23}$$

$$\hat{\Lambda}_c^z = (\,|w|+c)\Delta\hat{S}^z$$

而黏性谱半径(假设采用的是涡黏性湍流模型)为

$$\hat{\Lambda}_v^x = \max\left(\frac{4}{3\rho},\frac{\gamma}{\rho}\right)\left(\frac{\mu_L}{Pr_L}+\frac{\mu_T}{Pr_T}\right)\frac{(\Delta\hat{S}^x)^2}{\Omega} \tag{6.24}$$

变量 $\Delta\hat{S}^x$，$\Delta\hat{S}^y$ 和 $\Delta\hat{S}^z$ 分别表示控制体在 y-z、x-z 和 x-y 平面上的投影，由下式给出

$$\Delta\hat{S}^x = \frac{1}{2}\sum_{J=1}^{N_F}|S_x|_J$$

$$\Delta\hat{S}^y = \frac{1}{2}\sum_{J=1}^{N_F}|S_y|_J \tag{6.25}$$

$$\Delta\hat{S}^z = \frac{1}{2}\sum_{J=1}^{N_F}|S_z|_J$$

其中，S_x，S_y 和 S_z 表示面矢量 $\boldsymbol{S}=\boldsymbol{n}\cdot\Delta S$ 在 x，y 和 z 方向上的分量。

非结构网格的 CFL 数通常与结构网格相同。因此，表6.1和表6.2中 CFL 数的值仍然适用于非结构网格。而且采用当地时间步长技术(与结构网格相同)也可以加速收敛到稳态。如前所述，由式(6.17)可得模拟非定常流动所需的全局时间步长。

6.2 隐式时间推进方法

通过在式(6.2)中设置$\beta \neq 0$,可以得到各种隐式时间积分格式,其中$\omega = 0$的隐式格式最适合求解定常流动问题(非定常流动的求解见6.3节),此时,式(6.2)简化为

$$\frac{(\Omega M)_I}{\Delta t_I}\Delta \boldsymbol{W}_I^n = -\beta \boldsymbol{R}_I^{n+1} - (1-\beta)\boldsymbol{R}_I^n \tag{6.26}$$

隐式公式导出了$t+\Delta t$时刻未知流场变量的一组非线性方程组。式(6.26)的求解需要计算新时间层上的残差,即$\boldsymbol{R}^{n+1}$。由于$\boldsymbol{W}^{n+1}$未知,所以不能直接计算$\boldsymbol{R}^{n+1}$,但可以用当前时间层的残差对$\boldsymbol{R}_I^{n+1}$进行线性化,即

$$\boldsymbol{R}_I^{n+1} \approx \boldsymbol{R}_I^n + \left(\frac{\partial \boldsymbol{R}}{\partial \boldsymbol{W}}\right)_I \Delta \boldsymbol{W}^n \tag{6.27}$$

式中:$\partial \boldsymbol{R}/\partial \boldsymbol{W}$称为通量雅可比矩阵。值得注意的是,通量雅可比矩阵通常由用$\boldsymbol{R}^n$表示的空间离散的一种相当粗糙的近似得到,例如在高阶迎风离散情况下,通常仅基于一阶迎风格式计算通量雅可比矩阵,然而,为了获得最佳效率和鲁棒性,通量雅可比矩阵仍然应该能反映出空间离散的最重要特征。

如果将式(6.27)中$\boldsymbol{R}^{n+1}$线性化代入式(6.26)中,得到以下隐式格式:

$$\left[\frac{(\Omega M)_I}{\Delta t_I} + \beta\left(\frac{\partial \boldsymbol{R}}{\partial \boldsymbol{W}}\right)_I\right]\Delta \boldsymbol{W}^n = -\boldsymbol{R}_I^n \tag{6.28}$$

式(6.28)左侧方括号中的项称为隐式算子或系统矩阵,相应的,式(6.28)的右端项称为显式算子,解的空间精度仅取决于显式算子。

隐式算子为一个非对称大型稀疏块矩阵,其维数等于单元(格心格式)或网格点(格点格式)的总数。下面将进一步讨论结构网格和非结构网格的隐式算子的形式。如3.2节所述,质量矩阵$\boldsymbol{M}$可用单位矩阵代替,而不影响定常解。式(6.28)中的参数β一般设为1,从而得到一阶精度时间离散,$\beta = 1/2$时得到二阶精度时间离散,但不建议设置为二阶精度,因为$\beta = 1$的格式更加鲁棒,并且时间精度对于定常问题不重要。

对于刚性控制方程组,隐式算子中必须包含源项。这很自然地由式(6.27)的残差线性化可得到,从而产生与式(6.10)相同的公式。如文献[12]所示,如果$\partial \boldsymbol{Q}/\partial \boldsymbol{W}$的特征值都为负或零,则上述隐式格式(式(6.28))在任何时间步长都保持数值稳定。

线性方程组(6.28)的求解需要对隐式算子求逆,即对一个非常大的矩阵求逆,原则上,可以通过两种方式实现矩阵求逆。第一种方法是使用高斯消元法或

某个直接稀疏矩阵方法[21-22]，对矩阵直接求逆。然而，由于存储量过大，且计算量非常大，这种方法不适合实际问题[23]。

隐式算子矩阵求逆的第二种方法是迭代方法。3.2.2节中提到了使用最广泛的迭代方法，它大致可分为两类。第一类方法是将隐式算子分解成几个部分——这个过程称为因式分解，因子的构造方式使其比原来的隐式算子矩阵更容易求逆。第二类方法是采用Krylov子空间方法进行隐式算子矩阵求逆，通常通过设$\Delta t \to \infty$，将隐式时间推进格式（式(6.28)）转化为牛顿法，该方法因此被称为Newton-Krylov方法。

下面，将首先讨论隐式算子的矩阵结构，其次研究式(6.28)中通量雅可比矩阵$\partial \boldsymbol{Q}/\partial \boldsymbol{W}$计算的各种方法，此外还将详细介绍三种最流行的迭代方法，最后讨论隐式龙格-库塔法。

6.2.1 隐式算子的矩阵形式

根据式(6.27)，可将残差$\boldsymbol{R}^{n+1}$的线性化写成如下形式：

$$\boldsymbol{R}^{n+1} \approx \boldsymbol{R}^{n} + \sum_{m=1}^{N_{\mathrm{F}}} \left\{ \frac{\partial}{\partial \boldsymbol{W}} \left[(\boldsymbol{F}_{\mathrm{c}} - \boldsymbol{F}_{\mathrm{v}})_m \Delta S_m \right] \Delta \boldsymbol{W}^n \right\} - \frac{\partial (\Omega \boldsymbol{Q})}{\partial \boldsymbol{W}} \Delta \boldsymbol{W}^n \tag{6.29}$$

式中：N_{F}为控制体Ω的面数（参见式(4.2)或式(5.2)）。因此，通量雅可比为

$$\frac{\partial \boldsymbol{R}}{\partial \boldsymbol{W}} = \sum_{m=1}^{N_{\mathrm{F}}} \frac{\partial (\boldsymbol{F}_{\mathrm{c}})_m}{\partial \boldsymbol{W}} \Delta S_m - \sum_{m=1}^{N_{\mathrm{F}}} \frac{\partial (\boldsymbol{F}_{\mathrm{v}})_m}{\partial \boldsymbol{W}} \Delta S_m - \frac{\partial (\Omega \boldsymbol{Q})}{\partial \boldsymbol{W}} \tag{6.30}$$

需要强调的是，必须将通量雅可比视为作用于更新解$\Delta \boldsymbol{W}$的一个算子。如上所述，式(6.30)中的对流通量和黏性通量不一定与显式算子中的通量相同。

由于结构网格和非结构网格上的隐式算子之间存在显著差异，下面将分别探讨结构网格和非结构网格的隐式算子。

1. 结构网格的隐式算子

为了推导出式(6.28)中系统矩阵的形式，考虑如图6.1中的1D网格。另外还假设，空间离散基于二次控制体的格点格式（4.2.3节），采用简单的通量平均（见图4.8）计算通量。不考虑黏性通量时，残差为

$$\boldsymbol{R}_i^n = (\boldsymbol{F}_{\mathrm{c}})_{i+1/2} \Delta S_{i+1/2} + (\boldsymbol{F}_{\mathrm{c}})_{i-1/2} \Delta S_{i-1/2} - \Omega_i \boldsymbol{Q}_i \tag{6.31}$$

此外，式(6.30)中面$m=i+1/2$处对流通量的导数可表示为

$$\begin{aligned} \frac{\partial (\boldsymbol{F}_{\mathrm{c}})_{i+1/2}}{\partial \boldsymbol{W}} &= \frac{\partial}{\partial \boldsymbol{W}} \left\{ \frac{1}{2} \left[\boldsymbol{F}_{\mathrm{c}}(\boldsymbol{W}_{i+1}^n) + \boldsymbol{F}_{\mathrm{c}}(\boldsymbol{W}_i^n) \right] \right\} \\ &= \frac{1}{2} \left[(\boldsymbol{A}_{\mathrm{c}})_{i+1} + (\boldsymbol{A}_{\mathrm{c}})_i \right] \end{aligned} \tag{6.32}$$

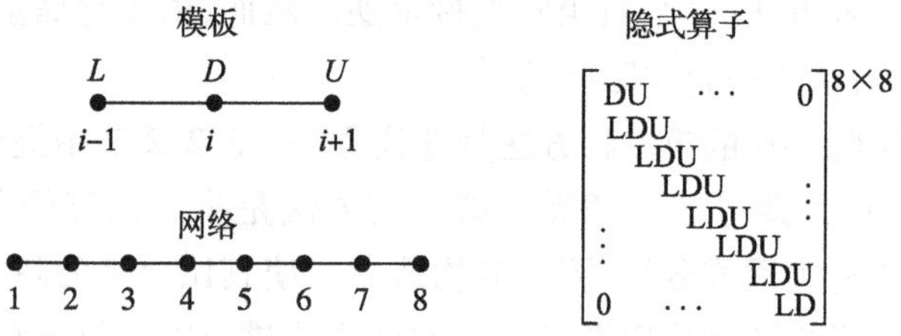

图 6.1 1D 结构网格和相关的三点模板隐式算子矩阵

式中:$\boldsymbol{A}_c$ 为对流通量雅可比矩阵(A.9 节)。因此,根据式(6.29)、式(6.31)和式(6.32),$t+\Delta t$ 时刻的残差近似为

$$\begin{aligned}\boldsymbol{R}_i^{n+1} \approx \boldsymbol{R}_i^n + \frac{1}{2}[(\boldsymbol{A}_c)_{i+1}\Delta\boldsymbol{W}_{i+1}^n + (\boldsymbol{A}_c)_i\boldsymbol{W}_i^n]\Delta S_{i+1/2} + \\ \frac{1}{2}[(\boldsymbol{A}_c)_{i-1}\Delta\boldsymbol{W}_{i-1}^n + (\boldsymbol{A}_c)_i\boldsymbol{W}_i^n]\Delta S_{i-1/2} - \frac{\partial(\Omega_i\boldsymbol{Q}_i)}{\partial\boldsymbol{W}}\end{aligned} \tag{6.33}$$

最后,对于 $\beta=1$ 和归并的质量矩阵,由式(6.26)推导出隐式格式:

$$\begin{aligned}\left\{\frac{\Omega_i}{\Delta t_i}\boldsymbol{I} + \frac{1}{2}[(\boldsymbol{A}_c)_i\Delta S_{i+1/2} + (\boldsymbol{A}_c)_i\Delta S_{i-1/2}] - \frac{\partial(\Omega_i\boldsymbol{Q}_i)}{\partial\boldsymbol{W}} + \right. \\ \left.\frac{1}{2}[(\boldsymbol{A}_c)_{i+1}\Delta S_{i+1/2}] + \frac{1}{2}[(\boldsymbol{A}_c)_{i-1}\Delta S_{i-1/2}]\right\}\Delta\boldsymbol{W}^n = -\boldsymbol{R}_i^n\end{aligned} \tag{6.34}$$

式中:$\boldsymbol{I}$ 表示单位矩阵。应注意的是,每个矩阵 $\boldsymbol{A}_c$ 的单位法向矢量在控制体的相关面积 ΔS 的同一侧计算。隐式算子包含与空间离散相同的 3 点模板(节点 $i-1$、i 和 $i+1$)。为了使系统矩阵更为直观,将隐式算子中与中心节点 i、下风节点($i+1$)和迎风节点($i-1$)相关的项分别表示为 $\boldsymbol{D}$、$\boldsymbol{U}$ 和 $\boldsymbol{L}$,即

$$\boldsymbol{D} \equiv \frac{\Omega_i}{\Delta t_i}\boldsymbol{I} + \frac{1}{2}[(\boldsymbol{A}_c)_i\Delta S_{i+1/2} + (\boldsymbol{A}_c)_i\Delta S_{i-1/2}] - \frac{\partial(\Omega_i\boldsymbol{Q}_i)}{\partial\boldsymbol{W}}$$

$$\boldsymbol{U} \equiv \frac{1}{2}(\boldsymbol{A}_c)_{i+1}\Delta S_{i+1/2}$$

$$\boldsymbol{L} \equiv \frac{1}{2}(\boldsymbol{A}_c)_{i-1}\Delta S_{i-1/2}$$

分别对图 6.1 中所有 8 个网格节点写出式(6.34),得到图 6.1 右侧显示的 8×8 的块三对角矩阵。每个块的 $\boldsymbol{L}$、$\boldsymbol{D}$ 和 $\boldsymbol{U}$ 代表 1D 情形的一个 3×3 矩阵(因为 1D 情形有三个守恒方程)。

同样的思路适用于多维情形,例如,如果 2D 空间离散采用图 6.2 所示的 5 点模板,则将得到一个块五对角矩阵,如图 6.2 的右侧所示。对节点进行排序,使得索引 i 的运行速度比索引 j 的要快(对应于 FORTRAN 中的 mat(i,j))。需

要注意的是,第二个非对角元素位于距离主对角元素 8 个元素处(i 方向的节点总数)。如果采用图 4.9(b)的 7 点模板进行空间离散,则 3D 情形的系统矩阵将变成块七对角矩阵。

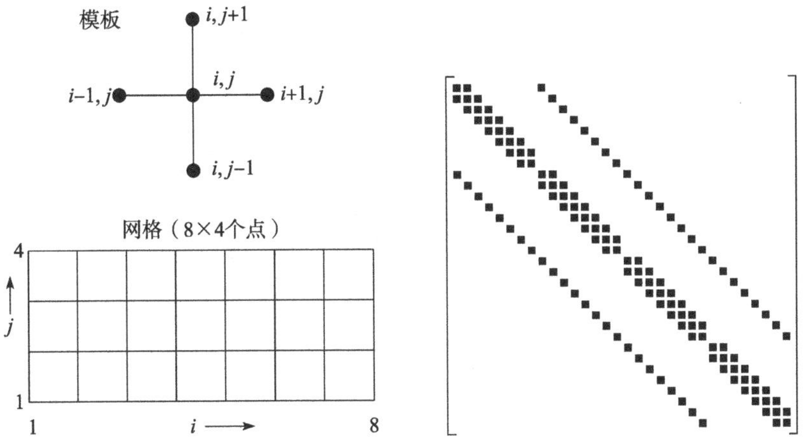

(a)　　　　　　　　(b)

图 6.2　2D 结构网格和与 5 点模板有关的隐式算子矩阵

(a) 结构网格;(b) 隐式算子矩阵。非零块矩阵用实心方块显示。

总之,对于结构网格,系统矩阵总是规则稀疏带状结构。应注意式(6.28)中的项

$$\frac{(\Omega \boldsymbol{M})_I}{\Delta t_I} \tag{6.35}$$

该项始终位于主对角线上,有时会造成一定的收敛困难,当时间步长变大,一些迭代的矩阵求逆方法(如高斯-赛德尔方法)可能会由于隐式算子矩阵对角占优减弱而导致求逆失败。

2. 非结构网格的隐式算子

从结构网格到非结构网格,系统矩阵的形式完全不同。这可以通过图 6.3 所示的小型非结构网格来说明,假设空间离散采用的是 5.2.1 节中介绍的格心格式,该格式只使用共面单元的流场变量(例如与 5.3.2 节和 5.3.3 节的一阶迎风格式类似),如单元 2 的模板包括单元 18、10 和 13,最终的系统矩阵显示在图 6.3(b)中。显然,由于网格单元(在中点-二次格式中为节点)通常按任意顺序编号,因此矩阵不可能具有规则形式,只有主对角线元素始终存在,它至少包含式(6.35),还可能包含源项的导数。

系统矩阵中非零元素的准随机分布是不受欢迎的,它使迭代求逆方法(如

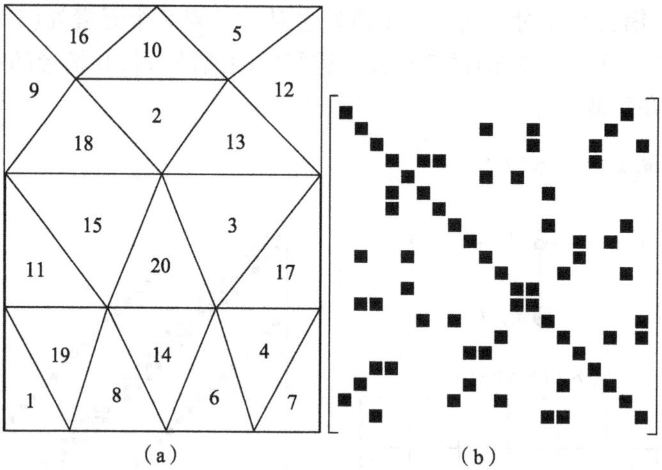

图 6.3　2D 非结构网格和相关的共面模板的隐式算子矩阵

(a) 非结构网格；(b) 隐式算子矩阵。实心方块表示非零块矩阵。

高斯-赛德尔方法）的收敛速度减慢，此外，ILU 等 Krylov 子空间方法的预处理技术也不能有效使用。因此开发了对单元（节点）重新排序的策略，从而大大降低了系统矩阵的带宽，即，使非零元素聚集在主对角线附近。最著名的重新排序策略是 RCM（Reverse-Cuthill-McKee）算法[24-25]。图 6.4 显示了将 RCM 算法应用于图 6.3 的示例网格时，最终的单元编号和系统矩阵，可以看出，矩阵的带宽显著减小，矩阵具有更规则的结构形式。

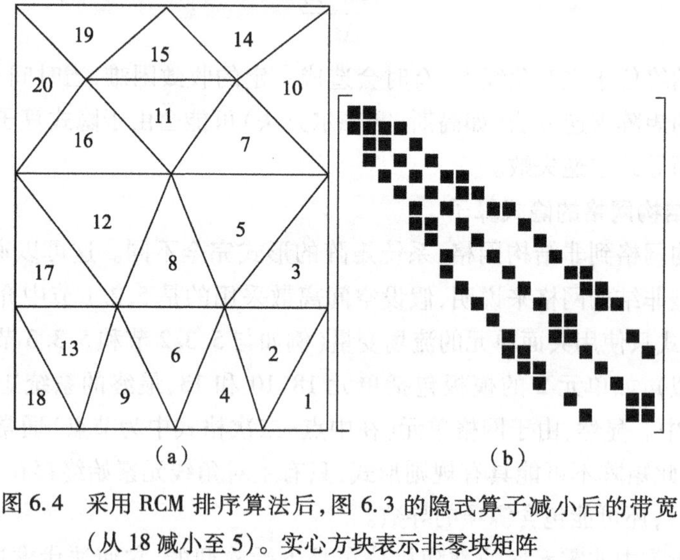

图 6.4　采用 RCM 排序算法后，图 6.3 的隐式算子减小后的带宽（从 18 减小至 5）。实心方块表示非零块矩阵

为了使缓存丢失最小化或允许数值格式矢量化,还开发了其他几种排序策略,文献[26]对这些排序策略进行了概述。

6.2.2 通量雅可比矩阵的计算

根据基本空间离散格式的类型,式(6.28)中的通量雅可比矩阵 $\partial \boldsymbol{R}/\partial \boldsymbol{W}$ 的解析计算可能变得非常复杂,甚至不可能得出结果。为了使这些概念更加清晰,我们将推导无黏流的通量雅可比矩阵,然后探讨向纳维-斯托克斯方程组的推广。

1. 中心格式

对于中心离散格式,很容易构造通量雅可比矩阵。正如我们在图6.1的例子中看到的,$\partial \boldsymbol{R}/\partial \boldsymbol{W}$ 由对流通量雅可比矩阵(式(6.34))组成,它可以解析地推导出来(见A.9节)。人工黏性通常以简化形式包含,不考虑非线性压力传感器(式(4.55))。6.2.3节将讨论有关人工黏性的问题。

2. 矢通量分裂格式

当使用一种矢通量分裂格式(4.3.2节)作为推导的基础时,通量雅可比矩阵的计算变得更加复杂,下面以Steger和Warming[27]的矢通量分裂格式为例进行说明。研究表明[28],对于隐式算子的各种迎风离散,Steger-Warming分裂比诸如van Leer的矢通量分裂格式(式(4.60))更好。

Steger-Warming矢通量分裂格式的基本思想是将对流通量分裂为正负两部分,即

$$\boldsymbol{F}_{\mathrm{c}}=\boldsymbol{F}_{\mathrm{c}}^{+}+\boldsymbol{F}_{\mathrm{c}}^{-} \tag{6.36}$$

其中正负通量定义为

$$\boldsymbol{F}_{\mathrm{c}}^{\pm}=\boldsymbol{A}_{\mathrm{SW}}^{\pm}\boldsymbol{W}=(\boldsymbol{T}\boldsymbol{\Lambda}^{\pm}\boldsymbol{T}^{-1})\boldsymbol{W} \tag{6.37}$$

式中:$\boldsymbol{A}_{\mathrm{SW}}^{\pm}$ 为正/负Steger-Warming通量分裂雅可比矩阵;$\boldsymbol{T}$ 表示右特征矢量矩阵;$\boldsymbol{T}^{-1}$ 表示左特征矢量矩阵;$\boldsymbol{\Lambda}^{\pm}$ 表示正/负特征值对角矩阵(见A.11节)。特征值矩阵定义为[27]

$$\boldsymbol{\Lambda}^{\pm}=\frac{1}{2}(\boldsymbol{\Lambda}_{\mathrm{c}}\pm|\boldsymbol{\Lambda}_{\mathrm{c}}|) \tag{6.38}$$

其中,$\boldsymbol{\Lambda}_{\mathrm{c}}$ 由式(A.84)给出。

利用式(6.36)中定义的对流通量分裂,得到通量雅可比矩阵与式(6.29)中更新 $\Delta \boldsymbol{W}^{n}$ 的乘积:

$$\frac{\partial \boldsymbol{R}_{I}}{\partial \boldsymbol{W}}\Delta \boldsymbol{W}^{n}=\sum_{m=1}^{N_{\mathrm{F}}}\left[\frac{\partial(\boldsymbol{F}_{\mathrm{c}}^{+}\Delta S)_{m}}{\partial \boldsymbol{W}_{\mathrm{L},m}}\Delta \boldsymbol{W}_{\mathrm{L},m}^{n}+\frac{\partial(\boldsymbol{F}_{\mathrm{c}}^{-}\Delta S)_{m}}{\partial \boldsymbol{W}_{\mathrm{R},m}}\Delta \boldsymbol{W}_{\mathrm{R},m}^{n}\right] \tag{6.39}$$

其中,$\Delta \boldsymbol{W}_{\mathrm{L},m}^{n}$ 和 $\Delta \boldsymbol{W}_{\mathrm{R},m}^{n}$ 分别表示在表面 m 处左右状态的更新。在结构网格上,面左右状态可以通过 MUSCL 方法进行计算(式(4.46));在非结构网格上,可以采用 5.3.3 节讨论的重构方法计算面左右状态。然而,随着精度的提高,模板变得更宽,导致系统矩阵的带宽更大,因此,为了降低数值复杂性,式(6.39)中通常只采用一阶精度近似。作为折中方案,以更高的精度重构面左右状态,但保留一阶方法的模板来计算导数[29]。

继续讨论式(6.39)中导数 $\partial \boldsymbol{F}_{\mathrm{c}}^{\pm}/\partial \boldsymbol{W}$ 的计算。在面 m 处的正通量为

$$\frac{\partial(\boldsymbol{F}_{\mathrm{c}}^{+}\Delta S)_m}{\partial \boldsymbol{W}_{\mathrm{L},m}}=\frac{\partial}{\partial \boldsymbol{W}_{\mathrm{L},m}}[(\boldsymbol{A}_{\mathrm{SW}}^{+}\boldsymbol{W})_{\mathrm{L},m}\Delta S_m]$$

根据式(6.37)和式(6.38),上式变为

$$\begin{aligned}\frac{\partial(\boldsymbol{F}_{\mathrm{c}}^{+}\Delta S)_m}{\partial \boldsymbol{W}_{\mathrm{L},m}}=&\frac{\Delta S_m}{2}[(\boldsymbol{A}_{\mathrm{c}})_{\mathrm{L},m}+|(\boldsymbol{A}_{\mathrm{c}})_{\mathrm{L},m}|]+\\&\frac{\Delta S_m}{2}\left[\frac{\partial(\boldsymbol{A}_{\mathrm{c}})_{\mathrm{L},m}}{\partial \boldsymbol{W}_{\mathrm{L},m}}+\frac{\partial|(\boldsymbol{A}_{\mathrm{c}})_{\mathrm{L},m}|}{\partial \boldsymbol{W}_{\mathrm{L},m}}\right]\boldsymbol{W}_{\mathrm{L},m}^{n}\end{aligned}\tag{6.40}$$

对于负通量,也可以得到类似于式(6.40)的表达式。如上所见,式(6.40)的第一项由对流通量雅可比矩阵组成(见 A.9 节),因此不存在什么困难。但第二项包含矩阵元素的导数,虽然可以通过手工计算或使用符号代数软件包解析地获得,但这将使得程序庞大,且计算效率低下[30]。或者,我们假设矩阵 $\boldsymbol{A}_{\mathrm{c}}$ 在当地为常数,因而式(6.40)中的第二项可以忽略,但是根据隐式方法的类型,这样做可能会严重限制 CFL 数的大小[31]。

还可以用其他方法来计算式(6.39)中的导数 $\partial \boldsymbol{F}_{\mathrm{c}}^{\pm}/\partial \boldsymbol{W}$,如源代码的自动微分(例如,使用 ADIFOR[32]、ADIC[33]、OpenAD[34]或 Adept[35]),或有限差分法(参见文献[23,30])。矢量 $\boldsymbol{F}$ 的第 i 个分量相对于因变量 $\boldsymbol{X}$ 的第 j 个分量的导数,可以近似为

$$\frac{\partial f_i}{\partial x_j}\approx\frac{f_i(\boldsymbol{X}+h_j\boldsymbol{e}^j)-f_i(\boldsymbol{X})}{h_j}\tag{6.41}$$

式中:$\boldsymbol{e}^j$为第 j 个标准基矢量。Dennis 和 Schnabel[36]建议步长 h_j 为

$$h_j=\sqrt{\epsilon}\max\{|x_j|,\mathrm{typ}x_j\}\,\mathrm{sign}(x_j)\tag{6.42}$$

式中,ϵ 为机器精度;$\mathrm{typ}x_j$ 为 x_j 的典型尺寸。

读者也可参考文献[37-38],了解关于雅可比矩阵高效数值计算的建议。

3. 通量差分裂格式

由 Roe 提出的通量差分分裂格式(见 4.3.3 节,式(4.91)),可以将通量雅

可比矩阵与式(6.29)中的更新 ΔW^n 的乘积写成

$$\frac{\partial \boldsymbol{R}_I}{\partial \boldsymbol{W}}\Delta \boldsymbol{W}^n = \sum_{m=1}^{N_F}\frac{\Delta S_m}{2}\Big\{(\boldsymbol{A}_c)_{L,m}\Delta \boldsymbol{W}_{L,m}^n+(\boldsymbol{A}_c)_{R,m}\Delta \boldsymbol{W}_{R,m}^n- \frac{\partial}{\partial \boldsymbol{W}_{L,m}}[\,|\boldsymbol{A}_{Roe}|_m(\boldsymbol{W}_{R,m}^n-\boldsymbol{W}_{L,m}^n)\,]\Delta \boldsymbol{W}_{L,m}^n- \frac{\partial}{\partial \boldsymbol{W}_{R,m}}[\,|\boldsymbol{A}_{Roe}|_m(\boldsymbol{W}_{R,m}^n-\boldsymbol{W}_{L,m}^n)\,]\Delta \boldsymbol{W}_{R,m}^n\Big\} \tag{6.43}$$

与矢通量分裂类似,式(6.43)包含对流通量雅可比矩阵以及 Roe 矩阵 $\boldsymbol{A}_{Roe}$ 的导数,文献[31]对导数进行了介绍。如上文所述,由于相应的公式非常复杂(参见文献[30]),因此,最好是对 $\partial \boldsymbol{F}_c/\partial \boldsymbol{W}$ 进行数值计算。不过,也可以假设 Roe 矩阵为当地常数,对式(6.43)进行简化[39]:

$$\frac{\partial \boldsymbol{R}_I}{\partial \boldsymbol{W}}\Delta \boldsymbol{W}^n \approx \sum_{m=1}^{N_F}\frac{\Delta S_m}{2}\{(\boldsymbol{A}_c)_{L,m}\Delta \boldsymbol{W}_{L,m}^n + (\boldsymbol{A}_c)_{R,m}\Delta \boldsymbol{W}_{R,m}^n - |\boldsymbol{A}_{Roe}|_m(\Delta \boldsymbol{W}_{R,m}^n - \Delta \boldsymbol{W}_{L,m}^n)\} \tag{6.44}$$

与 Steger-Warming 矢通量分裂格式相比,上面的近似线性化(式(6.44))仅使隐式格式的性能略微有所降低[31]。

4. 黏性流动

对于纳维-斯托克斯方程组,还必须考虑隐式算子中的黏性通量。导数 $\partial \boldsymbol{F}_v/\partial \boldsymbol{W}$,即式(6.30)中的黏性通量雅可比矩阵,通常不容易得到,由于黏性矢通量包含流动变量的导数,会出现其他复杂情况,因此必须使用有限差分法(式(6.41))计算黏性通量雅可比矩阵,或者使用简化公式。

对于纳维-斯托克斯方程组采用 TSL 近似的情况(参见 2.4.3 节和 A.6 节),通过假设动力学黏性系数和热传导系数为当地常数,可以解析地求得黏性通量雅可比矩阵。由 A.10 节讨论可知,式(6.29)中与黏性通量有关的项为

$$\frac{\partial(\boldsymbol{F}_v\Delta S)_m}{\partial \boldsymbol{W}}\Delta \boldsymbol{W}^n \approx [\,(\boldsymbol{A}_v^*)_{R,m}\Delta \boldsymbol{W}_{R,m}^n-(\boldsymbol{A}_v^*)_{L,m}\Delta \boldsymbol{W}_{L,m}^n\,]\Delta S_m \tag{6.45}$$

式中:$\boldsymbol{A}_v^*$ 表示式(A.71)或式(A.75)给出的黏性通量雅可比矩阵,但无空间算子 $\partial_\psi()$(参见式(A.74)和式(A.79))。使用所有变量的左侧或右侧状态计算雅可比矩阵(动力学黏性系数除外,它由算术平均值确定)。需要指出的是,如果以一阶精度计算左右侧状态,则式(6.45)使隐式算子为二阶中心差分近似。

6.2.3 ADI 方法

ADI 方法是最早的隐式迭代方法之一[40]。ADI 方法只能在结构网格上实现,它是基于式(6.28)中的隐式算子的近似分裂(或称为分解),分裂为两个(2D)或三个(3D)因式,每个因式只包含计算空间中某一特定方向的对流通量和黏性通量的线性化。3D 情况下的公式[40-41]为

$$\left\{\frac{\Omega}{\Delta t}\boldsymbol{I}+\frac{\partial[(\boldsymbol{F}_{\mathrm{c}}^{I}-\boldsymbol{F}_{\mathrm{v}}^{I})\Delta S^{I}]_{i+1/2}}{\partial \boldsymbol{W}}+\frac{\partial[(\boldsymbol{F}_{\mathrm{c}}^{I}-\boldsymbol{F}_{\mathrm{v}}^{I})\Delta S^{I}]_{i-1/2}}{\partial \boldsymbol{W}}\right\}$$
$$\left\{\frac{\Omega}{\Delta t}\boldsymbol{I}+\frac{\partial[(\boldsymbol{F}_{\mathrm{c}}^{J}-\boldsymbol{F}_{\mathrm{v}}^{J})\Delta S^{J}]_{j+1/2}}{\partial \boldsymbol{W}}+\frac{\partial[(\boldsymbol{F}_{\mathrm{c}}^{J}-\boldsymbol{F}_{\mathrm{v}}^{J})\Delta S^{J}]_{j-1/2}}{\partial \boldsymbol{W}}\right\} \tag{6.46}$$
$$\left\{\frac{\Omega}{\Delta t}\boldsymbol{I}+\frac{\partial[(\boldsymbol{F}_{\mathrm{c}}^{K}-\boldsymbol{F}_{\mathrm{v}}^{K})\Delta S^{K}]_{k+1/2}}{\partial \boldsymbol{W}}+\frac{\partial[(\boldsymbol{F}_{\mathrm{c}}^{K}-\boldsymbol{F}_{\mathrm{v}}^{K})\Delta S^{K}]_{k-1/2}}{\partial \boldsymbol{W}}-\frac{\partial(\Omega\boldsymbol{Q})}{\partial \boldsymbol{W}}\right\}\Delta\boldsymbol{W}^{n}=-\boldsymbol{R}^{n}$$

为简单起见,式(6.46)中省略了不必要的节点索引 i,j,k。此外,式(6.46)中的上标 I,J,K 用来标记矢通量或面积对应的计算空间的坐标,如 $\Delta S^{I}_{i+1/2,j,k}$ 对应于图 4.1(b)中的 ΔS_2。用单元索引代替节点索引,则式(6.46)中的格式可以应用于格心型离散。

式(6.46)中对流通量和黏性通量的导数可按照 6.2.2 节讨论的方法进行计算。ADI 方法传统上是与带人工耗散的中心格式耦合的,在这种情况下,类似于式(6.34)的公式对每个因式都成立(当然,源项的线性化只包含在一个因式中)。为了获得鲁棒高效的格式,需要在隐式算子中包含人工耗散项的线性化[42-44],通过把谱半径和耗散系数看作与 $\boldsymbol{W}$ 无关的量,可以简化人工耗散项的线性化,根据式(4.48),I 方向的因子可简化为

$$\left\{\frac{\Omega}{\Delta t}\boldsymbol{I}+\frac{\partial[(\boldsymbol{F}_{\mathrm{c}}^{I}-\boldsymbol{F}_{\mathrm{v}}^{I})\Delta S^{I}]_{i+1/2}}{\partial \boldsymbol{W}}+\frac{\partial[(\boldsymbol{F}_{\mathrm{c}}^{I}-\boldsymbol{F}_{\mathrm{v}}^{I})\Delta S^{I}]_{i-1/2}}{\partial \boldsymbol{W}}-\frac{\partial(\boldsymbol{D}_{\mathrm{IM}}^{I}\Delta S^{I})_{i+1/2}}{\partial \boldsymbol{W}}-\frac{\partial(\boldsymbol{D}_{\mathrm{IM}}^{I}\Delta S^{I})_{i-1/2}}{\partial \boldsymbol{W}}\right\}$$

其中,JST 格式的隐式人工耗散项(参见式(4.50))为

$$\frac{\partial(\boldsymbol{D}_{\mathrm{IM}}^{I})_{i+1/2}}{\partial \boldsymbol{W}}\Delta\boldsymbol{W}^{n}\approx\hat{\Lambda}^{\mathrm{S}}_{i+1/2}[(\epsilon_{\mathrm{IM}}^{(2)})_{i+1/2}(\Delta\boldsymbol{W}_{i+1}^{n}-\Delta\boldsymbol{W}_{i}^{n})-(\epsilon_{\mathrm{IM}}^{(4)})_{i+1/2}(\Delta\boldsymbol{W}_{i+2}^{n}-3\Delta\boldsymbol{W}_{i+1}^{n}+3\Delta\boldsymbol{W}_{i}^{n}-\Delta\boldsymbol{W}_{i-1}^{n})] \tag{6.47}$$

以类似的方式,可定义其他方向上的隐式耗散项。也可在隐式算子中只保留二阶差分[43-44],这使每个因式的矩阵从块五对角阵简化为块三对角阵(图 6.1),但是,如文献[44]所指出的,这样会限制格式的稳定性。

式(6.46)中的隐式算子求逆分三步(2D 情况下为两步)进行,即

$$
\begin{aligned}
(\boldsymbol{D}^I+\boldsymbol{L}^I+\boldsymbol{U}^I)\Delta\boldsymbol{W}^{(1)} &= -\boldsymbol{R}^n \\
(\boldsymbol{D}^J+\boldsymbol{L}^J+\boldsymbol{U}^J)\Delta\boldsymbol{W}^{(2)} &= \Delta\boldsymbol{W}^{(1)} \\
(\boldsymbol{D}^K+\boldsymbol{L}^K+\boldsymbol{U}^K)\Delta\boldsymbol{W}^{n} &= \Delta\boldsymbol{W}^{(2)}
\end{aligned}
\tag{6.48}
$$

其中,$\boldsymbol{D}$、$\boldsymbol{L}$ 和 $\boldsymbol{U}$ 分别表示对角阵、下三角阵和上三角阵。每一步都需要对块三对角阵或块五对角阵(如果式(6.47)中的 $\epsilon_{\mathrm{IM}}^{(4)}>0$)进行求逆,采用直接求解方法即可完成。为了减少计算量,Pulliam 和 Chaussee[45] 提出了 ADI 方法的对角化形式,将(由对流通量雅可比矩阵和黏性通量雅可比矩阵组成的)块矩阵转化为对角矩阵,只需对非块三对角或五对角矩阵进行求逆,从而大大节省了计算量和存储量[43]。尽管严格来说对角化只对欧拉方程组有效,但它也可用于黏性流动,在隐式算子中略去黏性通量(例如,如式(6.45))的线性化,或者用黏性特征值(式(6.19))来近似。

隐式算子的分裂引入了因式分解误差,它是式(6.28)中基准格式的隐式算子与因式分解运算之间的差异,对于 ADI 方法,该误差项由因子$(\Delta t)^N$ 表征,其中 N 为空间维数。该误差项导致 ADI 方法在 3D 情况下无法保持无条件稳定[46]。然而,如果隐式算子中包含了人工黏性的四阶差分,则稳定性会得到改善[44]。事实上,ADI 方法已成功用于各种 3D 问题的求解[43,47]。Rosenfeld 和 Yassour[48] 提出了在多块网格上 ADI 方法的一种令人关注的实现方法,该方法对块边界进行隐式处理。

采用与 6.1.4 节所述相同的方法,使用式(6.14)计算时间步长 Δt,最佳 CFL 数在 20~50 之间,具体取值取决于特定流动的物理特性。

6.2.4 LU-SGS 方法

隐式 LU-SGS 方法,又称 LU-SSOR 方法,因其数值复杂性低、内存需求适中(均与显式多步格式相当)而得到广泛应用。此外,LU-SGS 方法还易于在矢量计算机和并行计算机上实现,可以用于结构网格和非结构网格。

LU-SGS 方法源自 Jameson 和 Turkel[49] 的工作,他们将隐式算子分解成上下对角占优的因子。LU-SGS 方法本身是由 Yoon 和 Jameson[50-52] 作为一种松弛方法引入,用于求解式(6.28)中的无分裂隐式格式。Rieger 和 Jameson[53] 对其进一步发展并应用于 3D 黏性流动。此后,许多研究人员将 LU-SGS 方法应用于结构网格[31,54,56-63] 和非结构网格[64-71] 的黏性流动,它还常用于化学反应流动的模拟[72-78]。

LU-SGS 方法使用一阶精度 Steger-Warming 矢通量分裂格式(见 6.2.2 节)对式(6.29)中的对流通量进行线性化,无论显式算子如何离散,隐式算子的线性化始终保持不变。LU-SGS 方法将式(6.28)中的隐式算子分解为以下三部分:

$$(\boldsymbol{D}+\boldsymbol{L})\boldsymbol{D}^{-1}(\boldsymbol{D}+\boldsymbol{U})\Delta\boldsymbol{W}^n=-\boldsymbol{R}_I^n \tag{6.49}$$

上述因式的构造满足 $\boldsymbol{L}$ 为严格下三角矩阵，$\boldsymbol{U}$ 为严格上三角矩阵，而 $\boldsymbol{D}$ 为对角矩阵。值得注意的是，因式数量始终保持不变，与空间维数无关。

通过前扫描和后扫描，对式(6.49)中 LU-SGS 方法的系统矩阵求逆，即

$$\begin{aligned}(\boldsymbol{D}+\boldsymbol{L})\Delta\boldsymbol{W}^{(1)}&=-\boldsymbol{R}_I^n\\(\boldsymbol{D}+\boldsymbol{U})\Delta\boldsymbol{W}^n&=\boldsymbol{D}\Delta\boldsymbol{W}_I^{(1)}\end{aligned} \tag{6.50}$$

其中，$\boldsymbol{W}^{n+1}=\boldsymbol{W}^n+\Delta\boldsymbol{W}^n$。结构网格和非结构网格的算子 $\boldsymbol{L}$,$\boldsymbol{D}$ 和 $\boldsymbol{U}$ 以及求逆过程，在某些方面有所不同。因此，下面将分别讨论结构网格和非结构网格的 LU-SGS 方法。

1. 结构网格的 LU-SGS

在结构网格上，算子 $\boldsymbol{L}$、$\boldsymbol{D}$ 和 $\boldsymbol{U}$ 定义为（参见文献[50-52,55,73]）

$$\begin{aligned}\boldsymbol{L}=&(\boldsymbol{A}^++\boldsymbol{A}_\mathrm{v})_{i-1}\Delta S^I_{i-1/2}+(\boldsymbol{A}^++\boldsymbol{A}_\mathrm{v})_{j-1}\Delta S^J_{j-1/2}+\\&(\boldsymbol{A}^++\boldsymbol{A}_\mathrm{v})_{k-1}\Delta S^K_{k-1/2}\\\boldsymbol{U}=&(\boldsymbol{A}^--\boldsymbol{A}_\mathrm{v})_{i+1}\Delta S^I_{i+1/2}+(\boldsymbol{A}^--\boldsymbol{A}_\mathrm{v})_{j+1}\Delta S^J_{j+1/2}+\\&(\boldsymbol{A}^--\boldsymbol{A}_\mathrm{v})_{k+1}\Delta S^K_{k+1/2}\\\boldsymbol{D}=&\frac{\Omega}{\Delta t}\boldsymbol{I}+(\boldsymbol{A}^--\boldsymbol{A}_\mathrm{v})\Delta S^I_{i-1/2}+(\boldsymbol{A}^--\boldsymbol{A}_\mathrm{v})\Delta S^J_{j-1/2}+\\&(\boldsymbol{A}^--\boldsymbol{A}_\mathrm{v})\Delta S^K_{k-1/2}+(\boldsymbol{A}^++\boldsymbol{A}_\mathrm{v})\Delta S^I_{i+1/2}+\\&(\boldsymbol{A}^++\boldsymbol{A}_\mathrm{v})\Delta S^J_{j+1/2}+(\boldsymbol{A}^++\boldsymbol{A}_\mathrm{v})\Delta S^K_{k+1/2}-\frac{\partial(\Omega\boldsymbol{Q})}{\partial\boldsymbol{W}}\end{aligned} \tag{6.51}$$

为了更好的可读性，式(6.51)中只显示了与 i,j,k 不同的节点索引（格心格式中为网格单元索引）。ΔS 的上标 I,J,K 表示在计算空间中的方向。正/负通量雅可比矩阵 $\boldsymbol{A}^\pm$ 和黏性通量雅可比矩阵 $\boldsymbol{A}_\mathrm{v}$ 中的单位法向矢量的计算，与相关面的面积 ΔS 一样，在控制体的同一侧进行，需要注意的是，在此假设单位法向矢量指向控制体的外部，与之相反，在许多文献中，都假设控制体相对两侧的单位法向矢量指向同一方向。

式(6.51)中的黏性通量雅可比矩阵通过数值计算，或者用其 TSL 近似（对应于式(6.45)）代替，无论边界层的实际方向如何，在所有计算坐标中都可以采用 TSL 近似。另一种更进一步的简化方法，如文献[65]中所建议的那样，用黏性谱半径（式(6.19)）替换黏性通量雅可比矩阵，即 $\boldsymbol{A}_\mathrm{v}\Delta S\approx\hat{\Lambda}_\mathrm{v}$。

分裂的对流通量雅可比矩阵 $\boldsymbol{A}^\pm$ 是这样构造的，它使得矩阵 $\boldsymbol{A}^+$ 的特征值均为非负，而矩阵 $\boldsymbol{A}^-$ 的特征值均为非正。通常，矩阵 $\boldsymbol{A}^\pm$ 定义为[49]

$$\boldsymbol{A}^\pm\Delta S=\frac{1}{2}(\boldsymbol{A}_\mathrm{c}\Delta S\pm r_\mathrm{A}\boldsymbol{I}),\quad r_\mathrm{A}=\omega\hat{\Lambda}_\mathrm{c} \tag{6.52}$$

式中：$\boldsymbol{A}_{\mathrm{c}}$ 为对流通量雅可比矩阵（A.9 节）；$\hat{\Lambda}_{\mathrm{c}}$ 为对流通量雅可比矩阵的谱半径（由式(4.53)或式(6.15)给出）。请注意当忽略 $\boldsymbol{A}_{\mathrm{c}}$ 的导数时近似式(6.52)和式(6.40)之间的相似之处。式(6.52)中的系数 ω 表示超松弛参数，它还决定隐式格式的耗散量，进而会影响格式的收敛性，系数 ω 的可选择范围为 $1<\omega\leqslant 2$，较高的 ω 值会使 LU-SGS 格式的鲁棒性增加，但可能会减慢收敛到稳态的速度。式(6.52)中雅可比矩阵 $\boldsymbol{A}^{\pm}$ 的定义确保了其为对角占优系统矩阵，这对迭代求逆过程（式(6.50)）的效率和鲁棒性非常重要。

根据式(6.52)的分裂，结合平均的面矢量，可得简化的对角算子 $\boldsymbol{D}$，即

$$\boldsymbol{D}=\left[\frac{\Omega}{\Delta t}+\omega(\hat{\Lambda}_{\mathrm{c}}^{I}+\hat{\Lambda}_{\mathrm{c}}^{J}+\hat{\Lambda}_{\mathrm{c}}^{K})\right]\boldsymbol{I}+2(\boldsymbol{A}_{\mathrm{v}}^{I}\Delta S^{I}+\boldsymbol{A}_{\mathrm{v}}^{J}\Delta S^{J}+\boldsymbol{A}_{\mathrm{v}}^{K}\Delta S^{K})-\frac{\partial(\Omega \boldsymbol{Q})}{\partial \boldsymbol{W}} \tag{6.53}$$

对流通量雅可比矩阵的谱半径 $\hat{\Lambda}_{\mathrm{c}}$ 如式(6.15)所示。根据式(6.16)，分别在 I、J 或 K 方向上对面积和法向矢量进行平均。正如下面即将讨论到的，这种近似有助于显著减少运算量和内存需求。

LU-SGS 方法的一个显著特点是执行式(6.50)中的前扫描和后扫描的方式。对于 2D 情况，在计算空间中沿对角线 $i+j=$ 常数完成扫描，前扫描（式(6.50)的第一行）如图 6.5 所示。通过这种方式，$\boldsymbol{L}$ 和 $\boldsymbol{U}$ 算子中涉及的非对角项可以从扫描的前一部分获得。对于 3D 情况，在 $i+j+k=$ 常数平面上对隐式算子求逆，如图 6.6 所示。因此，LU-SGS 格式可写成

$$\begin{aligned}\boldsymbol{D}\Delta\boldsymbol{W}_{i,j,k}^{(1)}&=-\boldsymbol{R}_{i,j,k}^{n}-\boldsymbol{L}\Delta\boldsymbol{W}^{(1)}\\ \boldsymbol{D}\Delta\boldsymbol{W}_{i,j,k}^{n}&=\boldsymbol{D}\Delta\boldsymbol{W}_{i,j,k}^{(1)}-\boldsymbol{U}\Delta\boldsymbol{W}^{n}\end{aligned} \tag{6.54}$$

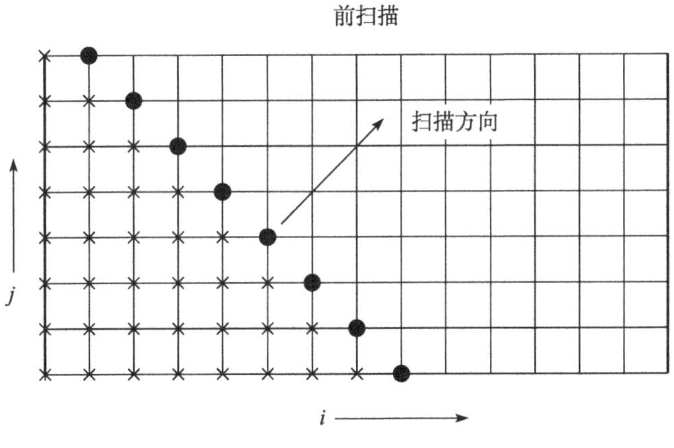

图 6.5　计算空间中 LU-SGS 方法的扫描方向

● 表示算子 D 当前求逆位置（$i+j=$常数的线）；× 表示 L 值已更新。

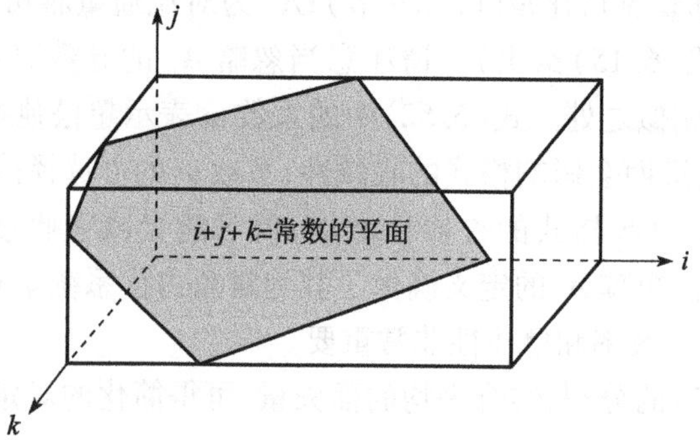

图 6.6 3D 隐式 LU-SGS 方法在计算空间中扫描的对角平面

由式(6.54)可见,唯一需要进行求逆的项是对角项 $\boldsymbol{D}$。因此,LU-SGS 方法将稀疏带状矩阵的求逆转化为块对角矩阵的求逆,此外,如果式(6.53)中的黏性通量雅可比矩阵用黏性谱半径近似,则算子 $\boldsymbol{D}$ 变成一个对角矩阵(除源项外)。因此,与其他隐式格式(如前面讨论的 ADI 方法)相比,LU-SGS 方法需要的计算量非常小。此外,对角算子的求逆可以针对对角平面的每个节点(单元)独立进行,这使得该方法易于矢量化。对角平面上节点/单元的索引可通过以下伪代码获得[79]

```
DO plane =1,nplanes
  DO k =1,kmax
    DO j =1,jmax
      DO i =1,imax
        IF (i +j +k =plane +2) store index
      ENDDO
    ENDDO
  ENDDO
ENDDO
```

对角平面数 $nplanes = imax + jmax + kmax - 2$。显然,可优化上述代码以获得更高的计算效率。

为了不对 $\boldsymbol{L}$ 和 $\boldsymbol{U}$ 中对流通量雅可比矩阵进行显式计算和存储,可用通量的泰勒级数展开来代替乘积 $\boldsymbol{A}^{\pm}\Delta\boldsymbol{W}^{n}$[53]。利用式(6.52),可写为

$$(\boldsymbol{A}^{\pm}\Delta S)\Delta\boldsymbol{W}^{n} \approx \frac{1}{2}(\Delta\boldsymbol{F}_{c}\Delta S \pm r_{A}\bar{I}\Delta\boldsymbol{W}^{n}) \tag{6.55}$$

其中,对流通量的修正量为

$$\Delta\boldsymbol{F}_{c} = \boldsymbol{F}_{c}^{n+1} - \boldsymbol{F}_{c}^{n} \tag{6.56}$$

由于沿着对角平面扫描，此时 $\boldsymbol{F}_{\mathrm{c}}^{n+1}$ 已知，所以可由式(6.55)给出对流通量的修正量简化形式，这进一步显著降低了 LU-SGS 方法的数值计算量。

可采用与 6.1.4 节所述相同的方法，使用式(6.14)计算时间步长 Δt。但是，如 Rieger 和 Jameson[53] 所述，式(6.49)中的隐式 LU-SGS 方法在 $\Delta t \to \infty$ 情况下是一种近似牛顿迭代，因此，在通常情况下，对于定常流动，实际使用的 CFL 数量级为 $10^4 \sim 10^6$，收敛性由超松弛参数 ω 控制。对于非定常流的模拟，可采用 6.3 节介绍的公式，另一种方法是使用文献[80]中描述的 LU-SGS 方法的改进版本。

2. 非结构网格的 LU-SGS

对于中点-二次格式，其算子为[65-67]

$$
\begin{aligned}
\boldsymbol{L} &= \sum_{j \in L(i)} \left[\boldsymbol{A}_j^{+} + (\boldsymbol{A}_{\mathrm{v}})_j\right] \Delta S_{ij} \\
\boldsymbol{U} &= \sum_{j \in U(i)} \left[\boldsymbol{A}_j^{-} - (\boldsymbol{A}_{\mathrm{v}})_j\right] \Delta S_{ij} \\
\boldsymbol{D} &= \frac{\Omega_i}{\Delta t_i}\boldsymbol{I} + \frac{\omega}{2}(\hat{\Lambda}_{\mathrm{c}})_i + \sum_{j=1}^{N_{\mathrm{F}}} (\boldsymbol{A}_{\mathrm{v}})_i \Delta S_{ij} - \frac{\partial(\Omega_i \boldsymbol{Q}_i)}{\partial \boldsymbol{W}}
\end{aligned}
\tag{6.57}
$$

其中，$L(i)$ 和 $U(i)$ 分别表示属于下三角矩阵和上三角矩阵的节点 i 的共线节点，ΔS_{ij} 表示与边 ij 相关面的面积(图 5.9)，N_{F} 表示控制体 Ω_i 的面数。对流通量的谱半径 $(\hat{\Lambda}_{\mathrm{c}})_i$ 由式(6.21)计算，黏性通量雅可比矩阵 $\boldsymbol{A}_{\mathrm{v}}$ 同样可用其谱半径近似[65]，在这种情况下，对角算子变为

$$
\boldsymbol{D} = \frac{\Omega_i}{\Delta t_i}\boldsymbol{I} + \frac{\omega}{2}(\hat{\Lambda}_{\mathrm{c}})_i + (\hat{\Lambda}_{\mathrm{v}})_i - \frac{\partial(\Omega_i \boldsymbol{Q}_i)}{\partial \boldsymbol{W}} \tag{6.58}
$$

其中，根据式(6.21)计算 $(\hat{\Lambda}_{\mathrm{v}})_i$。对于格心格式，可以得到类似于式(6.57)和式(6.58)的公式。主要的区别在于算子 $\boldsymbol{L}$ 和 $\boldsymbol{U}$ 的求和是在单元的面上进行，而不是在边上进行。

式(6.57)中的 $L(i)$ 和 $U(i)$ 的集合应满足与结构网格上的对角平面相同的功能。因此，有必要对节点(单元)分层[65]，以满足以下要求：

(1) 当前层的节点 i(单元 I)与以前更新的流动变量的层有关联，否则 LU-SGS 格式退化为雅可比迭代；

(2) 同一层中的节点(单元)互不相连，否则格式不能矢量化。

可以使用文献[65]介绍的过程为中点-二次格式生成网格层，文献[81-82]中提出了格心格式生成层的方法。

适当定义集合 $L(i)$ 和 $U(i)$，从而可以采取以下两步求逆过程[65-67]：

$$\begin{aligned} D\Delta W_i^{(1)} &= -R_i^n - \sum_{j\in L(i)} \frac{1}{2}[(\Delta F_c^{(1)})_j \Delta S_{ij} + (r_A^*)_j I \Delta W_j^{(1)}] \\ D\Delta W_i^n &= D\Delta W_i^{(1)} - \sum_{j\in U(i)} \frac{1}{2}[(\Delta F_c^n)_j \Delta S_{ij} - (r_A^*)_j I \Delta W_j^n] \end{aligned} \tag{6.59}$$

式中：黏性雅可比矩阵用其谱半径近似，而正/负雅可比矩阵根据式(6.55)进行线性化。此外，系数$(r_A^*)_j$定义为

$$(r_A^*)_j = \omega(|\boldsymbol{v}_j \cdot \boldsymbol{n}_{ij}| + c_j)\Delta S_{ij} + \frac{\Delta S_{ij}}{\|\boldsymbol{r}_j - \boldsymbol{r}_i\|_2}\left[\max\left(\frac{4}{3\rho_j}, \frac{\gamma_j}{\rho_j}\right)\left(\frac{\mu_L}{Pr_L} + \frac{\mu_T}{Pr_T}\right)_j\right] \tag{6.60}$$

式中：$\|\boldsymbol{r}_j - \boldsymbol{r}_i\|_2$为边$ij$的长度。

采用与显式格式（式(6.20)）相同的方法计算时间步长，但应忽略黏性特征值，因为黏性项已经包含在隐式算子中，所以不需要像显式格式那样减小时间步长。对于定常流动，CFL 数选择范围为$10^4 \sim 10^6$。利用超松弛参数ω可以调节 LU-SGS 方法的收敛速度和鲁棒性。

6.2.5 Newton-krylov 方法

首先，将式(6.28)给出的隐式格式改写为

$$J\Delta W^n = -R^n \tag{6.61}$$

其中，J表示隐式算子（系统矩阵）。正如我们已经看到的，J构成了一个大型稀疏矩阵，且通常为非对称矩阵。在前面的章节中讨论了将J分解成几个因式的两种方法，每个因式都比J本身更容易求逆，然而，由于因式分解误差（以及R^{n+1}的近似线性化），只能实现线性收敛到稳态。为了使非线性方程组的求解获得牛顿法的二次收敛性，必须满足四个条件：

(1) 残差线性化必须精确；

(2) 必须对J进行准确求逆；

(3) 时间步长必须满足$\Delta t \to \infty$；

(4) 在某种意义上，初始解必须接近最终解。

显然，必须克服的主要障碍是整个系统矩阵的线性化和求逆。

特别适合于求解大型线性方程组的一类迭代法是 Krylov 子空间法。人们提出了针对 CFD 中出现的矩阵求逆的几种方法，例如，CGS 方法[83]、Bi-CGSTAB 格式[84]或 TFQMR 法[85]。然而，最成功的 Krylov 子空间法是 GMRES 方法[86-87]，它最早由 Saad 和 Schulz 提出，一些研究人员对 GMRES 方法进行了改进和推广[88-91]。由于 GMRES 方法的流行，下面将重点介绍 GMRES 方法，但讨论的大部分内容也适用于其他 Krylov 子空间法。

1. GMRES 方法

正如在 3.2.2 节中提到的(另见 A.12 节),GMRES 方法使全局残差范数最小,即 $\| \boldsymbol{J}\Delta \boldsymbol{W}^n+\boldsymbol{R}^n \|$ 在 m 个标准正交矢量集(搜索方向)中最小,这些正交矢量范围为式(3.10)给出的 Krylov 子空间 K_m。GMRES 算法总结如下:

(1) 假设初始解为 $\Delta \boldsymbol{W}_0^n$,计算初始残差矢量 $\boldsymbol{r}_0=\boldsymbol{J}\Delta \boldsymbol{W}_0^n+\boldsymbol{R}^n$;

(2) 生成 m 个搜索方向(由 Gram-Schmidt 正交化生成);

(3) 求解全局残差范数最小化问题;

(4) 得到式(6.61)的近似解 $\Delta \boldsymbol{W}^n=\Delta \boldsymbol{W}_0^n+\boldsymbol{\gamma}_m$。

由于内存需求随搜索方向的数量线性增加,所以在实际中 m 的值限制在 10~40,这对于收敛解 $\Delta \boldsymbol{W}^n$ 可能不够,因此必须重新启动 GMRES 方法,即设置 $\Delta \boldsymbol{W}_0^n=\Delta \boldsymbol{W}^n$,计算 $\boldsymbol{r}_0$,并继续执行步骤 2。如文献[92]所指出的,如果全局残差的范数下降到指定公差以下,则应减少 m 值,而不是使用固定数量的搜索方向,这样可以在牛顿迭代的后面步骤中节省大量运算。

2. 通量雅可比矩阵的计算

GMRES 和其他 Krylov 子空间方法不需要显式计算和存储通量雅可比矩阵 $\partial \boldsymbol{R}/\partial \boldsymbol{W}$。这种思想是基于以下的发现,即这些方法仅依赖于形式为 $\boldsymbol{J}\Delta \boldsymbol{W}^n$ 的矩阵矢量积,而不需要显式求解矩阵 $\boldsymbol{J}$。通量雅可比矩阵与解的更新的乘积可用一个简单的有限差分近似为

$$\frac{\partial \boldsymbol{R}}{\partial \boldsymbol{W}}\Delta \boldsymbol{W}^n \approx \frac{\boldsymbol{R}(\boldsymbol{W}^n+h\Delta \boldsymbol{W}^n)-\boldsymbol{R}(\boldsymbol{W}^n)}{h} \tag{6.62}$$

它只需要对残差进行两次计算即可。为了使数值误差最小,必须谨慎选择步长 h[36],一种较好的步长 h 计算公式为[93]

$$h=\frac{\sqrt{\epsilon}}{\| \Delta \boldsymbol{W}^n \|_2^2}\max\{\,|d|,\mathrm{typ}[\boldsymbol{W}^n\cdot|\Delta \boldsymbol{W}^n|]\}\,\mathrm{sign}(d) \tag{6.63}$$

其中,ϵ 表示机器精度,d 为标量积 $\boldsymbol{W}^n\cdot\Delta \boldsymbol{W}^n$,$|\Delta \boldsymbol{W}^n|$ 为矢量 $\Delta \boldsymbol{W}^n$ 所有元素的绝对值,typ$U$ 表示 U 的典型大小。除了节省内存和运算外,有限差分近似还有一个更重要的优点,即可以很容易实现高阶残差 $\boldsymbol{R}^n$(包括边界条件、限制器、源项等)的精确线性化,从而可以以适中的代价实现牛顿方法的二次收敛性,因此,我们将这种方法称为 Newton-Krylov 方法[93-100]。

3. 预处理技术

Krylov 子空间方法的效率很大程度上取决于一个好的预处理器。预处理器旨在使系统矩阵 $\boldsymbol{J}$ 的特征值接近于 1,从而不需要求解式(6.61),而是根据

式(3.11)求解左预处理系统或右预处理系统。利用式(6.61)和式(6.62)以及条件 $\Delta t\to\infty$, Newton-Krylov 方法的左预处理为

$$\boldsymbol{P}_{\mathrm{L}}\frac{\boldsymbol{R}(\boldsymbol{W}^{n}+h\Delta\boldsymbol{W}^{n})-\boldsymbol{R}(\boldsymbol{W}^{n})}{h}=-\boldsymbol{P}_{\mathrm{L}}\boldsymbol{R}^{n}\tag{6.64}$$

右预处理为

$$\frac{\boldsymbol{R}(\boldsymbol{W}^{n}+h\boldsymbol{P}_{\mathrm{R}}\Delta\boldsymbol{W}^{*})-\boldsymbol{R}(\boldsymbol{W}^{n})}{h}=-\boldsymbol{R}^{n}\tag{6.65}$$

$$\boldsymbol{P}_{\mathrm{R}}^{-1}\Delta\boldsymbol{W}^{n}=\Delta\boldsymbol{W}^{*}$$

这两种预处理方法的主要区别在于:左预处理对残差 $\boldsymbol{R}^n$ 进行缩放,而右预处理则没有,当监测 Krylov 方法的收敛性时,必须记住这一点。

显然,预处理器应该尽可能接近系统矩阵的逆($\boldsymbol{P}_{\mathrm{L,R}}\approx\boldsymbol{J}^{-1}$),但另外,它应该是可逆的,并且求逆的计算量很小。因此,我们必须在 Krylov 方法的收敛速度和预处理矩阵求逆花费的时间之间找到一种最佳的折中方案。最成功的预处理方法之一是 ILU 因式分解法[101-102],它具有不同的填充层次(最常用的是零,记为 ILU(0))。对于黏性湍流流动,即对于刚性方程组[103],ILU 预处理器尤为有效。为了在非结构网格上获得良好的性能,需要对系统矩阵的元素进行重新排序,以减小矩阵带宽,在 6.2.1 节中我们已经讨论了 RCM 方法重新排序策略[24-25]。

ILU 预处理方法的一个严重缺陷是必须计算并存储矩阵 $\boldsymbol{J}$ 的元素(见 6.2.2 节)。因此,一些学者建议采用 LU-SGS 方法作为预处理方法[104-105],这样当使用式(6.62)时,完全不需要进行 $\boldsymbol{J}$ 的构造和存储。然而,文献[103]用几个 2D 实例证明了:就 CPU 时间而言,GMRES 方法与 LU-SGS 的结合不如 GMRES 与 ILU(0)相结合的方法。尽管如此,LU-SGS 方法(与多重网格相结合时最佳)仍然是一个很有吸引力的替代方法,特别是在 3D 情况下。

4. 启动问题

隐式 Newton-Krylov 方法的时间步长可为无穷大,但建议在牛顿迭代过程开始时使用较小的时间步长。因为在求解开始时,流动解一般都与稳态解(即非线性方程组 $\boldsymbol{R}(\boldsymbol{W})=0$ 的根)相差甚远,这可能会导致牛顿迭代失败。一种解决迭代失败的方法是切换进化松弛(switched evolution relaxation)技术[106],该方法保留式(6.61)中的 $\Omega/\Delta t$ 项,时间步长的计算方法与显式格式(式(6.14)或无黏性特征值的式(6.20))相同。随着残差的二范数的减小,CFL 数 σ 从一个小的初始值开始增加,即

$$\sigma^{n+1}=\sigma^{n}\frac{\|\boldsymbol{R}^{n-1}\|_2}{\|\boldsymbol{R}^{n}\|_2} \tag{6.66}$$

因此,迭代过程(式(6.61))的收敛最初是线性的,但对于大的 CFL 数,它接近牛顿法的二次收敛。时间项 $\Omega/\Delta t$ 的另一个影响是使 $\boldsymbol{J}$ 的对角占优增加(与 $\Omega/\Delta t$ 成反比),这将有助于迭代过程的稳定。文献[95]建议限制式(6.66)中 σ^{n+1} 的变化范围,使得 σ^{n+1}/σ^{n} 的范围为 0.1~2。

解决牛顿法启动问题的另一种方法为网格序列法,该方法在一系列较粗网格上获得初始解,然后将其插值到细网格上。还有一种方法是使用一种数值计算量小但鲁棒的迭代方法获得初始解,例如,以采用 LU-SGS 方法的多重网格方法开始,然后切换成以 LU-SGS 作为预处理器的 GMRES 方法。对于黏性湍流流动,这种方法值得特别关注,因为多重网格方法的收敛速度在初始阶段之后通常会减慢,但整体流场解已经接近稳态。

6.2.6　隐式龙格-库塔法

求解关于时间的常微分方程组(6.1)的一般 m 步方法可写为

$$\boldsymbol{W}_I^{n+1}=\boldsymbol{W}_I^{n}-\frac{\Delta t}{(\Omega\boldsymbol{M})_I}\sum_{k=1}^{m} b_k\boldsymbol{R}_I^{(k)} \tag{6.67}$$

其中,各步残差 $\boldsymbol{R}_I^{(k)}$ 是时间层和其他中间残差加权和的函数,即

$$\begin{aligned}\boldsymbol{R}_I^{(k)}&=\boldsymbol{R}_I\left(t^n+c_k\Delta t,\boldsymbol{W}_I^{n}-\frac{\Delta t}{(\Omega\boldsymbol{M})_I}\sum_{l=1}^{m} a_{kl}\boldsymbol{R}_I^{(l)}\right)\\&=\boldsymbol{R}_I(t^n+c_k\Delta t,\boldsymbol{W}_I^{(k)})\end{aligned} \tag{6.68}$$

系数 a_{kl},b_k 和 c_k 可采用 Butcher 表[15]的形式表示为

$$\begin{array}{c|cccc} c_1 & a_{11} & a_{12} & \cdots & a_{1m}\\ c_2 & a_{21} & a_{22} & \cdots & a_{2m}\\ \vdots & \vdots & \vdots & \ddots & \vdots\\ c_m & a_{m1} & a_{m2} & \cdots & a_{mm}\\ \hline & b_1 & b_2 & \cdots & b_m \end{array} \tag{6.69}$$

系数 c_k 定义中间时间层,a_{kl} 表示第 k 步的加权系数,b_k 为最后步的加权系数。这些系数决定了一个特定龙格-库塔法的精度、稳定性和效率。根据系数 a_{kl},b_k 和 c_k 的值,可以构造各种龙格-库塔法。例如,对于所有 $l\geqslant k$,设 $a_{kl}=0$,则得到纯显式方法,以下为经典四阶龙格-库塔法的系数:

$$
\begin{array}{c|cccc}
0 & 0 & 0 & 0 & 0 \\
1/2 & 1/2 & 0 & 0 & 0 \\
1/2 & 0 & 1/2 & 0 & 0 \\
1 & 0 & 0 & 1 & 0 \\
\hline
 & 1/6 & 1/3 & 1/3 & 1/6
\end{array}
\tag{6.70}
$$

另外，对于某些 $l \geqslant k$，即沿着表(6.69)的对角线或上三角矩阵，如果系数 $a_{kl} \neq 0$，则得到隐式龙格-库塔法。在这种情况下，第 k 步的中间残差，即式(6.68)中的 $\boldsymbol{R}_I^{(k)}$，也与同一步的解间接相关，甚至还可能与后面各步的解有关，CFD 应用中所采用的隐式龙格-库塔法通常避免与当前步之后的解耦合。如果 a_{kl} 仅在下三角矩阵中为非零，则该格式称为对角隐式龙格-库塔(diagonally implicit Runge-Kutta，DIRK)法；而如果所有对角系数 a_{kk} 均相等，则该格式称为单对角隐式龙格-库塔(singly diagonally implicit Runge-Kutta，SDIRK)法。

系数 a_{kl}，b_k 和 c_k 必须满足某些条件，首先，物理上合理的格式其系数应满足 $c_k = \sum_l a_{kl}$；其次，还要满足下面的阶数条件。

一阶精度：

$$
\sum_{k=1}^{m} b_k = 1 \tag{6.71}
$$

二阶精度：

$$
\sum_{k=1}^{m} b_k c_k = \frac{1}{2} \tag{6.72}
$$

三阶精度：

$$
\begin{aligned}
&\sum_{k=1}^{m} b_k c_k^2 = \frac{1}{3} \\
&\sum_{k=1}^{m} \sum_{l=1}^{m} b_k a_{kl} c_l = \frac{1}{6}
\end{aligned}
\tag{6.73}
$$

四阶精度：

$$
\begin{aligned}
&\sum_{k=1}^{m} b_k c_k^3 = \frac{1}{4} \\
&\sum_{k=1}^{m} \sum_{l=1}^{m} b_k c_k a_{kl} c_l = \frac{1}{8} \\
&\sum_{k=1}^{m} \sum_{l=1}^{m} b_k a_{kl} c_l^2 = \frac{1}{12}
\end{aligned}
$$

$$\sum_{k=1}^{m}\sum_{l=1}^{m}\sum_{n=1}^{m} b_k a_{kl} a_{ln} c_n = \frac{1}{24} \tag{6.74}$$

在满足上述阶数条件的情况下,可以对系数进行优化,以达到预期的精度、稳定性和阻尼特性。

对于隐式龙格-库塔法,式(6.68)中的中间残差 $\boldsymbol{R}_I^{(k)}$ 间接取决于尚待计算的残差(即 $l \geqslant k$ 的 $\boldsymbol{R}_I^{(l)}$)。因此,各步需要求解一组非线性方程组,通常用牛顿法来求解。下面以 DIRK 为例进行说明,第 k 步的中间解可以写成

$$\boldsymbol{W}_I^{(k)} = \boldsymbol{W}_I^n - \frac{\Delta t}{(\Omega \boldsymbol{M})_I}\left[\sum_{l=1}^{k-1} a_{kl}\boldsymbol{R}_I^{(l)} + a_{kk}\boldsymbol{R}_I^{(k)}\right] \tag{6.75}$$

式(6.75)中的隐式项 $\boldsymbol{R}_I^{(k)}$ 是 $\boldsymbol{W}_I^{(k)}$ 和 $(t^n + c_k\Delta t)$ 的函数,可用式(6.27)线性化处理,即

$$\boldsymbol{R}_I^{(p+1)} \approx \boldsymbol{R}_I^{(p)} + \boldsymbol{J}_I \Delta \boldsymbol{W}^{(p)} \tag{6.76}$$

其中,$\Delta \boldsymbol{W}^{(p)} = \boldsymbol{W}^{(p+1)} - \boldsymbol{W}^{(p)}$,$\boldsymbol{J} = \partial \boldsymbol{R}/\partial \boldsymbol{W}$ 为雅可比矩阵,将上式代入式(6.75),得到第 k 步的解,其表达式为

$$\left[\frac{(\Omega \boldsymbol{M})_I}{\Delta t} + a_{kk}\boldsymbol{J}_I\right]\Delta \boldsymbol{W}^{(p)} = \frac{(\Omega \boldsymbol{M})_I}{\Delta t}(\boldsymbol{W}_I^n - \boldsymbol{W}_I^{(p)}) - \sum_{l=1}^{k-1} a_{kl}\boldsymbol{R}_I^{(l)} - a_{kk}\boldsymbol{R}_I^{(p)} \tag{6.77}$$

该式在形式上与式(6.28)相同,可以用前面章节中的任何隐式方法进行求解。当收敛时,式(6.77)的解 $\boldsymbol{W}^{(p+1)}$ 与式(6.75)的第 k 步中间解近似。

隐式龙格-库塔法被用于这样的情况,即微分方程组(6.1)为刚性的,并且需要进行时间精确求解。方程组的刚性可能由湍流建模、高度拉伸的网格单元或真实气体效应等引起。Cooper 和 Sayfy[107-108] 提出的思路是将方程组分裂为刚性项和非刚性项,然后对刚性项进行隐式处理,同时对非刚性项进行显式积分,这种多步法称为显隐龙格-库塔法(additive Runge-Kutta scheme)。根据具体情况,与隐式处理所有项相比,这种分裂可使式(6.77)中的雅可比矩阵 $\boldsymbol{J}$ 更简单,从而大大减少计算量。Kennedy 和 Carpenter 在文献[109]中为此目的设计了优化的龙格-库塔法,表 6.3 和表 6.4 给出了耦合的显隐龙格-库塔法的系数,这两种方法都有六步,为时间四阶精度。需要注意的是,为了便于耦合,这两种方法具有相同的系数 b_k 和 c_k。这两种方法的第二组系数 $\hat{b}_k$(最后步的加权系数)定义了一种三阶方法,被称为“嵌入法”,用于计算积分误差,并用它来调节时间步长。时间误差计算如下:

$$\epsilon_I = \frac{\Delta t}{(\Omega \boldsymbol{M})_I}\sum_{k=1}^{m}(b_k - \hat{b}_k)\boldsymbol{R}_I^{(k)} \tag{6.78}$$

表 6.3　显式六步四阶和三阶精度($\hat{b}_k$)龙格-库塔法的系数[109]

0	0	0	0	0	0	0
$\frac{1}{2}$	$\frac{1}{2}$	0	0	0	0	0
$\frac{83}{250}$	$\frac{13861}{62500}$	$\frac{6889}{62500}$	0	0	0	0
$\frac{31}{50}$	$-\frac{116923316275}{2393684061468}$	$-\frac{2731218467317}{15368042101831}$	$\frac{9408046702089}{11113171139209}$	0	0	0
$\frac{17}{20}$	$-\frac{451086348788}{2902428689909}$	$-\frac{2682348792572}{7519795681897}$	$\frac{12662868775082}{11960479115383}$	$\frac{3355817975965}{11060851509271}$	0	0
1	$\frac{647845179188}{3216320057751}$	$\frac{73281519250}{8382639484533}$	$\frac{552539513391}{3454668386233}$	$\frac{3354512671639}{8306763924573}$	$\frac{4040}{17871}$	0
b_k	$\frac{82889}{524892}$	0	$\frac{15625}{83664}$	$\frac{69875}{102672}$	$-\frac{2260}{8211}$	$\frac{1}{4}$
$\hat{b}_k$	$\frac{4586570599}{29645900160}$	0	$\frac{178811875}{945068544}$	$\frac{814220225}{1159782912}$	$-\frac{3700637}{11593932}$	$\frac{61727}{225920}$

表 6.4　单对角隐式六步四阶和三阶精度($\hat{b}_k$)龙格-库塔法的系数[109]

0	0	0	0	0	0	0
$\frac{1}{2}$	$\frac{1}{4}$	$\frac{1}{4}$	0	0	0	0
$\frac{83}{250}$	$\frac{8611}{62500}$	$-\frac{1743}{31250}$	$\frac{1}{4}$	0	0	0
$\frac{31}{50}$	$\frac{5012029}{34652500}$	$-\frac{654441}{2922500}$	$\frac{174375}{388108}$	$\frac{1}{4}$	0	0
$\frac{17}{20}$	$\frac{15267082809}{155376265600}$	$-\frac{71443401}{120774400}$	$\frac{730878875}{902184768}$	$\frac{2285395}{8070912}$	$\frac{1}{4}$	0
1	$\frac{82889}{524892}$	0	$\frac{15625}{83664}$	$\frac{69875}{102672}$	$-\frac{2260}{8211}$	$\frac{1}{4}$
b_k	$\frac{82889}{524892}$	0	$\frac{15625}{83664}$	$\frac{69875}{102672}$	$-\frac{2260}{8211}$	$\frac{1}{4}$
$\hat{b}_k$	$\frac{4586570599}{29645900160}$	0	$\frac{178811875}{945068544}$	$\frac{814220225}{1159782912}$	$-\frac{3700637}{11593932}$	$\frac{61727}{225920}$

表 6.5~表 6.7 给出了读者可能感兴趣的其他对角隐式龙格-库塔法的系数，关于应用实例和得到的结果请参考文献[113-121]。

表 6.5　单对角隐式三步四阶精度龙格-库塔法的系数[110]

$(1+a)/2$	$(1+a)/2$	0	0
$1/2$	$-a/2$	$(1+a)/2$	0
$(1-a)/2$	$1+a$	$-(1+2a)$	$(1+a)/2$
	$1/(6a^2)$	$1-1/(3a^2)$	$1/(6a^2)$

参数 $a=(2/\sqrt{3})\cos(\pi/18)$，该方法具有良好的阻尼特性，即使在大时间步长下也如此。

表 6.6　对角隐式三步四阶精度龙格-库塔法的系数[111]

0.257820901066211	0.377847764031163	0.0	0.0
0.434296446908075	0.385232756462588	0.461548399939329	0.0
0.758519768667167	0.675724855841358	−0.061710969841169	0.241480233100410
	0.750869573741408	−0.362218781852651	0.611349208111243

对系数进行了优化，使该方法在大时间步长下也具有低色散和低耗散误差。但是，因为 $c_k \neq \sum_l a_{kl}$，所以该方法是物理不相容的。

表 6.7　对角隐式三步三阶精度龙格-库塔法优化后的系数[112]

1.559910257360900	1.559910257360900	0.0	0.0
14.78349948371908	1.656464402419667	13.12703508129941	0.0
14.37813978673026	1.658169529369301	1.672660289271657	11.04730996808931
	1.169960021844266	2.759712723699827	−2.929672745544100

6.3　非定常流动方法

目前，非定常流动现象的模拟在许多工程学科中非常重要。例如，叶轮机械中固定部件和旋转部件之间的相互作用、活塞式发动机、流固耦合、直升机空气动力学、气动声学、湍流的 DNS 或 LES、爆轰，等等。显然，对于所求解的问题，模拟的执行必须高效，且具有足够的精度。

对于某些非定常应用，时间尺度和空间尺度与特征值的比值相当，即由物理规定的 CFL 数为①的量级，显式方法是最好的选择。气动声学、DNS 和 LES 就属于此类情形，对于此类应用，通常采用显式龙格-库塔法。由于在这些应用中，全局物理现象的演变速度比解的局部变化慢得多，因此有必要在很长的物理时间内进行积分。为了精确地做到这一点，显式方法的时间分辨率必须是三阶或三阶以上，这需要采用与 6.1 节中介绍的方法不同的龙格-库塔法

(见文献[122-123])。

在其他情况下,和空间尺度与特征值的比值相比,物理时间尺度较大时(如颤振或转子-定子的相互作用),CFL 数可选择数百甚至数千而不会影响模拟的精度。在这种情况下,隐式方法更为合适。下面将讨论一种称为双时间步(dual time-stepping)方法的特殊技术,该方法常用于非定常流动模拟,它与 6.2.6 节的隐式龙格-库塔法的比较请参见文献[124-126]。

双时间步方法基于式(6.2)中基本非线性格式的二阶时间精度方法。设式(6.2)中的 $\beta=1$ 和 $\omega=1/2$,得

$$\frac{3(\Omega \boldsymbol{M})_I^{n+1}\boldsymbol{W}_I^{n+1}-4(\Omega \boldsymbol{M})_I^{n}\boldsymbol{W}_I^{n}+(\Omega \boldsymbol{M})_I^{n-1}\boldsymbol{W}_I^{n-1}}{2\Delta t}=-\boldsymbol{R}_I^{n+1} \tag{6.79}$$

式中:Δt 为全局物理时间步长;$\boldsymbol{M}$ 为质量矩阵。式(6.79)为式(6.1)中时间导数的三点后向差分(后向欧拉)近似。为了求解式(6.79)给出的非线性方程组,可使用牛顿法或时间推进法。时间推进法可写为

$$\frac{\partial}{\partial t^*}(\Omega_I^{n+1}\boldsymbol{W}_I^*)=-\boldsymbol{R}_I^*(\boldsymbol{W}^*) \tag{6.80}$$

式中:$\boldsymbol{W}_I^*$ 是 $\boldsymbol{W}_I^{n+1}$ 的近似;t^* 表示伪时间变量。需要注意的是,时间导数中无质量矩阵。非定常残差定义为

$$\boldsymbol{R}_I^*(\boldsymbol{W}^*)=\boldsymbol{R}_I(\boldsymbol{W}^*)+\frac{3}{2\Delta t}(\Omega \boldsymbol{M})_I^{n+1}\boldsymbol{W}_I^*-\boldsymbol{Q}_I^* \tag{6.81}$$

将式(6.80)中在时间推进过程中为常数的所有项都集中到源项中,即

$$\boldsymbol{Q}_I^*=\frac{2}{\Delta t}(\Omega \boldsymbol{M})_I^{n}\boldsymbol{W}_I^{n}-\frac{1}{2\Delta t}(\Omega \boldsymbol{M})_I^{n-1}\boldsymbol{W}_I^{n-1} \tag{6.82}$$

在网格运动和/或变形的情况下,控制体的新尺寸,即式(6.80)中的 Ω_I^{n+1} 必须满足几何守恒定律(详见 A.5 节)。

式(6.80)的定常解对应于新时间层上的流动变量,即 $\boldsymbol{W}^*=\boldsymbol{W}^{n+1}$。因为对于伪时间的定常状态,$\boldsymbol{R}_I^*=0$,因此满足式(6.79)。任何先前介绍的显式或隐式时间推进方法均可用于伪时间的方程组(6.80)的求解。下面将讨论双时间步方法在显式多步方法和隐式方法中的实现。

6.3.1 显式多步法的双时间步

Jameson[127]最早采用显式多步格式实现了双时间步方法,利用局部时间步长和多重网格技术进行加速。这种方法的显著优点在于物理时间步长不像显式方法那样受到限制,可以完全根据流动物理现象来选择时间步长,另外,仅源项 $\boldsymbol{Q}^*$ 需要额外的存储,因而该方法非常具有吸引力。求解伪时间问题

（式(6.80)）的 m 步显式格式为

$$
\begin{aligned}
\boldsymbol{W}_I^{(0)} &= (\boldsymbol{W}_L^*)^l \\
\boldsymbol{W}_I^{(1)} &= \boldsymbol{W}_I^{(0)} - \frac{\alpha_1 \Delta t_I^*}{\Omega_I^{n+1}} \boldsymbol{R}_I^*(\boldsymbol{W}^{(0)}) \\
\boldsymbol{W}_I^{(2)} &= \boldsymbol{W}_I^{(0)} - \frac{\alpha_2 \Delta t_I^*}{\Omega_I^{n+1}} \boldsymbol{R}_I^*(\boldsymbol{W}^{(1)}) \\
&\vdots \\
(\boldsymbol{W}_L^*)^{l+1} &= \boldsymbol{W}_I^{(0)} - \frac{\alpha_m \Delta t_I^*}{\Omega_I^{n+1}} \boldsymbol{R}_I^*(\boldsymbol{W}^{(m-1)})
\end{aligned}
\tag{6.83}
$$

其中，l 和 $l+1$ 分别表示真实的时间层和新的伪时间层。以 $(\boldsymbol{W}^*)^l=\boldsymbol{W}^n$ 或由以前的物理时间步插值得到的值启动时间推进过程（如文献[128]）

$$
(\boldsymbol{W}_I^*)^l = \boldsymbol{W}^n + \frac{3\boldsymbol{W}^n - 4\boldsymbol{W}^{n-1} + \boldsymbol{W}^{n-2}}{2} \tag{6.84}
$$

推进过程一直持续到 $(\boldsymbol{W}_I^*)^{l+1}$ 与 $\boldsymbol{W}_I^{n+1}$ 足够近似为止（通常当残差 $\boldsymbol{R}_I^*$ 下降 2 个或 3 个数量级时），然后执行下一个物理时间步。伪时间步长 Δt^* 的计算方法与 6.1.4 节中的方法相同。

Arnone 等[128]指出，当物理时间步长 Δt 与伪时间步长 Δt^* 相当或更小时，多步法（式(6.83)）变得不稳定。Melson 等[129]证明了不稳定是由式(6.81)中的项

$$
\frac{3}{2\Delta t}(\Omega \boldsymbol{M})_I^{n+1} \boldsymbol{W}_I^*
$$

引起的，当 Δt 较小时不稳定更显著。Melson 等建议采用隐式方式处理该项，因此，必须修正式(6.83)中的多步法，从而第 k 步变为[129]

$$
\begin{gathered}
\boldsymbol{W}_I^{(k)} = \boldsymbol{W}_I^{(0)} - \frac{\alpha_k \Delta t_I^*}{\Omega_I^{n+1}} \left[\bar{I} + \frac{3}{2\Delta t} \alpha_k \Delta t_I^* \boldsymbol{M}^{n+1} \right]^{-1} \cdot \\
\left[\boldsymbol{R}_I(\boldsymbol{W}^{(k-1)}) + \frac{3}{2\Delta t} (\Omega \boldsymbol{M})_I^{n+1} \boldsymbol{W}_I^{(k-1)} - \boldsymbol{Q}_I^* \right]
\end{gathered}
\tag{6.85}
$$

同样的方法也可应用于混合多步法（见 6.1.2 节）。式(6.85)对于任何物理时间步长都是稳定的[129]。

对于格心格式，可归并式(6.85)中的质量矩阵 $\boldsymbol{M}^{n+1}$（用单位矩阵代替），而不会降低解的精度，从而式(6.85)中的项

$$
\left[\boldsymbol{I} + \frac{3}{2\Delta t} \alpha_k \Delta t_I^* \boldsymbol{M}^{n+1} \right]^{-1}
$$

转变为标量值。然而,对于格点格式,必须考虑质量矩阵,否则,对于小的物理时间步长,多步法是不稳定的。为了规避昂贵的 $\boldsymbol{M}^{n+1}$ 矩阵的求逆, Venkatakrishnan[130]以及 Venkatakrishnan 和 Mavriplis[131]建议对式(6.85)进行如下修正:

$$\boldsymbol{W}_I^{(k)} = \boldsymbol{W}_I^{(0)} - \frac{\alpha_k \Delta t_I^*}{\Omega_I^{n+1}} \left[1 + \frac{3}{2\Delta t}\alpha_k \Delta t_I^* \beta\right]^{-1} \cdot \left[\boldsymbol{R}_I^*(\boldsymbol{W}^{(k-1)}) - \frac{3}{2\Delta t}\Omega_I^{n+1} \beta \boldsymbol{W}_I^{(k-1)}\right] \tag{6.86}$$

此时,参数 β 用于稳定时间推进格式。在实践中,选择 $\beta=2$ 就足以满足要求[130-131]。

伪时间解采用显式多步法的双时间步法应用非常广泛,当多步法采用局部时间步长(在时间 t^* 上)和多重网格技术进行加速时,计算效率最高,其原因是多重网格法很快(在几个循环内)就能收敛到式(6.80)的定常解。结构网格应用实例可参见文献[127-129],[132-134]。在文献[130-131,135]中介绍了该方法在非结构网格上的实现。

6.3.2 隐式方法的双时间步

在伪时间 t^* 内采用隐式方法求解式(6.80)的过程与6.2节中叙述的方法相同。首先,将式(6.80)写成非线性隐式格式,即

$$\frac{\partial}{\partial t^*}(\Omega_I^{n+1} \boldsymbol{W}_I^*) = -(\boldsymbol{R}_I^*)^{l+1} \tag{6.87}$$

其中, $l+1$ 是新的伪时间层。同样要注意的是,时间导数中无质量矩阵 $\boldsymbol{M}$。式(6.81)中定义的非定常残差可在伪时间步内进行线性化,即

$$(\boldsymbol{R}^*)^{l+1} \approx (\boldsymbol{R}^*)^l + \frac{\partial \boldsymbol{R}^*}{\partial \boldsymbol{W}^*}\Delta \boldsymbol{W}^* \tag{6.88}$$

其中, $\Delta\boldsymbol{W}^* = (\boldsymbol{W}^*)^{l+1} - (\boldsymbol{W}^*)^l$,而通量雅可比矩阵定义为

$$\frac{\partial \boldsymbol{R}^*}{\partial \boldsymbol{W}^*} = \frac{\partial \boldsymbol{R}}{\partial \boldsymbol{W}} + \frac{3}{2\Delta t}(\Omega \boldsymbol{M})^{n+1} \tag{6.89}$$

如果将上述线性化处理代入式(6.87),则得到无分裂隐式格式[136]为

$$\left[\left(\frac{1}{\Delta t_I^*} + \frac{3}{2\Delta t}\right)(\Omega \boldsymbol{M})_I^{n+1} + \left(\frac{\partial \boldsymbol{R}}{\partial \boldsymbol{W}}\right)_I\right]\Delta \boldsymbol{W}^* = -(\boldsymbol{R}_I^*)^l \tag{6.90}$$

6.2节中介绍的任何方法都可以用于方程组(6.90)的求解。时间精确隐式方法的详细讨论见文献[137],方法实现的实例读者可参见文献[136,138-139]等。

参考文献

[1] Jameson A, Schmidt W, Turkel E. Numerical solutions of the Euler equations by finite volume methods using Runge-Kutta time-stepping schemes. AIAA Paper 81-1259; 1981.

[2] van Leer B, Tai C-H, Powell KG. design of optimally smoothing multi-stage schemes for the Euler equations. AIAA Paper 89-1933, 1989.

[3] Tai C-H, Sheu J-H, van Leer B. Optimal multistage schemes for Euler equations with residual smoothing. AIAA J 1995; 33: 1008-16.

[4] Tai C-H, Sheu J-H, Tzeng P-Y. Improvement of explicit multistage schemes for central spatial discretization. AIAA J 1996; 34: 185-8.

[5] Martinelli L. Calculations of viscous flows with amultigrid method. PhD Thesis. Dept. of Mechanical and Aerospace Engineering, Princeton University; 1987.

[6] Mavriplis DJ, Jameson A. Multigrid solution of the Navier-Stokes equations on triangular meshes. AIAA J 1990; 28: 1415-25.

[7] Lafon A, Yee HC. On the numerical treatment of nonlinear source terms in reaction-convection equations. AIAA Paper 92-0419; 1992.

[8] Curtiss CF, Hirschfelder JO. Integration of stiff equations. Proc Natl Acad Sci USA 1952; 38.

[9] Bussing TRA. A finite volume method for the Navier-Stokes equations with finite rate chemistry. PhD Thesis. Dept. of Aeronautics and Astronautics, Massachusetts Inst. of Technology; 1985.

[10] Bussing TRA, Murman EM. Finite-volume method for the calculation of compressible chemically reacting flows. AIAA J 1988; 26: 1070-8.

[11] Kunz RF, Lakshminarayana B. Stability of explicit Navier-Stokes procedures using k-ε and k-ε/algebraic Reynolds stress turbulence models. J Comput Phys 1992; 103: 141-59.

[12] Jonas S, Frühauf HH, Knab O. Fully coupled approach to the calculation of nonequilibrium hypersonic flows using a Godunov-type method. In: 1st European computational fluid dynamics conf. Brussels, Belgium; September 7-11, 1992.

[13] Merci B, Steelant J, Vierendeels J, Riemslagh K, Dick E. Computational treatment of source terms in two-equation turbulence models. AIAA J 2000; 38: 2085-93.

[14] Rosenbrock HH. Some general implicit processes for the numerical solution of differential equations. Comput J 1963; 5: 329-30.

[15] Butcher JC. Implicit Runge-Kutta processes. Math Comput 1964; 18: 50-64.

[16] Courant R, Friedrichs KO, Lewy H. über die partiellen Differenzengleichungen der mathematischen Physik. Math Ann 1928; 100: 32-74. [Transl.: On the Partial Difference Equations of Mathematical Physics. IBM J 1967; 11: 215-34].

[17] Rizzi A, Inouye M. A time-split finite volume technique for three-dimensional blunt-body flow. AIAA Paper 73-0133; 1973.

[18] Müller B, Rizzi A. Runge-Kutta finite-volume simulation of laminar transonic flow over a NACA0012 airfoil using the Navier-Stokes equations. FFA TN 1986-60; 1986.

[19] Swanson RC, Turkel E, White JA. An effective multigrid method for high-speed flows. In: Proc. 5th Copper Mountain conf. on multigrid methods; 1991.

[20] Frink NT, Parikh P, Pirzadeh S. A fast upwind solver for the Euler equations on three-dimensional unstructured meshes. AIAA Paper 91-0102; 1991.

[21] George A, Liu JW. Computer solution of large sparse positive definite systems. Prentice Hall Series in Comput Math. Englewood Cliffs, NJ: Prentice Hall; 1981.

[22] Pothen A, Simon HD, Liou KP. Partitioning sparse matrices with eigenvectors of graphs. SIAM J Matrix Anal Appl 1990; 11: 430-52.

[23] Vanden KJ, Whitfield DL. Direct and iterative algorithms for the three-dimensional Euler equations. AIAA Paper 93-3378, 1993 [also AIAA J 1995; 33: 851-8].

[24] Cuthill E, McKee J. Reducing the bandwidth of sparse symmetric matrices. In: Proc. ACM 24th national conference; 1969. p. 157-61.

[25] Gibbs NE, Poole WG, Stockmeyer PK. An algorithm for reducing the bandwidth and profile of a sparse matrix. SIAM J Numer Anal 1976; 13: 236-50.

[26] Löner R. Some useful renumbering strategies for unstructured grids. Int J Numer Meth Eng 1993; 36: 3259-70.

[27] Steger JL, Warming RF. Flux vector splitting of the inviscid gasdynamic equations with application to finite-difference methods. J Comput Phys 1981; 40: 263-93.

[28] Liou MS, van Leer B. Choice of implicit and explicit operators for the upwind differencing method. AIAA Paper 88-0624; 1988.

[29] Barth TJ, Linton SW. An unstructured mesh Newton solver for compressible fluid flow and its parallel implementation. AIAA Paper 95-0221; 1995.

[30] Orkwis PD, Vanden KJ. On the accuracy of numerical versus analytical Jacobians. AIAA Paper 94-0176; 1994 [also AIAA J 1996; 34: 1125-9].

[31] Barth TJ. Analysis of implicit local linearization techniques for upwind and TVD algorithms. AIAA Paper 87-0595; 1987.

[32] Bischof C, Khademi P, Mauer A, Carle A. Adifor 2.0: automatic differentiation of Fortran 77 programs. IEEE Comput Sci Eng 1996; Fall: 18-32.

[33] Narayanan SHK, Norris B, Winnicka B. ADIC2: Development of a component source transformation system for differentiating C and C++. In: Int. conf. on computational science, ICCS; 2010.

[34] http://www.mcs.anl.gov/OpenAD.

[35] Hogan RJ. Fast reverse-mode automatic differentiation using expression templates in C++. ACM Trans Math Software 2014; 40: 26: 1-16 [see also: http://www.met.rdg.ac.uk/clouds/adept].

[36] Dennis JE, Schnabel RB. Numerical methods for unconstrained optimization and nonlinear equations. Englewood Cliffs, NJ: Prentice-Hall; 1983.

[37] Xu X, Richards BE. Simplified procedure for numerically approximate jacobian matrix generation in Newton's method for solving the Navier-Stokes equations. GU Aero Report 9320, Dept.

Aerospace Eng. ,University of Glasgow;1993.

[38] Curtis AR,Powell MD,Reid JK. On the estimation of sparse Jacobian matrices. J Inst Maths Appl 1974;13:117-19.

[39] Venkatakrishnan V. Preconditioned conjugate gradient methods for the compressible Navier-Stokes equations. AIAA Paper 90-0586;1990 [also AIAA J 1991;29:1092-100].

[40] Briley WR,McDonald H. Solution of the multi-dimensional compressible Navier-Stokes equations by a generalized implicit method. J Comput Phys 1977;24:372-97.

[41] Beam R,Warming RF. An implicit factored scheme for the compressible Navier-Stokes equations. AIAA J 1978;16:393-402.

[42] Steger JL. Implicit finite difference simulation of flow about arbitrary geometries with application to airfoils. AIAA J 1978;16:679.

[43] Pulliam TH,Steger JL. Recent improvements in efficiency,accuracy and convergence for implicit approximate factorization scheme. AIAA Paper 85-0360;1985.

[44] Pulliam TH. Artificial dissipation models for the Euler equations. AIAA J 1986;24:1931-40.

[45] Pulliam TH,Chaussee DS. A diagonal form of an implicit approximate factorization algorithm. J Comput Phys 1987;39:347-63.

[46] Lomax H. Some notes on finite difference methods. Lecture Notes. Stanford University;1986.

[47] Pulliam TH. Implicit methods in CFD. Numerical methods for fluid dynamics III. Oxford University Press;1988.

[48] Rosenfeld M,Yassour Y. The alternating direction multi-zone implicit method. J Comput Phys 1994;110:212-20.

[49] Jameson A, Turkel E. Implicit scheme and LU-decompositions. Math Comput 1981;37:385-97.

[50] Yoon S,Jameson A. A multigrid LU-SSOR scheme for approximate Newton-iteration applied to the Euler equations. NASA CR-17954;1986.

[51] Yoon S,Jameson A. Lower-upper symmetric-Gauss-Seidel method for the Euler and Navier-Stokes equations. AIAA Paper 87-0600,1987 [AIAA J 1988;26:1025-26].

[52] Yoon S,Jameson A. LU implicit schemes with multiple grids for the Euler equations. AIAA Paper 86-0105;1986 [also AIAA J 1987;7:929-35].

[53] Rieger H,Jameson A. Solution of steady 3-D compressible Euler and Navier-Stokes equations by an implicit LU scheme. AIAA Paper 88-0619;1988.

[54] Yoon S, Kwak D. 3-D incompressible Navier-Stokes solver using lower-upper symmetric-Gauss-Seidel algorithm. AIAA J 1991;29.

[55] Blazek J. Investigations of the implicit LU-SSOR scheme. DLR Research Report No. 93-51;1993.

[56] Pahlke K,Blazek J,Kirchner A. Time-accurate Euler computations for rotor flows. European forum:recent developments and applications in aeronautical CFD, Paper No. 15. Bristol, UK;1993.

[57] Yoon S,Kwak D. Multigrid convergence of an implicit symmetric relaxation scheme. AIAA

Paper 93-3357;1993.
[58] Blazek J. A multigrid LU-SSOR scheme for the solution of hypersonic flow problems. AIAA Paper 94-0062;1994.
[59] Candler GV, Wright MJ, McDonald JD. A data-parallel LU-SGS method for reacting flows. AIAA Paper 94-0410;1994.
[60] Stoll P, Gerlinger P, Brüggemann D. Domain decomposition for an implicit LU-SGS scheme using overlapping grids. AIAA Paper 97-1896;1997.
[61] Lee B-S, Lee DH. Data parallel symmetric Gauss-Seidel algorithm for efficient distributed computing using massively parallel supercomputers. AIAA Paper 97-2138;1997.
[62] Otero E, Eliasson P. Parameter investigation with line-implicit lower-upper symmetric Gauss-Seidel on 3D stretched grids. AIAA Paper 2014-2094;2014.
[63] Xu L, Weng P. High order accurate and low dissipation method for unsteady compressible viscous flow computation on helicopter rotor in forward flight. J Comput Phys 2014;258:470-88.
[64] Tomaro RF, Strang WZ, Sankar LN. An implicit algorithm for solving time dependent flows on unstructured grids. AIAA Paper 97-0333;1997.
[65] Sharov D, Nakahashi K. Reordering of 3-D hybrid unstructured grids for vectorized LU-SGS Navier-Stokes calculations. AIAA Paper 97-2102;1997.
[66] Kano S, Kazuhiro N. Navier-Stokes computations of HSCT off-design aerodynamics using unstructured hybrid grids. AIAA Paper 98-0232;1998.
[67] Sharov D, Nakahashi K. Low speed preconditioning and LU-SGS scheme for 3-D viscous flow computations on unstructured grids. AIAA Paper 98-0614;1998.
[68] Strang WZ, Tomaro RF, Grismer MJ. The defining methods of Cobalt60: a parallel, implicit, unstructured Euler/Navier-Stokes flow solver. AIAA Paper 99-0786;1999.
[69] Parsani M, Van den Abeele K, Lacor C, Turkel E. Implicit LU-SGS algorithm for high-order methods on unstructured grid with p-multigrid strategy for solving the steady Navier-Stokes equations. J Comput Phys 2010;229:828-50.
[70] Parsani M, Ghorbaniasl G, Lacor C. Analysis of the implicit LU-SGS algorithm for 3rd- and 4th-order spectral volume scheme for solving the steady Navier-Stokes equations. J Comput Phys 2011;230:7073-85.
[71] Haga T, Kuzuu K, Takaki R, Shima E. Assessment of an unstructured CFD solver for RANS simulation on body-fitted Cartesian grids. AIAA Paper 2014-0241;2014.
[72] Shuen S, Yoon S. Numerical study of chemically reacting flows using a lower-upper symmetric successive overrelaxation scheme. AIAA J 1989;27:1752-60.
[73] Shuen JS. Upwind differencing and LU factorization for chemical non-equilibrium Navier-Stokes equations. J Comput Phys 1992;99:233-50.
[74] Shuen JS, Chen KH, Choi Y. A coupled implicit method for chemical non-equilibrium flows at all speeds. J Comput Phys 1993;106:306-18.
[75] Palmer G, Venkapathy E. Comparison of nonequilibrium solution algorithms applied to chemically stiff hypersonic flows. AIAA J 1995;33:1211-19.

[76] Gerlinger P, Stoll P, Brüggemann D. An implicit multigrid method for the simulation of chemically reacting flows. J Comput Phys 1998;146:322-45.

[77] Yoon S, Jost G, Chang S. Parallelization of lower-upper symmetric Gauss-Seidel method for chemically reacting flow. AIAA Paper 2005-4627;2005.

[78] Chen B, Xu X, Cai G. A multi-code CFD solver for the efficient simulation of Ramjet/Scramjet inlet-engine coupled flowfields. AIAA Paper 2007-5414;2007.

[79] Janus JM. A matrix study of Ax=b for implicit factored methods. AIAA Paper 98-0112;1998.

[80] Yuan X, Daiguji H. A new LU-type implicit scheme for three-dimensional compressible Navier-Stokes equations. In: Hafez M, Oshima K, editors. 6th Int. symposium on CFD, vol. III. Lake Tahoe;1995 [Pergamon Press;1998. p. 1473-8].

[81] Soetrisno M, Imlay ST, Roberts DW. A zonal implicit procedure for hybrid structured-unstructured grids. AIAA Paper 94-0645;1994.

[82] Soetrisno M, Imlay ST, Roberts DW, Taflin DE. Development of a zonal implicit procedure for hybrid structured-unstructured grids. AIAA Paper 96-0167;1996.

[83] Sonneveld P. CGS, a fast Lanczos-type solver for nonsymmetric linear systems. SIAM J Sci Stat Comput 1989;10:36-52.

[84] Van der Vorst HA. BiCGSTAB: a fast and smoothly converging variant of Bi-CG for the solution of nonsymmetric linear systems. SIAM J Sci Stat Comput 1992;13:631-44.

[85] Freund RW. A transpose-free quasi-minimal residual algorithm for non-Hermitian linear systems. SIAM J Sci Comput 1993;14:470-82.

[86] Saad Y, Schulz MH. GMRES: a generalized minimum residual algorithm for solving nonsymmetric linear systems. SIAM J Sci Stat Comput 1986;7:856-69.

[87] Saad Y. Krylov subspace techniques, conjugate gradients, preconditioning and sparse matrix solvers. VKI Lecture Series;1994-05;1994.

[88] Walker HF. Implementation of the GMRES method using householder transformations. SIAM J Sci Comput 1988;9:152.

[89] Morgan RB. A restarted GMRES method augmented with eigenvectors. SIAM J Matrix Anal Appl 1995;16:1154-71.

[90] Chapman A, Saad Y. Deflated and augmented Krylov subspace techniques. Technical Report UMSI 95/181. Minnesota Supercomputer Institute;1995.

[91] Saad Y. Enhanced acceleration and reconditioning techniques. In: Hafez M, Oshima K, editors. CFD Review 1998. Singapore: World Scientific Publishing Co. ;1998. p. 478-87.

[92] Ajmani K, Ng, W-F, Liou M-S, Preconditioned conjugate gradient methods for low speed flow calculations. AIAA Paper 93-0881;1993.

[93] Brown PN, Saad Y. Hybrid Krylov methods for nonlinear systems of equations. SIAM J Sci Stat Comput 1990;11:450-81.

[94] Zingg D, Pueyo A. An efficient Newton-GMRES solver for aerodynamic computations. AIAA Paper 97-1955;1997.

[95] Gropp WD, Keyes DE, McInnes LC, Tidriri MD. Globalized Newton-Krylov-Schwarz algorithms

and software for parallel implicit CFD. ICASE Report No. 98-24;1998.

[96] Knoll DA, Keyes DE. Jacobian-free Newton-Krylov methods: a survey of approaches and applications. J Comput Phys 2004;193:357-97.

[97] Nejat A, Ollivier-Gooch C. A high-order accurate unstructured finite volume Newton-Krylov algorithm for inviscid compressible flows. J Comput Phys 2008;227:2582-609.

[98] Chisholm TT, Zingg DW. A Jacobian-free Newton-Krylov algorithm for compressible turbulent fluid flows. J Comput Phys 2009;228:3490-507.

[99] Lucas P, van Zuijlen AH, Bijl H. Fast unsteady flow computations with a Jacobian-free Newton-Krylov algorithm. J Comput Phys 2010;229:9201-15.

[100] Cai X-C, Gropp WD, Keyes DE, Tidriri MD. Newton-Krylov-Schwarz methods in CFD. In: Hebeker F-K, Rannacher R, Wittum G, editors. Proceedings of the International Workshop on the Navier-Stokes Equations;2013. p. 17-30.

[101] Meijerink JA, Van der Vorst HA. Guidelines for usage of incomplete decompositions in solving sets of linear equations as they occur in practical problems. J Comput Phys 1981;44:134-55.

[102] Hackbusch W. Iterative solution of large sparse systems of equations. New York: Springer Verlag;1994.

[103] Venkatakrishnan V, Mavriplis DJ. Implicit solvers for unstructured meshes. J Comput Phys 1993;105:83-91.

[104] Ajmani K, Liou M-S. Implicit conjugate gradient solvers on distributed-memory architectures. AIAA Paper 95-1695;1995.

[105] Luo H, Baum JD, Löner R. A fast, Matrix-free implicit method for compressible flows on unstructured grids. AIAA Paper 99-0936;1999.

[106] MulderW, van Leer B. Experiments with implicit upwind methods for the Euler equations. J Comput Phys 1985;59:232-46.

[107] Cooper GJ, Sayfy A. Additive methods for the numerical solution of ordinary differential equations. Math Comput 1980;35:1159-72.

[108] Cooper GJ, Sayfy A. Additive Runge-Kutta methods for stiff ordinary differential equations. Math Comput 1983;40:207-18.

[109] Kennedy CA, Carpenter MH. Additive Runge-Kutta schemes for convection-diffusion-reaction equations. Appl Numer Math 2003;44:139-81.

[110] Alexander R. Diagonally implicit Runge-Kutta methods for stiff ODE' s. SIAM J Numer Anal 1977;14:1006-21.

[111] Najafi-Yazdi A, Mongeau L. A low-dispersion and low-dissipation implicit Runge-Kutta scheme. J Comput Phys 2013;233:315-23.

[112] Nazari F, Mohammadian A, Charron M, Zadra A. Optimal high-order diagonally-implicit Runge-Kutta schemes for nonlinear diffusive systems on atmospheric boundary layer. J Comput Phys 2014;271:118-30.

[113] Zhong X. Additive semi-implicit Runge-Kutta schemes for computing high-speed nonequilibri-

um reactive flows. J Comput Phys 1996;128:19-31.
[114] Yoh JJ-I,Zhong X. Semi-implicit Runge-Kutta schemes for stiff multi-dimensional reacting flows. AIAA Paper 97-0803;1997.
[115] Yoh JJ-I,Zhong X. Low-Storage semi-implicit Runge-Kutta methods for reactive flow computations. AIAA Paper 98-0130;1998.
[116] Calvo MP,de Frutos J,Novo J. Linearly implicit Runge-Kutta methods for advection-reaction-diffusion equations. Appl Numer Math 2001;37:535-49.
[117] Dong H,Zhong X. Time-accurate simulations of hypersonic boundary layer stability and transition over blunt bodies using implicit parallel algorithms. AIAA Paper 2002-0156;2002.
[118] Calvo MP,Gerisch A. Linearly implicit Runge-Kutta methods and approximate matrix factorization. Appl Numer Math 2005;53:183-200.
[119] Kanevsky A,Carpenter MH,Gottlieb D,Hesthaven JS. Application of implicit-explicit high order Runge-Kutta methods to discontinuous-Galerkin schemes. J Comput Phys 2007;225:1753-81.
[120] Kazemi-Kamyab V,van Zuijlen AH,Bijl H. Analysis and application of high order implicit Runge-Kutta schemes for unsteady conjugate heat transfer: a strongly-coupled approach. J Comput Phys 2014;272:471-86.
[121] Langer S. Agglomeration multigrid methods with implicit Runge-Kutta smoothers applied to aerodynamic simulations on unstructured grids. J Comput Phys 2014;277:72-100.
[122] Tam CKW. Computational aeroacoustics:issues and methods. AIAA Paper 95-0677;1995.
[123] Tam CKW. Applied aero-acoustics:prediction methods. VKI Lecture Series 1996-04;1996.
[124] Bijl H,Carpenter MH,Vatsa VN,Kennedy CA. Implicit time integration schemes for the unsteady compressible Navier-Stokes equations: laminar flow. J Comput Phys 2002;179:313-29.
[125] Carpenter MH,Viken SA,Nielsen EJ. The efficiency of high order temporal schemes. AIAA Paper 2003-0086;2003.
[126] Jothiprasad G,Mavriplis DJ,Caughey DA. Higher-order time integration schemes for the unsteady Navier-Stokes equations on unstructured meshes. J Comput Phys 2003;191:542-66.
[127] Jameson A. Time-dependent calculations using multigrid with applications to unsteady flows past airfoils and wings. AIAA Paper 91-1596;1991.
[128] Arnone A,Liou MS,Povinelli LA. Multigrid time-accurate integration of Navier-Stokes equations. AIAA Paper 93-3361;1993.
[129] Melson ND,Sanetrik MD,Atkins HL. Time-accurate Navier-Stokes calculations with multigrid acceleration. In:Proc. 6th Copper Mountain conf. on multigrid methods;1993. p. 423-39.
[130] Venkatakrishnan V. Implicit schemes and parallel computing in unstructured grid CFD. ICASE Report,No. 95-28;1995.
[131] Venkatakrishnan V,Mavriplis DJ. Implicit method for the computation of unsteady flows on unstructured grids. J Comput Phys 1996;127:380-97.
[132] Alonso JJ, Jameson A. Fully-implicit time-marching aeroelastic solution. AIAA Paper

94-0056;1994.

[133] Belov A, Martinelli L, Jameson A. A new implicit algorithm with multigrid for unsteady incompressible flow calculations. AIAA Paper 95-0049;1995.

[134] Pierce NA, Alonso JJ. A preconditioned implicit multigrid algorithm for parallel computation of unsteady aeroelastic compressible flows. AIAA Paper 97-0444;1997.

[135] Singh KP, Newman JC, Baysal O. Dynamic unstructured method for flows past multiple objects in relative motion. AIAA J 1995;33:641-9.

[136] Dubuc L, Cantariti F, Woodgate M, Gribben B, Badcock KJ, Richards BE. Solution of the unsteady euler equations using an implicit dual-time method. AIAA J 1998;36:1417-24.

[137] Pulliam TH. Time accuracy and the use of implicit methods. AIAA paper 93-3360;1993.

[138] Bartels RE. An elasticity-based mesh scheme applied to the computation of unsteady three-dimensional spoiler and aeroelastic problems. AIAA Paper 99-3301;1999.

[139] Chassaing JC, Gerolymos GA, Vallet I. Reynolds-stress model dual-time-stepping computation of unsteady three-dimensional flows. AIAA J 2003;41:1882-94.

第7章
湍流建模

与层流相反,湍流的突出特征是,流体微团以一种无序的方式沿着复杂的不规则路径运动。强烈的无序运动使不同的流体层激烈地混合在一起,由于流体微团和固壁之间动量和能量交换的增加,在相同的条件下,相比层流,湍流导致更高的表面摩擦和热量交换。

虽然流动变量的无序波动具有确定性,但湍流模拟仍然是一个重要的问题。尽管现代超级计算机的性能很好,但用非定常纳维-斯托克斯方程组(2.19)直接模拟湍流(即直接数值模拟(direct numerical simulation,DNS)[1-16])仅适用于量级为 $10^4 \sim 10^5$ 的低 Re 数下相对简单的流动问题。DNS 需要与 $Re^{9/4}$ 成正比的网格量才足以分辨空间结构,需要的 CPU 时间与 Re^3 成正比,从而限制了它的广泛应用。因此,我们被迫用近似的方式考虑湍流的影响,人们发展了种类繁多的湍流模型,并且这方面研究还在持续推进中。现今的湍流模型主要有五类:

(1) 代数模型;

(2) 一方程模型;

(3) 多方程模型;

(4) 二阶封闭模型(雷诺应力模型);

(5) LES。

前三个模型属于一阶封闭模型,大多基于 Boussinesq 的涡黏性假设[17-18],但对于某些特定应用也基于非线性涡黏性公式。图 7.1 显示了湍流模型的分类概况,根据其复杂度降序排列。

我们应该注意到一个事实,即没有一个单一的湍流模型可以可靠地预测各种湍流流动,每个模型都有其优缺点。例如,如果一个特定的模型在附面层算例中表现很好,那么它可能在分离流动中表现糟糕,因此,关注模型是否能包括所研究流动的所有重要特征总是很重要的。计算量与特定应用精度需求之间的协

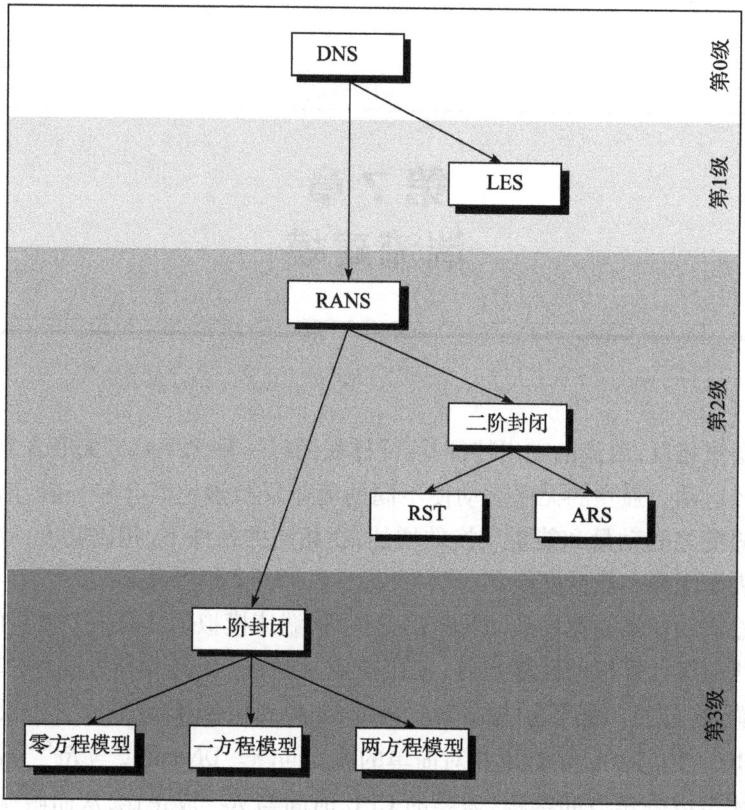

图 7.1　湍流模型的分级

调是需要关注的另一方面，在许多情况下，一个计算快捷的湍流模型能与更复杂的模型在某些全局量的预测上精度相当。

在后面几节中，我们首先介绍对控制方程组进行时间和质量平均得到的湍流基本方程组，其次介绍 Boussinesq 方法和非线性涡黏性方法，随后简要讨论 RST 方程，这是代数模型和 RST 模型的基础。在 7.2 节中，我们将介绍一些应用广泛的一方程和两方程一阶封闭模型。最后，由于 LES 及其相关方法在工程上应用越来越多，本章将对 LES 及其相关方法进行较为详细的讨论。

7.1　湍流基本方程组

首先，将控制方程组(2.19)重写为微分形式(见 A.1 节)，它在湍流建模的文献中经常使用，此外，这种表达形式紧凑、标记清晰。不过我们也将提供积分形式的湍流方程组示例。

在可压缩牛顿流体情况下，直角坐标系中不考虑源项的纳维-斯托克斯方程组的守恒形式为

$$
\begin{aligned}
&\frac{\partial\rho}{\partial t}+\frac{\partial}{\partial x_i}(\rho v_i)=0\\
&\frac{\partial}{\partial t}(\rho v_i)+\frac{\partial}{\partial x_j}(\rho v_j v_i)=-\frac{\partial p}{\partial x_i}+\frac{\partial\tau_{ij}}{\partial x_j}\\
&\frac{\partial}{\partial t}(\rho E)+\frac{\partial}{\partial x_j}(\rho v_j H)=\frac{\partial}{\partial x_j}(v_i\tau_{ij})+\frac{\partial}{\partial x_j}\left(k\frac{\partial T}{\partial x_j}\right)
\end{aligned}
\tag{7.1}
$$

其中，v_i 为速度分量（$v=[v_1,v_2,v_3]^{\mathrm{T}}$），$x_i$ 为坐标方向，张量符号在 A. 13 节中有详细解释。

式(7.1)中黏性应力张量 τ_{ij} 的分量定义为

$$\tau_{ij}=2\mu S_{ij}+\lambda\frac{\partial v_k}{\partial x_k}\delta_{ij}=2\mu S_{ij}-\left(\frac{2\mu}{3}\right)\frac{\partial v_k}{\partial x_k}\delta_{ij} \tag{7.2}$$

这里利用了斯托克斯假设（式(2.17)）。在笛卡儿坐标系下，式(7.2)等价于式(2.15)。式(7.2)中的第二项，即 $\partial v_k/\partial x_k$，对应于速度的散度，它在不可压缩流中消失。应变率张量的分量为

$$S_{ij}=\frac{1}{2}\left(\frac{\partial v_i}{\partial x_j}+\frac{\partial v_j}{\partial x_i}\right) \tag{7.3}$$

在此，定义旋转速率张量（速度梯度张量的反对称部分）为

$$\Omega_{ij}=\frac{1}{2}\left(\frac{\partial v_j}{\partial x_i}-\frac{\partial v_i}{\partial x_j}\right) \tag{7.4}$$

式(7.1)中的总能 E 和总焓 H 很容易从公式中得到

$$E=e+\frac{1}{2}v_iv_i \quad H=h+\frac{1}{2}v_iv_i \tag{7.5}$$

在笛卡儿坐标系下分别对应于式(2.6)和式(2.12)。

对于不可压缩流，可以将式(7.1)简化为

$$
\begin{aligned}
&\frac{\partial v_i}{\partial x_i}=0\\
&\frac{\partial v_i}{\partial t}+v_j\frac{\partial v_i}{\partial x_j}=-\frac{1}{\rho}\frac{\partial p}{\partial x_i}+\nu\nabla^2 v_i\\
&\frac{\partial T}{\partial t}+v_j\frac{\partial T}{\partial x_j}=k\nabla^2 T
\end{aligned}
\tag{7.6}
$$

其中，$\nu=\mu/\rho$ 为运动学黏度系数，∇^2 为拉普拉斯算子。在没有弹性效应的情况下，温度 T 的方程与质量守恒和动量方程解耦。

7.1.1 雷诺平均

第一种近似处理湍流的方法由雷诺在 1895 年提出。该方法将流场变量分解为平均值和脉动两部分，然后由控制方程组(7.1)求解对工程应用来说最感兴趣的平均值。首先考虑不可压缩流动，将式(7.1)中的速度分量和压力用式(7.7)代替[19]

$$v_i = \bar{v}_i + v_i' \quad p = \bar{p} + p' \tag{7.7}$$

其中，$\overline{(\cdot)}$表示平均值，$(\cdot)'$表示脉动量。平均值由平均过程得到，雷诺平均有三种不同形式：

(1) 时间平均——适于稳态湍流（统计上稳定的湍流）：

$$\bar{v}_i = \lim_{T\to\infty} \frac{1}{T} \int_t^{t+T} v_i \mathrm{d}t \tag{7.8}$$

因此，平均值 $\bar{v}_i$ 不随时间变化，只随空间变化，这种情况见图 7.2。在实际中，$T\to\infty$ 意味着积分时间间隔 T 要比湍流脉动特征时间尺度大。

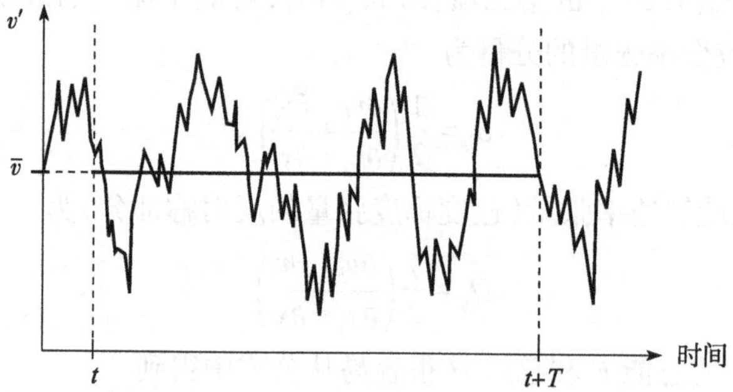

图 7.2　雷诺平均——湍流速度脉动 v' 与统计均值 $\bar{v}$ 示意图

(2) 空间平均——适用于均匀湍流：

$$\bar{v}_i = \lim_{\Omega\to\infty} \frac{1}{\Omega} \int_\Omega v_i \mathrm{d}\Omega \tag{7.9}$$

其中，Ω 是控制体体积。在这种情况下，$\bar{v}_i$ 在空间上是不变的，但允许在时间上变化。

(3) 系综平均——适用于一般湍流：

$$\bar{v}_i = \lim_{N\to\infty} \frac{1}{N} \sum_{m=1}^{N} v_i \tag{7.10}$$

在此，平均值 $\bar{v}_i$ 仍然是时间和空间坐标的函数。

对于这三种方法，脉动量的平均值均为 0，即 $\overline{v_i'} = 0$，但是可以很容易看出

$\overline{v_i'v_i'}\neq 0$。如果湍流的两个速度分量是相关的，那么对于 $\overline{v_i'v_j'}$ 也是如此。

如果湍流是稳态且均匀的，则这三种平均形式是等效的，这被称为遍历假设。

7.1.2 Favre（质量）平均

在密度不恒定的情况下，建议对式(7.1)中的某些量采用密度（质量）加权或 Favre 分解[20-21]，而不是采用雷诺平均。否则，由于额外的与密度脉动有关的相关项，平均控制方程组将变得相当复杂。最方便的方法是对密度和压力采用雷诺平均，对其他变量如速度、内能、焓和温度采用 Favre 平均。速度分量的 Favre 平均，可从以下关系式中得到[20-21]

$$\tilde{v}_i=\frac{1}{\bar{\rho}}\lim_{T\to\infty}\frac{1}{T}\int_t^{t+T}\rho v_i\mathrm{d}t \tag{7.11}$$

其中，$\bar{\rho}$ 表示雷诺平均的密度。因此，Favre 分解如下：

$$v_i=\tilde{v}_i+v_i'' \tag{7.12}$$

其中，$\tilde{v}_i$ 和 v_i'' 分别为速度分量 v_i 的均值和脉动。同样，脉动量的均值为零，即 $\widetilde{v_i''}=0$。此外，如果两个脉动量是相关的，那么它们乘积的平均值就不是零，例如 $\widetilde{v_i''v_i''}\neq 0$，一般情况下 $\widetilde{v_i''v_j''}\neq 0$。

对于 Favre 和雷诺平均的混合，可以推导出下列关系：

$$\widetilde{\rho v_i}=\bar{\rho}\tilde{v}_i,\quad \overline{\rho v_i''}=0,\quad 但\overline{v_i''}\neq 0 \tag{7.13}$$

以下各节将使用到这些关系。

7.1.3 雷诺平均纳维-斯托克斯方程组

对不可压缩纳维-斯托克斯方程组(7.6)进行时间平均（式(7.8)）或者系综平均（式(7.10)），可以得到如下的质量和动量守恒方程：

$$\begin{gathered}\frac{\partial\bar{v}_i}{\partial x_i}=0\\ \rho\frac{\partial\bar{v}_i}{\partial t}+\rho\bar{v}_j\frac{\partial\bar{v}_i}{\partial x_j}=-\frac{\partial\bar{p}}{\partial x_i}+\frac{\partial}{\partial x_j}(\bar{\tau}_{ij}-\rho\overline{v_i'v_j'})\end{gathered} \tag{7.14}$$

这样就得到 RANS 方程组。式(7.14)在形式上与纳维-斯托克斯方程组(7.6)相同，只是增加了一项：

$$\boldsymbol{\tau}_{ij}^{\mathrm{R}}=-\rho\overline{v_i'v_j'}=-\rho(\overline{v_iv_j}-\bar{v}_i\bar{v}_j) \tag{7.15}$$

它构成了雷诺应力张量，表示由湍流脉动引起的动量传输。层流黏性应力根据式(7.2)和式(7.3)使用雷诺平均的速度分量计算，即

$$\overline{\tau}_{ij}=2\mu\overline{S}_{ij}=\mu\left(\frac{\partial\overline{v}_i}{\partial x_j}+\frac{\partial\overline{v}_j}{\partial x_i}\right) \tag{7.16}$$

3D 问题中，雷诺应力张量由九个分量组成：

$$\rho\overline{v_i'v_j'}=\begin{bmatrix}\rho\overline{(v_1')^2} & \rho\overline{v_1'v_2'} & \rho\overline{v_1'v_3'}\\ \rho\overline{v_2'v_1'} & \rho\overline{(v_2')^2} & \rho\overline{v_2'v_3'}\\ \rho\overline{v_3'v_1'} & \rho\overline{v_3'v_2'} & \rho\overline{(v_3')^2}\end{bmatrix} \tag{7.17}$$

然而，由于在表达式中 v_i' 和 v_j' 可以互换，雷诺应力张量只包含 6 个独立分量。正应力求和之后除以密度定义了湍动能，即

$$K=\frac{1}{2}\overline{v_i'v_i'}=\frac{1}{2}\left[\overline{(v_1')^2}+\overline{(v_2')^2}+\overline{(v_3')^2}\right] \tag{7.18}$$

基于 RANS 方程组的湍流建模的基本问题是，找到 6 个附加关系式以封闭式(7.14)，我们将在 7.1.5~7.1.7 节介绍基本方法。

7.1.4 Favre 和雷诺平均纳维-斯托克斯方程组

在湍流建模中，通常假定 Morkovin 假设[22]有效，它表明如果 $\rho'\ll\overline{\rho}$，边界层的湍流结构不会受到密度脉动的显著影响。这对于马赫数 5 以内的壁面边界流动一般是正确的，然而在高超声速或可压缩自由剪切流中，必须考虑密度的脉动，对于燃烧流动或有明显传热过程的流动同样如此。

对密度和压力进行雷诺平均(式(7.8)或式(7.14))，可压缩纳维-斯托克斯方程组中的其他流动变量进行 Favre 平均(式(7.11))，得到如下的控制方程组：

$$\begin{aligned}&\frac{\partial\overline{\rho}}{\partial t}+\frac{\partial}{\partial x_i}(\overline{\rho}\widetilde{v}_i)=0\\&\frac{\partial}{\partial t}(\overline{\rho}\widetilde{v}_i)+\frac{\partial}{\partial x_j}(\overline{\rho}\widetilde{v}_j\widetilde{v}_i)=-\frac{\partial\overline{p}}{\partial x_i}+\frac{\partial}{\partial x_j}(\widetilde{\tau}_{ij}-\overline{\rho}\widetilde{v_i''v_j''})\\&\frac{\partial}{\partial t}(\overline{\rho}\widetilde{E})+\frac{\partial}{\partial x_j}(\overline{\rho}\widetilde{v}_j\widetilde{H})=\frac{\partial}{\partial x_j}\left(k\frac{\partial\widetilde{T}}{\partial x_j}-\overline{\rho}\widetilde{v_j''h''}+\widetilde{\tau_{ij}v_i''}-\overline{\rho}\widetilde{v_j''K}\right)+\\&\qquad\frac{\partial}{\partial x_j}\left[\widetilde{v}_i(\widetilde{\tau}_{ij}-\overline{\rho}\widetilde{v_i''v_j''})\right]\end{aligned} \tag{7.19}$$

这就是 Favre 和雷诺平均纳维-斯托克斯方程组。与雷诺平均类似，动量和能量方程中的黏性应力张量扩展为 Favre 平均雷诺应力张量，即

$$\tau_{ij}^{F}=-\bar{\rho}\widetilde{v_i''v_j''} \tag{7.20}$$

其形式类似于式(7.17),但采用 Favre 平均而不是雷诺平均。层流(分子)黏性应力张量 τ_{ij} 的分量由式(7.2)利用 Favre 平均的速度分量计算。

如果我们采用 Favre 平均的湍动能定义,即

$$\bar{\rho}\widetilde{K}=\frac{1}{2}\bar{\rho}\widetilde{v_i''v_i''} \tag{7.21}$$

那么式(7.19)中的总能可以表示为

$$\bar{\rho}\widetilde{E}=\bar{\rho}\widetilde{e}+\frac{1}{2}\bar{\rho}\widetilde{v}_i\widetilde{v}_i+\frac{1}{2}\bar{\rho}\widetilde{v_i''v_i''}=\bar{\rho}\widetilde{e}+\frac{1}{2}\bar{\rho}\widetilde{v}_i\widetilde{v}_i+\bar{\rho}\widetilde{K} \tag{7.22}$$

总焓定义为

$$\bar{\rho}\widetilde{H}=\bar{\rho}\widetilde{h}+\frac{1}{2}\bar{\rho}\widetilde{v}_i\widetilde{v}_i+\frac{1}{2}\bar{\rho}\widetilde{v_i''v_i''}=\bar{\rho}\widetilde{h}+\frac{1}{2}\bar{\rho}\widetilde{v}_i\widetilde{v}_i+\bar{\rho}\widetilde{K} \tag{7.23}$$

Favre 和雷诺平均纳维-斯托克斯方程组(式(7.19))中的每一项的物理意义如下[23]:

$\frac{\partial}{\partial x_j}\left(k\frac{\partial\widetilde{T}}{\partial x_j}\right)$ ——热的分子扩散

$\frac{\partial}{\partial x_j}(\bar{\rho}\widetilde{v_j''h''})$ ——热的湍流输运

$\frac{\partial}{\partial x_j}(\widetilde{\tau_{ij}v_i''})$ ——$\widetilde{K}$ 的分子扩散

$\frac{\partial}{\partial x_j}(\bar{\rho}\widetilde{v_j''K})$ ——$\widetilde{K}$ 的湍流输运

$\frac{\partial}{\partial x_j}(\widetilde{v}_i\widetilde{\tau}_{ij})$ ——分子应力做的功

$\frac{\partial}{\partial x_j}(\widetilde{v}_i\tau_{ij}^{F})$ ——Favre 平均雷诺应力做的功

$\widetilde{K}$ 的分子扩散和湍流输运常被忽略,这对于跨声速和超声速流动是合理的近似。为了封闭 Favre 和雷诺平均方程组(式(7.19)),还必须提供 Favre 和雷诺平均应力张量(式(7.20))的 6 个分量和湍流热通量的 3 个分量。我们将在下面几节中讨论三种基本方法。

7.1.5　涡黏性假设

湍流建模最重要的贡献之一是 Boussinesq 在 1877 年提出的涡黏性假

设[17-18]，他的观点基于这样的观察：湍流中的动量传递是由含能大尺度涡的混合主导的。Boussinesq 假设假定湍流剪切应力与平均应变率呈线性关系，就像在层流中一样，其比例系数就是涡黏性。对于雷诺平均不可压缩流动（式(7.14)），Boussinesq 假设可表示为

$$\tau_{ij}^{\mathrm{R}}=-\rho\overline{v_i'v_j'}=2\mu_{\mathrm{T}}\overline{S}_{ij}-\frac{2}{3}\rho K\delta_{ij} \tag{7.24}$$

这里 $\overline{S}_{ij}$ 表示雷诺平均的应变率张量（式(7.3)或式(7.16)），K 是湍动能（$K=(1/2)\overline{v_i'v_i'}$），$\mu_{\mathrm{T}}$ 为涡黏性系数。与分子黏性系数 μ 不同，涡黏性系数 μ_{T} 不表征流体的物理特性，而是局部流动条件的函数，此外 μ_{T} 还受到流动历史效应的强烈影响。

在可压缩 Favre 和雷诺平均纳维-斯托克斯方程组(7.19)中，Boussinesq 涡黏性假设表示为

$$\tau_{ij}^{\mathrm{F}}=-\bar{\rho}\widetilde{v_i''v_j''}=2\mu_{\mathrm{T}}\widetilde{S}_{ij}-\left(\frac{2\mu_{\mathrm{T}}}{3}\right)\frac{\partial\widetilde{v}_k}{\partial x_k}\delta_{ij}-\frac{2}{3}\bar{\rho}\widetilde{K}\delta_{ij} \tag{7.25}$$

这里 $\widetilde{S}_{ij}$ 和 $\widetilde{K}$ 分别代表 Favre 平均的应变率张量和湍动能，注意它与式(7.2)的相似性。式(7.24)和式(7.25)中的 $(2/3)\rho K\delta_{ij}$ 项需要指定以便正确获得 τ_{ij}^{R} 和 τ_{ij}^{F} 的迹。这意味着在 $\overline{S}_{ii}=0$（连续方程）或 $\widetilde{S}_{ii}=0$ 的情况下，必有

$$\tau_{ii}^{\mathrm{R}}=-2\rho K \quad 或 \quad \tau_{ii}^{\mathrm{F}}=-2\bar{\rho}\widetilde{K}$$

这样才能满足式(7.18)或式(7.21)中湍动能的关系式。然而，$(2/3)\rho K\delta_{ij}$ 这一项经常被忽略，特别是在简单的湍流模型（如代数模型）中。

湍流热通量的建模通常采用基于经典雷诺比拟[24]的近似。因此，可以写为

$$\bar{\rho}\widetilde{v_j''h''}=-k_{\mathrm{T}}\frac{\partial\widetilde{T}}{\partial x_j} \tag{7.26}$$

其中，湍流热传导系数 k_{T} 定义为

$$k_{\mathrm{T}}=c_p\frac{\mu_{\mathrm{T}}}{Pr_{\mathrm{T}}} \tag{7.27}$$

在式(7.27)中，c_p 表示比定压热容，Pr_{T} 是湍流 Pr 数，在流场中通常假定为常数（空气中 $Pr_{\mathrm{T}}=0.9$）。

通过对控制方程组(2.19)或控制方程组(7.1)的雷诺平均（和 Favre 平均）形式应用涡黏性假设，式(2.15)或式(7.2)的黏性应力张量中的动力学黏性系数 μ 可以简单地替换为层流和湍流分量的累加，即

$$\mu=\mu_{\mathrm{L}}+\mu_{\mathrm{T}} \tag{7.28}$$

对于层流黏性系数 μ_{L}，可以通过如式(2.30)的萨瑟兰公式计算。此外，根

据式(7.26)的雷诺比拟,式(2.24)或式(7.2)中的热传导系数 k 由下式计算:

$$k=k_{\mathrm{L}}+k_{\mathrm{T}}=c_p\left(\frac{\mu_{\mathrm{L}}}{Pr_{\mathrm{L}}}+\frac{\mu_{\mathrm{T}}}{Pr_{\mathrm{T}}}\right) \tag{7.29}$$

至少从工程角度来看,Boussinesq 的涡黏性概念是非常有吸引力的,因为它只需要确定 μ_{T}(式(7.24)或式(7.25)中的 $(2/3)\rho K\delta_{ij}$ 项需要的湍动能 K 或者是湍流模型的副产品,或者被简单地省略掉)。一旦获得了 μ_{T},就可以通过引入平均流动变量并在层流黏性中加入 μ_{T},很容易地将纳维-斯托克斯方程组(2.19)推广到式(7.1)来模拟湍流,因此,Boussinesq 方法是大量一阶湍流封闭模型的基础。然而,也有一些应用场景下 Boussinesq 假设不再有效(例如文献[23,p.214]或文献[25,p.111]),包括:

(1) 平均应变率突然变化的流动;

(2) 流线显著弯曲的流动;

(3) 旋转和分层的流动;

(4) 管道和涡轮机械中的二次流;

(5) 分离和再附的边界层流动。

涡黏性方法的局限是由湍流与平均应变场之间的平衡假设,以及独立于系统旋转引起的。在湍流模型中使用适当的修正项可以显著改善预测结果[26-27],进一步可以通过应用非线性涡黏性模型来提高预测的精度,这些模型将在 7.1.6 节中介绍。

7.1.6　非线性涡黏性

为了消除湍流与平均应变率平衡假设的限制,Lumley[28-29] 提出用应变张量和旋转张量的更高阶乘积对线性 Boussinesq 方法进行推广,可以看作泰勒级数展开。根据 Lumley 的思想,学者们提出了许多非线性涡黏性模型,如文献[30-37]。

下面介绍 Shih 等[31] 提出的一种方法,在广义的涡黏性公式中加入三阶项,特别适合于旋转流动。正如在文献[25,p.194]中已经指出的,三次项是提高精度所必需的。雷诺应力 τ_{ij}^{R} 可表示为[31-32]

$$\begin{aligned}\rho\overline{v_i'v_j'}=&\frac{2}{3}\rho K\delta_{ij}-C_1\frac{\rho K^2}{\epsilon}2S_{ij}^*-C_3\frac{\rho K^3}{\epsilon^2}[\overline{S}_{ik}\overline{\Omega}_{kj}-\overline{\Omega}_{ik}\overline{S}_{kj}]-\\&C_4\frac{\rho K^4}{\epsilon^3}[(\overline{S}_{ik})^2\overline{\Omega}_{kj}-\overline{\Omega}_{ik}(\overline{S}_{kj})^2]+\\&C_5\frac{\rho K^4}{\epsilon^3}\left[\overline{\Omega}_{ik}\overline{S}_{km}\overline{\Omega}_{mj}-\frac{1}{3}\overline{\Omega}_{kl}\overline{S}_{lm}\overline{\Omega}_{mk}\delta_{ij}+I_sS_{ij}^*\right]\end{aligned} \tag{7.30}$$

其中，$\overline{S}_{ij}$，$\overline{\Omega}_{ij}$ 的定义见式(7.3)和式(7.4)。此外

$$I_s = \frac{1}{2}[\overline{S}_{kk}\overline{S}_{ll} - (\overline{S}_{kk})^2]S_{ij}^*$$
$$S_{ij}^* = \overline{S}_{ij} - \frac{1}{3}\overline{S}_{kk}\delta_{ij} \tag{7.31}$$

湍动能 K 和耗散率 ϵ 的数值由低 Re 数 K-ϵ 湍流模型得到（见7.2.2节）。式(7.30)中的参数 C_1 至 C_5 在文献[31-32]中给出。

与线性涡黏性方法相比，非线性模型的计算量只增加一点，但大幅改善了对某些复杂湍流的预测能力。

7.1.7 雷诺应力输运方程

可以通过时间平均（二阶矩）来推导雷诺应力的精确方程，即

$$\overline{v_i' N(v_j) + v_j' N(v_i)} = 0 \tag{7.32}$$

其中，$N(v_i)$ 为纳维-斯托克斯算子，即

$$N(v_i) = \rho\frac{\partial v_i}{\partial t} + \rho v_j\frac{\partial v_i}{\partial x_j} + \frac{\partial p}{\partial x_i} - \mu\nabla^2 v_i \tag{7.33}$$

将式(7.32)与式(7.33)结合，得到如下的不可压缩流动 RST 方程[38]：

$$\frac{\partial \tau_{ij}^{\mathrm{R}}}{\partial t} + \overline{v}_k\frac{\partial \tau_{ij}^{\mathrm{R}}}{\partial x_k} = P_{ij} + \Pi_{ij} - \epsilon_{ij} - \frac{\partial C_{ijk}}{\partial x_k} + \mu\nabla^2\tau_{ij}^{\mathrm{R}} \tag{7.34}$$

可压缩流动的情况可以参考文献[23, p. 179]或文献[39-40]。式(7.34)中湍动能的生成项 P_{ij}、压力-应变项 Π_{ij}、耗散项 ϵ_{ij} 和三阶扩散项 C_{ijk} 分别定义为

$$P_{ij} = -\tau_{ik}^{\mathrm{R}}\frac{\partial \overline{v}_j}{\partial x_k} - \tau_{jk}^{\mathrm{R}}\frac{\partial \overline{v}_i}{\partial x_k}$$
$$\Pi_{ij} = \overline{p'\left(\frac{\partial v_i'}{\partial x_j} + \frac{\partial v_j'}{\partial x_i}\right)} = 2\overline{p'S_{ij}'} \tag{7.35}$$
$$\epsilon_{ij} = 2\mu\overline{\frac{\partial v_i'}{\partial x_k}\frac{\partial v_j'}{\partial x_k}}$$
$$C_{ijk} = \rho\overline{v_i'v_j'v_k'} + \overline{p'v_i'}\delta_{jk} + \overline{p'v_j'}\delta_{ik}$$

其中，S_{ij}' 表示应变率张量的脉动部分。C_{ijk} 的第一部分是湍流脉动的三阶相关项，表示对流脉动驱动的输运，其余两部分是压力输运项（压力-速度相关）。

可以看到，精确的雷诺应力方程包含了新的未知高阶相关项，例如 $\overline{v_i'v_j'v_k'}$。因此，式(7.34)只能用经验模型来封闭，这是由纳维-斯托克斯方程组的非线性

造成的。二阶封闭模型(雷诺应力模型)为求解式(7.34)提供了必要的框架,可以在文献[41-45]中找到实现的示例。

7.2 一阶封闭

在雷诺/Favre平均纳维-斯托克斯方程组中,一阶封闭是近似雷诺应力最简单的方法,它们基于7.1.5节的Boussinesq假设或7.1.6节的非线性涡黏性模型,因此,湍流模型的任务是计算涡黏性系数μ_T。

在众多一阶封闭模型中,我们选择了三个应用广泛、代表当前技术水平的模型,这三种模型都可以在结构网格和非结构网格上轻松实现。首先,讨论由Spalart和Allmaras提出的一方程模型。其次,介绍著名的K-ϵ两方程模型。最后,讨论Menter提出的K-ω切应力输运(shear-stress transportation, SST)两方程模型。这类湍流模型在各种算例中的详细比较可以参见文献[46]。

在7.2.1节中,密度和速度分量表示雷诺/Favre平均之后的变量,为了书写方便省略了相应的符号。

7.2.1 Spalart-Allmaras一方程模型

Spalart-Allmaras一方程湍流模型[47]采用关于涡黏性变量$\tilde{\nu}$的输运方程,它是基于经验、量纲分析和伽利略不变性而提出的,使用2D混合层、尾流和平板边界层的结果进行了校准。Spalart-Allmaras模型能相对合理地预测逆压梯度下的湍流流动,此外,它还能够在用户指定位置提供从层流到湍流的光滑过渡。Spalart-Allmaras模型有多个有利的数值特性。它是"局部的"模型,这意味着某一点的方程不依赖于其他点上的解,因此可以很容易地在结构多块或非结构网格上实现。它还具有鲁棒性,能够快速收敛到稳态解,并且在近壁区域只需要中等尺度的网格分辨率。

1. 微分形式

使用张量符号的Spalart-Allmaras湍流模型形式为[47]

$$
\begin{aligned}
\frac{\partial \tilde{\nu}}{\partial t}+\frac{\partial}{\partial x_j}(\tilde{\nu} v_j)= & C_{b1}(1-f_{t2})\tilde{S}\tilde{\nu}+ \\
& \frac{1}{\sigma}\left\{\frac{\partial}{\partial x_j}\left[(\nu_L+\tilde{\nu})\frac{\partial \tilde{\nu}}{\partial x_j}\right]+C_{b2}\frac{\partial \tilde{\nu}}{\partial x_j}\frac{\partial \tilde{\nu}}{\partial x_j}\right\}- \\
& \left[C_{w1}f_w-\frac{C_{b1}}{\kappa^2}f_{t2}\right]\left(\frac{\tilde{\nu}}{d}\right)^2+f_{t1}\|\Delta \boldsymbol{v}\|_2^2
\end{aligned} \tag{7.36}
$$

右端项分别表示涡黏性的生成、守恒扩散、非守恒扩散、近壁湍流破坏、生成的转捩衰减和湍流转捩源项。此外 $\nu_L = \mu_L/\rho$ 表示层流运动学黏性系数，d 表示距最近壁面的距离（壁面距离的计算可参考文献[48]）。式(7.28)~式(7.29)中的湍流涡黏性系数由下式得到

$$\mu_T = f_{v1}\rho\tilde{\nu} \tag{7.37}$$

生成项通过下式计算：

$$\begin{aligned} &\tilde{S} = S + \frac{\tilde{\nu}}{\kappa^2 d^2} f_{v2} \\ &f_{v1} = \frac{\chi^3}{\chi^3 + C_{v1}^3}, \quad f_{v2} = 1 - \frac{\chi}{1+\chi f_{v1}} \\ &\chi = \frac{\tilde{\nu}}{\nu_L} \end{aligned} \tag{7.38}$$

其中，S 表示平均旋转率的大小，即

$$S = \sqrt{2\Omega_{ij}\Omega_{ij}}$$

这里 Ω_{ij} 由式(7.4)给出。为了防止数值求解上的问题，$\tilde{S}$ 不能为零或负值，可以采取的一种简单限制是 $\tilde{S} = \max(\tilde{S}, 0.3S)$，文献[49]给出了更精细的方法。

控制涡黏性破坏的项为

$$\begin{aligned} &f_w = g\left(\frac{1+C_{w3}^6}{g^6 + C_{w3}^6}\right)^{1/6} \\ &g = r + C_{w2}(r^6 - r) \\ &r = \min\left(\frac{\tilde{\nu}}{\tilde{S}\kappa^2 d^2}, 10\right) \end{aligned} \tag{7.39}$$

模化层流-湍流转捩效应的函数为

$$\begin{aligned} &f_{t1} = g_t C_{t1} \exp\left(-C_{t2}\frac{\omega_t^2}{\Delta U^2}(d^2 + g_t^2 d_t^2)\right) \\ &f_{t2} = C_{t3}\exp(-C_{t4}\chi^2), \quad g_t = \min[0.1, \|\Delta \boldsymbol{v}\|_2/(\omega_t \Delta x_t)] \end{aligned} \tag{7.40}$$

其中，ω_t 表示壁面转捩处（需要用户指定）的涡量，$\|\Delta \boldsymbol{v}\|_2$ 表示转捩点和当地流场位置之间速度差量的 2 范数，d_t 是离最近转捩点的距离，Δx_t 是转捩点沿壁面的间距。

式(7.36)~式(7.40)中的常数定义为

$$\begin{aligned} &C_{b1} = 0.1355 \qquad C_{b2} = 0.622 \\ &C_{v1} = 7.1 \qquad \sigma = 2/3 \qquad \kappa = 0.41 \\ &C_{w1} = C_{b1}/\kappa^2 + (1+C_{b2})/\sigma \qquad C_{w2} = 0.3 \qquad C_{w3} = 2 \\ &C_{t1} = 1 \qquad C_{t2} = 2 \qquad C_{t3} = 1.2 \qquad C_{t4} = 0.5 \end{aligned} \tag{7.41}$$

为了考虑非平衡效应对生成项的影响,有学者提出将 C_{b1} 表示为应变率的函数[50]。

文献[47]指出,将式(7.36)中的扩散项:

$$\frac{1}{\sigma}\left\{\frac{\partial}{\partial x_j}\left[(\nu_{\mathrm{L}}+\widetilde{\nu})\frac{\partial\widetilde{\nu}}{\partial x_j}\right]+C_{b2}\frac{\partial\widetilde{\nu}}{\partial x_j}\frac{\partial\widetilde{\nu}}{\partial x_j}\right\}$$

替换为

$$\frac{1+C_{b2}}{\sigma}\frac{\partial}{\partial x_j}\left[(\nu_{\mathrm{L}}+\widetilde{\nu})\frac{\partial\widetilde{\nu}}{\partial x_j}\right]-\frac{C_{b2}}{\sigma}(\nu_{\mathrm{L}}+\widetilde{\nu})\nabla^2\widetilde{\nu} \tag{7.42}$$

会更加方便,这样做回避了$(\partial\widetilde{\nu}/\partial x_j)^2$ 项的离散难题。

2. 积分形式

将 Spalart-Allmaras 湍流模型(式(7.36))转化到有限体积法的框架下,得到如下方程:

$$\frac{\partial}{\partial t}\int_{\Omega}\widetilde{\nu}\mathrm{d}\Omega+\oint_{\partial\Omega}(F_{\mathrm{c,T}}-F_{\mathrm{v,T}})\mathrm{d}S=\int_{\Omega}Q_{\mathrm{T}}\mathrm{d}\Omega \tag{7.43}$$

其中,Ω 表示控制体,$\partial\Omega$ 是控制体的面,$\mathrm{d}S$ 是 Ω 的面微元。对流通量定义为

$$F_{\mathrm{c,T}}=\widetilde{\nu}V \tag{7.44}$$

其中,V 是逆变速度(式(2.22))。对流通量通常使用一阶迎风格式离散。黏性通量为

$$F_{\mathrm{v,T}}=n_x\tau_{xx}^{\mathrm{T}}+n_y\tau_{yy}^{\mathrm{T}}+n_z\tau_{zz}^{\mathrm{T}} \tag{7.45}$$

其中,n_x,n_y 和 n_z 是单位法向矢量的分量。黏性正应力为

$$\begin{aligned}\tau_{xx}^{\mathrm{T}}&=\frac{1}{\sigma}(\nu_{\mathrm{L}}+\widetilde{\nu})\frac{\partial\widetilde{\nu}}{\partial x}\\ \tau_{yy}^{\mathrm{T}}&=\frac{1}{\sigma}(\nu_{\mathrm{L}}+\widetilde{\nu})\frac{\partial\widetilde{\nu}}{\partial y}\\ \tau_{zz}^{\mathrm{T}}&=\frac{1}{\sigma}(\nu_{\mathrm{L}}+\widetilde{\nu})\frac{\partial\widetilde{\nu}}{\partial z}\end{aligned} \tag{7.46}$$

式(7.43)中的源项为

$$\begin{aligned}Q_{\mathrm{T}}=&C_{b1}(1-f_{t2})\widetilde{S}\widetilde{\nu}+\frac{C_{b2}}{\sigma}\left[\left(\frac{\partial\widetilde{\nu}}{\partial x}\right)^2+\left(\frac{\partial\widetilde{\nu}}{\partial y}\right)^2+\left(\frac{\partial\widetilde{\nu}}{\partial z}\right)^2\right]-\\ &\left[C_{w1}f_w-\frac{C_{b1}}{\kappa^2}f_{t2}\right]\left(\frac{\widetilde{\nu}}{d}\right)^2+f_{t1}\|\Delta\overline{\boldsymbol{v}}\|_2^2\end{aligned} \tag{7.47}$$

也可以使用式(7.42)建议的非守恒扩散形式。模型常数在式(7.41)中定义。

3. 初始和边界条件

$\tilde{\nu}$ 的初始条件通常取为 $\tilde{\nu}=0.1\nu_L$，入流边界或者远场边界同样使用该值，然而，当转捩项 f_{t1}（通常还包括 f_{t2}）被忽略，即完全湍流情况下，入流边界或者远场边界的 $\tilde{\nu}$ 通常被设为 ν_L 的 3~5 倍。在出流边界，$\tilde{\nu}$ 简单地由计算域内部外插得到，对于固壁，设定 $\tilde{\nu}=0$ 是合理的，此时 $\mu_T=0$。

7.2.2 K-ϵ 两方程模型

K-ϵ 湍流模型可能是应用最广泛的两方程涡黏性模型，它求解湍动能 K 的方程和湍流耗散率 ϵ 的方程。K-ϵ 模型的历史根源可以回溯到 Chou[51] 的工作，在 20 世纪 70 年代，学者们提出了该模型的各种形式，最重要的贡献来自 Jones 和 Launder[52-53]、Launder 和 Sharma[54] 以及 Launder 和 Spalding[55]。

K-ϵ 模型需要加入衰减函数用以确保模型从黏性底层直到壁面一直合理有效，衰减函数的目的是保证在壁面处 K 和 ϵ 有正确的渐进性质，即

$$\text{当 } y\to 0 \text{ 时，}\quad K\sim y^2 \quad \frac{\epsilon}{K}\sim\frac{2\nu}{y^2} \tag{7.48}$$

其中，y 表示壁面法向坐标。此外，雷诺应力具有如下性质（参见文献[23, p. 138-139]）

$$\text{当 } y\to 0,\quad i\neq j \text{ 时，}\quad \boldsymbol{\tau}_{ij}^{\mathrm{R}}\sim y^3 \tag{7.49}$$

包含衰减函数的 K-ϵ 模型也称为低 *Re* 数模型。使用最广泛的衰减函数公式是 Jones 和 Launder[52]、Launder 和 Sharma[54]，Lam 和 Bremhorst[56]，以及 Chien[57] 提出的。读者可以在文献[58]中了解到七个不同低 *Re* 数 K-ϵ 模型的比较。

K-ϵ 模型的数值求解比前面讨论的 Spalart-Allmaras 模型（7.2.1 节）要更困难。特别是衰减函数导致湍流方程源项的刚性很强，以及（为了解析黏性底层）壁面附近必要的高分辨率网格，要求至少使用点隐式格式，或者更好的，使用全隐式时间推进格式。文献[59]给出了 K-ϵ 模型显式时间离散的有用建议，文献[60-74]给出了在结构和非结构网格上实现 K-ϵ 的示例。最后强调一点，在存在逆压梯度的流动中，K-ϵ 模型的准确度会降低[23,58]，文献[75]讨论了在各种压力梯度流动中，模型参数是如何影响模型准确度的。

1. 微分形式

低 Re 数 K-ϵ 模型可表示为

$$\frac{\partial\rho K}{\partial t}+\frac{\partial}{\partial x_j}(\rho v_j K)=\frac{\partial}{\partial x_j}\left[\left(\mu_L+\frac{\mu_T}{\sigma_K}\right)\frac{\partial K}{\partial x_j}\right]+\tau_{ij}^{\mathrm{F}}S_{ij}-\rho\epsilon$$

$$\frac{\partial \rho \epsilon^*}{\partial t}+\frac{\partial}{\partial x_j}(\rho v_j \epsilon^*)=\frac{\partial}{\partial x_j}\left[\left(\mu_{\mathrm{L}}+\frac{\mu_{\mathrm{T}}}{\sigma_\epsilon}\right)\frac{\partial \epsilon^*}{\partial x_j}\right]+C_{\epsilon 1} f_{\epsilon 1}\frac{\epsilon^*}{K}\tau_{ij}^{\mathrm{F}} S_{ij}- C_{\epsilon 2} f_{\epsilon 2}\rho\frac{(\epsilon^*)^2}{K}+\phi_\epsilon \tag{7.50}$$

右端项分别表示守恒扩散、涡黏性的生成和耗散。ϕ_ϵ 表示显式壁面项。Favre 平均的湍流应力 τ_{ij}^{F} 由式(7.25)给出，应变率张量 S_{ij} 由式(7.3)给出。式(7.28)和式(7.29)中的湍流涡黏性系数由下式计算得到

$$\mu_{\mathrm{T}}=C_\mu f_\mu \rho\frac{K^2}{\epsilon^*} \tag{7.51}$$

根据式(7.24)或式(7.25)，在涡黏性系数的计算中也用到了湍动能。ϵ^* 与湍流耗散率 ϵ 有关，具体为

$$\epsilon=\epsilon_w+\epsilon^* \tag{7.52}$$

其中，ϵ_w 项表示壁面处的耗散率。式(7.52)中的定义极大地简化了壁面边界条件的应用(见下文)。

不同 K-ϵ 模型之间的模型常数、近壁衰减函数以及壁面项是不同的。在此，我们选择 Launder-Sharma 模型，因为它在广泛的应用中都给出了良好的结果[58]。对于 Launder-Sharma 模型，常数和湍流 Pr 数由文献[54]给出

$$\begin{gathered}C_\mu=0.09 \quad C_{\epsilon 1}=1.44 \quad C_{\epsilon 2}=1.92\\ \sigma_K=1.0 \quad \sigma_\epsilon=1.3 \quad Pr_{\mathrm{T}}=0.9\end{gathered} \tag{7.53}$$

此外，近壁衰减函数为

$$\begin{aligned}f_\mu&=\exp\left(\frac{-3.4}{(1+0.02Re_{\mathrm{T}})^2}\right)\\ f_{\epsilon 1}&=1\\ f_{\epsilon 2}&=1-0.3\exp(Re_{\mathrm{T}}^2)\end{aligned} \tag{7.54}$$

其中，$Re_{\mathrm{T}}=\rho K^2/(\epsilon^*\mu_{\mathrm{L}})$ 表示湍流 Re 数。最后，显式壁面项 ϕ_ϵ 和 ϵ_w 定义为

$$\begin{aligned}\phi_\epsilon&=2\mu_{\mathrm{T}}\frac{\mu_{\mathrm{L}}}{\rho}\left(\frac{\partial^2 v_s}{\partial y_n^2}\right)^2\\ \epsilon_w&=\frac{2\mu_{\mathrm{L}}}{\rho}\left(\frac{\partial\sqrt{K}}{\partial y_n}\right)^2\end{aligned} \tag{7.55}$$

其中，v_s 为平行于壁面的速度，y_n 为垂直于壁面的坐标。为了避免需要事前掌握距壁面距离和方位，通常根据文献[66,76]中的笛卡儿张量来计算壁面项和 ϵ_w，即

$$\phi_\epsilon = 2\mu_{\mathrm{T}}\frac{\mu_{\mathrm{L}}}{\rho}\left(\frac{\partial^2 v_i}{\partial x_j \partial x_k}\right)^2$$
$$\epsilon_w = \frac{2\mu_{\mathrm{L}}}{\rho}\left(\frac{\partial\sqrt{K}}{\partial x_j}\right)^2 \tag{7.56}$$

而不是通过式(7.55)计算。

2. 积分形式

对于表面微元为 dS 的控制体 Ω,积分形式的低 Re 数 K-ϵ 湍流模型为

$$\frac{\partial}{\partial t}\int_\Omega \boldsymbol{W}_{\mathrm{T}}\mathrm{d}\Omega + \oint_{\partial\Omega}(\boldsymbol{F}_{\mathrm{c,T}} - \boldsymbol{F}_{\mathrm{v,T}})\mathrm{d}S = \int_\Omega \boldsymbol{Q}_{\mathrm{T}}\mathrm{d}\Omega \tag{7.57}$$

守恒变量为

$$\boldsymbol{W}_{\mathrm{T}} = \begin{bmatrix}\rho K \\ \rho\epsilon^*\end{bmatrix} \tag{7.58}$$

对流通量为

$$\boldsymbol{F}_{\mathrm{c,T}} = \begin{bmatrix}\rho K V \\ \rho\epsilon^* V\end{bmatrix} \tag{7.59}$$

其中,V 为逆变速度(见式(2.22))。黏性通量矢量为

$$\boldsymbol{F}_{\mathrm{v,T}} = \begin{bmatrix}n_x\tau_{xx}^K + n_y\tau_{yy}^K + n_z\tau_{zz}^K \\ n_x\tau_{xx}^\epsilon + n_y\tau_{yy}^\epsilon + n_z\tau_{zz}^\epsilon\end{bmatrix} \tag{7.60}$$

其中,湍流黏性正应力为

$$\tau_{xx}^K = \left(\mu_{\mathrm{L}} + \frac{\mu_{\mathrm{T}}}{\sigma_K}\right)\frac{\partial K}{\partial x}, \cdots$$
$$\tau_{xx}^\epsilon = \left(\mu_{\mathrm{L}} + \frac{\mu_{\mathrm{T}}}{\sigma_\epsilon}\right)\frac{\partial \epsilon^*}{\partial x}, \cdots \tag{7.61}$$

在式(7.60)中,n_x, n_y, n_z 表示面 $\partial\Omega$ 处的单位外法向矢量的分量。源项通过下式计算:

$$\boldsymbol{Q}_{\mathrm{T}} = \begin{bmatrix}P - \rho\epsilon \\ (C_{\epsilon1}f_{\epsilon1}P - C_{\epsilon2}f_{\epsilon2}\rho\epsilon^*)\dfrac{\epsilon^*}{K} + \phi_\epsilon\end{bmatrix} \tag{7.62}$$

其中,P 表示湍动能的生成项,定义为

$$P = \tau_{xx}^{\mathrm{F}}\frac{\partial u}{\partial x} + \tau_{yy}^{\mathrm{F}}\frac{\partial v}{\partial y} + \tau_{zz}^{\mathrm{F}}\frac{\partial w}{\partial z} + \tau_{xy}^{\mathrm{F}}\left(\frac{\partial u}{\partial y} + \frac{\partial v}{\partial x}\right) + \tau_{xz}^{\mathrm{F}}\left(\frac{\partial u}{\partial z} + \frac{\partial w}{\partial x}\right) + \tau_{yz}^{\mathrm{F}}\left(\frac{\partial v}{\partial z} + \frac{\partial w}{\partial y}\right) \tag{7.63}$$

Favre 平均的湍流应力 τ_{ij}^{F} 由式(7.25)给出。基于 Lauder-Sharma 模型的模型常数、近壁衰减函数和壁面项的定义参见式(7.52)~式(7.56)。湍流涡黏性系数通过式(7.51)得到。

3. 初始和边界条件

初始化 K 和 ϵ^* 的最简单的方法是使用它们各自的自由来流值。更好的替代方案是在固壁处指定 K 和 ϵ^* 的型面,它可以通过与湍流平板边界层进行比拟[60]得到。但这需要事前知道距壁面的距离,这点在非结构网格上可能不太容易得到。

如果使用了式(7.52)的变换,固壁处适当的边界条件是 $K=0$ 和 $\epsilon^*=0$,这意味着壁面处 $\mu_T=0$。在入流边界处,K 和 ϵ^* 可以根据湍流强度和长度尺度的关系计算,即

$$(T_u)_\infty = \frac{\sqrt{\dfrac{2}{3}K_\infty}}{\|\boldsymbol{v}_\infty\|_2} \quad (l_T)_\infty = \frac{C_\mu K_\infty^{3/2}}{\epsilon_\infty^*} \tag{7.64}$$

这里假设 $\epsilon_\infty^* = \epsilon_\infty$。在叶轮机械中,$(l_T)_\infty$ 的取值范围为 $10^{-3} \sim 10^{-2}$ 的叶片平均径向间距[77]。在出流边界处,K 和 ϵ^* 由计算域内部外插得到。

4. 壁面函数

低 Re 数模型在壁面处需要非常密的网格,其标准条件是第一个网格节点(或单元体心)离壁面的距离应满足 $y^+ \leqslant 1$。为了降低湍流方程的刚性和节省网格点/单元的数量,通常采用 $10 \leqslant y^+ \leqslant 100$ 的较粗网格,在这种情况下,忽略掉式(7.50)或式(7.57)中 K-ϵ 模型的衰减函数($f_\mu = f_{\epsilon 1} = f_{\epsilon 2} = 1$;$\epsilon_w = 0$)和壁面项($\phi_\epsilon = 0$),我们称之为高 Re 数湍流模型。显然,第一个网格节点(单元体心)和壁面之间的区域需要使用壁面函数桥接。壁面函数传递了靠近壁面处网格节点(单元体心)上的 K 和 ϵ^* 值,在壁面本身和第一层网格节点(单元)处,不再求解湍流方程。学者们提出了各种样式的壁面函数,一般都是基于壁面对数率,一个例子是 Spalding 提出的函数[78],它模化了黏性底层、过渡区以及对数层。文献[79-85]或文献[67]介绍了高 Re 数模型的实现过程。

只要网格不是太粗,对于附着边界层,壁面函数的使用可以得到相当准确的预测结果。它还使得可以使用纯显式的时间推进格式求解湍流方程。然而,对于分离流动,壁面函数的使用变得非常可疑。

7.2.3 Menter SST 两方程模型

Menter 的 K-ω SST 湍流模型[86-87]融合了 Wilcox 的 K-ω 模型[23,88]和高 Re 数

K-ϵ 模型（转化为 K-ω 形式）。SST 模型试图结合两种模型的优点。它在边界层的底层采用 K-ω 模型，因为 K-ω 模型不需要衰减函数，与 K-ϵ 模型相比，在相近的准确度下，数值稳定性明显更高，此外，在边界层对数区，也使用 K-ω 模型，它在逆压流动和可压缩流动中，优于 K-ϵ 模型。另外，在边界层尾流区域采用 K-ϵ 模型，因为 K-ω 模型对 ω 的自由来流值非常敏感[89]，K-ϵ 模型也应用于自由剪切层，因为它在尾迹、射流和混合层的预测精度上有一定的折中。

SST 湍流模型的一个显著特征是修正后的湍流涡黏性函数，目的是提高对强逆压梯度流动和压力诱导边界层分离的预测精度。这一修正考虑了湍流切应力的输运，根据 Bradshaw 的观察，主切应力与湍动能成正比。

SST 模型的一个缺点是必须明确知道到最近壁面的距离，这点在多块结构网格或非结构网格中都需要专门准备，壁面距离的计算可参见文献[48]。SST 湍流模型的应用示例见文献[90-97]。

1. 微分形式

湍动能和湍流比耗散的输运方程的微分形式为[86]

$$
\begin{aligned}
&\frac{\partial \rho K}{\partial t}+\frac{\partial}{\partial x_j}(\rho v_j K)=\frac{\partial}{\partial x_j}\left[(\mu_{\mathrm{L}}+\sigma_K\mu_{\mathrm{T}})\frac{\partial K}{\partial x_j}\right]+\tau_{ij}^{\mathrm{F}}S_{ij}-\beta^*\rho\omega K\\
&\frac{\partial \rho\omega}{\partial t}+\frac{\partial}{\partial x_j}(\rho v_j\omega)=\frac{\partial}{\partial x_j}\left[(\mu_{\mathrm{L}}+\sigma_\omega\mu_{\mathrm{T}})\frac{\partial \omega}{\partial x_j}\right]+\frac{C_\omega\rho}{\mu_{\mathrm{T}}}\tau_{ij}^{\mathrm{F}}S_{ij}-\\
&\qquad\qquad \beta\rho\omega^2+2(1-f_1)\frac{\rho\sigma_{\omega2}}{\omega}\frac{\partial K}{\partial x_j}\frac{\partial\omega}{\partial x_j}
\end{aligned}
\tag{7.65}
$$

式(7.65)右端项分别表示守恒扩散、涡黏性的生成和耗散项。此外，ω 方程的最后一项描述了交叉扩散。Favre 平均的湍流应力 τ_{ij}^{F} 由式(7.25)给出，应变率张量 S_{ij} 由式(7.3)给出。式(7.28)～式(7.29)中的湍流涡黏性系数通过下式计算[89]：

$$
\mu_{\mathrm{T}}=\frac{a_1\rho K}{\max(a_1\omega,f_2\|\mathrm{curl}\boldsymbol{v}\|_2)}
\tag{7.66}
$$

湍流涡黏性的这一定义保证了在逆压梯度边界层中，当 K 的产生大于其耗散 ω（因此 $a_1\omega<\|\mathrm{curl}\boldsymbol{v}\|_2$）时，Bradshaw 假设（即 $\tau=a_1\rho K$，切应力正比于湍动能）是满足的。

式(7.65)中的函数 f_1，用于混合边界层内 K-ω 模型的系数和自由剪切层及自由来流区域内变换后的 K-ϵ 模型的系数，定义为

$$
\begin{aligned}
&f_1=\tanh(\arg_1^4)\\
&\arg_1=\min\left[\max\left(\frac{\sqrt{K}}{0.09\omega d},\frac{500\mu_{\mathrm{L}}}{\rho\omega d^2}\right),\frac{4\rho\sigma_{\omega2}K}{CD_{K\omega}d^2}\right]
\end{aligned}
\tag{7.67}
$$

其中，d 为到最近壁面的距离，$CD_{K\omega}$ 是式(7.65)中交叉扩散项的正数部分，即

$$CD_{K\omega}=\max\left(2\frac{\rho\sigma_{\omega 2}}{\omega}\frac{\partial K}{\partial x_j}\frac{\partial \omega}{\partial x_j},10^{-20}\right) \tag{7.68}$$

式(7.66)中的辅助函数 f_2 为

$$\begin{aligned} &f_2=\tanh(\arg_2^2)\\ &\arg_2=\max\left(\frac{2\sqrt{K}}{0.09\omega d},\frac{500\mu_L}{\rho\omega d^2}\right)\end{aligned} \tag{7.69}$$

模型常数为

$$a_1=0.31 \quad \beta^*=0.09 \quad \kappa=0.41 \tag{7.70}$$

最终 SST 湍流模型的系数 $\beta, C_\omega, \sigma_K$ 和 σ_ω 是通过将 K-ω 模型的系数 ϕ_1 与变换形式后的 K-ϵ 模型系数 ϕ_2 混合得到的。相应的关系为

$$\phi=f_1\phi_1+(1-f_1)\phi_2 \tag{7.71}$$

内层模型(K-ω 模型)的系数为

$$\begin{aligned} &\sigma_{K1}=0.85 \quad \sigma_{\omega 1}=0.5 \quad \beta_1=0.075\\ &C_{\omega 1}=\beta_1/\beta^*-\sigma_{\omega 1}\kappa^2/\sqrt{\beta^*}=0.533\end{aligned} \tag{7.72}$$

外层模型(K-ϵ)的系数为

$$\begin{aligned} &\sigma_{K2}=1.0 \quad \sigma_{\omega 2}=0.856 \quad \beta_2=0.0828\\ &C_{\omega 2}=\beta_2/\beta^*-\sigma_{\omega 2}\kappa^2/\sqrt{\beta^*}=0.440\end{aligned} \tag{7.73}$$

SST 湍流模型的积分形式原则上与 7.2.2 节中的 K-ϵ 模型的积分形式相对应，因此，在此不再重复。

2. 边界条件

固壁边界处湍动能和比耗散的边界条件为

$$K=0 \quad \omega=10\frac{6\mu_L}{\rho\beta_1(d_1)^2} \tag{7.74}$$

其中，d_1 是第一个网格节点(或体心)到壁面的距离。网格需要加密直至满足 $y^+<3$。

对于入流边界，建议取下列的自由来流值：

$$\omega_\infty=C_1\frac{\|\boldsymbol{v}_\infty\|_2}{L} \quad (\mu_T)_\infty=(\mu_L)_\infty 10^{-C_2} \quad K_\infty=\frac{(\mu_T)_\infty}{\rho_\infty}\omega_\infty \tag{7.75}$$

其中，L 表示计算域的长度，$1\leqslant C_1\leqslant 10$，$2\leqslant C_2\leqslant 5$。在出流边界，$K$ 和 ω 的值由流场内部外插得到。

7.3 大涡模拟

1963 年,Smagorinsky 在气象学[98]（大气环流）中已经采用了 LES 方法。LES 的首次工程应用(湍流槽流)是由 Deardorff 于 1970 年完成的[99],他的方法后来被 Schumann 推广和改进[100]。20 世纪 80 年代,湍流模拟的研究重点从 LES 转向 DNS,然而,一些重要的工作仍在开展,如 Bardina 等[101]、Moin 和 Kim[102]。对 LES 的兴趣在 20 世纪 90 年代初又重新回归[103-109]。如今,LES 越来越多地应用于物理和几何复杂的与工程相关的流动中,例如燃烧室流动,具体示例可参考文献[110-128],当然,这一趋势得到了低成本、功能强大的计算服务器的支持。此外,工程师还经常面临标准湍流模型无力解决的流动问题,在某些情况下,平均流动的频率与湍流脉动量级相同,因此时间平均法变得不再有效,人们不得不求助于 LES 或 DNS。

LES 基于小尺度湍流结构比大涡结构更具有普遍特征这一观察结果,主要思路是计算大的、含能结构对动量和能量传输的贡献,模化数值格式无法解析的小尺度结构的影响。由于小尺度结构更具有均匀性和普遍性,可以预期亚格子尺度模型比 RANS 方程组的湍流模型形式更简单。

LES 表征了 3D、时间相关的控制方程组的解。与基于 RANS 方程组的湍流建模相比,LES 在流向($50\leqslant x^+\leqslant 150$)和横流方向($15\leqslant z^+\leqslant 40$)也要求高分辨率的网格,然而,LES 在计算上比 DNS 廉价得多,解析边界层外层所需的网格点(单元)数量与 $Re^{0.4}$ 成正比[129],在黏性底层必须提高到 $Re^{1.8}$,因此与 DNS 所需网格的 $Re^{9/4}$ 相比,LES 可以应用于 Re 数至少高一个量级的问题。为了进一步减少网格分辨率要求,LES 可与近似壁面模型(7.3.4 节)或 RANS 模型(7.3.5 节)结合使用,这两种方法都允许在合理的计算成本下对工程问题开展 LES。

精确地分辨湍流高波数脉动需要空间离散格式在波数空间中具有相应的性质(参见文献[130]或文献[131]),因此经常使用谱方法,然而,谱方法仅适用于具有(准)周期边界、几何上简单的计算域,这样,有限差分或有限体积空间离散变得越来越流行。其中,中心差分格式比迎风格式更适合,其原因是,由于固有的数值衰减,迎风格式(无论精度的高低)在湍流谱的很大部分上耗散了太多的能量[132-133],可在文献[134-136]中找到关于各种离散方法数值误差的讨论。

在非结构网格上实现 LES 方法[137-147]是一个特殊的挑战。然而,它能够处理高度复杂的几何外形、运动边界,能够使用动态网格自适应技术,现有的研究主题包括在混合网格上发展数值高效的高阶空间离散格式。

感兴趣的读者可以在文献[25,p. 269-336]和文献[148-152]或文献[113]中找到 LES 方法的介绍。文献[153-155]概述了 LES 的研究现状。

7.3.1 空间滤波

LES 基于空间滤波操作，将任意流动变量 U 分解为滤波(大尺度、可分辨)部分 $\overline{U}$ 和亚滤波(不可分辨)部分 U'，即

$$U=\overline{U}+U' \tag{7.76}$$

空间 $\boldsymbol{r}_0$ 处滤波后的变量定义为

$$\overline{U}(\boldsymbol{r}_0,t)=\int_{\Omega}U(\boldsymbol{r},t)G(\boldsymbol{r}_0,\boldsymbol{r},\Delta)\,\mathrm{d}\boldsymbol{r} \tag{7.77}$$

其中，Ω 为整个流动区域，G 为滤波函数，$\boldsymbol{r}$ 为位置矢量。滤波函数决定了小尺度的结构和大小，它依赖于空间坐标差量 $\boldsymbol{r}_0-\boldsymbol{r}$ 和滤波宽度 $\Delta=(\Delta_1\Delta_2\Delta_3)^{1/3}$，其中 Δ_i 为第 i 个空间坐标方向的滤波宽度。常用的滤波函数为(图 7.3)

(1) 盒式滤波器：

$$G=\begin{cases}1/\Delta^3, & |(x_0)_i-x_i|\leqslant\Delta_i/2\\0, & \text{其他}\end{cases} \tag{7.78}$$

(2) 谱截断滤波器：

$$G=\prod_{i=1}^{3}\frac{\sin\left(\dfrac{\pi}{\Delta_i}[(x_0)_i-x_i]\right)}{\pi[(x_0)_i-x_i]} \tag{7.79}$$

(3) 高斯滤波器：

$$G=\left(\frac{6}{\pi\Delta^2}\right)^{3/2}\exp\left(\frac{-6\|\boldsymbol{r}_0-\boldsymbol{r}\|_2^2}{\Delta^2}\right) \tag{7.80}$$

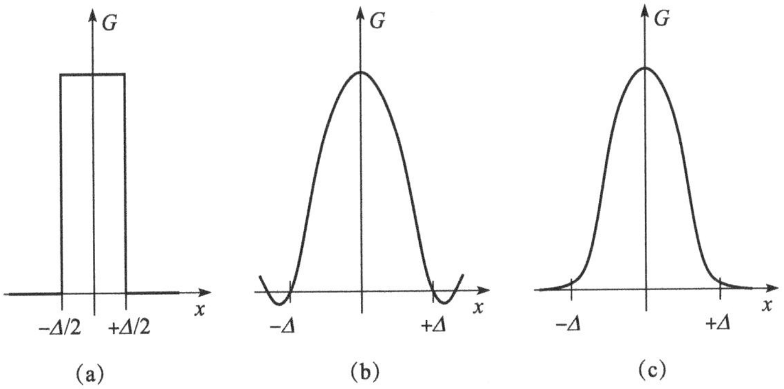

图 7.3　物理空间中的 LES 滤波函数

(a) 盒式；(b) 谱空间截断；(c) 高斯。

盒式和高斯滤波器对大尺度脉动以及滤波尺度以下的小尺度脉动都进行了光滑,截断滤波器只影响截断波数以下的尺度,在实践中,高斯滤波器总是与谱截断滤波器结合使用。文献[156-157]提出了适用于单元尺寸变化的网格的滤波器。

7.3.2 滤波控制方程组

将式(7.76)~式(7.77)定义的空间滤波,应用于纳维-斯托克斯方程组,以消除小尺度湍流脉动,将滤波宽度Δ和滤波函数视为自由参数。实际上,通常不是对控制方程组进行显式的滤波操作,而是由网格和离散误差来定义滤波器G,关于显式滤波的讨论请参见文献[158-159]。

由于处理方法的不同,下面将分别讨论不可压缩纳维-斯托克斯方程组(7.6)和可压缩纳维-斯托克斯方程组(7.1)。

1. 不可压缩纳维-斯托克斯方程组

对于牛顿流体的不可压缩流动,滤波后的控制方程组(7.6)形式如下:

$$\begin{aligned}&\frac{\partial \bar{v}_i}{\partial x_i}=0\\&\frac{\partial \bar{v}_i}{\partial t}+\frac{\partial}{\partial x_j}(\bar{v}_i\bar{v}_j)=-\frac{1}{\rho}\frac{\partial \bar{p}}{\partial x_i}+\nu\nabla^2\bar{v}_i-\frac{\partial \tau_{ij}^{S}}{\partial x_j}\end{aligned} \tag{7.81}$$

其中,ν为运动学黏性系数。式(7.81)描述了大尺度的含能运动的时空演化。对流项的非线性产生了亚格子尺度应力(subgrid-scale stress,SGS)张量:

$$\tau_{ij}^{S}=\overline{v_iv_j}-\bar{v}_i\bar{v}_j \tag{7.82}$$

它描述了不可分辨尺度的影响,需要对SGS张量进行模化处理以封闭方程(见7.3.3节)。

SGS张量可以分解为三个部分[160],即

$$\tau_{ij}^{S}=L_{ij}+C_{ij}+\tau_{ij}^{SR} \tag{7.83}$$

各个部分的物理意义如下:

$$L_{ij}=\overline{\bar{v}_i\bar{v}_j}-\bar{v}_i\bar{v}_j \tag{7.84}$$

是Leonard应力项,表示产生小尺度湍流的大尺度涡之间的相互作用。这一项只能由滤波速度场v_i显式地计算。此外,交叉应力项

$$C_{ij}=\overline{\bar{v}_iv_j'}-\overline{v_i'\bar{v}_j} \tag{7.85}$$

描述了大尺度涡和小尺度涡之间的相互作用。最后

$$\tau_{ij}^{SR}=\overline{v_i'v_j'} \tag{7.86}$$

是SGS雷诺应力张量,它反映了小尺度结构之间的相互作用。上面关于

式(7.83)的分解不再使用,主要是因为 L_{ij} 和 C_{ij} 不满足伽利略不变性①。

2. 可压缩纳维-斯托克斯方程组

如果要将 LES 应用于可压缩流动,必须将 Favre 平均(7.1.2 节)与空间滤波一起作用于式(7.1),否则滤波纳维-斯托克斯方程组将包含密度和其他变量(如速度或温度)之间的乘积。因此,式(7.1)中的速度分量、能量和温度分解为

$$U=\widetilde{U}+U'' \tag{7.87}$$

空间 $\boldsymbol{r}_0$ 处滤波后的变量为

$$\widetilde{U}(\boldsymbol{r}_0,t)=\frac{\overline{\rho U}}{\overline{\rho}}=\frac{1}{\overline{\rho}}\int_D \rho(\boldsymbol{r},t)U(\boldsymbol{r},t)G(\boldsymbol{r}_0,\boldsymbol{r},\Delta)\,\mathrm{d}\boldsymbol{r} \tag{7.88}$$

其中,$\overline{(\cdot)}$表示根据式(7.77)对变量进行滤波。Favre-滤波后的纳维-斯托克斯方程组(7.1)为[149,153]

$$\begin{aligned}
&\frac{\partial\overline{\rho}}{\partial t}+\frac{\partial}{\partial x_j}(\overline{\rho}\widetilde{v}_j)=0\\
&\frac{\partial(\overline{\rho}\widetilde{v}_i)}{\partial t}+\frac{\partial(\overline{\rho}\widetilde{v}_j\widetilde{v}_i)}{\partial x_j}+\frac{\partial\overline{p}}{\partial x_i}-\frac{\partial\hat{\sigma}_{ij}}{\partial x_j}=-\frac{\partial\tau_{ij}^{\mathrm{SF}}}{\partial x_j}+\frac{\partial}{\partial x_j}(\overline{\sigma}_{ij}-\hat{\sigma}_{ij})\\
&\frac{\partial(\overline{\rho}\widetilde{e})}{\partial t}+\frac{\partial(\overline{\rho}\widetilde{v}_j\widetilde{e})}{\partial x_j}+\frac{\partial\hat{q}}{\partial x_j}+\overline{p}\widetilde{S}_{kk}-\hat{\sigma}_{ij}\widetilde{S}_{ij}=-A-B-C+D
\end{aligned} \tag{7.89}$$

其中,各项分别表示:

$A=\frac{\partial}{\partial x_j}[\overline{\rho}(\widetilde{v_j e}-\widetilde{v}_j\widetilde{e})]$ ——亚格子尺度热通量的散度

$B=\frac{\partial}{\partial x_j}[\overline{q}_j-\hat{q}_j]$ ——SGS 热扩散的散度

$C=[\overline{pS_{kk}}-\overline{p}\widetilde{S}_{kk}]$ ——SGS 压力膨胀

$D=[\overline{\sigma_{ij}S_{ij}}-\hat{\sigma}_{ij}\widetilde{S}_{ij}]$ ——SGS 黏性耗散

以及

$$\begin{aligned}
&\overline{\sigma}_{ij}=\overline{2\mu S_{ij}}+\overline{\left(\mu_{\mathrm{B}}-\frac{2\mu}{3}\right)\delta_{ij}S_{kk}}\\
&\hat{\sigma}_{ij}=2\widetilde{\mu}\widetilde{S}_{ij}+\left(\widetilde{\mu}_{\mathrm{B}}-\frac{2\widetilde{\mu}}{3}\right)\delta_{ij}\widetilde{S}_{kk}\\
&\widetilde{S}_{ij}=\frac{1}{2}\left(\frac{\partial\widetilde{v}_i}{\partial x_j}+\frac{\partial\widetilde{v}_j}{\partial x_i}\right)
\end{aligned} \tag{7.90}$$

---

① 伽利略不变性是指数学性质在所有彼此相对匀速平移的参考系下是不变的。

$$\bar{q}_j = -k\frac{\partial \bar{T}}{\partial x_j} \quad \tilde{q}_j = -\tilde{k}\frac{\partial \tilde{T}}{\partial x_j}$$

在式(7.89)~式(7.90)中，e 表示单位质量内能，$\widetilde{S}_{ij}$ 为 Favre -滤波的应变率张量，$\tau_{ij}^{\mathrm{SF}}=\bar{\rho}(\widetilde{v_i v_j}-\tilde{v}_i\tilde{v}_j)$ 为 Favre 平均的亚格子应力。此外 μ，μ_{B} 和 k 分别代表分子黏性系数、体积黏性系数和热传导系数。最后 $\tilde{\mu}$，$\tilde{\mu}_{\mathrm{B}}$ 和 $\tilde{k}$ 是在滤波后的温度 $\widetilde{T}$ 下对应的值。

式(7.89)的右端包含了需要模化的项。在动量方程中，SGS 应力 $\boldsymbol{\tau}_{ij}^{\mathrm{SF}}$ 可以近似得到，但第二项($\bar{\sigma}_{ij}-\hat{\sigma}_{ij}$)经常忽略。在能量方程中，A 项可以由 SGS 应力表示[161]，B 项可以忽略，C 和 D 项可以用文献[162]的方式模化。

7.3.3 亚格子尺度建模

亚格子尺度模型的主要任务是模拟大尺度和亚格子尺度之间的能量输运。一般来说，能量是从大尺度传递到小尺度(湍流级串过程)，因此，亚格子尺度模型必须提供足够的能量耗散方式。然而，在某些情况下，能量也会从小尺度传递到大尺度——称为能量反级串过程，显然，模型也应该考虑到这种影响，文献[163]讨论了能量反级串模型。

学者们提出了各种亚格子尺度模型，这项研究现在仍在继续。现有的模型可以分为两种基本的类型。第一类是显式模化亚格子应力 τ_{ij}^{S} 的方法，其先决条件是空间离散引起的数值耗散必须远小于亚格子尺度耗散，大多数显式 SGS 模型基于涡黏性概念，下面将对此进行解释。此外，我们将阐述 Smagorinsky 模型，它构成了所有亚格子尺度模型的基础，我们还将简要讨论动态亚格子尺度模型的基础。最近出版的文献[164]对六种不同的显式亚格子尺度模型进行了比较。

第二类是通过对对流通量的恰当离散隐式地模化 SGS 应力的方法(因此忽略 τ_{ij}^{S} 项)，这些模型被称为单调集成的 LES(monotonically integrated LES, MILES)。这种方法首先由 Boris 等[165]提出，并被 Grinstein 和 Fureby[166-167]所倡导，它的先决条件是数值耗散能够正确地模化亚格子尺度耗散。就像在文献[168-169]中所证明的那样，这并不容易实现，然而，MILES 在各种流动问题上还是取得了一定的成功[170-177]。

1. 涡黏性模型

这些显式模型能够表征小尺度的全局耗散效应，但不能呈现能量交换的局部细节。在不可压缩流动中，涡黏性模型将 SGS 应力与大尺度应变率 $\bar{S}_{ij}$ 联系起来：

$$\tau_{ij}^{\mathrm{S}}-\frac{\delta_{ij}}{3}\tau_{kk}^{\mathrm{S}}=-2\nu_{\mathrm{T}}\bar{S}_{ij} \tag{7.91}$$

其中,应变率 $\overline{S}_{ij}$ 由式(7.3)使用滤波后的速度分量得到。涡黏性系数 ν_T 一般通过代数关系式计算,这样会节省计算量。SGS 应力的各向同性部分(τ_{kk}^{S})要么添加到滤波后的压力中[178],要么模化[179],要么忽略不计。

与式(7.91)相似的关系也适用于可压缩纳维-斯托克斯方程组。Favre 平均的 SGS 应力张量分量近似为

$$\tau_{ij}^{SF}-\frac{\delta_{ij}}{3}\tau_{kk}^{SF}=-2\bar{\rho}\nu_T\widetilde{S}_{ij}+\left(\frac{2\bar{\rho}\nu_T}{3}\right)\frac{\partial\widetilde{v}_k}{\partial x_k}\delta_{ij} \tag{7.92}$$

应变率张量 $\widetilde{S}_{ij}$ 的分量由式(7.90)计算。

2. Smagorinsky SGS 模型

Smagorinsky 模型[98]基于平衡假设,即小尺度完全且瞬时地将它们从大尺度接收到的所有能量耗散掉。代数模型采用如下的形式:

$$\nu_T=(C_s\Delta)^2\,|\overline{S}| \tag{7.93}$$

其中,$|\overline{S}|=(2\overline{S}_{ij}\overline{S}_{ij})^{1/2}$ 为应变率张量的幅值。C_s 为 Smagorinsky 常数,Lilly[180]发现其理论值为 $C_s=0.18$。然而,Smagorinsky 常数取决于流动的类型,例如在剪切流中 C_s 必须降低到大约 0.1。通常选择式(7.93)中的滤波尺度 Δ 为平均网格大小的两倍,即 $\Delta=2(\Delta x_1\Delta x_2\Delta x_3)^{1/3}$。

考虑到壁面附近小尺度增长的减缓,涡黏性系数 ν_T 也必须减小。因此,根据 van Driest 衰减将式(7.93)中的 Smagorinsky 模型修改为

$$\nu_T=[C_s\Delta(1-e^{-y^+/25})]^2\,|\overline{S}| \tag{7.94}$$

其中,y^+表示无量纲的壁面距离(壁面距离的计算可见文献[48])。

Smagorinsky 模型数值成本低且易于实现。然而,它有几个严重的缺点:

(1) 在平均剪切的层流区,其耗散性太大;

(2) 在靠近壁面和层流-湍流过渡区需要特别指定;

(3) 参数 C_s 不是唯一定义的;

(4) 没有考虑能量反级串过程。

由于这些缺点,人们提出了各种其他方法(如文献[149]),非常流行的是下面的动态模型。

3. 动态 SGS 模型

动态 SGS 模型采用与 Smagorinsky 模型(式(7.93))相同的关系式来计算式(7.91)或(7.92)中的涡黏性系数 ν_T,不同之处在于(先验调节的)Smagorinsky 常数不再是一个定值,而是在空间和时间动态演化的参数。因此:

$$\nu_T=C_d(\boldsymbol{r},t)\Delta^2\,|\overline{S}| \tag{7.95}$$

参数 C_d 根据湍流最小尺度的能量含量计算。为此,Germano 等[181]提出采用第二个滤波器,即测试滤波器 $\hat{\Delta}$。测试滤波器的滤波宽度必须大于应用于控制方程组的滤波器的尺度 Δ(通常 $\hat{\Delta}=2\Delta$)。对滤波方程组再使用测试滤波器进行二次滤波,得到亚测试尺度应力 τ_{ij}^{ST},即

$$\tau_{ij}^{ST}=\widehat{\overline{v_i v_j}}-\hat{\overline{v}}_i\hat{\overline{v}}_j \tag{7.96}$$

Germano 等式[182]将亚测试尺度应力和 SGS 应力 τ_{ij}^{S}(式(7.83))联系了起来,得到

$$\hat{L}_{ij}=\tau_{ij}^{ST}-\hat{\tau}_{ij}^{S}=\widehat{\overline{v}_i\overline{v}_j}-\hat{\overline{v}}_i\hat{\overline{v}}_j \tag{7.97}$$

其中,$\hat{L}_{ij}$ 表示测试滤波器作用下的 Leonard 应力,它表征尺度介于滤波宽度 Δ 和测试滤波宽度 $\hat{\Delta}$ 之间的湍流结构对雷诺应力的贡献。

如果用式(7.91)和式(7.95)的涡黏性方法表示式(7.97)中的亚测试尺度应力和 SGS 应力,可以得到

$$\hat{L}_{ij}-\frac{\delta_{ij}}{3}L_{kk}=-2C_d M_{ij} \tag{7.98}$$

其中

$$M_{ij}=\hat{\Delta}^2\,|\hat{\overline{S}}|\,\hat{\overline{S}}_{ij}-[\Delta^2\,|\overline{S}|\,\overline{S}_{ij}]^{\wedge} \tag{7.99}$$

记号$[\ ]^{\wedge}$表示方括号中的项是经过测试滤波器滤波的。参数 C_d 可以使用 Lilly[183]提出的最小二乘最小化方法从式(7.98)中推导出,即

$$C_d(\boldsymbol{r},t)=-\frac{1}{2}\frac{L_{ij}M_{ij}}{M_{mn}M_{mn}} \tag{7.100}$$

由于在式(7.98)中将参数 C_d 拿到了测试滤波器之外,所以式(7.100)在数学上是不相容的。因此,在实际操作中,需要对式(7.100)中的分子和分母在齐次方向上做系综平均,即

$$C_d(\boldsymbol{r},t)=-\frac{1}{2}\frac{\langle L_{ij}M_{ij}\rangle}{\langle M_{mn}M_{mn}\rangle} \tag{7.101}$$

Ghosal 等[184]、Carati 等[185]、Piomelli 和 Liu[186]、Held[113]等提出了改进的动态 SGS 模型。

7.3.4 壁面模型

高 Re 数($Re>10^6$)下壁面约束流动 LES 的计算成本仍然过高,原因是需要巨大的网格点(单元)数来适当地解析壁面边界层。为了降低成本,可以通过指定外部流动速度和壁面应力之间的关系来模拟壁面边界层,这种方法与 RANS 模拟中使用的壁面函数非常相似,它的基本假设是近壁与外区仅存在微弱的相

互作用,这得到了文献[187-188]的支持。

早期壁面模型的实现是基于这样的假设:壁面边界层内的动力学特性是通用的,因此可以用广义壁面律来近似。这些模型基本上都采用对数律(见文献[100,189-191])。Balaras 等[192]最近提出了一种新的分区方法,在两层模型框架下,求解滤波纳维-斯托克斯方程组(7.81)直到壁面外第一个网格点,从这一点到壁面,在加密的嵌入网格上求解 2D 边界层方程,然后使用嵌入网格上的解指定壁面切应力,作为 LES 的边界条件。Balaras 等[192]的分区方法允许在 $20<y^+<100$ 区域内设置第一个网格点,这样就显著减少了网格量和计算时间。该方法成功地应用到槽道湍流、方管和旋转槽道,后来,它还用于分离流动的 LES,得到了令人鼓舞的结果[193-196]。文献[197]对各种壁面模型进行了概述和评估。

7.3.5 分离涡模拟

尽管前面提到的壁面模型有助于大大减少网格点(单元)的数量,LES 对于复杂的工程外形仍然成本过大。为此,Spalart 最近提出了另一种方法,即分离涡模拟(detached eddy simulation,DES),旨在模拟高 *Re* 数下的大分离流动[198-199]。该方法代表了 RANS 和 LES 之间的混合,主要思想是采用高度拉伸的网格和 RANS 模型(主要是 7.2.1 节中的 Spalart-Allmaras 模型或 7.2.3 节中的 Menter SST 模型)来解析附着边界层,在壁面以外区域使用 LES 和各向同性网格捕捉脱体的 3D 旋涡,因此,DES 试图在单一框架下结合这两种方法的优点。

对于 Spalart-Allmaras 湍流模型,将式(7.36)和式(7.38)~式(7.40)中的壁面距离 d 替换为 DES 长度尺度,即

$$l=\min(d,C_{\mathrm{DES}}\Delta) \tag{7.102}$$

即可实现 DES。DES 长度尺度取决于控制体的最大维尺度,即 $\Delta=\max(\Delta x,\Delta y,\Delta z)$。常数 C_{DES} 在一定程度上取决于流动类型,对于均匀湍流,$C_{\mathrm{DES}}=0.65$ 是最优值[200],另外,跨声速和超声速射流中推荐使用 $C_{\mathrm{DES}}=0.1$[201]。式(7.102)中长度尺度 l 的定义确保在边界层内,由于 $d<C_{\mathrm{DES}}\Delta$,因而 $l=d$,DES 恢复成原始的 RANS 模型;另外,在边界层外 $l=C_{\mathrm{DES}}\Delta$,Spalart-Allmaras 模型充当了 LES 的单方程 SGS 模型(参见式(7.91)和式(7.24))。在对控制方程组进行时间积分时,需要调整全局时间步长,以解析分离涡的时间尺度。这通常意味着在边界层区域时间步长将远远超过显式格式的稳定边界,因此,采用时间精确的隐式格式更有效,如 6.3 节中介绍的双时间步方法。更多关于 DES 方法的细节和模拟示例可在上述参考文献或文献[202-209]中找到。

参考文献

[1] Liu C, Liu Z. Multigrid methods and high order finite difference for flow in transition. AIAA Paper 93-3354; 1993.

[2] Olejniczak DJ, Weirs VG, Liu J, Candler GV. Hybrid finite-difference methods for DNS of compressible turbulent boundary layers. AIAA Paper 96-2086; 1996.

[3] Cook AW, Riley JJ. Direct numerical simulation of a turbulent reactive plume on a parallel computer. J Comput Phys 1996; 129: 263-83.

[4] Pinelli A, Vacca A, Quarteroni A. A spectral multidomain method for the numerical simulation of turbulent flows. J Comput Phys 1997; 136: 546-58.

[5] Zhong X. High-order finite-difference schemes for numerical simulation of hypersonic boundary-layer transition. J Comput Phys 1998; 144: 662-709.

[6] Lange M, Riedel U, Warnatz J. Parallel DNS of turbulent flames with detailed reaction schemes. AIAA Paper 98-2979; 1998.

[7] Hernandez G, Brenner G. Boundary conditions for direct numerical simulations of free jets. AIAA Paper 99-0287; 1999.

[8] Ladeinde F, Liu W, O' Brien EE. DNS evaluation of chemistry models for turbulent, reacting, and compressible mixing layers. AIAA Paper 99-0413; 1999.

[9] Freund JB. Acoustic sources in a turbulent jet—a direct numerical simulation study. AIAA Paper 99-1858; 1999.

[10] Rizzetta DP, Visbal MR. Application of a high-order compact difference scheme to large-eddy and direct numerical simulation. AIAA Paper 99-3714; 1999.

[11] Luchini P, Quadrio M. A low-cost parallel implementation of direct numerical simulation of wall turbulence. J Comput Phys 2006; 211: 551-71.

[12] Martín MP, Candler GV. A parallel implicit method for the direct numerical simulation of wall-bounded compressible turbulence. J Comput Phys 2006; 215: 153-71.

[13] Martín MP, Taylor EM, Wu M, Weirs VG. A bandwidth-optimized WENO scheme for the effective direct numerical simulation of compressible turbulence. J Comput Phys 2006; 220: 270-89.

[14] Salvadore F, Bernardini M, Botti M. GPU accelerated flow solver for direct numerical simulation of turbulent flows. J Comput Phys 2013; 235: 129-42.

[15] Khajeh-Saeed A, Perot JB. Direct numerical simulation of turbulence using GPU accelerated supercomputers. J Comput Phys 2013; 235: 241-57.

[16] Marxen O, Magin TE, Shaqfeh E, Iaccarino G. A method for the direct numerical simulation of hypersonic boundary-layer instability with finite-rate chemistry. J Comput Phys 2013; 255: 572-89.

[17] Boussinesq J. Essai sur la théorie des eaux courantes. Mem Pres Acad Sci 1877; XXIII: 46.

[18] Boussinesq J. Theorie de l' écoulement tourbillonant et tumulteur des liquides dans les lits rectiligues. Comptes Rendus Acad Sci 1896; CXXII: 1293.

[19] Reynolds O. On the dynamical theory of incompressible viscous fluids and the determination of the criterion. Philos Trans R Soc Lond A 1895;186:123-64.

[20] Favre A. Equations des gaz turbulents compressibles, part 1: formes générales. J Mécan 1965; 4:361-90.

[21] Favre A. Equations des gaz turbulents compressibles, part 2: méthode des vitesses moyennes; methode des vitesses moyennes pondérées par la masse volumique. J Mécan 1965;4:391-21.

[22] Morkovin MV. Effects of compressibility on turbulent flow. In: Favre A, editor. The mechanics of turbulence. New York: Gordon and Breach; 1964.

[23] Wilcox DC. Turbulence modeling for CFD. La Canada, CA: DCW Industries, Inc. ; 1993.

[24] Reynolds O. On the extent and action of the heating surface for steam boilers. Proc Manchester Lit Philos Soc 1874;14:7-12.

[25] Hallbäck M, Henningson DS, Johansson AV, Alfredsson PH, editors. Turbulence and transition modelling. ERCOFTAC Series, vol. 2. Dordrecht, The Netherlands: Kluwer Academic Publishers; 1996.

[26] Spalart P, Shur M. On the sensitization of turbulence models to rotation and curvature. Aerospace Sci Technol 1997;5:297-302.

[27] Shur M, Strelets M, Travin A, Spalart P. Turbulence modeling in rotating and curved channels: assessment of the Spalart-Shur correction term. AIAA Paper 98-0325; 1998.

[28] Lumley JL. Toward a turbulent constitutive equation. J Fluid Mech 1970;41:413-34.

[29] Lumley JL. Computational modeling of turbulent flows. Adv Appl Mech 1978;18:123-76.

[30] Craft TJ, Launder BE, Suga K. A non-linear eddy-viscosity model including sensitivity to stress anisotropy. In: Proc. 10th symp. turbulent shear flows, Paper 23:19. Pennsylvania State University; 1995.

[31] Shih TH, Zhu J, Liou WW, Chen K-H, Liu N-S, Lumley J. Modeling of turbulent swirling flows. In: Proc. 11th symposium on turbulent shear flows. Grenoble, France; 1997 [also NASA TM-113112; 1997].

[32] Chen K-H, Liu N-S. Evaluation of a non-linear turbulence model using mixed volume unstructured grids. AIAA Paper 98-0233, 1998.

[33] Goldberg U, Peroomian O, Chakravarthy S. Application of the kε-R turbulence model to wall-bounded compressive flows. AIAA Paper 98-0323; 1998.

[34] Abdel Gawad AF, Abdel Latif OE, Ragab SA, Shabaka IM. Turbulent flow and heat transfer in rotating non-circular ducts with nonlinear k-ε model. AIAA Paper 98-0326; 1998.

[35] Yoshizawa A, Fujiwara H, Hamba F, Nisizima S, Kumagai Y. Nonequilibrium turbulent-viscosity model for supersonic free-shear layer/wall-bounded flows. AIAA J 2003;41:1029-36.

[36] Tucker PG, Liu Y, Chung YM, Jouvray A. Computation of an unsteady complex geometry flow using novel non-linear turbulence models. Int J Numer Methods Fluids 2003;43:979-1001.

[37] Yang X, Ma H. Computation of strongly swirling confined flows with cubic eddy-viscosity turbulence models. Int J Numer Methods Fluids 2003;43:1355-70.

[38] Hinze JO. Turbulence. 2nd ed. New York: McGraw-Hill; 1975.

[39] Bai XS. On the modeling of turbulent combustion at low Mach numbers. PhD Thesis. Stock-

holm, Sweden: Royal Inst. of Technology, Dept. of Mechanics/Applied CFD; 1994.

[40] Adumitroaie V, Ristorcelli JR, Taulbee DB. Progress in Favre-Reynolds stress closures for compressible flows. ICASE Report No. 98-21; 1998.

[41] Kobayashi MH, Marques NPC, Pereira JCF. Critical evaluation of near wall treatment for turbulent Reynolds stress models to predict aerodynamic stall. AIAA Paper 97-1826; 1997.

[42] Rizzetta DP. Evaluation of algebraic Reynolds-stress models for separated high-speed flows. AIAA Paper 97-2125; 1997.

[43] Chassaing JC, Gerolymos GA, Vallet I. Reynolds-stress model dual-time-stepping computation of unsteady three-dimensional flows. AIAA J 2003; 41: 1882-94.

[44] Ben Nasr N, Gerolymos GA, Vallet I. Low-diffusion approximate Riemann solvers for Reynolds-stress transport. J Comput Phys 2014; 268: 186-35.

[45] Mor-Yossef Y. Unconditionally stable time marching scheme for Reynolds stress models. J Comput Phys 2014; 276: 635-64.

[46] Bardina JE, Huang PG, Coakley TJ. Turbulence modeling validation, testing, and development. NASA TM-110446; 1997.

[47] Spalart SR, Allmaras SA. A one-equation turbulence model for aerodynamic flows. AIAA Paper 92-0439; 1992 [also in La Recherche Aerospatiale 1994; 1: 5-21].

[48] Tucker PG. Differential equation based length scales to improve DES and RANS simulations. AIAA Paper 2003-3968; 2003.

[49] Allmaras, S. R.; Johnson, F. T.; Spalart, P. R.: Modifications and Clarifications for the Implementation of the Spalart-Allmaras Turbulence Model. ICCFD7-1902, 7th Int. Conf. on Comput. Fluid Dynamics, 2012.

[50] Rung T, Bunge U, Schatz M, Thiele F. Restatement of the Spalart-Allmaras eddy-viscosity model in strain-adaptive formulation. AIAA J 2003; 41: 1396-9.

[51] Chou PY. On velocity correlations and the solutions of the equations of turbulent fluctuations. Quart Appl Math 1945; 3: 38-54.

[52] Jones WP, Launder BE. The prediction of laminarization with a two-equation model of turbulence. Int J Heat Mass Transfer 1972; 15: 301-14.

[53] Jones WP, Launder BE. The prediction of low-Reynolds-number phenomena with a two-equation model of turbulence. Int J Heat Mass Transfer 1973; 16: 1119-30.

[54] Launder BE, Sharma BI. Application of the energy dissipation model of turbulence to the calculation of flow near a spinning disc. Lett Heat Mass Transfer 1974; 1: 131-8.

[55] Launder BE, Spalding B. The numerical computation of turbulent flows. Comput Methods Appl Mech Eng 1974; 3: 269-89.

[56] Lam CKG, Bremhorst KA. Modified form of Kε model for predicting wall turbulence. ASME J Fluid Eng 1981; 103: 456-60.

[57] Chien K-Y. Predictions of channel and boundary-layer flows with a low-Reynolds-number turbulence model. AIAA J 1982; 20: 33-38.

[58] Patel VC, Rodi W, Scheurer G. turbulence models for near-wall and low Reynolds number flows: a review. AIAA J 1985; 23: 1308-19.

[59] Kunz RF, Lakshminarayana B. stability of explicit Navier-Stokes procedures using K-ϵ and K-ϵ/algebraic Reynolds stress turbulence models. J Comput Phys 1992;103:141-59.
[60] Gerolymos GA. Implicit multiple-grid solution of the compressible Navier-Stokes equations using K-ϵ turbulence closure. AIAA J 1990;28:1707-17.
[61] Mavriplis DJ, Martinelli L. Multigrid solution of compressible turbulent flow on unstructured meshes using a two-equation model. AIAA Paper 91-0237;1991.
[62] Turner MG, Jennions IK. An investigation of turbulence modeling in transonic fans including a novel implementation of an implicit K-ϵ turbulence model. J Turbomach 1993;115:249-60.
[63] Grasso F, Falconi D. On high speed turbulence modeling of shock-wave boundary-layer interaction. AIAA Paper 93-0778;1993.
[64] Koobus B. An implicit method for turbulent boundary layers simulation. INRIA Report No. 2450, December 1994.
[65] Sondak DL. Parallel implementation of the K-ϵ turbulence model. AIAA Paper 94-0758;1994.
[66] Gerolymos GA, Vallet I. Implicit computation of three-dimensional compressible Navier-Stokes equations using K-ϵ closure. AIAA J 1996;34:1321-30.
[67] Luo H, Baum JD, Löhner R. Computation of compressible flows using a two-equation turbulence model on unstructured grids. AIAA Paper 97-0430;1997.
[68] Papp J, Ghia K. Study of turbulent compressible mixing layers using two-equation turbulence models including an RNG K-ϵ model. AIAA paper 1998-320;1998.
[69] Wang Q, Massey SJ, Abdol-Hamid KS, Frink NT. Solving Navier-Stokes equations with advanced turbulence models on three-dimensional unstructured grids. AIAA Paper 99-0156;1999.
[70] Sinha K, Candler G, Martin M. Assessment of the K-ϵ turbulence model for compressible flows using direct simulation data. AIAA Paper 2001-730;2001.
[71] Van Maele K, Merci B, Dick E. Comparative study of K-ϵ turbulence models in inert and reacting swirling flow. AIAA Paper 2003-3744;2003.
[72] Ferreira VG, Mangiavacchi N, Tomé MF, Castelo A, Cuminato JA, McKee S. Numerical simulation of turbulent free surface flow with two-equation K-ϵ eddy-viscosity models. Int J Numer Methods Fluids 2004;44:347-75.
[73] Brinckman K, Rodebaugh G, Dash S. Towards a unified K-epsilon turbulence model for the practical analysis of aeropropulsive flows. AIAA Paper 2013-392;2013.
[74] Sinha K, Balasridhar SJ. Conservative formulation of the K-ϵ turbulence model for shock-turbulence interaction. AIAA J 2013;51:1872-82.
[75] Edeling WN, Cinnella P, Dwight RP, Bijl H. Bayesian estimates of parameter variability in the K-ϵ turbulence model. J Comput Phys 2014;258:73-94.
[76] Lakshminarayana B. Turbulence modeling for complex shear flows. AIAA J 1986;24:1900-17.
[77] Kunz RF, Lakshminarayana B. Explicit Navier-Stokes computation of cascade flows using the K-ϵ turbulence model. AIAA J 1992;30:13-22.
[78] Spalding DB. A single formula for the "law of the wall". ASME J Appl Mech 1961;28:455-58.
[79] Viegas JR, Rubesin MW. Wall-function boundary conditions in the solution of the Navier-Stokes equations for complex compressible flows. AIAA Paper 83-1694;1983.

[80] Abdol-Hamid KS, Lashmanan B, Carlson JR. Application of Navier-Stokes code PAB3D with k-ϵ turbulence models to attached and separated flows. NASA TR-3480; 1995.
[81] Frink N. Assessment of an unstructured-grid method for predicting 3-D turbulent viscous flows. AIAA Paper 96-0292; 1996.
[82] Mohammadi B, Pironneau O. Unsteady separated turbulent flows computation with wall-laws and k-ϵ model. Comput Methods Appl Eng 1997; 148: 393-405.
[83] Grotjans H, Menter FR. Wall functions for general application CFD codes. In: Proc. fourth European CFD conf, 7-11 Sept. Athens, Greece; 1998.
[84] Kalitzin G, Medic G, Iaccarino G, Durbin P. Near-wall behavior of RANS turbulence models and implications for wall functions. J Comput Phys 2005; 204: 265-91.
[85] Knopp T, Alrutz T, Schwamborn D. A grid and flow adaptive wall-function method for RANS turbulence modeling. J Comput Phys 2006; 220: 19-40.
[86] Menter FR. Two-equation eddy-viscosity turbulence models for engineering applications. AIAA Paper 93-2906; 1993 [also AIAA J 1994; 32: 1598-605].
[87] Menter FR, Rumsey LC. Assessment of two-equation turbulence models for transonic flows. AIAA Paper 94-2343; 1994.
[88] Wilcox DC. Reassessment of the scale-determining equation for advanced turbulence models. AIAA J 1988; 26: 1299-1310.
[89] Menter FR. Influence of freestream values on K-ω turbulence model predictions. AIAA J 1992; 30: 1651-9.
[90] Hellsten A, Laine S. Extension of the K-ω-SST turbulence model for flows over rough surfaces. AIAA Paper 97-3577; 1997 [also AIAA J 1998; 36: 1728-29].
[91] Hellsten A. Some improvements in Menter' s K-ω SST turbulence model. AIAA Paper 98-2554; 1998.
[92] Forsythe JR, Hoffmann KA, Suzen YB. Investigation of modified Menter' s two-equation turbulence model for supersonic applications. AIAA Paper 99-0873; 1999.
[93] Goodheart K, Dykas S, Schnerr G. Numerical insights into the solution of transonic flow test cases using the Wilcox K-ω, EASM and SST models. AIAA Paper 2003-1141; 2003.
[94] Steed R. High lift CFD simulations with an SST-based predictive laminar to turbulent transition model. AIAA paper 2011-864; 2011.
[95] Murman S. Evaluating modified diffusion coefficients for the SST turbulence model using benchmark tests. AIAA Paper 2011-3571; 2011.
[96] Geisbauer S. Numerical spoiler wake investigations at the borders of the flight envelope. AIAA Paper 2011-3811; 2011.
[97] Schoenawa S, Hartmann R. Discontinuous Galerkin discretization of the Reynolds-averaged Navier-Stokes equations with the shear-stress transport model. J Comput Phys 2014; 262: 194-216.
[98] Smagorinsky J. General circulation experiments with the primitive equations. Mon Weather Rev 1963; 91: 99-165.
[99] Deardorff JW. A numerical study of three-dimensional turbulent channel flow at large Reynolds numbers. J Fluid Mech 1970; 41: 453-80.

[100] Schumann U. Subgrid-scale model for finite-difference simulations of turbulent flows in plane channels and annuli. J Comput Phys 1975;18:376-404.

[101] Bardina J, Ferziger JH, Reynolds WC. Improved subgrid models for large eddy simulation. AIAA Paper 80-1357;1980.

[102] Moin P, Kim J. Numerical investigation of turbulent channel flow. J Fluid Mech 1982;118:341-77.

[103] Piomelli U, Zang TA, Speziale CG, Hussaini MY. On the large-eddy simulation of transitional wall-bounded flows. ICASE Report No. 89-55;1989.

[104] Hussaini MY, Speziale CG, Zang TA. Discussion of "The potential and limitations of direct and large-eddy simulations." ICASE Report No. 89-61;1989.

[105] Erlebacher G, Hussaini MY, Speziale CG, Zang TA. Toward the large-eddy simulation of compressible turbulent flows. ICASE Report No. 90-76;1990.

[106] Erlebacher G, Hussaini MY, Speziale CG, Zang TA. On the large-eddy simulation of compressible isotropic turbulence. In: 12th International conf. on numerical methods in fluid dynamics, Oxford, England; Lecture Notes in Physics, vol. 371. Berlin: Springer; 1990.

[107] Schumann U. Direct and large-eddy simulation of turbulence—summary of the state of the art 1993. VKI Lecture Series 1993-02. Von Karman Institute: Rhode-St-Genèse, Belgium; 1993.

[108] Jones WP, Wille M. Large eddy simulation of a jet in a cross-flow. In: 10th Symposium on turbulent shear flows, vol. 1. The Pennsylvania State University, PA, August 14-16;1995.

[109] Lund TS, Moin P. Large eddy simulation of a boundary layer on a concave surface. In: 10th symposium on turbulent shear flows, vol. 1. The Pennsylvania State University, PA, August 14-16;1995.

[110] Calhoon WH, Menon S. Subgrid modeling for reacting large eddy simulations. AIAA Paper 96-0516;1996.

[111] Cook AW, Riley JJ, Kosaly G. A laminar flamelet approach to subgrid-scale chemistry in turbulent flows. Combust Flame 1997;109:332-41.

[112] Wu Y, Cao S, Yang J, Ge L, Guilbaud M, Soula V. Turbulent flow calculation through a water turbine draft tube by using the Smagorinsky model. In: ASME fluids engineering division summer meeting (FEDSM), Paper No. FEDSM98-4864. Washington DC;1998.

[113] Held J. Large eddy simulations of separated compressible flows around 3-D configurations. PhD Thesis. Lund, Sweden: Department of heat and power engineering/fluid mechanics, Lund Institute of Technology;1998.

[114] Kim W-W, Menon S, Mongia HC. Large eddy simulations of reacting flows in a dump combustor. AIAA Paper 98-2432;1998.

[115] Ducros F, Ferrand V, Nicoud F, Weber C, Darracq D, Gacherieu C, Poinsot T. Large-eddy simulation of the shock/turbulence interaction. J Comput Phys 1999;152:517-49.

[116] Jaberi FA. Large eddy simulation of turbulent premixed flames via filtered mass density function. AIAA Paper 99-0199;1999.

[117] Han J, Lin Y-L, Arya SP, Proctor FH. Large eddy simulation of aircraft wake vortices in a homogeneous atmospheric turbulence: vortex decay and descent. AIAA Paper 99-0756;1999.

[118] Bui TT. A parallel, finite-volume algorithm for large-eddy simulation of turbulent flows. AIAA Paper 99-0789; 1999.

[119] Kravchenko AG, Moin P, Shariff K. B-Spline method and zonal grids for simulations of complex turbulent flows. J Comput Phys 1999; 151: 757-89.

[120] Cook AW. A consistent approach to large eddy simulation using adaptive mesh refinement. J Comput Phys 1999; 154: 117-33.

[121] Pascarelli A, Piomelli U, Candler GV. Multi-block large-eddy simulations of turbulent boundary layers. J Comput Phys 2000; 157: 256-79.

[122] Kurose R, Makino H. Large eddy simulation of a solid-fuel jet flame. Combust Flame 2003; 135: 1-16.

[123] Di Mare F, Jones WP, Menzies KR. Large eddy simulation of a model gas turbine combustor. Combust Flame 2004; 137: 278-94.

[124] Roux S, Lartigue G, Poinsot T, Meier U, Bérat C. Studies of mean and unsteady flow in a swirled combustor using experiments, acoustic analysis, and large eddy simulations. Combust Flame 2005; 141: 40-54.

[125] Trouve A, Wang Y. Large eddy simulation of compartment fires. Int J Comput Fluid Dynam 2010; 24: 449-66.

[126] Knopp T, Zhang X, Kessler R, Lube G. Enhancement of an industrial finite-volume code for large-eddy-type simulation of incompressible high Reynolds number flow using near-wall modelling. Comput Methods Appl Mech Eng 2010; 199: 890-902.

[127] Granet V, Vermorel O, Lacour C, Enaux B, Dugué V, Poinsot T. Large-eddy simulation and experimental study of cycle-to-cycle variations of stable and unstable operating points in a spark ignition engine. Combust Flame 2012; 159: 1562-75.

[128] Viazzo S, Poncet S, Serre E, Randriamampianina A, Bontoux P. High-order large eddy simulations of confined rotor-stator flows. Flow Turbul Combust 2012; 88: 63-75.

[129] Chapman DR. Computational aerodynamics development and outlook. AIAA J 1979; 17: 1293-313.

[130] Tam CKW, Webb JC. Dispersion-relation-preserving finite difference schemes for computational acoustics. J Comput Phys 1993; 107: 262-81.

[131] Tam CKW. Applied aero-acoustics: prediction methods. VKI Lecture Series 1996-04. Rhode-St-Genèse, Belgium: Von Karman Institute; 1996.

[132] Mittal R, Moin P. Suitability of upwind-biased finite difference schemes for large-eddy simulation of turbulent flows. AIAA J 1997; 35: 1415-17.

[133] Garnier E, Mossi M, Sagaut P, Comte P, Deville M. On the use of shock-capturing schemes for large-eddy simulation. J Comput Phys 1999; 153: 273-311.

[134] Kravchenko AG, Moin P. On the effect of numerical errors in large eddy simulations of turbulent flow. J Comput Phys 1997; 131: 310-22.

[135] Park N, Yoo JY, Choi H. Discretization errors in large eddy simulation: on the suitability of centered and upwind-biased compact difference schemes. J Comput Phys 2004; 198: 580-616.

[136] Denaro FM. What does finite volume-based implicit filtering really resolve in Large-eddy simulations? J Comput Phys 2011;230:3849-83.

[137] Miet P, Laurence D, Nitrosso B. Large eddy simulation with unstructured grids and finite elements. In: 10th Symposium on turbulent shear flows, vol. 2. The Pennsylvania State University, PA, August 14-16, 1995.

[138] Jansen K. Large-eddy simulation of flow around a NACA 4412 Airfoil using unstructured grids. Center for Turbulence Research, NASA Ames/Stanford University, Annual Research Briefs; 1996. p. 225-32.

[139] Chalot F, Marquez B, Ravachol M, Ducros F, Nicoud F, Poinsot T. A consistent finite element approach to large eddy simulation. AIAA Paper 98-2652; 1998.

[140] Okong' o N, Knight D. Compressible large eddy simulation using unstructured grids: channel and boundary layer flows. AIAA Paper 98-3315; 1998.

[141] Urbin G, Knight D, Zheltovodov AA. Compressible large eddy simulation using unstructured grid: supersonic turbulent boundary layer and compression corner. AIAA Paper 99-0427; 1999.

[142] Marsden AL, Vasilyev OV, Moin P. Construction of commutative filters for LES on unstructured meshes. J Comput Phys 2002; 175: 584-603.

[143] Camarri S, Salvetti MV. Koobus B, Dervieux A. A low-diffusion MUSCL scheme for LES on unstructured grids. Comput Fluids 2004; 33: 1101-29.

[144] Selle L, Lartigue G, Poinsot T, Koch R, Schildmacher K-U, Krebs W, et al. Compressible large eddy simulation of turbulent combustion in complex geometry on unstructured meshes. Combust Flame 2004; 137: 489-505.

[145] Koobus B, Farhat C. A variational multiscale method for the large eddy simulation of compressible turbulent flows on unstructured meshes—application to vortex shedding. Comput Methods Appl Mech Eng 2004; 193: 1367-83.

[146] Ciardi M, Sagaut P, Klein M, Dawes WN. A dynamic finite volume scheme for large-eddy simulation on unstructured grids. J Comput Phys 2005; 210: 632-55.

[147] Wang G, Duchaine F, Papadogiannis D, Duran I, Moreau S, Gicquel LYM. An overset grid method for large eddy simulation of turbomachinery stages. J Comput Phys 2014; 274: 333-55.

[148] Ferziger JH. Direct and large eddy simulation of turbulence. In: Vincent A, editor. Numerical methods in fluid mechanics, vol. 16. Centre de Recherches Mathématiques Université de Montréal, Proceedings and Lecture Notes; 1998. p. 53-97.

[149] Piomelli U. Large-eddy simulation of turbulent flows. VKI Lecture Series 1998-05. Rhode-St-Genèse, Belgium: Von Karman Institute; 1998.

[150] Mossi M. Simulation of benchmark and industrial unsteady compressible turbulent fluid flows. PhD Thesis, No. 1958, École Polytechnique Fédérale de Lausanne, Département de Génie Mécanique, Switzerland; 1999.

[151] Sagaut P. Large eddy simulation for incompressible flows: an introduction. 3rd ed. Berlin: Springer; 2005.

[152] Garnier E, Adams N, Sagaut P. Large eddy simulation for compressible flows. Berlin: Springer; 2009.

[153] Piomelli U. Large-eddy simulation: present state and future directions. AIAA Paper 98-0534; 1998.

[154] Pitsch H. Large-eddy simulation of turbulent combustion. Annu Rev Fluid Mech 2006; 38:453-482.

[155] Piomelli U, Benocci C, van Beeck JP. Large eddy simulation and related techniques. VKI Lecture Series 2010-04. Rhode-St-Genèse, Belgium: Von Karman Institute; 2010.

[156] Vasilyev OV, Lund TS. A General theory of discrete filtering for LES in complex geometry. Annual Research Briefs, Center for Turbulence Research, NASA Ames/Stanford Univ.; 1997. p. 67-82.

[157] Vasilyev OV, Lund TS, Moin P. A general class of commutative filters for les in complex geometries. J Comput Phys 1998; 146:82-104.

[158] Lund TS, Kaltenbach H-J. Experiments with explicit filtering for les using a finite-difference method. Annual Research Briefs, Center for Turbulence Research, NASA Ames/Stanford Univ.; 1995. p. 91-105.

[159] Lund TS. On the use of discrete filters for large eddy simulation. Annual Research Briefs, Center for Turbulence Research, NASA Ames/Stanford Univ.; 1997. p. 83-95.

[160] Leonard A. Energy cascade in large-eddy simulations of turbulent fluid flows. Adv Geophys 1974; 18:237-48.

[161] Moin P, Squires KD, Cabot WH, Lee S. A dynamic subgrid-scale model for compressible turbulence and scalar transport. Phys Fluids 1991; 3:2746-57.

[162] Vreman B, Geurts B, Kuerten H, Broeze J, Wasistho B, Streng M. Dynamic subgrid-scale models for LES of transitional and turbulent compressible flow id 3-D shear layers. In: 10th symposium on turbulent shear flows, vol. 1. The Pennsylvania State University, PA, August 14-16; 1995.

[163] Domaradzki JA, Saiki EM. Backscatter models for large-eddy simulations. Theor Comput Fluid Dyn 1997; 9:75-83.

[164] Lenormand E, Sagaut P, Phuoc LT, Comte P. Subgrid-scale models for large-eddy simulations of compressible wall bounded flows. AIAA J 2000; 38:1340-50.

[165] Boris JP, Grinstein FF, Oran ES, Kolbe RJ. New insights into large eddy simulations. Fluid Dyn Res 1992; 10:199-228.

[166] Fureby C, Grinstein FF. Monotonically integrated large eddy simulation of free shear flows. AIAA J 1999; 37:544-56.

[167] Grinstein FF, Fureby C. Recent progress on MILES for high Reynolds number flows. J Fluids Eng 2002; 124:848-61.

[168] Garnier E, Mossi M, Sagaut P, Comte P, Deville M. On the use of shock-capturing schemes for large-eddy simulation. J Comput Phys 1999; 153:273-311.

[169] Domaradzki JA, Radhakrishnan S, Xiao Z, Smolarkiewicz PK. Effective eddy viscosities in implicit modeling of high Reynolds number flows. AIAA Paper 2003-4103; 2003.

[170] Fureby C, Grinstein FF. Monotonically integrated large eddy simulation of free shear flows. AIAA J 1999;37:544-56.

[171] Fureby C. Large eddy simulation of rearward-facing step flow. AIAA J 1999;37:1401-10.

[172] Okong' o N, Knight D, Zhou G. Large eddy simulations using an unstructured grid compressible Navier-Stokes algorithm. Int J Comput Fluid Dynam 2000;13:303-26.

[173] Yan H, Knight D, Zheltovodov AA. Large-eddy simulation of supersonic flat-plate boundary layers using the monotonically integrated large-eddy simulation (MILES) technique. J Fluids Eng 2002;124:868-75.

[174] Fureby C, Grinstein FF. Large eddy simulation of high-Reynolds-number free and wall-bounded flows. J Comput Phys 2002;181:68-97.

[175] Raverdy B, Mary I, Sagaut P. High-resolution large-eddy simulation of flow around low-pressure turbine blade. AIAA J 2003;41:390-7.

[176] Grinstein FF, Fureby C. Implicit large eddy simulation of high-re flows with flux-limiting schemes. AIAA Paper 2003-4100;2003.

[177] Hickel S, Adams NA, Domaradzki JA. An adaptive local deconvolution method for implicit LES. J Comput Phys 2006;213:413-36.

[178] Rogallo RS, Moin P. Numerical simulation of turbulent flows. Annu Rev Fluid Mech 1984;16:99-137.

[179] Vreman AW. Direct and large-eddy simulation of the compressible turbulent mixing layer. PhD Thesis. The Netherlands: Dept. of Applied Mathematics, University of Twente Enschede; 1995.

[180] Lilly DK. The representation of small-scale turbulence in numerical simulation experiments. In: Proc. IBM sci. comp. symp. on environmental sciences, New York; November 14-16, 1967.

[181] Germano M, Piomelli U, Moin P, Cabot WH. A dynamic subgrid-scale eddy viscosity model. Phys Fluids 1991;3:1760-5.

[182] Germano M. Turbulence: the filtering approach. J Fluid Mech 1992;238:325-36.

[183] Lilly DK. A proposed modification of the Germano subgrid-scale closure method. Phys Fluids 1992;4:633-5.

[184] Ghosal S, Lund TS, Moin P, Akselvoll K. A dynamic localization model for large-eddy simulation of turbulent flows. J Fluid Mech 1995;286:229-55.

[185] Carati D, Ghosal S, Moin P. On the representation of backscatter in dynamic localization models. Phys Fluids 1995;7:606.

[186] Piomelli U, Liu J. Large-eddy simulation of rotating channel flows using a localized dynamic model. Phys Fluids 1995;7:839-48.

[187] Chapman DR, Kuhn GD. The limiting behavior of turbulence near a wall. J Fluid Mech 1986;70:265-92.

[188] Brooke JW, Hanratty TJ. Origin of turbulence-producing eddies in a channel flow. Phys Fluids 1993;5:1011-22.

[189] Grötzbach G. Direct numerical and large eddy simulation of turbulent channel flows. In: Cheremisinoff NP, editor. Encyclopedia of fluid mechanics. West Orange: Gulf Publishing; 1987,

p. 1337.
[190] Mason PJ, Callen NS. On the magnitude of the subgrid-scale eddy coefficient in large-eddy simulations of turbulent channel flow. J Fluid Mech 1986;162:439-62.
[191] Piomelli U, Ferziger J, Moin P, Kim J. New approximate boundary conditions for large eddy simulations of wall-bounded flows. Phys Fluids 1989;1:1061-68.
[192] Balaras E, Benocci C, Piomelli U. Two-layer approximate boundary condition for large-eddy simulations. AIAA J 1996;34:1111-19.
[193] Cabot W. Large-eddy simulations with wall models. Annual Research Briefs, Center for Turbulence Research, NASA Ames/Stanford Univ.; 1995. p. 41-50.
[194] Cabot W. Near-wall models in large eddy simulations of flow behind a backward-facing step. Annual Research Briefs, Center for Turbulence Research, NASA Ames/Stanford Univ.; 1996. p. 199-210.
[195] Cabot W, Moin P. Approximate wall boundary conditions in the large-eddy simulation of high Reynolds number flow. Flow Turbulence Combust 1999;63:269-91.
[196] Wang M, Moin P. Wall modelling in LES of trailing-edge flow. In: Proc. 2nd int. symposium on turbulence and shear flow phenomena, vol. 2. Stockholm, Sweden; 2001. p. 165-70.
[197] Piomelli U, Balaras E. Wall-layer models for large-eddy simulations. Prog Aerosp Sci 2008; 44:437-46.
[198] Spalart PR. Strategies for turbulence modelling and simulations. Int J Heat Fluid Flow 2000; 21:252-63.
[199] Spalart PR. Trends in turbulence treatments. AIAA Paper 2000-2306; 2000.
[200] Strelets M. Detached eddy simulation of massively separated flows. AIAA Paper 2001-0879; 2001.
[201] Mani M. Hybrid turbulence models for unsteady flow simulation. J Aircraft 2004;41:110-18.
[202] Forsythe J, Squires K, Wurtzler K, Spalart PR. Detached eddy simulation of fighter aircraft at high alpha. AIAA Paper 2002-0591; 2002.
[203] Forsythe J, Squires K, Wurtzler K, Spalart PR. Prescribed spin of the F-15E using detached-eddy simulation. AIAA Paper 2003-839; 2003.
[204] Schlüter JU. Consistent boundary conditions for integrated RANS/LES simulations: LES inflow conditions. AIAA Paper 2003-3971; 2003.
[205] Hamba F. A hybrid RANS/LES simulation of turbulent channel flow. Theor Comput Fluid Dyn 2003;16:387-403.
[206] Georgiadis NJ, Alexander JID, Reshotko E. Hybrid Reynolds-averaged Navier-Stokes/large-eddy simulations of supersonic turbulent mixing. AIAA J 2003;41:218-29.
[207] Schlüter JU, Pitsch H. Antialiasing filters for coupled Reynolds-averaged/large-eddy simulations. AIAA J 2005;43:608-15.
[208] Kawai S, Fujii K. Analysis and prediction of thin-airfoil stall phenomena with hybrid turbulence methodology. AIAA J 2005;43:953-61.
[209] Gopalan H, Heinz S, Stöllinger MK. A unified RANS-LES model: computational development, accuracy and cost. J Comput Phys 2013;249:249-74.

第8章 边界条件

任何数值模拟都只能考虑真实物理域或系统的一部分,域的截断产生了人为边界,必须在此指定特定物理量的值。此外,流场中的固壁构成了物理域的天然边界。需要特别注意边界条件的数值处理,不恰当的实现可能导致对真实系统的模拟不准确,也会对求解格式的稳定性和收敛速度产生不利影响。

在欧拉方程组和纳维-斯托克斯方程组的数值求解中,一般会遇到以下几种边界条件:

(1) 固壁;

(2) 外流中的远场;

(3) 内流中的入流和出流;

(4) 注入边界;

(5) 对称;

(6) 坐标切割和周期边界;

(7) 网格块间的边界。

下面将详细介绍这些边界条件的数值处理方法,读者可参阅3.4节了解更多边界条件(如壁面或自由面的热辐射)的相关文献。

8.1 虚拟单元的概念

在讨论边界条件之前,应该首先提及虚拟单元或虚拟点的概念。这种方法在结构网格中非常流行,同时,虚拟单元在非结构网格中也有一定的优势。虚拟单元是物理域之外网格单元或网格点的附加层,图8.1给出了2D结构网格虚拟单元的示意图,可以看到,计算域所有边界都被两层虚拟单元包围(用虚线标记)。虚拟单元(或点)通常不会像域内网格一样生成(除非是在网格块的界面上)，这些单元只是虚拟的，尽管它们也与体积或面矢量这样的几何量相关联。

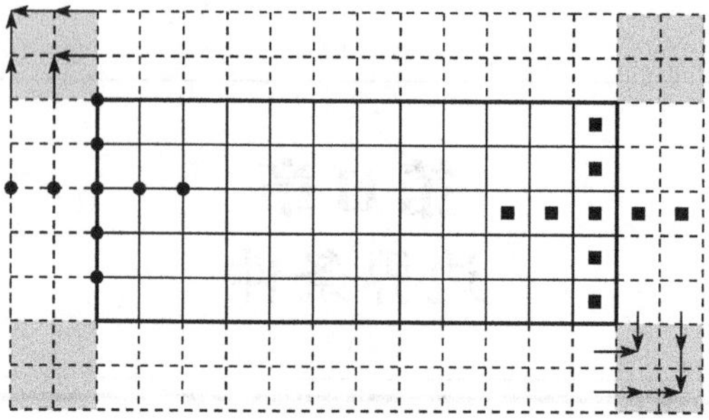

图 8.1　2D 计算域(粗线)周围的两层虚拟单元(虚线)。实心圆点代表二阶格点(二次)格式的标准模板，实心矩形代表二阶格心格式的模板(见 4.3 节)

虚拟单元的目的是简化边界上通量、梯度、耗散等数值项的计算，通过虚拟单元，可将空间离散格式的模板扩展到物理边界。图 8.1 所示，边界处可以采用与物理域内相同的离散格式，从而我们可以对所有物理网格点用相同的方法求解控制方程组，这使得离散格式更容易实现。此外，结构网格的所有网格点都可以在单个循环中访问，这在循环并行化或矢量化中具有显著的优势，当然，这样做的前提条件是为虚拟单元(点)上的守恒变量和几何量恰当赋值，虚拟单元层数必须能够完全覆盖物理域之外的模板单元。虚拟单元(点)上的守恒变量通过边界条件设置，几何量通常取自边界处相应的控制体，在多个网格块间的边界处(图 3.4)，所有的流动变量和几何量都是从相邻的网格块传输过来。

对于图 8.1 中灰色阴影的虚拟单元(及其相关联的点)，如果没有相邻的网格块，如何对其赋值并不是很明确。标准的交叉类型离散格式的模板不需要这些单元上的值，然而，这些单元(点)对于梯度的计算(黏性通量，见 4.4 节)或多重网格的转换算子(见 9.4 节)是重要的。解决这个问题最简单的方法是将其赋值为相邻“规则”虚拟单元的平均值，如图 8.1 中箭头所示。然而，这种方法处理固壁或对称边界的效果不尽人意，在这些情况下，最好将物理边界扩展到虚拟单元层(请参阅本身附带的结构 2D 代码)。

8.2　固壁

8.2.1　无黏流动

在无黏流动的情况下，流体滑过壁面。因为没有摩擦力，速度矢量与壁面相切，也就是说没有垂直于壁面的流动，即

$$\text{在壁面上} \quad \boldsymbol{v} \cdot \boldsymbol{n} = 0 \tag{8.1}$$

其中,$\boldsymbol{n}$ 为壁面的单位法向矢量。因此,壁面处的逆变速度 V(式(2.22))为零,对流通量(式(2.21))缩减为只包含压力,即

$$(\boldsymbol{F}_{\mathrm{c}})_{\mathrm{w}} = \begin{bmatrix} 0 \\ n_x p_{\mathrm{w}} \\ n_y p_{\mathrm{w}} \\ n_z p_{\mathrm{w}} \\ 0 \end{bmatrix} \tag{8.2}$$

其中,p_{w} 是壁面压力。

1. 结构网格的格心格式

格心格式中,在单元体心处计算压力。然而,式(8.2)中要求在边界单元的界面上计算 p_{w},这可以很容易地从域内通过外插得到。如图 8.2 所示,可以简单地设 $p_{\mathrm{w}}=p_2$,更高的精度可以通过使用两点外插公式:

$$p_{\mathrm{w}} = \frac{1}{2}(3p_2 - p_3) \tag{8.3}$$

或者三点外插公式得到

$$p_{\mathrm{w}} = \frac{1}{8}(15p_2 - 10p_3 + 3p_4) \tag{8.4}$$

考虑到网格的拉伸,可以采用到壁面的距离代替插值中的常系数[1]。

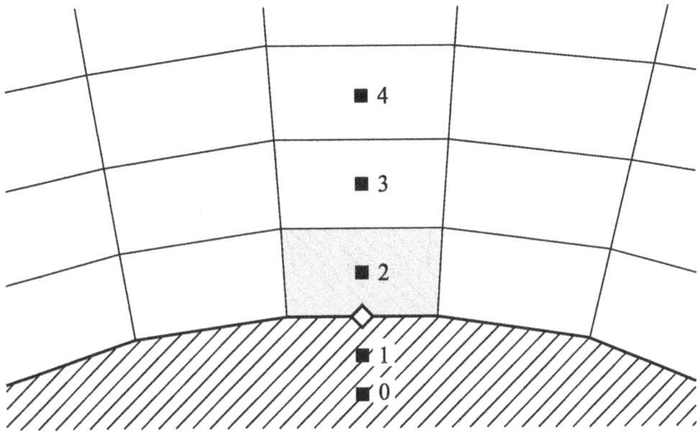

图 8.2　格心格式的固壁边界条件。虚拟单元表示为 0 和 1,对流通量(式(8.2))计算的位置用菱形表示

式(8.3)~式(8.4)的外插算法没有考虑网格和壁面的几何信息。Rizzi[2] 提出了法向动量关系法,该方法基于"在无黏流动中固壁代表了一条流线"这一

性质，将式(8.1)中的零法向流动条件沿壁面流线进行微分，并将微分结果代入动量方程得到

$$\rho \boldsymbol{v} \cdot (\boldsymbol{v} \cdot \nabla) \boldsymbol{n} = \boldsymbol{n} \cdot \nabla p \tag{8.5}$$

式(8.5)将密度、速度、壁面几何信息和压力的法向导数联系起来。现有结果证明法向动量关系可以给出非常准确的结果[1]，然而，在复杂几何情况下，法向动量关系的数值求解可能会出现问题，具体的算法实现和精度比较可查阅文献[1]。

虚拟单元上守恒变量的值可从域内部线性外插得到，即

$$\begin{aligned} \boldsymbol{W}_1 &= 2\boldsymbol{W}_2 - \boldsymbol{W}_3 \\ \boldsymbol{W}_0 &= 3\boldsymbol{W}_2 - 2\boldsymbol{W}_3 \end{aligned} \tag{8.6}$$

式(8.6)中的下标与图8.2对应。如果空间离散格式使用到虚拟单元（例如耗散因子的计算），对流通量的计算必须与式(8.2)一致。

2. 结构网格的格点格式

对于重叠控制体的格点离散格式（见4.2.2节），式(8.1)的边界条件的实现非常简单，根据式(8.2)计算壁面的对流通量，壁面压力 p_w 通过节点平均得到，如式(4.26)中的2D情况或式(4.27)中的3D情况，分布公式（2D情况下的式(4.30)）现在只占据两个（3D情况下为四个）单元（对比图4.4和图8.3）。

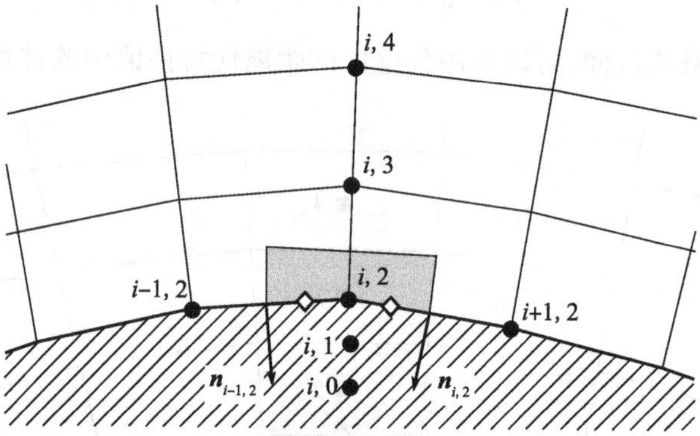

图8.3　结构格点（二次控制体）格式的固壁边界条件。虚拟点记为$(i,0)$和$(i,1)$，对流通量（式(8.2)）计算的位置用菱形标记，与图4.6相对应

对于二次控制体的格点格式（见4.2.3节），可以采用几种不同的方法。一种方法是将式(8.2)中的条件分别应用于壁面上控制体的每个壁面，因此，根据图8.3可以得到

$$
(\boldsymbol{F}_{c,w}\Delta S)_{i,2}=\begin{bmatrix}0\\(n_x)_{i-1,2}(p_w)_{i-1/4}\\(n_y)_{i-1,2}(p_w)_{i-1/4}\\(n_z)_{i-1,2}(p_w)_{i-1/4}\\0\end{bmatrix}\frac{\Delta S_{i-1,2}}{2}+\begin{bmatrix}0\\(n_x)_{i,2}(p_w)_{i+1/4}\\(n_y)_{i,2}(p_w)_{i+1/4}\\(n_z)_{i,2}(p_w)_{i+1/4}\\0\end{bmatrix}\frac{\Delta S_{i,2}}{2} \tag{8.7}
$$

式(8.7)中的压力$(p_w)_{i-1/4}$和$(p_w)_{i+1/4}$可以通过线性插值得到,例如:

$$
(p_w)_{i+1/4}=\frac{1}{4}[3p_{i,2}+p_{i+1,2}] \tag{8.8}
$$

对应的3D公式将在非结构中点-二次格式一节介绍(式(8.11))。

另一种可能的实现方法是在相应的壁面节点处(即图8.3中的节点$(i,2)$)直接施加式(8.2)的条件。壁面压力p_w简单地设为$p_{i,2}$,单位法向矢量通过将所有共享节点$(i,2)$的壁面小面的法向矢量进行平均得到,这种方法需要对速度矢量进行修正。在通过时间推进格式更新流场后,将壁面处的速度矢量投影到切向平面[3-4],即

$$
(\boldsymbol{v}_{i,2})_{corr}=\boldsymbol{v}_{i,2}-[\boldsymbol{v}_{i,2}\cdot(\boldsymbol{n}_{av})_{i,2}]\cdot(\boldsymbol{n}_{av})_{i,2} \tag{8.9}
$$

其中,$\boldsymbol{n}_{av}$是平均的单位法向矢量。这样,流动就与壁面相切。

为了给虚拟点赋值,可以使用类似于式(8.6)的关系将守恒变量从内场外插到虚拟点(式(2.20))。

3. 非结构网格的格心格式

对于非结构网格的格心格式,可以使用与结构网格类似的方法实现式(8.1)中的固壁边界条件。如果边界单元是四边形、六面体或棱柱(壁面为三角形),压力可通过式(8.3)外插到壁面,邻近单元(图8.2中的单元3)可以根据5.2.1节中的基于面的数据结构获得。对于三角形或四面体单元,文献[5-6]建议使用一层虚拟单元,虚拟单元上的速度矢量通过镜像壁面边界单元的速度矢量得到,例如,在图8.2中的虚拟单元1上,速度为

$$
\boldsymbol{v}_1=\boldsymbol{v}_2-2V_2\boldsymbol{n} \tag{8.10}
$$

其中,$V_2=u_2n_x+v_2n_y+w_2n_z$是逆变速度,$\boldsymbol{n}=[n_x,n_y,n_z]^T$表示壁面处的单位法向矢量。虚拟单元的压力和密度设置为相应边界单元上的值(即$p_w=p_2$)

4. 非结构网格中点-二次格式

对于非结构网格中点-二次离散格式,边界条件(式(8.1))需要更多的关注。图8.4和图8.5分别给出了2D和3D的情况。式(8.2)中的对流通量在位于壁面的控制体的每个壁面上分别计算,这与之前介绍的结构网格二次控制体格点格式中的第一种方法相同。对于四边形单元(如图8.4中的1-3-4-5),压力

根据式(8.8)插值得到。在六面体、三棱柱或金字塔单元情况下,控制体的面是四边形(如图 8.5 中的面 1-4-5-6),插值公式为

$$p_{\mathrm{w}}=\frac{1}{16}[9p_1+3p_4+3p_6+p_5] \tag{8.11}$$

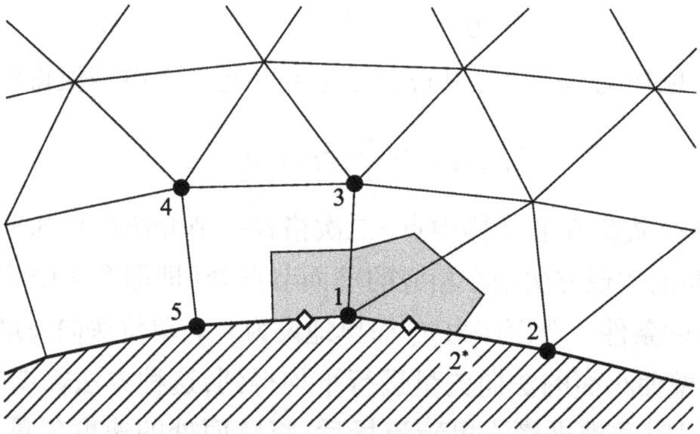

图 8.4　2D 非结构二次控制体混合网格的固壁边界条件。对流通量(式(8.2))计算的位置用菱形标记

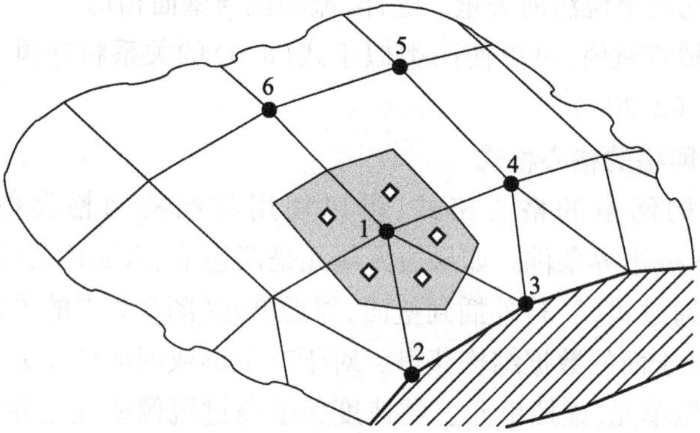

图 8.5　3D 非结构混合网格的固壁边界条件。对流通量(式(8.2))计算的位置用菱形标记

如果边界单元是四面体或三棱柱(2D 情况下为三角形),应采用类似有限元的方法计算壁面压力[7]。例如,在图 8.4 中的壁面段 1-2 处,面 1-2* 处的压力通过下式计算:

$$p_{\mathrm{w}}=\frac{1}{6}[5p_1+p_2] \tag{8.12}$$

对于四面体的情况,以图 8.5 中壁面 1-2-3 为例,压力通过下式计算:

$$p_w=\frac{1}{8}[6p_1+p_2+p_3] \tag{8.13}$$

需要强调的是,这种有限元类型的插值方法对于准确地流动求解很重要[7]。

8.2.2 黏性流动

对于流过固壁的黏性流体,假定壁面与流体之间的相对速度为零,因而称为无滑移边界条件。在壁面静止的情况下,笛卡儿速度分量为

$$\text{在壁面上 } u=v=w=0 \tag{8.14}$$

无滑移条件有两个基本后果。首先,不需要求解壁面处的动量方程,在格点格式中利用到了这一点。其次,无滑移壁面的对流通量同样由式(8.2)给出,式(2.24)中的项简化为 $\boldsymbol{\Theta}=k\nabla T$,因此,对流通量中的壁面压力的计算与无黏流动中的方法相同,但虚拟单元(点)的处理方式不同。

1. 格心格式

利用虚拟单元可以简化式(8.14)中无滑移边界条件的实现。在绝热壁(没有热流穿过壁面)情况下,可以设为(图 8.2)

$$\begin{aligned}&\rho_1=\rho_2 \quad E_1=E_2\\&u_1=-u_2 \quad v_1=-v_2 \quad w_1=-w_2\end{aligned} \tag{8.15}$$

对于单元 0 和 3 也类似。该方法同时适用于结构网格和非结构网格[6]。

如果已知壁面温度,速度分量仍如式(8.15)一样取反得到。虚拟单元上的温度利用指定的壁温从内场线性外插得到。由于壁面处压力法向梯度为零,将虚拟单元上的压力指定为边界单元的值,即 $p_0=p_1=p_2$。虚拟单元上的密度和总能根据插值后的量计算得到。

2. 格点格式

由于动量方程不需要求解,因此壁面处对流通量(式 8.2)为零。式(2.23)中的黏性通量为能量方程只提供了壁面处温度的法向导数。对于绝热壁面,$\nabla T_w\cdot\boldsymbol{n}$ 为零,因此无须计算壁面上的对流通量和黏性通量。动量方程的残差应设为零,以防止在壁面节点产生非零速度分量。

在给定壁面温度的情况下,可以根据理想气体假设直接设置壁面(图 8.3 中的节点$(i,2)$)的总能,即

$$(\rho E)_{i,2}=\frac{c_p}{\gamma}\rho_{i,2}T_w \tag{8.16}$$

式中:T_w 为给定的壁面温度。动量和能量方程的残差必须归零,同样的策略也适用于非结构网格。

非绝热壁面的另一种处理方法,它在某些应用中更鲁棒,在壁面处根本不求解控制方程组。密度和能量都直接指定:

$$\rho_{i,2}=\frac{p_{i,3}}{T_w R} \quad (\rho E)_{i,2}=\frac{p_{i,3}}{\gamma} \tag{8.17}$$

式(8.17)假定了壁面处压力的法向梯度为零,因此有 $p_{i,2}=p_{i,3}$。既然所有的守恒变量都是指定的,在节点$(i,2)$处所有方程的残差都需设为零。这种方法也可以在非结构网格上使用,不过,在三角形或四面体上单元上,压力的外插需要额外的运算。

如果壁面是绝热的,则虚拟点上的值如下获得:

$$\begin{aligned}&\rho_{i,1}=\rho_{i,3} \quad E_{i,1}=E_{i,3}\\&u_{i,1}=-u_{i,3} \quad v_{i,1}=-v_{i,3} \quad w_{i,1}=-w_{i,3}\end{aligned} \tag{8.18}$$

对于节点 0 和 4 亦是如此。如果壁面温度给定,虚拟单元上的温度从内场外插得到,即

$$T_{i,1}=2T_w-T_{i,3} \quad T_{i,0}=3T_w-2T_{i,3} \tag{8.19}$$

公式中的下标如图 8.3 所示。速度分量根据式(8.18)取反得到,密度和能量使用插值的温度值和压力 $p_{i,3}$ 计算。

8.3 远场

翼型、机翼、汽车和其他构型的外流数值模拟,必须在一个有界区域内进行,因此需要人工的远场边界条件。远场边界条件的数值实现必须满足两个基本要求。首先,与无限域相比,计算域的截断对流动解应该没有明显的影响。其次,任何向外传播的扰动不得反射回流场[9]。由于亚声速和跨声速流动问题的椭圆性质,它们对远场边界条件特别敏感,不合适的实现会导致收敛到稳态的速度显著减慢,还可能影响数值解的精度。人们发展了各种方法,能够在人工边界处吸收向外传播的扰动[10-15]。关于不同的无反射边界条件的综述可见文献[16]。

接下来的两节将讨论 Whitfield 和 Janus[12] 提出的特征变量的概念以及升力体远场边界条件的推广。

8.3.1 特征变量的概念

根据对流通量雅可比矩阵特征值的符号(A.11 节式(A.84)或式(A.88)),信息沿着特征方向传出或进入计算域。例如,在亚声速入流情况下,有四个流入特征

(3D 情况下)和一个流出特征(式(A.88)中的 Λ_5),亚声速出流则与之相反。根据 Kreiss[17] 的 1D 理论,边界处从外部施加的条件数量应该等于流入特征的数量。剩余条件应该由域内流场解确定。

Whitfield 和 Janus[12] 的方法基于边界法向的欧拉方程组(2.44)的特征形式,在结构网格和非结构网格上的大量流动算例中都有很好的表现,它不仅适用于远场边界,也适用于无黏固壁边界(见 8.2.1 节)。

远场边界的两种基本流动情况如图 8.6 所示,流动可以进入或者离开计算域,根据当地马赫数,需要处理四种不同类型的远场边界条件:

(1) 超声速入流;

(2) 超声速出流;

(3) 亚声速入流;

(4) 亚声速出流。

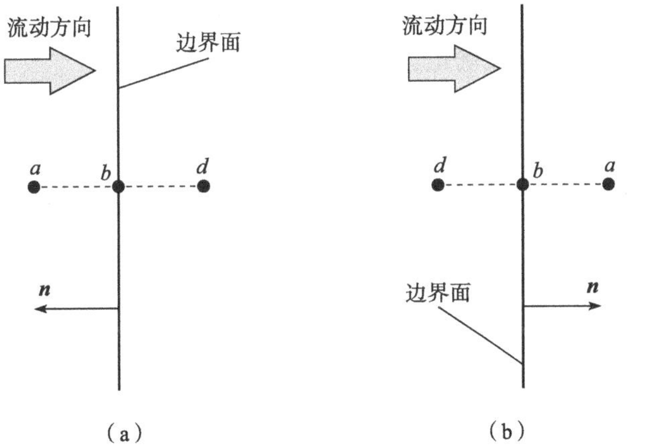

图 8.6　远场边界:入流(a)和出流(b)。位置 a 在计算域外,b 在边界上,d 在域内。单位法向矢量 $\boldsymbol{n}=[n_x, n_y, n_z]^{\mathrm{T}}$ 指向域外

1. 超声速入流

对于超声速入流,所有特征值具有相同的符号。由于流动进入物理域,边界上(图 8.6 中的 b 点)的守恒变量仅由自由来流决定。因此

$$\boldsymbol{W}_b = \boldsymbol{W}_a \tag{8.20}$$

$\boldsymbol{W}_a$ 的值由给定的马赫数 M_∞ 和两个流动角(迎角、侧滑角)确定。

2. 超声速出流

这种情况下,所有的特征值也有相同的符号。然而,流动离开物理域,边界上所有守恒变量都必须由域内解来确定。可以简单设置为

$$W_b = W_d \tag{8.21}$$

这对应于常数外插。也可以采用线性外插，但稳定性较差。

3. 亚声速入流

此时四个特征进入物理域，一个特征离开物理域。因此，四个特征变量取值由自由来流确定，一个特征变量从物理域内部外插得到，可采用如下的边界条件[12]：

$$\begin{aligned}
p_b &= \frac{1}{2}\{p_a+p_d-\rho_0 c_0[n_x(u_a-u_d)+n_y(v_a-v_d)+n_z(w_a-w_d)]\} \\
\rho_b &= \rho_a+(p_b-p_a)/c_0^2 \\
u_b &= u_a-n_x(p_a-p_b)/(\rho_0 c_0) \\
v_b &= v_a-n_y(p_a-p_b)/(\rho_0 c_0) \\
w_b &= w_a-n_z(p_a-p_b)/(\rho_0 c_0)
\end{aligned} \tag{8.22}$$

其中，ρ_0 和 c_0 表示参考状态，参考状态通常设置为内点的状态（图 8.6 中的点 d）。点 a 的值由自由来流确定。

4. 亚声速出流

此时，四个流动变量（密度和三个速度分量）必须从物理域内部外插得到，剩下的第五个变量（压力）由外部指定。根据文献[12]，远场边界的原始变量为

$$\begin{aligned}
p_b &= p_a \\
\rho_b &= \rho_d+(p_b-p_d)/c_0^2 \\
u_b &= u_d+n_x(p_d-p_b)/(\rho_0 c_0) \\
v_b &= v_d+n_y(p_d-p_b)/(\rho_0 c_0) \\
w_b &= w_d+n_z(p_d-p_b)/(\rho_0 c_0)
\end{aligned} \tag{8.23}$$

其中，p_a 为指定的静压。

虚拟单元/节点上的物理量可以通过 b 和 d 线性外插得到。

8.3.2 升力体修正

上述特征远场边界条件假定零环量，这对于亚声速或跨声速流动中的升力体是不正确的。因此，远场边界必须设置在离物体非常远的位置，否则，流场解会不准确。如果在自由来流中引入单涡（3D 情况下的马蹄涡）效应，则远场的距离可以显著缩短（一个量级）。假设涡心位于升力体上，涡的强度正比于物体产生的升力。下面介绍 2D 和 3D 中涡修正的具体实现。

1. 2D 涡修正

在此,我们介绍的方法由 Usab 和 Murman[18] 提出。在可压缩流动下,修正的自由来流速度分量由下式给出:

$$
\begin{aligned}
u_\infty^* &= u_\infty + \left(\frac{\Gamma\sqrt{1-M_\infty^2}}{2\pi d}\right)\frac{1}{1-M_\infty^2\sin^2(\theta-\alpha)}\sin\theta \\
v_\infty^* &= v_\infty + \left(\frac{\Gamma\sqrt{1-M_\infty^2}}{2\pi d}\right)\frac{1}{1-M_\infty^2\sin^2(\theta-\alpha)}\cos\theta
\end{aligned}
\tag{8.24}
$$

其中,Γ 为环量,(d,θ) 为远场点的极坐标,α 为迎角,M_∞ 为自由来流马赫数。环量根据 Kutta-Joukowsky 定理由下式计算:

$$
\Gamma = \frac{1}{2}\|\boldsymbol{v}_\infty\|_2 a C_\mathrm{L} \tag{8.25}
$$

在式(8.25)中,a 表示翼型弦长,C_L 为由表面压力积分得到的升力系数。式(8.24)中的极坐标为

$$
\begin{aligned}
d &= \sqrt{(x-x_\mathrm{ref})^2+(y-y_\mathrm{ref})^2} \\
\theta &= \tan\left(\frac{y-y_\mathrm{ref}}{x-x_\mathrm{ref}}\right)
\end{aligned}
\tag{8.26}
$$

其中,x_ref 和 y_ref 是参考点的坐标(涡的位置——例如,1/4 弦长处)。

修正的自由来流压力 p_∞^* 为

$$
p_\infty^* = \left[p_\infty^{(\gamma-1)/\gamma} + \left(\frac{\gamma-1}{\gamma}\right)\frac{\rho_\infty(\|\boldsymbol{v}_\infty\|_2^2-\|\boldsymbol{v}_\infty^*\|_2^2)}{2p_\infty^{1/\gamma}}\right]^{\gamma/(\gamma-1)} \tag{8.27}
$$

其中,$\|\boldsymbol{v}_\infty^*\|_2^2=(u_\infty^*)^2+(v_\infty^*)^2$。修正的自由来流密度由状态方程得到

$$
\rho_\infty^* = \rho_\infty\left(\frac{p_\infty^*}{p_\infty}\right)^{1/\gamma} \tag{8.28}
$$

将修正后的变量 u_∞^*,v_∞^*,p_∞^* 和 ρ_∞^* 代入式(8.22)或式(8.23)中,代替 u_a,v_a,p_a 和 ρ_a。

上述涡修正方法(式(8.24)~式(8.28))只对亚声速流严格成立,然而,自由来流条件修正在跨声速流动中也是有益的,这点从图 8.7 中升力系数与远场边界距离的关系可以证实。远场半径设置为 5、20、50 和 99 倍的弦长,可以看到,没有涡修正时,模拟结果与远场距离强相关,而涡修正使得远场位于大约 20 倍弦长距离时模拟仍然足够精确,这显著减少了网格单元/点的数量。文献[19]证实,如果在涡修正中使用更高阶的项,远场边界可以设置在大约 5 倍弦长的位置,且没有明显的精度损失。

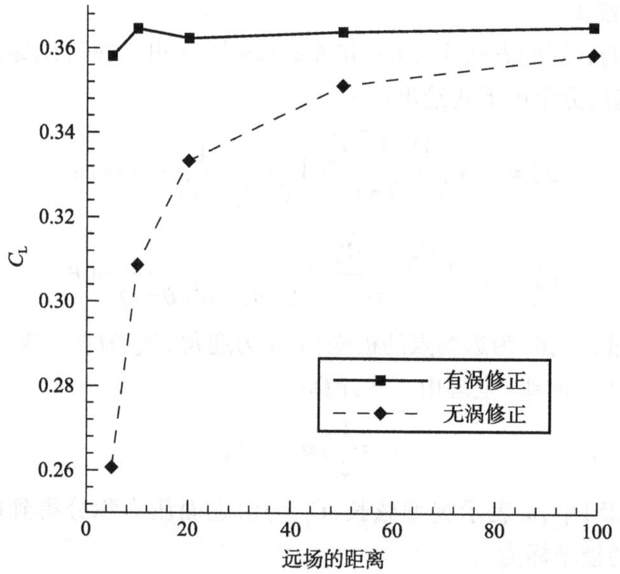

图 8.7　远场边界距离和单涡修正对升力系数的影响：NACA0012 翼型，$M_\infty=0.8$，$\alpha=1.25°$

2. 3D 涡修正

机翼对远场边界的影响可以用马蹄涡近似表示。在可压缩流动情况下，修正的自由来流速度分量由以下关系式求得[20-21]：

$$
\begin{aligned}
u_\infty^* &= u_\infty + \frac{\Gamma\beta^2}{2\pi}A \\
v_\infty^* &= v_\infty - \frac{\Gamma}{2\pi}\left[\frac{z+l}{(z+l)^2+y^2}B - \frac{z-l}{(z-l)^2+y^2}C + \frac{x\beta^2}{x^2+y^2\beta^2}A\right] \\
w_\infty^* &= w_\infty + \frac{\Gamma}{2\pi}\left[\frac{y}{(z+l)^2+y^2}B - \frac{y}{(z-l)^2+y^2}C\right]
\end{aligned}
\tag{8.29}
$$

其中，Γ 为环量，(x,y,z) 为远场点的笛卡儿坐标，l 表示半展长。在式(8.29)中假定气流方向是正 x 方向，机翼沿 z 轴放置。式(8.29)中的 A，B 和 C 分别为

$$
\begin{aligned}
A &= \frac{z+l}{\sqrt{\psi_+}} - \frac{z-l}{\sqrt{\psi_-}} \\
B &= 1 + \frac{x}{\sqrt{\psi_+}} \\
C &= 1 + \frac{x}{\sqrt{\psi_-}}
\end{aligned}
\tag{8.30}
$$

其中

$$\psi_+ = x^2 + \beta^2 (z+l)^2 + \gamma^2 \beta^2$$
$$\psi_- = x^2 + \beta^2 (z-l)^2 + \gamma^2 \beta^2 \tag{8.31}$$
$$\beta = \sqrt{1 - M_\infty^2}$$

其中，M_∞ 为自由来流马赫数。环量 Γ 用式(8.25)计算，a 表示平均弦长。修正的压力 p_∞^* 和密度 ρ_∞^* 分别通过式(8.27)~式(8.28)得到。式(8.22)或式(8.23)中的 u_a, v_a, w_a, p_a 和 ρ_a 使用相应的修正量 $u_\infty^*, v_\infty^*, w_\infty^*, p_\infty^*$ 和 ρ_∞^* 代替。

式(8.29)中修正速度分量 v_∞^* 和 w_∞^* 在涡线与出流边界相交的位置为无穷大。在这些位置，有

$$z = +l, \quad y = 0$$
$$z = -l, \quad y = 0$$

且 $x = x_{\text{farf}}$。为了避免数值奇异性，文献[21]建议将式(8.29)中的

$$(z+l)^2 + y^2$$
$$(z-l)^2 + y^2$$

限制在 1/4 翼展，即 $l/2$ 之内，这样，在涡线 $z = l$ 和 $z = -l$ 的 $l/2$ 距离内减小了速度 v_∞^* 和 w_∞^* 的修正量。

文献[21]给出的数值结果表明，如果采用式(8.29)中的涡修正，升力和阻力系数对远场距离的灵敏度会降低，到远场边界的距离为 $7l$ 就足以得到精确的结果。

8.4　进口/出口边界

对于纳维-斯托克斯方程组，人们发展了各种方法来实现数值进口和出口(也称为开口)边界条件[22-26]，在此，我们集中讨论为叶轮机应用而发展的方法。文献[27-30]提出了无反射进口和出口边界，Giles[31]、Hirsch 和 Verhoff[32]提出了欧拉方程组的无反射边界条件，用于处理物体与进口或出口之间距离较短的情况。

在某些情况下，进口和出口边界上速度、压力和温度梯度也是周期性的，比如在热交换器模拟[33]中就会遇到这种类型的流动，文献[34-36]实现了槽道 LES 模拟的周期性进/出口边界条件。

1. 亚声速进口

通常做法是指定总压、总温和两个流动角。一个特征变量必须从流域内部插值，可以使用向外传播的黎曼不变量[30]，定义为

$$R^{-}=\boldsymbol{v}_{d}\cdot\boldsymbol{n}-\frac{2c_{d}}{\gamma-1} \tag{8.32}$$

其中,下标 d 表示域内状态(图 8.6(a))。黎曼不变量用来确定边界处的绝对速度或声速。通过实践发现选择声速会使计算更稳定,特别是对于低马赫数流动。因此设定:

$$c_{b}=\frac{-R^{-}(\gamma-1)}{(\gamma-1)\cos^{2}\theta+2}\left\{1+\cos\theta\sqrt{\frac{[(\gamma-1)\cos^{2}\theta+2]c_{0}^{2}}{(\gamma-1)(R^{-})^{2}}-\frac{\gamma-1}{2}}\right\} \tag{8.33}$$

其中,θ 为相对于边界的流动角,c_0 为滞止声速。因此有

$$\cos\theta=-\frac{\boldsymbol{v}_{d}\cdot\boldsymbol{n}}{\|\boldsymbol{v}_{d}\|_{2}} \tag{8.34}$$

和

$$c_{0}^{2}=c_{d}^{2}+\frac{\gamma-1}{2}\|\boldsymbol{v}_{d}\|_{2}^{2} \tag{8.35}$$

这里 $\|\boldsymbol{v}_d\|_2$ 表示内点 d 处的合速度(图 8.6(a)),式(8.34)中的单位法向矢量 $\boldsymbol{n}$ 设为指向计算域外。

边界处的其他变量,例如静温、压力、密度或绝对速度等的计算方法如下:

$$\begin{aligned}
T_{b}&=T_{0}\left(\frac{c_{b}^{2}}{c_{0}^{2}}\right)\\
p_{b}&=p_{0}\left(\frac{T_{b}}{T_{0}}\right)^{\gamma/(\gamma-1)}\\
\rho_{b}&=\frac{p_{b}}{RT_{b}}\\
\|\boldsymbol{v}_{b}\|_{2}&=\sqrt{2c_{p}(T_{0}-T_{b})}
\end{aligned} \tag{8.36}$$

式中:T_0 和 p_0 为给定的总温和总压;R 和 c_p 分别为气体常数和比定压热容。进口速度分量根据指定的两个(2D 情况下是一个)流动角对 $\|\boldsymbol{v}_b\|_2$ 分解得到的。

2. 亚声速出口

在叶轮机械中,通常在出口处指定静压。亚声速出口边界可以用类似于式(8.23)中的出口方法来处理,只有环境压力 p_a 用给定出口静压代替。

虚拟单元(点)上的流动变量可以通过边界和内部点 d 的状态线性外插得到。

3. 超声速进口和出口

在这种情况下,所有特征要么传入,要么传出。因此,所有守恒变量都必须像式(8.20)一样在进口处指定,在出口处则根据式(8.21)外插。

8.5 注入边界

根据给定的质量流量 $\dot{m}$ 和注入温度 T_{inj}，计算出注入速度和其他流动变量。在此假定流量垂直于边界注入，对流通量根据式(2.21)计算，逆变速度 V 设为注入速度，即

$$V=V_{\text{inj}}=\frac{\dot{m}}{\rho_{\text{inj}}} \tag{8.37}$$

速度分量根据面法向矢量计算(图 8.8)：

$$\boldsymbol{v}_{\text{inj}}=-\boldsymbol{n}V_{\text{inj}} \tag{8.38}$$

式(8.37)中的密度 ρ_{inj} 是 T_{inj} 和边界处压力 p_b 的函数，在理想气体的情况下有

$$\rho_{\text{inj}}=\frac{p_b}{RT_{\text{inj}}} \tag{8.39}$$

对于图 8.8(a)所示的格心格式，可以将 p_b 设置为等于边界单元内压力，即 $p_b=p_2$。对于中点-二次格点格式，可以使用式(8.8)、式(8.11)~式(8.13)这样的公式，由组成特定面的点(即图 8.8(b)中的$(i,2)$和$(i+1,2)$)插值得到 p_b。虚拟单元(点)中的值可以通过常数或线性外插得到。

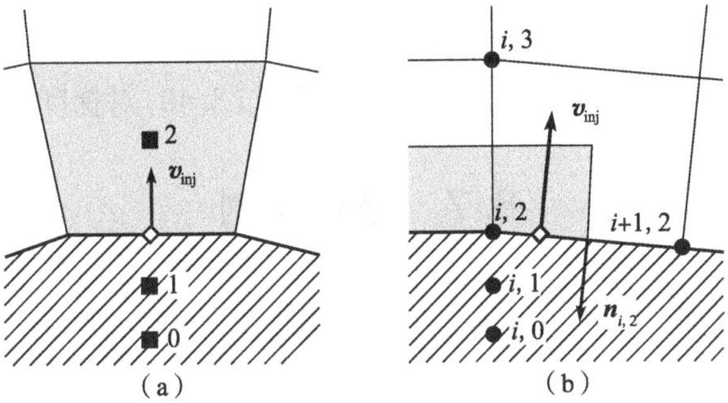

图 8.8 注入边界条件

(a) 结构格心格式；(b) 二次控制体格点格式。虚拟单元(点)表示为 1 和 0，对流通量计算的位置用菱形标记。

8.6 对称面

如果流动关于线或平面对称，第一个必须满足的条件是没有流过边界的通量，这等价于要求垂直于对称边界的速度为零，此外，下列梯度也必须为零：

（1）标量的法向梯度；

（2）切向速度的法向梯度；

（3）法向速度的切向梯度（因为 $\boldsymbol{v}\cdot\boldsymbol{n}=0$）。

这些条件可以写为

$$\begin{aligned}\boldsymbol{n}\cdot\nabla U&=0\\ \boldsymbol{n}\cdot\nabla(\boldsymbol{v}\cdot\boldsymbol{t})&=0\\ \boldsymbol{t}\cdot\nabla(\boldsymbol{v}\cdot\boldsymbol{n})&=0\end{aligned}\tag{8.40}$$

其中，U 表示标量，$\boldsymbol{t}$ 为与对称面相切的矢量。

1. 格心格式

采用虚拟单元可以大大简化对称边界条件的实现。使用反射单元的概念获取虚拟单元上的流场变量，虚拟单元上的密度或压力等标量设置等于镜像对称的内部单元上的值，即

$$U_1=U_2\quad U_0=U_3\tag{8.41}$$

下标记号与图 8.2 对应。速度分量关于边界镜像对称，如式（8.10）所示，虚拟单元上法向速度的法向梯度等于对面内部单元的值，但符号相反。

2. 格点格式（二次控制体）

可以采用两种不同方法来处理。第一种方法是通过在边界上镜像网格来构造控制体缺失的一半，通量和梯度的计算跟内点一样，使用镜像的流场变量。第二种方法是计算半个控制体的通量（但不越过边界），与对称面垂直的残差分量设为零，过程中还需要修正这些面的法向矢量（如图 8.4 中的点 2^*），包括移除所有垂直于对称面的法向矢量分量，以及根据式（8.40）对梯度进行修正。

8.7 坐标切割

这种类型的边界条件只出现在结构网格中。坐标切割代表一种人工的、非物理的边界，它是由具有不同计算坐标但物理位置相同的网格点组成的线（3D 情况下是平面），这意味着网格是折叠的，可以接触到自身。正如将在 11.1.1 节中看到的，坐标切割表现为 C 型（图 11.5）或 O 型网格拓扑（图 11.9）。流动变量及其梯度通过切割面必须保持连续。

实现坐标切割边界的最佳方法是使用虚拟单元（点），如图 8.9 所示，由图可见，这时虚拟层不是虚拟的，而是与切割面另一边的网格重合。因此，虚拟单元（格心格式）或虚拟点（顶点格式）上的物理量直接从对面单元（点）获得。在格心格式中，边界单元（图 8.9（a）阴影部分）的切割面上的通量计算与内部流场完全相同。

对于格点格式,可以用两种不同的方法处理坐标切割。第一种方法是在切割面生成一个完整的控制体(另一部分在图 8.9(b)中以虚线表示),利用虚拟单元,通量可以使用与内点相同的方式计算。如果实现正确,点 2(上面网格部分)和点 5(下面网格部分)的流场变量将相等。第二种方法是针对控制体的每一半分别计算通量,然后将图 8.9(b)中点 2 和点 5 的残差累加,需要注意的是,点 2 和点 5 的半控制体体积也需要累加。

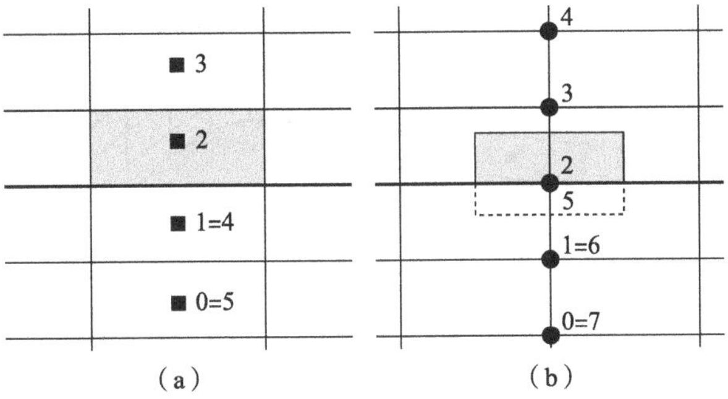

图 8.9 坐标切割边界(粗线)

(a) 格心格式;(b) 二次控制体格式。虚拟单元(点)编号为 0 和 1。

8.8 周期边界

在实际应用中存在流场相对于一个或多个坐标方向具有周期性的情况,在这种情况下,只在一个重复区域内模拟就足够了,该区域与其他物理域的正确相互作用通过周期边界条件来确保。

在此,我们将周期边界分为两种基本类型。第一种是平移周期,即一个周期边界可以仅通过坐标平移就可转换为另一个周期边界。第二种类型是旋转周期,它通过坐标旋转得到,因此称为旋转周期。

下面将介绍格心格式和格点格式中周期边界条件的实现,还将考虑旋转周期的情况。关于周期边界处理的更多细节可以查阅文献[37-38]。

1. 格心格式

利用虚拟单元可以简单地实现周期边界条件。考虑图 8.10 中的叶轮机示例,该构型在垂直方向上具有周期性,阴影单元 1 和 2 分别位于周期的上下边界。根据周期条件,第一层虚拟单元对应于对面周期边界的边界单元,第二层虚拟单元与实际单元的第二层相关联,依此类推。因此,虚拟单元中的所有标量(密度、压力等)都直接从对应的实际单元中获得,即

$$U_{1'} = U_1 \quad U_{2'} = U_2 \tag{8.42}$$

同样的关系也适用于平移周期下的矢量（速度、梯度）。旋转周期边界需要对矢量进行修正，这将在后面进行讨论。

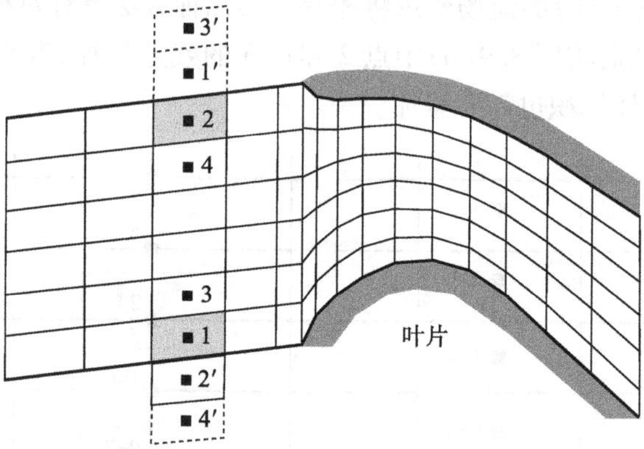

图 8.10　2D 非结构格心格式的周期边界（粗线）。虚拟单元（虚线）由相应的实际单元标号加单引号表示

2. 格点格式（二次控制体）

这种情况如图 8.11 所示。一种处理方法是沿着阴影控制体的面进行通量积分，然后对图中点 1 和点 2 的残差进行累加，得到完整的通量。因此

$$\boldsymbol{R}_{1,\text{sum}} = \boldsymbol{R}_1 + \boldsymbol{R}_{2'} \quad \text{和} \quad \boldsymbol{R}_{2,\text{sum}} = \boldsymbol{R}_2 + \boldsymbol{R}_{1'} \tag{8.43}$$

点 1 和点 2 的半控制体（图 8.11 阴影部分）体积也必须累加。对于平移周期，对面边界的残差保持不变，即 $\boldsymbol{R}_{1'} = \boldsymbol{R}_1$ 和 $\boldsymbol{R}_{2'} = \boldsymbol{R}_2$，因此 $\boldsymbol{R}_{1,\text{sum}} = \boldsymbol{R}_{2,\text{sum}}$。旋转周期边界需要在使用式（8.43）之前转换动量方程。

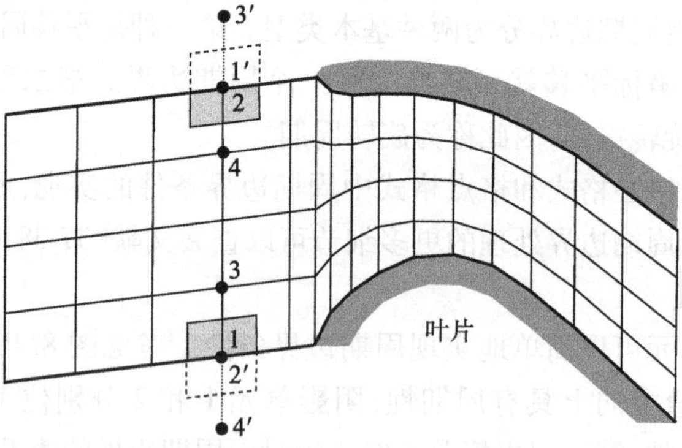

图 8.11　2D 非结构二次控制体格点格式的周期边界（粗线）。控制体的虚拟部分（虚线）由对面边界相应控制体标号加单引号表示，虚拟点 3′和 4′亦是如此

3. 旋转周期性

旋转周期条件基于坐标系的旋转，因此，所有矢量，如速度或标量的梯度，都必须相应地进行变换。压力或密度等标量关于坐标旋转具有不变性，因此它们保持不变。假设旋转轴平行于 x 轴（图 8.12），则旋转矩阵为

$$\boldsymbol{R}=\begin{bmatrix}1 & 0 & 0\\ 0 & \cos\phi & -\sin\phi\\ 0 & \sin\phi & \cos\phi\end{bmatrix} \tag{8.44}$$

其中，周期边界 A 和 B 之间的夹角 ϕ 顺时针为正。因此，从边界 A 到 B（图 8.10 中的单元 1′、2′和图 8.11 中的点 1′~4′）的速度矢量变换为

$$\boldsymbol{v}_B=\boldsymbol{R}\boldsymbol{v}_A \tag{8.45}$$

容易证明速度的 x 分量不因旋转而改变，因此有 $u_B=u_A$。所有流场变量的梯度以类似方式转换。

如上所述，在格点格式下，按照式（8.43）进行累加前必须修正动量方程组的残差。使用旋转矩阵（式（8.44））可以得到

$$\boldsymbol{R}_{B,\mathrm{sum}}^{u,v,w}=\boldsymbol{R}_B^{u,v,w}+\boldsymbol{R}\boldsymbol{R}_A^{u,v,w} \tag{8.46}$$

式（8.46）中的上标 u,v,w 分别表示三个动量方程。

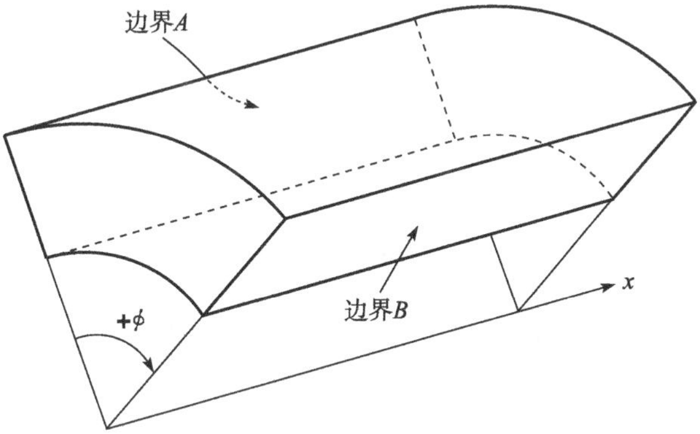

图 8.12　旋转周期边界（A 和 B）。假设旋转轴与 x 轴重合

8.9　网格块边界

在 3.1 节讨论结构网格的空间离散过程中我们看到，通常不可能在几何复杂区域内生成单块网格（图 3.4），一般有两种解决这个问题的方法，分别是多块网格方法和重叠网格技术。下面将介绍多块网格方法的基本实现过程，更全面的讨论可查阅文献［39-45］，文献［46］对多块方法作了非常有用的介绍。关于

重叠网格技术的细节，在此没有讨论，读者可参阅文献[47-53]。

在多块技术中，物理域被分割成一定数量的虚拟部分，因此，计算域也被划分成相同数量的块。一般情况下，特定块上的物理解取决于一个或多个相邻块中的流动，因此必须设计一种数据结构，允许块之间有效地交换信息。如果使用不同的处理器来求解网格块中的控制方程组，在进程间通信时也需要该数据结构。

数据结构的第一部分由块边界的编号组成。图 8.13 显示了一种特定的编号格式，该编号策略可以总结如下：

边界 1： $i = \mathrm{IBEG}$

边界 2： $i = \mathrm{IEND}$

边界 3： $j = \mathrm{JBEG}$

边界 4： $j = \mathrm{JEND}$

边界 5： $k = \mathrm{KBEG}$

边界 6： $k = \mathrm{KEND}$

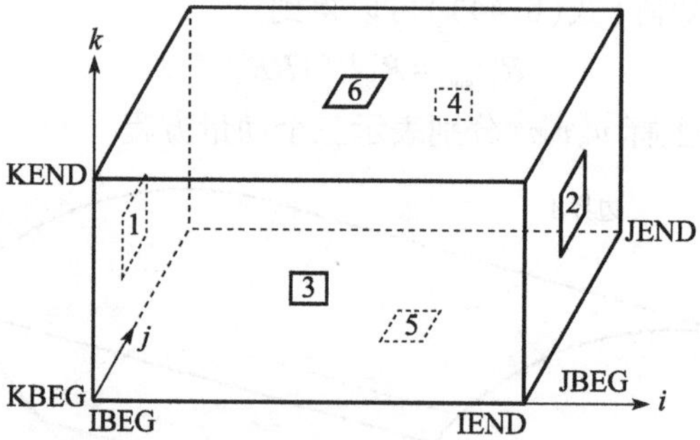

图 8.13　计算空间边和块边界的编号方案

所有网格块需使用相同的编号方案。计算空间中网格点的索引 i,j,k 定义在下列范围内：

$$\mathrm{IBEG} \leqslant i \leqslant \mathrm{IEND}$$
$$\mathrm{JBEG} \leqslant j \leqslant \mathrm{JEND}$$
$$\mathrm{KBEG} \leqslant k \leqslant \mathrm{KEND}$$

格心格式所需的单元索引 I,J,K 也以类似的方式定义。由于多块方法通常使用虚拟单元/点实现，实际单元/点将在每个范围的开始或结束有一定的偏移量（图 8.1）。

每个块的边界可被分成若干个互不重叠的小面，这使得可以在同一块边界上指定不同的边界条件，如图 8.14 所示。为了唯一地识别每个小面，需要存储相应块和块边界的编号，此外，还必须存储小面的原点、高度和宽度。为此，

图 8.14 中使用了 L1BEG,L1END,L2BEG 和 L2END 坐标。建议根据循环方向定位小面的坐标系,即,如果我们考虑 i 坐标,j 和 k 将是第一个和第二个循环方向。对于 j 坐标,循环方向分别为 k 和 i。由于图 8.14 中的小面位于 j=JBEG 边界上,所以 l_1 坐标是 k 方向,l_2 是 i 方向。使用循环方向能够对每个小面的方位有唯一的定义。

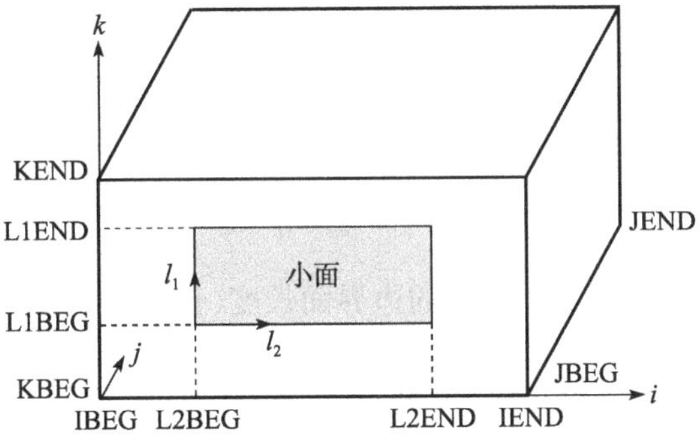

图 8.14 计算空间中边界面的坐标。该面有自身的局部坐标系 l_1,l_2

数据结构还应包含网格块间的交界面信息,确保数据可以在这些小面间交换(在此假设块之间只通过交界面通信)。为此,有必要通过相邻块和小面的编号扩展上述小面的数据结构,此外,还需要对小面之间的方位进行编码。

两个块之间流场变量的交换如图 8.15 所示,该过程包括两个步骤。第一步,将与邻近小面的虚拟层重叠的计算域部分的变量写入到自身的虚拟单元/点(图 8.15 中的 A'和 B')中或者临时存储区中,对所有块均如此操作。第二步,A'和 B'中的数据在两个块之间交换,这意味着 A'被写入到 B 块的虚拟层,B'被写入到 A 块的虚拟层。如果两个小面的方向不同,必须对数据进行相应的转换。如果网格线在块交界面上不匹配,则需要进一步的操作,具体可参见文献[54-56]。

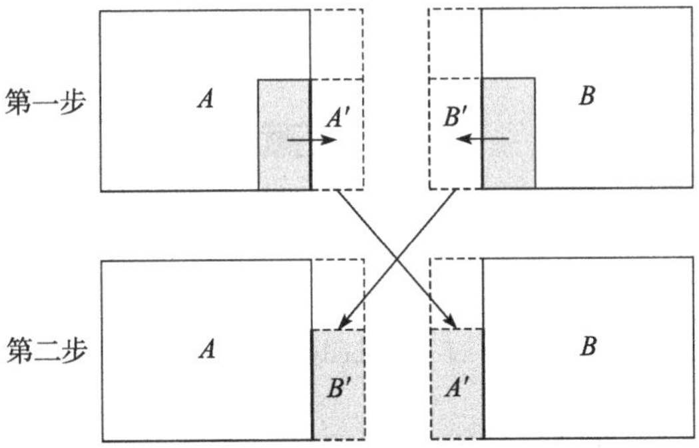

图 8.15 两个块 A 和 B 之间流动变量的交换(在阴影区域 A',B')。虚拟层用虚线表示

8.10 非结构网格边界处的流场变量梯度

在5.3.4节中介绍过，在中点-二次格式中，流场变量梯度的计算需要额外关注。如果使用式(5.50)中的Green-Gauss方法在三角形或四面体网格上计算梯度，域边界（对称或周期边界除外）的贡献必须用类似式(8.12)或式(8.13)的方式来计算，而不能用算术平均，否则梯度将不准确。根据图8.4中的记号，面1-2对边界节点1的贡献为

$$\frac{1}{6}(5U_1+U_2)\boldsymbol{n}_{12}\frac{\Delta S_{12}}{2}$$

其中，ΔS_{12}是节点1和节点2之间的边界面长度（因此需要减半）。与式(8.13)对应，图8.5中三角形面1-2-3对节点1的贡献为

$$\frac{1}{8}(6U_1+U_2+U_3)\boldsymbol{n}_{123}\frac{\Delta S_{123}}{3}$$

其中，$\Delta S_{123}/3$是三角形1-2-3中的灰色区域。在混合网格上，使用基于虚拟边的最小二乘方法更为合适[8]（图5.15）。

格心格式在对称或周期边界上不需要特殊处理，其实现与8.6节或8.8节中讨论的通量实现相同。如果使用最小二乘法计算梯度，这也适用于中点-二次格式，唯一需要做的额外工作是如8.6节所述，将某些梯度置零（参见式(8.40)）。

如果在中点-二次格式中使用Green-Gauss方法（式(5.50)），需要修正控制体中接触对称边界的那些面的法向矢量（如图8.4中的点2*），这可以通过将这些面的矢量（它们都垂直于对称面）的所有分量都设为零来实现。最后，按照8.6节讨论的方式对梯度进行修正。在周期边界处，边界两侧的梯度和体积需要求和，如8.8节中通量一样（式(8.43)），对于旋转周期，梯度还需要使用式(8.44)的旋转矩阵进行变换。

参考文献

[1] Kroll N, Jain RK. Solution of two-dimensional Euler equations—experience with a finite volume code. DFVLR-FB 87-41; 1987.

[2] Rizzi A. Numerical implementation of solid-body boundary conditions for the Euler equations. ZAMM 1978; 58: 301-4.

[3] Hall MG. Cell vertex multigrid scheme for solution of the Euler equations. Proc. Conf. on Numerical Methods for Fluid Dynamics. Reading, UK; 1985.

[4] Koeck C. Computation of three-dimensional flow using the Euler equations and a multiple-grid scheme. Int J Numer Meth Fluids 1985;5:483-500.

[5] Frink NT, Parikh P, Pirzadeh S. A fast upwind solver for the Euler equations on three-dimensional unstructured meshes. AIAA Paper 91-0102;1991.

[6] Frink NT. Recent progress toward a three-dimensional Navier-Stokes solver. AIAA Paper 94-0061;1994.

[7] Luo H, Baum JD, Löhner R. An improved finite volume scheme for compressible flows on unstructured grids. AIAA Paper 95-0348;1995.

[8] Haselbacher A, Blazek J. On the accurate and efficient discretisation of the Navier-Stokes equations on mixed grids. AIAA Paper 99-3363;1999 [also AIAA J 2000;38:2094-102].

[9] Mazaheri K, Roe PL. Numerical wave propagation and steady-state solutions: soft wall and outer boundary conditions. AIAA J 1997;35:965-75.

[10] Engquist B, Majda A. Absorbing boundary conditions for numerical simulation of waves. Math Comput 1977;31:629-51.

[11] Bayliss A, Turkel E. Far field boundary conditions for compressible flow. J Comput Phys 1982;48:182-99.

[12] Whitfield DL, Janus JM. Three-dimensional unsteady Euler equations solution using flux vector splitting. AIAA Paper 84-1552;1984.

[13] Gustafsson B. Far field boundary conditions for time-dependent hyperbolic systems. Center for Large Scale Sci. Comput., CLaSSiC-87-16. Stanford University;1987.

[14] Thompson KW. Time dependent boundary conditions for hyperbolic systems. J Comput Phys 1987;68:1-24.

[15] Hayder ME, Hu FQ, Hussaini MY. Towards perfectly absorbing boundary conditions for Euler equations. ICASE Report No. 97-25;1997.

[16] Givoli D. Non-reflecting boundary conditions. J Comput Phys 1991;94:1-29.

[17] Kreiss HO. Initial boundary value problems for hyperbolic systems. Commun Pure Appl Math 1970;23:277-98.

[18] Usab WJ, Murman EM. Embedded mesh solution of the Euler equation using a multiple-grid method. AIAA Paper 83-1946;1983.

[19] Giles MB, Drela M, Thompkins WT. Newton solution of direct and inverse transonic Euler equations. AIAA Paper 85-1530;1985.

[20] Klunker EB, Harder KC. Notes on linearized subsonic wing theory. Unpublished.

[21] Radespiel R. A cell-vertex multigrid method for the Navier-Stokes equations. NASA TM-101557;1989.

[22] Papanastasiou TC, Malamataris N, Ellwood K. A new outflow boundary condition. Int J Numer Meth Fluids 1992;14:587-608.

[23] Sani RL, Gresho PM. Résumé and remarks on the open boundary condition minisymposium. Int J Numer Meth Fluids 1994;18:983-1008.

[24] Baum M, Poinsot T, Thevenin D. Accurate boundary conditions for multicomponent reactive flows. J Comput Phys 1994;116:247-61.
[25] Griffiths DF. The 'no boundary condition' outflow boundary condition. Int J Numer Meth Fluids 1997;24:393-411.
[26] Renardy M. Imposing 'no' boundary condition at outflow: why does it work? Int J Numer Meth Fluids 1997;24:413-17.
[27] Rudy DH, Strikwerda JC. A nonreflective outflow boundary condition for subsonic Navier-Stokes calculations. J Comput Phys 1980;36:55-70.
[28] Yokota JW, Caughey DA. An L-U implicit multigrid algorithm for the three-dimensional Euler equations. AIAA Paper 87-0453;1987.
[29] Yokota JW. Diagonally inverted lower-upper factored implicit multigrid scheme for the three-dimensional Navier-Stokes equations. AIAA J 1989;28:1642-9.
[30] Holmes DG. Inviscid 2D solutions on unstructured, adaptive grids. Numerical Methods for Flows in Turbomachinery, Von Karman Institute, Rhode-St-Genèse, Belgium: VKI Lecture Series 1989-06, 1989.
[31] Giles MB. Non-reflecting boundary conditions for Euler equation calculations. AIAA Paper 89-1942;1989.
[32] Hirsch Ch, Verhoff A. Far field numerical boundary conditions for internal and cascade flow computations. AIAA Paper 89-1943;1989.
[33] Patankar SV, Liu CH, Sparrow EM. Fully developed flow and heat transfer in ducts having streamwise-periodic variations of cross-sectional area. ASME J Heat Transfer 1977;99:180-86.
[34] Deschamps V. Simulations numériques de la turbulence inhomogène incompressible dans un écoulement de canal plan. ONERA, note technique 1988-5;1988.
[35] Mossi M. Simulation of benchmark and industrial unsteady compressible turbulent fluid flows. PhD Thesis, No. 1958, École Polytechnique Fédérale de Lausanne, Département de Génie Mécanique, Switzerland;1999. p. 136-7.
[36] Lenormand E, Sagaut P, Phuoc LT, Comte P. Subgrid-scale models for large-eddy simulations of compressible wall bounded flows. AIAA J 2000;38:1340-50.
[37] Chung H-T, Baek J-H. Influence of trailing-edge grid structure on Navier-Stokes computation of turbomachinery cascade flow. Int J Numer Meth Fluids 1992;15:883-94.
[38] Segal G, Vuik K, Kassels K. On the implementation of symmetric and antisymmetric periodic boundary conditions for incompressible flow. Int J Numer Meth Fluids 1994;18:1153-1165.
[39] Rossow C-C. Efficient computation of inviscid flow fields around complex configurations using a multiblock multigrid method. Commun Appl Numer Meth 1992;8:735-47.
[40] Jacquotte OP. Generation, optimization and adaptation of multiblock grids around complex configurations in computational fluid dynamics. Int J Numer Meth Eng 1992;34:443-454.
[41] Szmelter J, Marchant MJ, Evans A, Weatherill NP. Two-dimensional Navier-Stokes equations with adaptivity on structured meshes. Comput Meth Appl Mech Eng 1992;101:355-68.

[42] Marchant MJ, Weatherill NP. The construction of nearly orthogonal multiblock grids for compressible flow simulation. Commun Numer Meth Eng 1993;9:567-78.

[43] Jenssen CB. Implicit multiblock Euler and Navier-Stokes calculations. AIAA J 1994;32:1808-14.

[44] De-Nicola C, Pinto G, Tognaccini R. On the numerical stability of block structured algorithms with applications to 1-D advection-diffusion problems. Comput Fluids 1995;24:41-54.

[45] Enander R, Sterner E. Analysis of internal boundary conditions and communication strategies for multigrid multiblock methods. Report No. 191, Dept. Sci. Computing, Uppsala University, Sweden;1997.

[46] Rizzi A, Eliasson P, Lindblad I, Hirsch C, Lacor C, Haeuser J. The engineering ofmultiblock/multigrid software for Navier-Stokes flows on structured meshes. Comput Fluids 1993;22:341-67.

[47] Benek JA, Buning PG, Steger JL. A 3-D chimera grid embedding technique. AIAA Paper 85-1523;1985.

[48] Buning PG, Chu IT, Obayashi S, Rizk YM, Steger JL. Numerical simulation of the integrated space shuttle vehicle in ascent. AIAA Paper 88-4359-CP;1988.

[49] Chesshire G, Henshaw WD. Composite overlapping meshes for the solution of partial differential equations. J Comput Phys 1990;90:1-64.

[50] Pearce DG, Stanley SA, Martin EW, Gomez RJ, Le Beau GJ, Buning PG, et al. Development of a large scale chimera grid system for the space shuttle launch vehicle. AIAA Paper 93-0533;1993.

[51] Kao H-J, Liou M-S, Chow C-Y. Grid adaptation using chimera composite overlapping meshes. AIAA J 1994;32:942-9.

[52] Henshaw WD, Schwendeman DW. An adaptive numerical scheme for high-speed reactive flow on overlapping grids. J Comput Phys 2003;191:420-47.

[53] Xu L, Weng P. High order accurate and low dissipation method for unsteady compressible viscous flow computation on helicopter rotor in forward flight. J Comput Phys 2014;258:470-88.

[54] Rai MM. A conservative treatment of zonal boundaries for Euler equation calculations. J Comput Phys 1986;62:472-503.

[55] Kassies A, Tognaccini R. Boundary conditions for Euler equations at internal block faces of multi-block domains using local grid refinement. AIAA Paper 90-1590;1990.

[56] Peng Y-F, Mittal R, Sau A, Hwang RR. Nested Cartesian grid method in incompressible viscous fluid flow. J Comput Phys 2010;229:7072-101.

第9章 加速收敛技术

为了加速定常流场控制方程组(2.19)的求解,人们开发了各种加速技术,这些加速技术同样适用于双时间步格式(见6.3节)的内迭代。本章将讨论以下几种加速收敛技术:

(1) 局部时间步长;

(2) 焓阻尼;

(3) 残差光顺;

(4) 多重网格;

(5) 低速预处理。

其中,局部时间步长、焓阻尼和低速预处理是基于对常微分方程组(6.1)的修改,而残差光顺和多重网格是对求解过程的改进。除残差光顺外,其他方法都可应用于显式(见6.1节)和隐式(见6.2节)时间推进格式,而残差光顺方法是专门针对显式时间推进格式发展的(见6.1.1和6.1.2节)。文献[1-2]概述了几种加速方法。

本章还介绍另一种能提高计算效率的技术:并行技术。虽然并行技术主要是关于计算机代码的设计,但对所使用的数值方法也有影响,鉴于当前计算机硬件的发展趋势,这是另一种重要的加速技术,9.6节将讨论几种可能的并行方法。

9.1 局部时间步长

当使用局部时间步长技术时,对离散控制方程组(6.1)进行积分的每个控制体都尽可能采用最大的时间步长,局部时间步长值 Δt_I 根据式(6.14)、式(6.18)、式(6.20)或式(6.22)计算。相比于全局时间步长,虽然局部时间步长技术推进求解过程的每一步瞬态解不是时间精确解,但是流场收敛到稳态的速度大大加快。图9.1为机翼无黏亚声速绕流的局部时间步长方法和全局时间步长方法的效率对比,该流场计算通过显式多步法(见6.1.1节)使流动达到稳态,由图可知,使用全局时间步长会导致收敛到稳态的速度减缓,相比于定常无黏流场计算,定常黏性流场计算采用局部时间步长法通常可以节省更多的CPU时间。

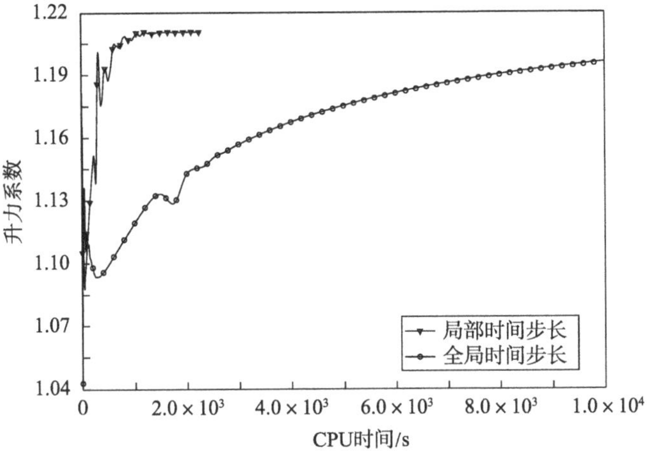

图9.1　非结构网格上,无黏亚声速流动采用和不采用局部时间步长时升力系数收敛历程的比较

9.2　焓阻尼

在某些情况下,整个流场的总焓H(式(2.12))为常数,例如在没有热源和外力的情况下,由欧拉方程组控制的外流场就属于此类情况,可以利用这种物理性质来减少计算量。

第一种方法是指定流场中的总焓值,从而欧拉方程组(2.44)可以略去能量方程,这样可以节省内存和CPU时间。

Jameson提出了另一种求解势流方程的方法——焓阻尼法[3]。该方法利用总焓H与其自由来流焓值H_∞之间的差定义一个附加的强迫项,从而可以大大加快收敛到稳态的速度。将焓阻尼应用于欧拉方程组(2.44),得到如下的校正形式[4]:

$$\frac{\partial}{\partial \boldsymbol{t}}\int_\Omega \boldsymbol{W}\mathrm{d}\Omega + \oint_{\partial\Omega}\boldsymbol{F}_\mathrm{c}\mathrm{d}S = \int_\Omega(\boldsymbol{Q} - \boldsymbol{Q}_\mathrm{ED})\mathrm{d}\Omega \tag{9.1}$$

其中,强迫项$\boldsymbol{Q}_\mathrm{ED}$由下式给出:

$$\boldsymbol{Q}_{\mathrm{ED}}=\vartheta\begin{bmatrix}\rho(H-H_{\infty})\\ \rho u(H-H_{\infty})\\ \rho v(H-H_{\infty})\\ \rho w(H-H_{\infty})\\ \rho(H-H_{\infty})\end{bmatrix} \tag{9.2}$$

阻尼系数 ϑ 是一个小量，需要根据经验确定。空间离散格式，特别是人工耗散都必须遵循 $H=H_{\infty}$ 为离散方程的有效解，于是源项 $\boldsymbol{Q}_{\mathrm{ED}}$ 的增量才不会改变最终的定常解。

焓阻尼作为每次流场解更新后的一个附加步骤执行。例如，在显式 m 步时间推进格式（见 6.1.1 节或 6.1.2 节）中，阻尼步为

$$\boldsymbol{W}_I^{n+1}=\frac{1}{1+\vartheta(H_I^{(m)}-H_{\infty})}\boldsymbol{W}_I^{(m)} \tag{9.3}$$

而能量方程则变为

$$(\rho E)_I^{n+1}=\frac{1}{1+\vartheta}[(\rho E)_I^{(m)}-\vartheta p_I^{(m)}] \tag{9.4}$$

在式(9.3)中，$\boldsymbol{W}_I^{(m)}$ 表示 m 步显式时间推进格式的最终解。文献[1]中 NACA0012 翼型的跨声速绕流（$M_{\infty}=0.8, \alpha=1.25°$）数值实验证实，焓阻尼法可减少大约一半的步数使流场达到稳态。

9.3 残差光顺

显式多步时间推进格式（见 6.1.1 节和 6.1.2 节）的最大 CFL 数和收敛性在一定程度上受步系数优化的影响[5-7]。Jameson 和 Baker[8] 引入残差光顺技术，其目的是使显式方法具有隐式方法的特征，从而大幅增加最大允许的 CFL 数。残差光顺的另一个目的是更好地抑制残差的高频误差分量，这对多重网格方法的成功应用也特别重要。

残差光顺可通过显式、隐式或混合方式实现[9-10]，它通常应用于显式时间推进方法（式(6.5)和式(6.7)）的每一步，在更新解 $\boldsymbol{W}^{(k)}$ 之前，将先前计算的残差 $\boldsymbol{R}^{(k)}$ 替换为光顺后的残差 $\boldsymbol{R}^*$。下面将讨论应用较为广泛的隐式残差光顺（implicit residual smoothing, IRS）技术在结构和非结构网格上的实现。

9.3.1 结构网格的中心 IRS

3D 中心 IRS 的标准形式为

$$
\begin{aligned}
-\epsilon^{I}\boldsymbol{R}^{*}_{I-1,J,K}+(1+2\epsilon^{I})\boldsymbol{R}^{*}_{I,J,K}-\epsilon^{I}\boldsymbol{R}^{*}_{I+1,J,K}&=\boldsymbol{R}_{I,J,K}\\
-\epsilon^{J}\boldsymbol{R}^{**}_{I,J-1,K}+(1+2\epsilon^{J})\boldsymbol{R}^{**}_{I,J,K}-\epsilon^{J}\boldsymbol{R}^{**}_{I,J+1,K}&=\boldsymbol{R}^{*}_{I,J,K}\\
-\epsilon^{K}\boldsymbol{R}^{***}_{I,J,K-1}+(1+2\epsilon^{K})\boldsymbol{R}^{***}_{I,J,K}-\epsilon^{K}\boldsymbol{R}^{***}_{I,J,K+1}&=\boldsymbol{R}^{**}_{I,J,K}
\end{aligned}
\tag{9.5}
$$

其中，$\boldsymbol{R}^{*}_{I,J,K}$、$\boldsymbol{R}^{**}_{I,J,K}$ 和 $\boldsymbol{R}^{***}_{I,J,K}$ 分别表示沿 I,J 和 K 方向光顺后的残差，参数 ϵ^{I}，ϵ^{J} 和 ϵ^{K} 表示三个计算坐标方向的光顺系数。式(9.5)中的隐式算子类似于二阶中心差分，因此称为中心隐式残差光顺(central implicit residual smoothing，CIRS)。式(9.5)中的隐式残差方程组采用三对角矩阵求逆的 Thomas 算法求解。

光顺系数通常定义为对流通量雅可比矩阵谱半径的函数[11]，其目的是在每个坐标方向上仅进行基于稳定性和较好的误差抑制所必要的光顺。文献[12]建议的适用于 2D 的光顺系数为

$$
\begin{aligned}
\epsilon^{I}&=\max\left\{\frac{1}{4}\left[\left(\frac{\sigma^{*}}{\sigma}\frac{1}{1+\psi r}\right)^{2}-1\right],0\right\}\\
\epsilon^{J}&=\max\left\{\frac{1}{4}\left[\left(\frac{\sigma^{*}}{\sigma}\frac{1}{1+\psi/r}\right)^{2}-1\right],0\right\}
\end{aligned}
\tag{9.6}
$$

其中，σ^{*}/σ 表示残差光顺后与光顺前的 CFL 数比值，变量 r 表示对流谱半径之比(式(4.53))，即 $r=\hat{\Lambda}^{J}_{c}/\hat{\Lambda}^{I}_{c}$，参数 $\psi\approx0.125$ 确保光顺算子的线性稳定性。

3D 情况下，可以使用类似于式(9.6)的表达式计算光顺系数[12]，即

$$
\begin{aligned}
\epsilon^{I}&=\max\left\{\frac{1}{4}\left[\left(\frac{\sigma^{*}}{\sigma}\frac{1}{1+\psi(r^{JI}+r^{KI})}\right)^{2}-1\right],0\right\}\\
\epsilon^{J}&=\max\left\{\frac{1}{4}\left[\left(\frac{\sigma^{*}}{\sigma}\frac{1}{1+\psi(r^{KJ}+r^{IJ})}\right)^{2}-1\right],0\right\}\\
\epsilon^{K}&=\max\left\{\frac{1}{4}\left[\left(\frac{\sigma^{*}}{\sigma}\frac{1}{1+\psi(r^{IK}+r^{JK})}\right)^{2}-1\right],0\right\}
\end{aligned}
\tag{9.7}
$$

其中

$$
r^{JI}=\hat{\Lambda}^{J}_{c}/\hat{\Lambda}^{I}_{c}\quad r^{KI}=\hat{\Lambda}^{K}_{c}/\hat{\Lambda}^{I}_{c}
$$

参数 ψ 的值通常取 0.0625。

最大 CFL 数比值 σ^{*}/σ 与光顺系数的值和空间离散方法的类型有关。对于中心格式(见 4.3.1 节)，该比值按下式确定：

$$
\frac{\sigma^{*}}{\sigma}\leqslant\sqrt{1+4\epsilon}
\tag{9.8}
$$

在实际应用中，σ^{*}/σ 的值可近似达到 2(当 $\epsilon=0.8$ 时)，更高的 σ^{*}/σ 比值会降低时间推进方法对误差的抑制作用。对于迎风空间离散，不存在式(9.8)这样的简单条件，因为 σ^{*}/σ 的最大值还与各步系数有关，但通常情况下，无黏

流计算的 CFL 数(见表 6.1 和表 6.2),可提高至 2 倍左右,而黏性流计算的 CFL 数可提高至 5 倍(使用表 6.2 中的混合方法)。图 9.2 和图 9.3 显示了在结构网格上 CIRS 对无黏流场计算和黏性流场计算收敛特性的影响,为了考虑 CIRS 带来的额外计算量,图中比较的是 CPU 时间。如图所示,采用 CIRS 后的求解时间显著减少,特别是对于黏性流场的求解而言,当 CIRS 与多重网格技术相结合时,可以节省更多的求解时间(见 9.4 节)。

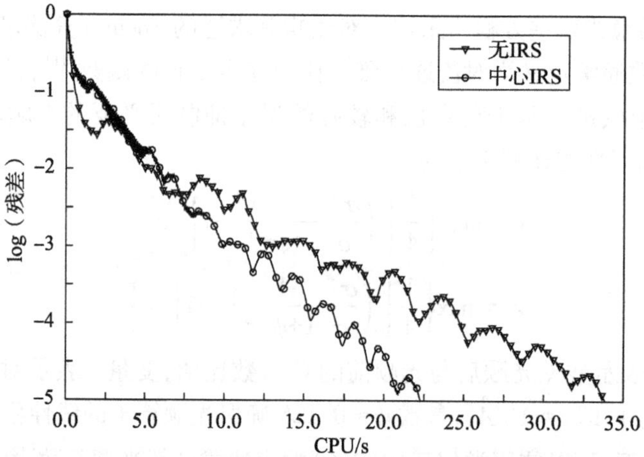

图 9.2 结构网格上无黏跨声速流采用和不采用 CIRS 时的收敛历程比较

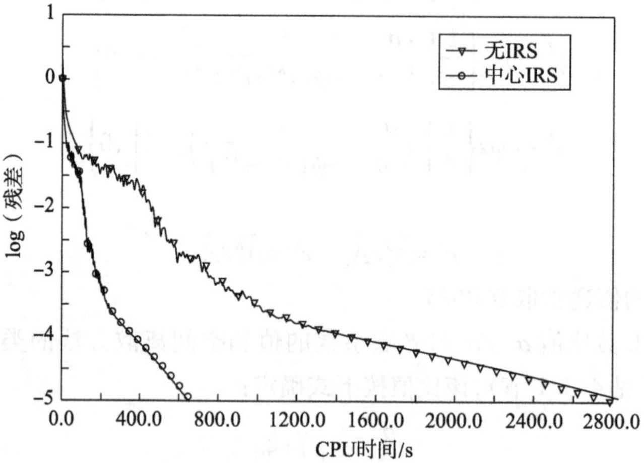

图 9.3 结构网格上黏性亚声速流采用和不采用 CIRS 时的收敛历程比较

借助 CIRS 技术,可以抵消式(6.18)中黏性谱半径 $\hat{\Lambda}_v$ 产生的对时间步长的限制，根据式(6.14)计算最大时间步长，不需要考虑$\hat{\Lambda}_v$，在黏性谱半径占主导的流动区域，采用较大系数$\epsilon^I,\epsilon^J,\epsilon^K$进行光顺。基于黏性谱半径的光顺系数可由下式计算[12-13]：

$$\begin{aligned}\epsilon_v^I &= C\left(\frac{\sigma^*}{\sigma}\right)\frac{\hat{\Lambda}_v^I}{\hat{\Lambda}_c^I+\hat{\Lambda}_c^J+\hat{\Lambda}_c^K}\\ \epsilon_v^J &= C\left(\frac{\sigma^*}{\sigma}\right)\frac{\hat{\Lambda}_v^J}{\hat{\Lambda}_c^I+\hat{\Lambda}_c^J+\hat{\Lambda}_c^K}\\ \epsilon_v^K &= C\left(\frac{\sigma^*}{\sigma}\right)\frac{\hat{\Lambda}_v^K}{\hat{\Lambda}_c^I+\hat{\Lambda}_c^J+\hat{\Lambda}_c^K}\end{aligned} \tag{9.9}$$

其中，常数$C\approx5/4$，然后取式(9.7)和式(9.9)中的系数的最大值作为最终的光顺系数。

9.3.2 非结构网格的中心 IRS

在非结构网格上，采用拉普拉斯算子(见式(5.24))对残差进行CIRS。因此，控制体I光顺后的残差$\boldsymbol{R}_I^*$由以下隐式关系获得[14]：

$$\boldsymbol{R}_I^* + \sum_{J=1}^{N_A}\epsilon(\boldsymbol{R}_I^* - \boldsymbol{R}_J^*) = \boldsymbol{R}_I \tag{9.10}$$

上式中求和运算包括所有N_A个相邻控制体。使用雅可比迭代求解式(9.10)。光顺系数的有效值范围为$0.3\leqslant\epsilon\leqslant0.8$，利用该范围的$\epsilon$值，可以将CFL数增加2~5倍(从而增大时间步长)。由于矩阵的对角占优性质，雅可比迭代大约只需两步即可收敛。

9.3.3 结构网格的迎风 IRS

前面讨论的CIRS方法对亚声速和跨声速流动都能达到令人满意的效果，在黏性主导的区域也有帮助。然而，CIRS与迎风离散格式相结合时，表现出较差的误差阻尼特性，此外，在强激波情况下，由CIRS加速的多步时间推进方法的鲁棒性会受到影响。因此，发展了迎风IRS(upwind implicit residual smoothing, UIRS)方法[15-16]，该方法特别适合于高马赫数流动。

与CIRS方法相比，UIRS方法考虑了对流特征值$\overline{\Lambda}_c$的符号，仅在欧拉方程组的特征方向上(即dx/dt=常数$=\Lambda_c$)光顺残差，从而避免对上游残差产生非物理影响。UIRS方法需要将残差转换为特征变量(见附录A.11节)，这样，就可以根据对应特征值的符号对残差矢量的每个分量单独进行光顺。对于残差的第

1 个分量,1D 隐式算子定义为[15-16]

$$
\begin{aligned}
-\epsilon^I(\boldsymbol{R}^*)_{I-1}^l+(1+\epsilon^I)(\boldsymbol{R}^*)_I^l=(\boldsymbol{R}^c)_I^l, \quad \Lambda_c^l\geqslant 0 \\
(1+\epsilon^I)(\boldsymbol{R}^*)_I^l-\epsilon^I(\boldsymbol{R}^*)_{I+1}^l=(\boldsymbol{R}^c)_I^l, \quad \Lambda_c^l<0
\end{aligned}
\tag{9.11}
$$

其中,$\boldsymbol{R}^c$ 表示转换为特征变量的残差,即

$$
\boldsymbol{R}^c=\boldsymbol{T}^{-1}\boldsymbol{R} \tag{9.12}
$$

使用 Thomas 算法求解三对角(如果 Λ_c^l 符号不变,则为双对角)方程组得到光顺后的残差 $\boldsymbol{R}^*$,然后残差 $\boldsymbol{R}^*$ 被转换回守恒变量,之后可对解进行更新。

UIRS 最大的特点是,它使多步法具有非常好的阻尼特性,特别是在与迎风离散相结合的情况下,对于较高的光顺系数,抑制解误差的能力保持不变,甚至还有所提高。文献[15-16]证明了对于 1D 欧拉方程组,像 $\epsilon=500$ 和 $\sigma^*=1000$ 等参数值会得到稳定且非常快速的显式多步法(另参见源代码目录 analysis/mstage)。但是在多维情况下实现 UIRS 还存在一定的问题,由于对流通量雅可比不能在所有坐标方向上同时对角化(见附录 A.11 节),因此必须对每个计算坐标方向分别执行式(9.12)的变换和式(9.11)的残差光顺,坐标分裂导致最大光顺系数降至 $2\leqslant\epsilon\leqslant 6$。尽管如此,与 CIRS 相比,收敛到稳态的速度仍将大大提高[16-17],特别是与多重网格相结合时[16,18],收敛速度和鲁棒性方面将得到最大程度的提高。

在多维情况下,光顺系数根据特征值进行标定。例如,2D 情况可以使用下式计算:

$$
\begin{aligned}
\epsilon^I=\epsilon\cdot\min\left[\frac{\Lambda_c^I}{\Lambda_c^J},1\right] \\
\epsilon^J=\epsilon\cdot\min\left[\frac{\Lambda_c^J}{\Lambda_c^I},1\right]
\end{aligned}
\tag{9.13}
$$

光顺方法的 CFL 数与系数 ϵ 之间的关系为

$$
\frac{\sigma^*}{\sigma}\leqslant 1+C\epsilon \tag{9.14}
$$

常数 C 取决于空间离散的类型,中心格式时 $C=1$,一阶或二阶迎风格式时 $C=2$。

为了避免将残差转换成特征变量的数值计算,建议采用简化的 UIRS 方法[16-18]。在 i 方向上可写成

$$
\begin{aligned}
-\epsilon^I\boldsymbol{R}_{I-1}^*+(1+\epsilon^I)\boldsymbol{R}_I^*=\boldsymbol{R}_I, \quad & M^I>1 \\
-\epsilon^I\boldsymbol{R}_{I-1}^*+(1+2\epsilon^I)\boldsymbol{R}_I^*-\epsilon^I\boldsymbol{R}_{I+1}^*=\boldsymbol{R}_I, \quad & |M^I|\leqslant 1 \\
(1+\epsilon^I)\boldsymbol{R}_I^*-\epsilon^I\boldsymbol{R}_{I+1}^*=\boldsymbol{R}_I, \quad & M^I<-1
\end{aligned}
\tag{9.15}
$$

马赫数 M 是投影到特定计算坐标方向(此处为 I 方向)的速度。由于运算

量少,简化的 UIRS 特别适合于 3D 流动问题,文献[16]证明了其达到稳态所需的 CPU 时间只有 CIRS 的一半。另一个实例如图 9.4 所示,由图可见,UIRS 有助于使多重网格方法稳定,并可加快其收敛速度,采用 CIRS 的方法无法得到完全收敛的解,如果不采用迎风延拓,该方法将不起作用。

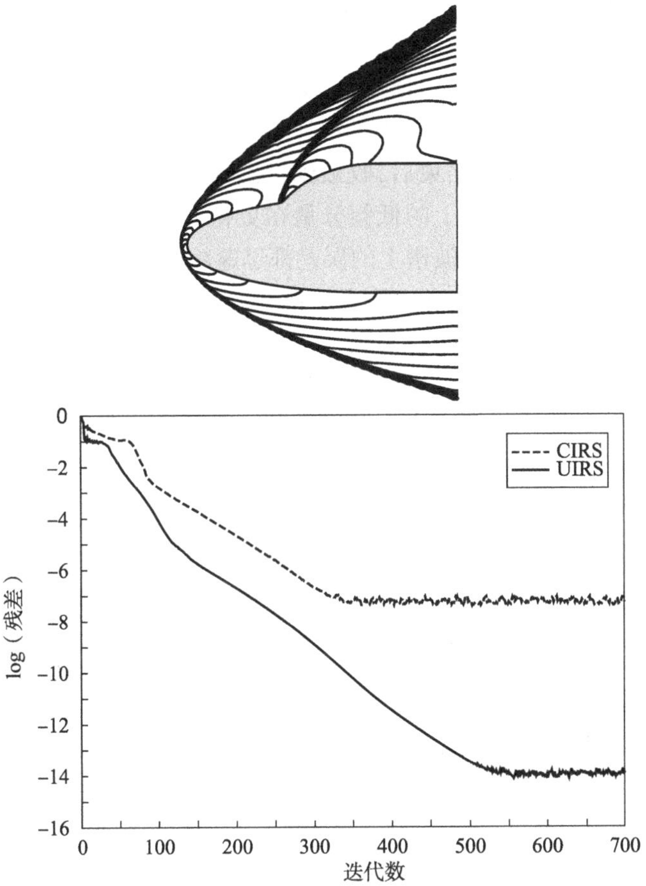

图 9.4 在 $M_\infty=20$、$\alpha=10°$时常规钝体高超声速绕流 CIRS 和 UIRS 的多重网格收敛比较

9.4 多重网格

多重网格方法是一种非常强大的加速技术(图 3.7)。它通过在一系列逐步粗化的网格上求解控制方程组,然后合并粗网格解的更新,并将其添加到最细网格的解中。多重网格方法最初由 Brandt[19] 发展用于椭圆型偏微分方程组,后来由 Jameson 应用到欧拉方程组[20-22],之后,多重网格法被用于求解纳维-斯托克

斯方程组[11-13,23-32]。多重网格法可用于显式和隐式时间推进格式[33-39]。当前研究的目标是显著提高双曲型流动问题多重网格法的效率[40-43]。

多重网格法的基本思想是利用粗网格使最细网格上的解更快地达到稳态,它的成功主要取决于如下两方面:

(1)(因为控制体更大)粗网格可采用更大的时间步长,同时减少计算量。由于确定新解的工作主要是在粗网格上,因此收敛速度更快,且计算时间减少;

(2)大多数显式和隐式时间推进法或迭代格式能有效减少求解误差中的高频分量(见10.3节),但是通常几乎无法抑制低频误差分量,这导致在初始阶段(此阶段消除了最大误差)结束后,收敛到稳态的速度减慢,多重网格法能帮助解决这个问题——最细网格上的低频分量在更粗的网格上变成高频分量,从而能够被逐次抑制,最终,整个波谱上的误差都迅速减小,收敛速度显著加快。

因此,可以看出,多重网格法的成功很大程度上取决于时间推进或迭代格式对高频误差分量的很好抑制。

代数多重网格(algebraic multigrid,AMG)方法为几何(基于网格的)多重网格提供了一种替代方法[44-53]。AMG是专门为隐式格式开发的,它直接对系统矩阵(左端算子)进行操作。AMG的基本思想是应用粗化矩阵来减少隐式算子的维数,从而减少方程组的数量,然后对代表粗层级的简化系统进行求解,得到细层级解的修正。粗化矩阵的构建使得耦合最强的方程组(即系统矩阵中最大的非对角阵)累加在一起,因而粗层级的生成完全由流动问题的物理特性决定,而不是由网格决定。AMG的优点是不需要构造或存储粗网格拓扑结构,这对于复杂的非结构网格尤为有利。

9.4.1 多重网格的基本循环

在应用几何多重网格法之前,必须生成粗网格。标准的做法是在所有坐标方向均匀地对网格进行粗化,而Mulder[54]提出了一种称为半粗化的方法,该方法仅在一个方向上对网格进行粗化,从一种层级的粗网格变为另一种层级的粗网格。半粗化方法特别适用于控制方程组在一个空间方向为刚性的流动问题,边界层中垂直于壁面的方向就属于这类情形,文献[26,55-57]中报道了半粗化方法在纳维-斯托克斯方程组中的应用。

根据关系式(6.1),细网格上的离散控制方程组为

$$\frac{\mathrm{d}}{\mathrm{d}t}\boldsymbol{W}_h = -\frac{1}{\Omega_h}\boldsymbol{R}_h \tag{9.16}$$

在下面的章节中,根据网格线的间距(对于非结构网格为控制体的特征尺寸),最细网格用下标 h 表示。从已知解 $\boldsymbol{W}_h^n$ 开始,采用合适的迭代格式,经过一

个时间步后得到新解 $\boldsymbol{W}_h^{n+1}$，用此新解计算出新的残差 $\boldsymbol{R}_h^{n+1}$。为了改进解 $\boldsymbol{W}_h^{n+1}$，使用一套粗网格，需要执行以下三个步骤：

1. 将解和残差转换到粗网格上

通过插值方法，将解转换到粗网格

$$\boldsymbol{W}_{2h}^{(0)}=\hat{I}_h^{2h}\boldsymbol{W}_h^{n+1} \tag{9.17}$$

其中，下标 $2h$ 表示粗网格①，$\hat{I}_h^{2h}$ 为插值算子。残差也必须传递到粗网格，以便可以抹平其低频误差分量，为此，采用守恒型转换算子，这意味着，当控制体尺寸增大时，残差值也必须增加同样的量。为了在粗网格上保持细网格的求解精度，还需要细网格的残差，为此，构造了一个称为强迫函数[19,22]的源项，作为从细网格转换的残差与在粗网格上利用初始解 $\boldsymbol{W}_{2h}^{(0)}$ 计算的残差之间的差值，即

$$(\boldsymbol{Q}_{\mathrm{F}})_{2h}=I_h^{2h}\boldsymbol{R}_h^{n+1}-\boldsymbol{R}_{2h}^{(0)} \tag{9.18}$$

式中：I_h^{2h} 表示将残差从细网格转换到粗网格的限制算子。这类多重网格法称为完全近似存储法（FAS）[19]。FAS 法特别适用于非线性方程组，因为它通过再离散，将系统的非线性特性代入粗层网格中。

2. 计算粗网格上的新解

采用与细网格相同的方法，计算粗网格上的解。将式（9.18）中的强迫函数叠加到粗网格的残差中，即

$$(\boldsymbol{R}_{\mathrm{F}})_{2h}=\boldsymbol{R}_{2h}+(\boldsymbol{Q}_{\mathrm{F}})_{2h} \tag{9.19}$$

因此，式（6.1）中的时间推进格式可以写成

$$\frac{\mathrm{d}}{\mathrm{d}t}\boldsymbol{W}_{2h}=-\frac{1}{\Omega_{2h}}(\boldsymbol{R}_{\mathrm{F}})_{2h}=-\frac{1}{\Omega_{2h}}[\boldsymbol{R}_{2h}+(\boldsymbol{Q}_{\mathrm{F}})_{2h}] \tag{9.20}$$

根据式（6.5），对于显式多步格式（见 6.1.1 节），则为

$$\begin{aligned}\boldsymbol{W}_{2h}^{(k)}&=\boldsymbol{W}_{2h}^{(0)}-\alpha_k\frac{\Delta t_{2h}}{\Omega_{2h}}[\boldsymbol{R}_{2h}^{(k-1)}+(\boldsymbol{Q}_{\mathrm{F}})_{2h}],\quad k=1,2,\cdots,m\\ \boldsymbol{W}_{2h}^{n+1}&=\boldsymbol{W}_{2h}^{(m)}\end{aligned} \tag{9.21}$$

必须注意的是，在第一次迭代（式（9.21）中的时间步）中，$(\boldsymbol{R}_{\mathrm{F}})_{2h}$ 与从细网格传递的残差相同（即，由式（9.18）可得 $(\boldsymbol{R}_{\mathrm{F}})_{2h}=I_h^{2h}\boldsymbol{R}_h^{n+1}$），这确保粗网格上的解取决于细网格上的残差，从而保持了细网格的精度。

由于粗网格不会影响细网格解的精度，所以对于粗网格上的空间离散，一阶格式就足够保证计算精度。与高阶格式相比，粗网格上的一阶精确离散方法具有鲁棒性强、阻尼性好、计算量小等优点。

---

① 符号 $2h$ 不必从严格的几何意义上理解。在非结构网格上，细网格和粗网格控制体的特征尺寸的比值通常不为 2，半粗化方法同样如此。

3. 将解从粗网格插值到细网格

在粗网格上执行一个或几个时间步(迭代)后,计算相对于初始插值解(式(9.17))的修正。粗网格修正为

$$\delta \boldsymbol{W}_{2h}=\boldsymbol{W}_{2h}^{n+1}-\boldsymbol{W}_{2h}^{(0)} \tag{9.22}$$

将粗网格修正插值到细网格,以改进细网格的解。因此,细网格上的新解为

$$\boldsymbol{W}_{h}^{+}=\boldsymbol{W}_{h}^{n+1}+I_{2h}^{h}\delta \boldsymbol{W}_{2h} \tag{9.23}$$

式中:I_{2h}^{h} 称为延拓算子。

9.4.2 多重网格策略

上述基本多重网格方法仅由一层粗网格组成,如果存在多层粗网格,则重复步骤 1 和步骤 2,直到最粗网格。需要注意的是,粗网格上的强迫函数是由式(9.19)的受限制的修正残差构成。例如,在粗网格 $4h$ 上,由下式得到强迫函数,即

$$(\boldsymbol{Q}_{\mathrm{F}})_{4h}=I_{2h}^{4h}(\boldsymbol{R}_{\mathrm{F}})_{2h}^{n+1}-\boldsymbol{R}_{4h}^{(0)}=I_{2h}^{4h}[\boldsymbol{R}_{2h}^{n+1}+(\boldsymbol{Q}_{\mathrm{F}})_{2h}]-\boldsymbol{R}_{4h}^{(0)} \tag{9.24}$$

通过这种方式,最细网格的残差可以控制所有粗网格上的求解精度。在最粗网格上执行给定数量的时间步之后,可以连续重复步骤 3,直到再次回到最细网格,这个过程称为锯齿形循环或 V 循环(图 9.5(a))。当然,在粗网格上也可以进行更多的循环,这种策略称为 W 循环,如图 9.5(b)所示,它常用于跨声速流动。在超声速和高超声速流动情况下,V 循环被证实更有效率。

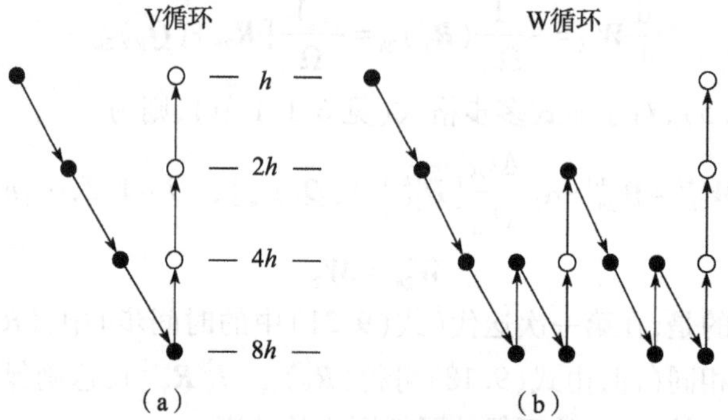

图 9.5 多重网格的循环类型

● 表示限制前的时间步;○ 表示延拓后的时间步。

1. 时间步数

限制前和延拓后的最佳时间步数取决于时间推进方法的类型。对于显式多步方法(见 6.1.1 节或 6.1.2 节),通常只在残差限制前执行一个时间步,而延拓

后不执行时间步。然而,在将粗网格校正累加到细网格解 W_h^{n+1}(式(9.23))之前,对粗网格校正(式(9.22))进行光顺可提高多重网格法的鲁棒性,光顺方法可以采用 9.3 节中所述的中心隐式光顺方法(采用常系数)。

另一种应用广泛的时间推进方法——隐式 LU-SGS 方法(见 6.2.4 节),需要在限制前进行两次迭代,以获得最佳的多重网格效率[35]。延拓后的时间步数取决于空间离散格式。对于中心格式(见 4.3.1 节),不需要时间步数[35-37],但可对解的修正进行光顺,而如果采用迎风离散,则需要在延拓后进行一个时间步,这种(2,1)格式对各种流动条件,在鲁棒性和计算时间方面都被证明是最优的[36-37]。

2. 起始网格

应该指出的是,在实践中,多重网格法并非直接从最细网格开始,而是从一个粗网格开始执行多次多重网格循环,然后将近似解插值到下一层级更细网格(延拓也使用相同的算子),再执行几个循环,将解再次插值到下一层级更细的网格,依此类推,直到最细网格。通过这种方式,只需适度的计算量,就可以在最细网格上获得很好的初始解,这种非常有效的方法称为完全多重网格法(FMG)[19]。

3. 转换算子精度

限制算子(式(9.18))和延拓算子(式(9.23))必须满足一定的精度要求,即[58]

$$m_R + m_P > m_E \tag{9.25}$$

其中,m_R 和 m_P 分别表示由限制算子和延拓算子进行精确插值的多项式的次数加 1,例如在线性插值的情况下,m_R 或 m_P 等于 2。而 m_E 表示控制方程组的阶数,对于欧拉方程组 $m_E = 1$,对于纳维-斯托克斯方程组 $m_E = 2$。如果违背式(9.25)的条件,则限制算子和/或延拓算子引入的附加误差将会干扰细网格解,这种多重网格方法将收敛缓慢,甚至发散。

9.4.3　在结构网格上的实现

在结构网格上实现多重网格非常简单,只要在每个坐标方向上删除每两条网格线的一条,即可很容易地生成粗网格。因此,网格线的间距为 $2h$、$4h$ 等,这就保证了与最细网格相比,粗网格上的计算量较低。文献[11-13,20-22,26,33-37]中提供了几个有代表性的例子。

如图 9.6 所示,对于格心格式和格点格式,必须以不同的方式定义解的插值算子、残差的限制算子和校正延拓算子。例如,在两层连续的网格上,每两个网格点有一个公共网格点,而网格单元中心则始终位于不同的位置。因此我们将分别讨论格心型和格点型有限体积格式的转换算子的标准形式。

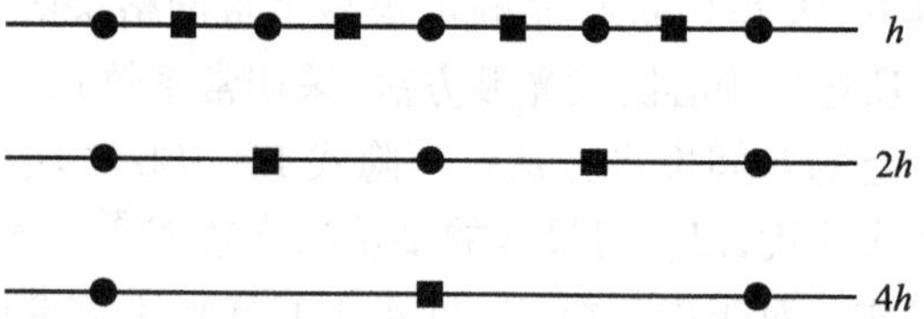

图 9.6　1D 情况的细网格(h)和两层粗网格($2h$,$4h$)示意图。圆圈表示网格点，矩形表示单元中心

除了(将在下文介绍的)对称的、纯几何定义的限制算子和延拓算子外，文献[16]第 4 章中还提出了迎风偏置形式的算子，迎风限制算子和延拓算子两者均考虑了流动方程组的特点，提高了高超声速流动中多重网格法的鲁棒性。为了节省篇幅，我们将仅讨论用于格点型空间离散的迎风延拓算子的实现。

1. 格心格式的转换算子

利用体积加权插值，将解从细网格转换到粗网格。2D 情况下，式(9.17)变为(图 9.7(a))

$$
(\boldsymbol{W}_{2h}^{(0)})_{I,J}=\frac{(\boldsymbol{W}_h^{n+1})_{I,J}\Omega_{I,J}+(\boldsymbol{W}_h^{n+1})_{I+1,J}\Omega_{I+1,J}+(\boldsymbol{W}_h^{n+1})_{I,J+1}\Omega_{I,J+1}}{\Omega_{I,J}+\Omega_{I+1,J}+\Omega_{I,J+1}+\Omega_{I+1,J+1}}+\frac{(\boldsymbol{W}_h^{n+1})_{I+1,J+1}\Omega_{I+1,J+1}}{\Omega_{I,J}+\Omega_{I+1,J}+\Omega_{I,J+1}+\Omega_{I+1,J+1}} \tag{9.26}
$$

3D 情况采用类似的转换算子，其体积总和为形成一个粗网格单元的 8 个细网格控制体。

限制算子定义为一个粗网格控制体中所包含所有网格单元的残差之和。因此，2D 情况的限制算子为(图 9.7(a))

$$
(I_h^{2h}\boldsymbol{R}_h^{n+1})_{I,J}=(\boldsymbol{R}_h^{n+1})_{I,J}+(\boldsymbol{R}_h^{n+1})_{I+1,J}+(\boldsymbol{R}_h^{n+1})_{I,J+1}+(\boldsymbol{R}_h^{n+1})_{I+1,J+1} \tag{9.27}
$$

同理可得 3D 情况的限制算子。注意，必须将式(9.27)中位于物理域之外的残差 $\boldsymbol{R}_h^{n+1}$ 设为零。

粗网格修正公式(9.23)的延拓可以通过两种不同的方式进行。第一种方式，将 $\delta\boldsymbol{W}_{2h}$ 均匀分配到周围所有网格单元中心(图 9.7(b))，则得到零阶延拓算子，如

$$
(\boldsymbol{W}_h^{+})_{I,J+1}=(\boldsymbol{W}_h^{n+1})_{I,J+1}+(\delta\boldsymbol{W}_{2h})_{I,J} \tag{9.28}
$$

第二种方式可使多重网格法更快收敛，包括两个步骤：步骤一，类似于(下文中的)格点格式，将 $\delta\boldsymbol{W}_{2h}$ 插值到网格节点；步骤二，对节点值进行平均得到细网格单元中心的值。根据图 9.7(c)，可推导出以下最终关系式：

$$(I_{2h}^{b}\delta W_{2h})_{I,J}=\frac{1}{16}[9(\delta W_{2h})_{I,J}+3(\delta W_{2h})_{I-1,J}+3(\delta W_{2h})_{I,J-1}+(\delta W_{2h})_{I-1,J-1}] \tag{9.29}$$

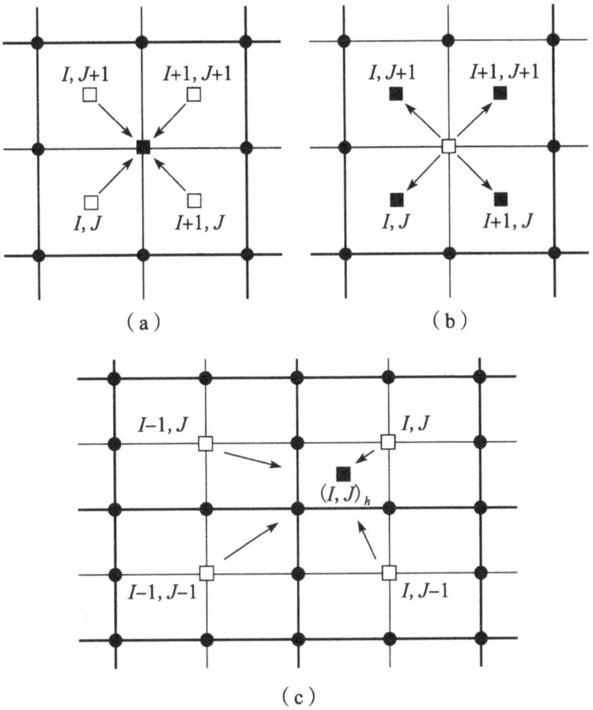

图 9.7 2D 结构网格格心格式的转换算子

(a) 解的插值和残差的限制；(b)、(c) 粗网格修正的延拓。实心圆=网格点；实心矩形=插值对象的格心；矩形=插值来源的格心；粗线=粗网格；细线=细网格。

采用类似方法可得到相应的 3D 表达式

$$\begin{aligned}(I_{2h}^{b}\delta W_{2h})_{I,J,K}=\frac{1}{64}[&27(\delta W_{2h})_{I,J,K}+9(\delta W_{2h})_{I-1,J,K}+\\&9(\delta W_{2h})_{I,J-1,K}+9(\delta W_{2h})_{I,J,K-1}+\\&3(\delta W_{2h})_{I-1,J-1,K}+3(\delta W_{2h})_{I-1,J,K-1}+\\&3(\delta W_{2h})_{I,J-1,K-1}+(\delta W_{2h})_{I-1,J-1,K-1}\end{aligned} \tag{9.30}$$

2. 格点格式的转换算子

由于细网格和粗网格之间有共同的节点，因此通过简单赋值就可将解从细网格转换到粗网格，即

$$(W_{2h}^{(0)})_{i,j,k}=(W_{h}^{n+1})_{i,j,k} \tag{9.31}$$

标准中心限制算子表示组成一个粗网格单元的所有 4 个(3D 为 8 个)细网格单元的节点的线性插值。根据图 9.8,2D 限制残差计算公式为

$$
\begin{aligned}
(I_h^{2h}\boldsymbol{R}_h^{n+1})_{i,j} = (\boldsymbol{R}_h^{n+1})_{i,j} + \frac{1}{2}[&(\boldsymbol{R}_h^{n+1})_{i-1,j} + (\boldsymbol{R}_h^{n+1})_{i+1,j} + \\
&(\boldsymbol{R}_h^{n+1})_{i,j-1} + (\boldsymbol{R}_h^{n+1})_{i,j+1}] + \\
&\frac{1}{4}[(\boldsymbol{R}_h^{n+1})_{i-1,j-1} + (\boldsymbol{R}_h^{n+1})_{i+1,j-1} + \\
&(\boldsymbol{R}_h^{n+1})_{i-1,j+1} + (\boldsymbol{R}_h^{n+1})_{i+1,j+1}]
\end{aligned} \tag{9.32}
$$

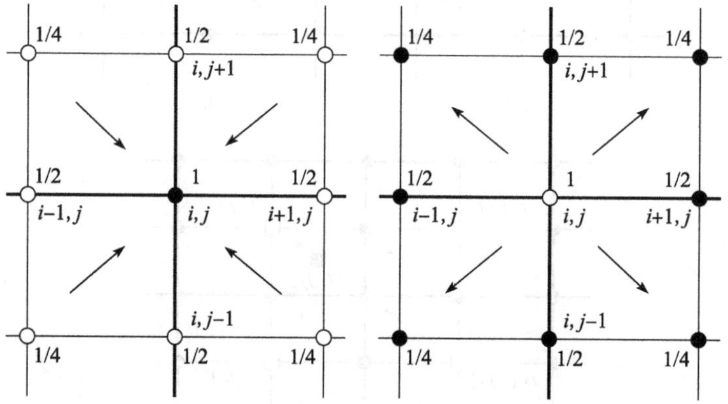

（a）　　　　（b）

图 9.8　2D 结构网格格点格式的插值因子

(a) 限制算子;(b) 延拓算子。实心圆=插值对象点;空心圆=插值来源点;粗线=粗网格;细线=细网格。点(i,j)为两层网格的公共点。

3D 限制残差计算公式为

$$
(I_h^{2h}\boldsymbol{R}_h^{n+1})_{i,j,k} = (\boldsymbol{R}_h^{n+1})_{i,j,k} + \frac{1}{2}A + \frac{1}{4}B + \frac{1}{8}C \tag{9.33}
$$

式中各因子为

$$
\begin{aligned}
A = &(\boldsymbol{R}_h^{n+1})_{i+1} + (\boldsymbol{R}_h^{n+1})_{i-1} + (\boldsymbol{R}_h^{n+1})_{j+1} + (\boldsymbol{R}_h^{n+1})_{j-1} + (\boldsymbol{R}_h^{n+1})_{k+1} + (\boldsymbol{R}_h^{n+1})_{k-1} \\
B = &(\boldsymbol{R}_h^{n+1})_{i+1,j+1} + (\boldsymbol{R}_h^{n+1})_{i-1,j+1} + (\boldsymbol{R}_h^{n+1})_{i+1,j-1} + (\boldsymbol{R}_h^{n+1})_{i-1,j-1} + \\
&(\boldsymbol{R}_h^{n+1})_{i+1,k+1} + (\boldsymbol{R}_h^{n+1})_{i-1,k+1} + (\boldsymbol{R}_h^{n+1})_{i+1,k-1} + (\boldsymbol{R}_h^{n+1})_{i-1,k-1} + \\
&(\boldsymbol{R}_h^{n+1})_{j+1,k+1} + (\boldsymbol{R}_h^{n+1})_{j-1,k+1} + (\boldsymbol{R}_h^{n+1})_{j+1,k-1} + (\boldsymbol{R}_h^{n+1})_{j-1,k-1} \\
C = &(\boldsymbol{R}_h^{n+1})_{i+1,j+1,k+1} + (\boldsymbol{R}_h^{n+1})_{i-1,j+1,k+1} + (\boldsymbol{R}_h^{n+1})_{i-1,j-1,k+1} + (\boldsymbol{R}_h^{n+1})_{i+1,j-1,k+1} + \\
&(\boldsymbol{R}_h^{n+1})_{i+1,j+1,k-1} + (\boldsymbol{R}_h^{n+1})_{i-1,j+1,k-1} + (\boldsymbol{R}_h^{n+1})_{i-1,j-1,k-1} + (\boldsymbol{R}_h^{n+1})_{i+1,j-1,k-1}
\end{aligned} \tag{9.34}
$$

上式中,只显示了下标与 i,j,k 不同的项。值得注意的是,在限制之前,所有

物理边界和虚拟点处的残差 $\boldsymbol{R}_h^{n+1}$ 必须设为零。

粗网格修正的延拓可通过对粗网格点的一个循环来实现，在该循环中，使用与限制算子相同的权重，将值 $(\delta\boldsymbol{W}_{2h})_{i,j,k}$ 分配到细网格点上（图 9.8(b)），将每个点的贡献值进行累加，从而得到细网格的完整的转换修正值。

3. 迎风延拓（格点格式）

原则上，迎风延拓可以用特征变量或守恒变量表示。文献[16]提出了一种特别有效的基于守恒变量的迎风延拓方法，该方法根据马赫数和速度方向，采用式(9.23)中修正量的迎风偏置插值，它的计算量非常小，但可显著提高多重网格法在高速流动计算中的鲁棒性[16]。

将解修正量 $\delta\boldsymbol{W}_{2h}$ 插值到细网格应分两步完成。第一步，两层网格共有的点 A、B、C、D（图 9.9）处的修正量直接传递到细网格。第二步，将修正量插值到只属于细网格的点 e,f,e',f' 和 g，插值取决于对应点处马赫数的符号和绝对值，在这种情况下，使用逆变速度计算马赫数，例如点 e 处的马赫数为

$$M_e = \frac{\boldsymbol{n} \cdot \boldsymbol{v}_e}{c} \tag{9.35}$$

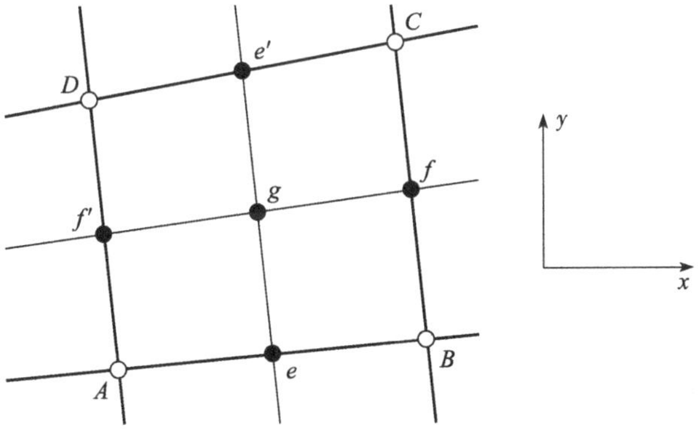

图 9.9　2D 迎风延拓。点 A、B、C、D 为粗网格（粗线表示）和细网格（细线表示）所共有的点。点 e,f,e',f',g 只属于细网格

对控制体的面矢量（A-B 方向）求平均，或对点 A 到点 B 的矢量进行归一化，可得到式(9.35)中的法向矢量 $\boldsymbol{n}$，然后根据如下准则将校正量传递到点 e：

$$(I_{2h}^{h}\delta\boldsymbol{W}_{2h})_e = \begin{cases} (\delta\boldsymbol{W}_{2h})_A, & M_e > 1 \\ \dfrac{1}{2}[(\delta\boldsymbol{W}_{2h})_A + (\delta\boldsymbol{W}_{2h})_B], & |M_e| \leqslant 1 \\ (\delta\boldsymbol{W}_{2h})_B, & M_e < -1 \end{cases} \tag{9.36}$$

点 f 和点 f' 的解修正量转换采用同样的过程。迎风延拓有助于使网格之间

的信息交换更符合物理实际。另外，将修正量插值到点 g 比较困难，在文献[16]中，采用对周围点 $A\sim D$ 的值进行简单平均

$$(I_{2h}^{h}\delta\boldsymbol{W}_{2h})_{g}=\frac{1}{4}[(\delta\boldsymbol{W}_{2h})_{A}+(\delta\boldsymbol{W}_{2h})_{B}+(\delta\boldsymbol{W}_{2h})_{C}+(\delta\boldsymbol{W}_{2h})_{D}] \tag{9.37}$$

在此，其他类型的迎风加权插值可能会更加合适。迎风延拓可以以类似的方式推广到3D情况，虽然方法进行了适当的简化，但在许多测试算例中获得了令人满意的结果[16]。

9.4.4 在非结构网格上的实现

与结构网格相比，非结构网格中粗网格的构建要复杂得多，主要问题是如何从一组没有特定顺序的元素（网格单元）构建一个均匀粗化的网格。此外，另一个需要注意的问题是粗网格与细网格的单元体积之比必须保持在一定范围内（2D大约为4，3D大约为8）。解决这一问题的比较合适的方法是AMG方法[44-53]，在9.4节开头简要讨论了该方法，但是几何多重网格的应用领域仍然十分广泛，因此，在此我们将重点介绍几何多重网格方法。

粗网格的生成方法主要有三种：

(1) 非嵌套网格方法；

(2) 拓扑方法；

(3) 控制体融合方法。

文献[59-60]对上述方法进行了综述。

标准的限制算子和延拓算子是基于纯几何定义的插值，Leclerq 和 Stoufflet[61]提出了适合于强激波流动的迎风转换算子，该迎风限制算子/延拓算子将残差/修正量转换为特征变量，经过迎风偏置插值后，限制值/延拓值再被转换回物理变量。文献[16]介绍了计算量更小的迎风多重网格法（另可参考式(9.36)）。

1. 非嵌套网格方法

非结构网格粗化最明显的思路是生成一系列完全独立、逐渐粗化网格[62-67]，这些粗网格不必包含任何公共节点，因此称之为非嵌套网格。但重要的是，所有粗网格必须保留主要的几何特征（前后缘、机头等），然而要实现这一点并不容易，尤其是对于较为复杂的几何构型，因此基于非嵌套网格的多重网格目前很少使用。

2. 拓扑方法

一种特殊的拓扑方法是应用基于图的算法，从细网格中删除某些节点，然后对其余节点重新进行三角剖分[68-69]。与非嵌套网格方法相反，因为这种网格序列包含公共节点，网格之间的插值变得更加容易，但是该方法同样存在非嵌套网

格在几何一致性方面的缺点。

另一种拓扑方法采用网格细分技术[29,70-71]，从粗网格开始，通过单元剖分生成更细的网格。该方法既可以应用于整个物理域，也可以仅应用于局部区域（如边界层）。它的缺点是最细网格的质量在很大程度上取决于初始粗网格和细化过程，通过边交换可以部分解决这个问题[72-73]。

另一种粗网格生成的思路是基于边消除技术[74]，它最初是针对四面体网格的无黏流场计算发展的，文献[75]进一步将边消除技术推广到混合网格的黏性流场求解。

3. 融合多重网格法

融合多重网格法是一种对非结构网格有效的方法，它首先由 Lallemand[76]、Lallemand 等[77]和 Koobus 等[78]提出，后来，融合多重网格法被许多学者采用[31,79-87]。该方法将细网格的控制体与其相邻控制体融合生成粗网格，生成的粗网格由逐渐变大的不规则多面体单元组成，如图 9.10 所示。融合技术保留了边界面的完全离散，与之前讨论的所有非结构多重网格法相比，具有明显优势。但需要指出的是，到目前为止，融合多重网格的实现大多是采用中点-二次格点格式（见 5.2.2 节），文献[79]介绍了融合多重网格在格心格式（见 5.2.1 节）中的应用。

4. 基于体融合的粗网格生成

格点格式的体融合按以下步骤进行：

(1) 建立一个种子点列表。种子点是选定的用于融合周围控制体的网格点，种子点列表可包含构成近似最大独立集合的点[81]，或简单地包含当前网格层的所有点；

(2) 循环遍历所有种子点；

(3) 如果种子点是未融合的，则融合其所有未融合的最近的相邻点（由一条边连接）；

(4) 检查粗化率（即在一个粗网格控制体中包含多少个细网格控制体），如果粗化率小于 4(3D 为 8)，则增加（与其他种子点无关联的）已经融合的最近的相邻控制体，直到达到最佳粗化率，文献[27]提出首先对间隔为 2 的相邻控制体进行融合，即这两个已经融合的控制体之间至少有 2 个(3D 为 3 个)最近的相邻点；

(5) 如果列表中仍有种子点，转到步骤(2)；

(6) 消除孤立控制体。这些控制体是无法融合的单个控制体，因为它们不存在未融合的相邻控制体，消除孤立控制体的方法是将它与粗化率最小的相邻控制体融合，这样使得粗网格层级的控制体的大小分布更为规则[31]。

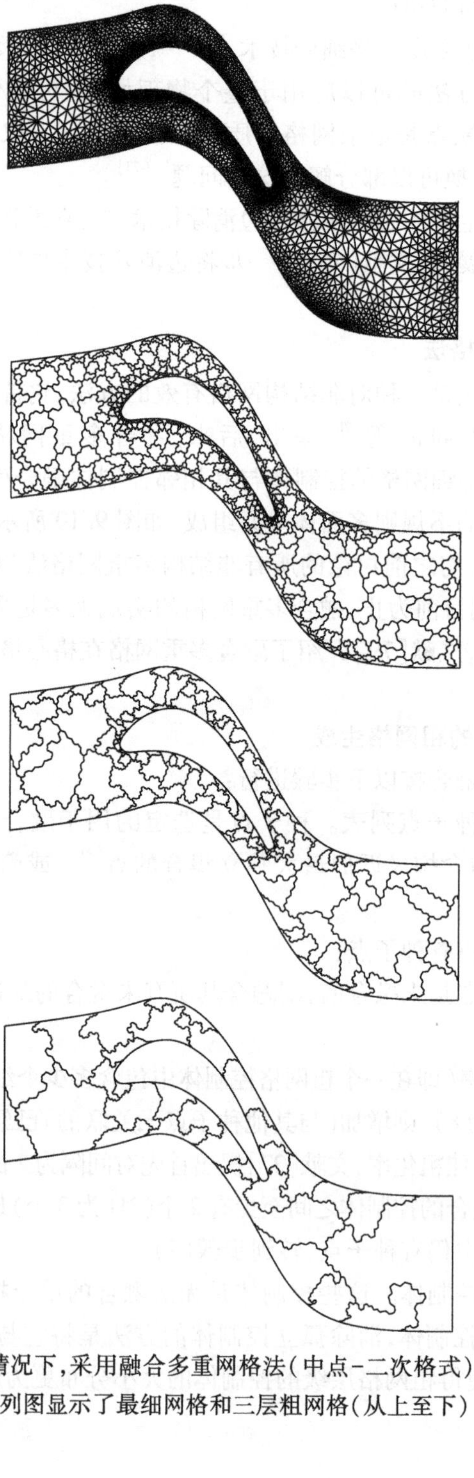

图 9.10　2D 情况下，采用融合多重网格法（中点-二次格式）生成粗网格。
该序列图显示了最细网格和三层粗网格（从上至下）

重复上述步骤,直到生成所有粗网格。

控制体融合必须从边界开始,以保持网格的各向同性。在 3D 情况下,可能需要用户干预,根据边界形状来指定融合方向。为了克服这一困难,Okamoto 等[83]提出了另一种称为全局粗化的算法。它采用基于边染色的全局剖分格式,使用这个剖分格式首先生成独立的边的集合,其次对于这个边的集合中的每条边,将共享它的所有控制体进行融合。重复这个步骤,直到达到规定的粗化率。该方法不需要指定初始种子点或融合方向,它还可以处理任意网格单元类型。此外,Moulitsas 和 Karypis 开发了一种方法[88-89],采用一种多级范式对给定的粗网格进行优化。

5. 融合多重网格的问题

对于无黏流场,融合多重网格的实现不存在什么困难。欧拉方程组一般离散为单个控制体面上的通量,从这方面来说,控制体的形状(实际使用的是平均面矢量)的复杂程度并不重要。此外,一阶精度空间格式只需要知道相邻控制体中的流场变量,这些信息在粗网格中很容易获得。

对于黏性流场,在任意形状的控制体上扩散通量的离散不能直接进行,其问题在于面心处梯度的计算(参见文献[31]中的讨论)。另一个更加严重的问题与所需的延拓算子的精度有关,式(9.25)中的不等式建议取 $m_P=2$,因为标准限制算子(残差的和)仅使 $m_R=1$。但是,在粗网格上构造线性插值并不容易。

为了避免存在一阶精度延拓算子的问题,Mavriplis[59]提议使用常数延拓(即,包含在融合粗网格控制体内的细网格的所有点都获得相同的解修正),并对黏性通量采用附加的缩放(缩放因子为 $0.5^{(\text{level}-1)}$;同时应用于梯度)。但是,这种方法无法得到最优的多重网格效率。

Haselbacher[31]建议在粗网格上仍保持细网格的黏性通量离散,并施加像在细网格上一样的边界条件。此外,他提出了一种分段线性延拓算子,对于标量 U,有

$$I_{2h}^{h}\delta U_{2h}=(\delta U_{2h})_i+\Psi_i(\nabla\delta U_{2h})_i\cdot\boldsymbol{r}_{i,j} \tag{9.38}$$

式(9.38)中的梯度 $(\nabla\delta U_{2h})_i$ 使用式(5.55)中介绍的线性最小二乘重构来计算,根据 5.3.5 节介绍的 Barth-Jespersen 限制器计算限制器 Ψ_i 的值。

9.5　低速预处理

在低亚声速马赫数范围,当流速大小与声速相比很小时,控制方程组(2.19)的对流项数值求解具有很强的刚性。我们可以用下面的例子来说明

这一点，在 3D 情况下，有如下的 5 个特征值

$$\begin{aligned}(\Lambda_c)_{1,2,3} &= V \quad (\text{对流模式})\\ (\Lambda_c)_{4,5} &= V \pm c \quad (\text{声速模式})\end{aligned} \tag{9.39}$$

式中：V 为逆变速度；c 为声速。控制方程组的刚性（进行时间推进时）由特征条件数决定。特征条件数的定义是最大特征值与最小特征值之比：

$$C_N = \frac{|(\Lambda_c)_{\max}|}{|(\Lambda_c)_{\min}|} = \frac{|V| + c}{|V|} = \frac{M+1}{M} \tag{9.40}$$

局部时间步长受速度最快的波（即 $(\Lambda_c)_4$）限制，在一个时间步内，最慢波只传输了单元宽度的一小部分：$\Lambda_{\min}\Delta t \approx (\Lambda_{\min}/\Lambda_{\max})h = h/C_N$。因此，大的条件数 C_N（即 $M \to 0$ 的情形）使波的传播效率降低，进而使收敛到定常状态的速度减慢[90]。此外，文献[91-92]也证明，可压缩流动的格式存在大量的人工耗散，对于马赫数趋近于零的情况，这些耗散不能被正确标定。因此，在低马赫数流动区域，这种空间离散的计算精度会受到损害[93]。

如果整个流场的速度较低（$M<0.2$），则可压缩性效应可以忽略，可使用不可压缩方程组求解，不可压缩纳维-斯托克斯方程组可以用众所周知的基于压力的格式求解[94-99]。另一种方法则是应用人工压缩性（或伪压缩性）方法[100-106]。但是也存在如下的流动情况，使用单独的不可压缩方法难以求解：

（1）包含很大低速区域的高速流动。强收缩喷管的上游亚声速流动就属于此类流动。

（2）由热源引起的密度变化的可压缩低速流动。这发生在表面传热或体积热添加（燃烧模拟）。

（3）不同马赫数的可压缩流和不可压缩流同时存在的流动问题，称为全速度流动，带推进系统的高升力构型和 V/STOL 机动中会出现这种情形。

上述流动情况下需应用可压缩控制方程组。为了在低马赫数下高效、准确地求解这些问题，可以采用预处理方法，预处理方法的优点是它能够提供一种适用于所有马赫数的求解方法。下面将推导预处理控制方程组。

9.5.1 预处理方程组的推导

以 1D 欧拉方程组为例进行说明，它的微分形式可写成

$$\frac{\partial \boldsymbol{W}}{\partial t} + \frac{\partial \boldsymbol{F}_c}{\partial x} = 0 \tag{9.41}$$

式中：$\boldsymbol{W} = [\rho, \rho u, \rho E]^T$ 为守恒变量；$\boldsymbol{F}_c$ 表示对流通量。为了理解预处理对（对计算时间步长很重要的）谱半径和（对迎风耗散很重要的）对流通量雅可比的影

响,将欧拉方程组(9.41)改写为准线性形式:

$$\frac{\partial \boldsymbol{W}}{\partial t}+\boldsymbol{A}_{\mathrm{c}} \frac{\partial \boldsymbol{W}}{\partial x}=0 \tag{9.42}$$

其中,$\boldsymbol{A}_{\mathrm{c}}=\partial \boldsymbol{F}_{\mathrm{c}}/\partial \boldsymbol{W}$ 为对流通量雅可比(见 A.2 节)。

低速预处理的思想是对控制方程组(9.41)进行转换,使新方程组在低马赫数(即低于 $M\approx0.2$)时具有更好的性质。首先,要使对流特征值变为相同量级,以限定特征条件数(式(9.40)),从而消除 $M\to0$ 时的数值刚性。其次,希望通过改变数值耗散的标度来提高精度。最后,与基于压力的格式处理一样,将压力和速度耦合。通过转换,由一组不同的流动变量取代守恒变量 $\boldsymbol{W}$,流动变量有各种选择[107],但最常用的是压力、速度分量和温度。将式(9.42)中的准线性形式转换成新变量 $\boldsymbol{W}_{\mathrm{p}}$,得

$$\boldsymbol{P} \frac{\partial \boldsymbol{W}_{\mathrm{p}}}{\partial t}+\boldsymbol{A}_{\mathrm{c}}\boldsymbol{P} \frac{\partial \boldsymbol{W}_{\mathrm{p}}}{\partial x}=0 \tag{9.43}$$

其中,$\boldsymbol{P}=\partial \boldsymbol{W}/\partial \boldsymbol{W}_{\mathrm{p}}$ 表示从新变量 $\boldsymbol{W}_{\mathrm{p}}$ 到守恒变量 $\boldsymbol{W}$ 的转换矩阵。引入一个新的通量雅可比:

$$\boldsymbol{A}_{\mathrm{c,p}}=\boldsymbol{A}_{\mathrm{c}}\boldsymbol{P}=\frac{\partial \boldsymbol{F}_{\mathrm{c}}}{\partial \boldsymbol{W}_{\mathrm{p}}} \tag{9.44}$$

则式(9.43)可写为

$$\boldsymbol{P} \frac{\partial \boldsymbol{W}_{\mathrm{p}}}{\partial t}+\boldsymbol{A}_{\mathrm{c,p}} \frac{\partial \boldsymbol{W}_{\mathrm{p}}}{\partial x}=0 \tag{9.45}$$

如果用一个合适的预处理矩阵 $\boldsymbol{\Gamma}$(稍后有详细说明)代替时间导数前的 $\boldsymbol{P}$,则预处理方程组(9.45)变为

$$\boldsymbol{\Gamma} \frac{\partial \boldsymbol{W}_{\mathrm{p}}}{\partial t}+\boldsymbol{A}_{\mathrm{c,p}} \frac{\partial \boldsymbol{W}_{\mathrm{p}}}{\partial x}=0 \tag{9.46}$$

或者写成

$$\frac{\partial \boldsymbol{W}_{\mathrm{p}}}{\partial t}+\boldsymbol{\Gamma}^{-1}\boldsymbol{A}_{\mathrm{c,p}} \frac{\partial \boldsymbol{W}_{\mathrm{p}}}{\partial x}=0 \tag{9.47}$$

现在,可以使用任何标准的空间和时间离散格式,求解采用新变量 $\boldsymbol{W}_{\mathrm{p}}$ 的方程组(9.47)。另一种方法是将预处理方程组(9.47)转换回守恒变量 $\boldsymbol{W}$,得

$$\boldsymbol{P}^{-1} \frac{\partial \boldsymbol{W}}{\partial t}+\boldsymbol{\Gamma}^{-1}\boldsymbol{A}_{\mathrm{c,p}}\boldsymbol{P}^{-1} \frac{\partial \boldsymbol{W}}{\partial x}=0 \tag{9.48}$$

其中,$\boldsymbol{P}^{-1}=\partial \boldsymbol{W}_{\mathrm{p}}/\partial \boldsymbol{W}$ 为从守恒变量到新变量的转换矩阵。为方便起见,式(9.48)可写成

$$\frac{\partial \boldsymbol{W}}{\partial t}+\boldsymbol{P\Gamma}^{-1}\boldsymbol{A}_{\mathrm{c,p}}\boldsymbol{P}^{-1}\frac{\partial \boldsymbol{W}}{\partial x}=0 \tag{9.49}$$

或等价于

$$\frac{\partial \boldsymbol{W}}{\partial t}+\boldsymbol{P\Gamma}^{-1}\boldsymbol{A}_{\mathrm{c}}\frac{\partial \boldsymbol{W}}{\partial x}=0 \tag{9.50}$$

式(9.49)和式(9.50)中的项 $\boldsymbol{P\Gamma}^{-1}$ 称为守恒变量预处理矩阵。

可以看出,采用守恒变量的预处理方程组(9.49)~式(9.50)具有以下特点:

(1) 空间导数乘以 $\boldsymbol{P\Gamma}^{-1}$;

(2) 非定常方程组与式(9.42)中的原始形式不同,因此解不再是时间精确的;

(3) 定常解(即 $\partial \boldsymbol{W}/\partial t=0$)保持不变;

(4) 预处理系统的特征值和特征矢量对应于矩阵 $\boldsymbol{P\Gamma}^{-1}\boldsymbol{A}_{\mathrm{c}}=\boldsymbol{P\Gamma}^{-1}\boldsymbol{A}_{\mathrm{c,p}}\boldsymbol{P}^{-1}$ 的特征值和特征矢量,因此与式(9.41)中原系统的特征值和特征矢量不同。

现在的主要任务是找到一组合适的变量 $\boldsymbol{W}_{\mathrm{p}}$ 和矩阵 $\boldsymbol{\Gamma}$,使预处理系统(式(9.50))的对流特征值尽可能接近于相等,同样重要的是,在 $M\to 0$ 时矩阵必须仍是有定义的,此外,对于更高的马赫数,最好能将预处理系统转换为原始系统。

在9.5.3节介绍转换矩阵和预处理矩阵之前,我们先讨论低速预处理在流场解算器中的实现。

9.5.2 预处理在流场解算器中的实现

由于我们求解的是控制方程组的积分形式,所以将式(9.50)写成

$$\frac{\partial}{\partial t}\int_{\Omega}\boldsymbol{W}\mathrm{d}\Omega+\boldsymbol{P\Gamma}^{-1}\oint_{\partial\Omega}(\boldsymbol{F}_{\mathrm{c}}-\boldsymbol{F}_{\mathrm{v}})\mathrm{d}S=\boldsymbol{P\Gamma}^{-1}\int_{\Omega}(\boldsymbol{Q})\mathrm{d}\Omega \tag{9.51}$$

值得注意的是,对于非定常流动,式(9.51)是不守恒的。因此,为了获得时间精确解,必须采用双时间步方法(见6.3节)。

使用显式多步格式(见6.1节)求解预处理控制方程组(9.51)的步骤如下:

(1) 根据矩阵 $\boldsymbol{P\Gamma}^{-1}\boldsymbol{A}_{\mathrm{c}}$ 的谱半径计算新的时间步长;

(2) 利用谱半径(如JST格式)或预处理通量雅可比矩阵的特征值和特征矢量(如Roe迎风格式)计算人工耗散,人工耗散可以用守恒变量或原始变量表示[107];

(3) 计算对流通量(无变化或基于新变量 $\boldsymbol{W}_{\mathrm{p}}$);

(4) 对耗散和对流通量求和。根据人工耗散的形式,将总的残差或仅对流项乘以守恒变量预处理矩阵,即 $\boldsymbol{P\Gamma}^{-1}$;

(5) 残差乘以 $\alpha_k \Delta t/\Omega$;

(6) 执行隐式残差光顺 IRS(可选);

(7) 旧的守恒变量 $\boldsymbol{W}^{(0)}$ 减去残差,得到新的守恒变量 $\boldsymbol{W}^{n+1}$;

(8) 更新边界条件(请注意,所有基于特征变量的边界条件,如入流、出流、远场等,都需要根据 $\boldsymbol{P\Gamma}^{-1}\boldsymbol{A}_c$ 进行变化)。

对于隐式格式,按照类似步骤求解预处理控制方程组,只是省略了步骤(5)和步骤(6),且守恒变量以不同的方式进行更新(见 6.2 节)。还应注意,由于雅可比矩阵为 $\boldsymbol{P\Gamma}^{-1}\boldsymbol{A}_c$,隐式算子中必须包含预处理。

预处理标量耗散格式(JST)的形式为(见式(4.50))

$$\begin{aligned}\boldsymbol{D}_{I+1/2}=\hat{\Lambda}^{S}_{I+1/2}(\boldsymbol{\Gamma P}^{-1})_{I+1/2}[&\epsilon^{(2)}_{I+1/2}(\boldsymbol{W}_{I+1}-\boldsymbol{W}_I)-\\ &\epsilon^{(4)}_{I+1/2}(\boldsymbol{W}_{I+2}-3\boldsymbol{W}_{I+1}+3\boldsymbol{W}_I-\boldsymbol{W}_{I-1})]\end{aligned} \tag{9.52}$$

请注意,如果完整残差是在之后乘以 $\boldsymbol{P\Gamma}^{-1}$ 的,则需要 $\boldsymbol{\Gamma P}^{-1}$ 项,因为谱半径 $\hat{\Lambda}^S$ 已包含了预处理矩阵(从而与式(4.53)不同)。还可以将式(9.52)中的 $\boldsymbol{P}^{-1}$ 和 $\boldsymbol{W}$ 合并成 $\boldsymbol{W}_p$,则可以将预处理标量耗散格式用原始变量表示为

$$\begin{aligned}\boldsymbol{D}_{I+1/2}=\hat{\Lambda}^{S}_{I+1/2}[&\epsilon^{(2)}_{I+1/2}(\boldsymbol{W}_{p,I+1}-\boldsymbol{W}_{p,I})-\\ &\epsilon^{(4)}_{I+1/2}(\boldsymbol{W}_{p,I+2}-3\boldsymbol{W}_{p,I+1}+3\boldsymbol{W}_{p,I}-\boldsymbol{W}_{p,I-1})]\end{aligned} \tag{9.53}$$

重要的是要认识到,在这种情况下,只有对流通量乘以 $\boldsymbol{\Gamma}^{-1}$,总残差则乘以 $\boldsymbol{P}$。式(9.53)中的非守恒形式比式(9.52)略简单,但对于内流或包含激波的全速度流动,存在精度方面的问题。

预处理 Roe 迎风格式可以写成(见式(4.91))

$$\begin{aligned}(\boldsymbol{F}_c)_{I+1/2}=\frac{1}{2}[&\boldsymbol{F}_c(\boldsymbol{W}_R)+\boldsymbol{F}_c(\boldsymbol{W}_L)-\\ &(\boldsymbol{\Gamma P}^{-1})_{I+1/2}|\boldsymbol{A}_{Roe}|_{I+1/2}(\boldsymbol{W}_R-\boldsymbol{W}_L)]\end{aligned} \tag{9.54}$$

其中,$|\boldsymbol{A}_{Roe}|=\boldsymbol{T}_{c,p}|\boldsymbol{\Lambda}_{c,p}|\boldsymbol{T}^{-1}_{c,p}$,$(\boldsymbol{T}_{c,p})$,$(\boldsymbol{T}^{-1}_{c,p})$ 和 $(\boldsymbol{\Lambda}_{c,p})$ 为矩阵 $\boldsymbol{P\Gamma}^{-1}\boldsymbol{A}_c=\boldsymbol{P\Gamma}^{-1}\boldsymbol{A}_{c,p}\boldsymbol{P}^{-1}$ 的左、右特征矢量以及特征值。$(I+1/2)$ 处的流动变量值由式(4.89)中的 Roe 平均计算得到,将总残差乘以 $\boldsymbol{P\Gamma}^{-1}$,同样,式(9.54)中的预处理 Roe 格式可以用原始变量 $\boldsymbol{W}_p$ 表示为

$$\begin{aligned}(\boldsymbol{F}_c)_{I+1/2}=\frac{1}{2}[&\boldsymbol{F}_c(\boldsymbol{W}_{p,R})+\boldsymbol{F}_c(\boldsymbol{W}_{p,L})-\\ &\boldsymbol{\Gamma}_{I+1/2}|\boldsymbol{A}_{Roe,p}|_{I+1/2}(\boldsymbol{W}_{p,R}-\boldsymbol{W}_{p,L})]\end{aligned} \tag{9.55}$$

构成 $\boldsymbol{A}_{Roe,p}$ 的特征值和特征矢量由矩阵 $\boldsymbol{\Gamma}^{-1}\boldsymbol{A}_{c,p}$ 确定(见式(9.47))。$\boldsymbol{\Gamma}^{-1}\boldsymbol{A}_{c,p}$ 的特征值与 $\boldsymbol{P\Gamma}^{-1}\boldsymbol{A}_c$ 的特征值(即 $\boldsymbol{\Lambda}_{c,p}$)相同,但两者的特征矢量不同,因此式(9.55)中的 Roe 矩阵可写成 $|\boldsymbol{A}_{Roe,p}|=\boldsymbol{T}_p|\boldsymbol{\Lambda}_{c,p}|\boldsymbol{T}_p^{-1}$。需要指出的是,在这

种情况下,残差必须乘以 $\boldsymbol{P\Gamma}^{-1}$,以便将其转换为守恒变量。关于低马赫数下迎风格式的讨论参见文献[108-109]。

9.5.3 预处理的矩阵形式

在各种原始变量 $\boldsymbol{W}_{\mathrm{p}}$ 的选择中,最常用的是

$$\boldsymbol{W}_{\mathrm{p}}=[p,u,v,w,T]^{\mathrm{T}} \tag{9.56}$$

后面的讨论将限于 $\boldsymbol{W}_{\mathrm{p}}$ 这种特定形式。下面将给出一般流体与完全气体的转换矩阵和预处理矩阵以及矩阵的特征值和左右特征矢量。

1. 转换矩阵

对于一般流体,由守恒变量到原始变量的转换矩阵 $\boldsymbol{P}^{-1}=\partial\boldsymbol{W}_{\mathrm{p}}/\partial\boldsymbol{W}$ 由下式给出[110]:

$$\boldsymbol{P}^{-1}=\begin{bmatrix} \dfrac{\rho h_T+\rho_T(H-q^2)}{a_1} & \dfrac{\rho_T u}{a_1} & \dfrac{\rho_T v}{a_1} & \dfrac{\rho_T w}{a_1} & -\dfrac{\rho_T}{a_1} \\ -\dfrac{u}{\rho} & \dfrac{1}{\rho} & 0 & 0 & 0 \\ -\dfrac{v}{\rho} & 0 & \dfrac{1}{\rho} & 0 & 0 \\ -\dfrac{w}{\rho} & 0 & 0 & \dfrac{1}{\rho} & 0 \\ \dfrac{1-\rho_p(H-q^2)-\rho h_p}{a_1} & -\dfrac{\rho_p u}{a_1} & -\dfrac{\rho_p v}{a_1} & -\dfrac{\rho_p w}{a_1} & \dfrac{\rho_p}{a_1} \end{bmatrix} \tag{9.57}$$

式中

$$\begin{aligned} q^2 &= \|\boldsymbol{v}\|_2^2=u^2+v^2+w^2 \\ a_1 &= \rho\rho_p h_T+\rho_T(1-\rho h_p) \end{aligned} \tag{9.58}$$

由原始变量到守恒变量的转换矩阵,即 $\boldsymbol{P}=\partial\boldsymbol{W}/\partial\boldsymbol{W}_{\mathrm{p}}$ 为[110]

$$\boldsymbol{P}=\begin{bmatrix} \rho_p & 0 & 0 & 0 & \rho_T \\ \rho_p u & \rho & 0 & 0 & \rho_T u \\ \rho_p v & 0 & \rho & 0 & \rho_T v \\ \rho_p w & 0 & 0 & \rho & \rho_T w \\ \rho_p H+\rho h_p-1 & \rho u & \rho v & \rho w & \rho_T H+\rho h_T \end{bmatrix} \tag{9.59}$$

式(9.57)~式(9.59)中的密度和焓关于压力和温度的导数可以写成以下形式:

$$
\begin{aligned}
\rho_p &= \rho\alpha_p \\
\rho_T &= -\rho\alpha_T \\
h_p &= \frac{1-\alpha_T T}{\rho} \\
h_T &= c_p
\end{aligned} \tag{9.60}
$$

式中：α_p 和 α_T 分别为定压和定温下的可压缩系数。声速可以由下式计算：

$$
c^2 = \frac{\rho h_T}{a_1} \tag{9.61}
$$

对于完全气体（见 2.4.1 节），式（9.60）中的压缩系数为 $\alpha_p = 1/p$ 和 $\alpha_T = 1/T$。在此情况下，由式（9.57）表示的守恒变量到原始变量的变换矩阵 $\boldsymbol{P}^{-1} = \partial \boldsymbol{W}_\mathrm{p}/\partial \boldsymbol{W}$ 可写成

$$
\boldsymbol{P}^{-1} = \begin{bmatrix}
(\gamma-1)\frac{q^2}{2} & (1-\gamma)u & (1-\gamma)v & (1-\gamma)w & \gamma-1 \\
-\frac{u}{\rho} & \frac{1}{\rho} & 0 & 0 & 0 \\
-\frac{v}{\rho} & 0 & \frac{1}{\rho} & 0 & 0 \\
-\frac{w}{\rho} & 0 & 0 & \frac{1}{\rho} & 0 \\
\frac{1}{\rho}\left[\frac{\gamma q^2}{2c_p} - T\right] & -\frac{\gamma u}{c_p\rho} & -\frac{\gamma v}{c_p\rho} & -\frac{\gamma w}{c_p\rho} & \frac{\gamma}{c_p\rho}
\end{bmatrix} \tag{9.62}
$$

对于完全气体，原始变量到守恒变量的变换矩阵 $\boldsymbol{P} = \partial \boldsymbol{W}/\partial \boldsymbol{W}_\mathrm{p}$ 为

$$
\boldsymbol{P} = \begin{bmatrix}
\frac{\rho}{p} & 0 & 0 & 0 & -\frac{\rho}{T} \\
\frac{\rho u}{p} & \rho & 0 & 0 & -\frac{\rho u}{T} \\
\frac{\rho v}{p} & 0 & \rho & 0 & -\frac{\rho v}{T} \\
\frac{\rho w}{p} & 0 & 0 & \rho & -\frac{\rho w}{T} \\
\frac{\rho E}{p} & \rho u & \rho v & \rho w & -\frac{\rho q^2}{2T}
\end{bmatrix} \tag{9.63}
$$

其中，q^2 定义如式（9.58）所示。

2. 预处理矩阵

对于欧拉方程组,预处理矩阵的构造相对容易。van Leer 等[111]提出的公式实现了最低可达到的特征条件数,文献[91]对此方法进行了详细讨论。但纳维–斯托克斯方程组的预处理更为复杂,其原因是黏性项引起了复杂的波速,使预处理系统难以分析。目前,得到大家认可的黏性流动预处理器为分别由 Choi 和 Merkle[112-113]、Turkel[107,114-116]、Lee 和 van Leer[117-118]、Lee[92]、Jorgenson 和 Pletcher[119]、Weiss 和 Smith[48,120]等提出的预处理器,应用实例可参见文献[110,121-135]。

1) Weiss 和 Smith 提出的预处理器

对于一般流体,Weiss 和 Smith[48,120]提出的预处理矩阵 $\boldsymbol{\Gamma}$ 的形式与式(9.59)中的 $\boldsymbol{P}$ 相同,用合适的预处理参数 θ 代替 ρ_p。这同样适用于预处理矩阵的逆矩阵 $\boldsymbol{\Gamma}^{-1}$,它类似于式(9.57)中的 $\boldsymbol{P}^{-1}$。因此,式(9.58)中的参数 a_1 变为

$$\alpha_1^{\Gamma}=\rho\theta h_T+\rho_T(1-\rho h_p) \tag{9.64}$$

对于完全气体,预处理矩阵 $\boldsymbol{\Gamma}$ 可表示为

$$\boldsymbol{\Gamma}=\begin{bmatrix} \theta & 0 & 0 & 0 & -\dfrac{\rho}{T} \\ \theta u & \rho & 0 & 0 & -\dfrac{\rho u}{T} \\ \theta v & 0 & \rho & 0 & -\dfrac{\rho v}{T} \\ \theta w & 0 & 0 & \rho & -\dfrac{\rho w}{T} \\ \theta H-1 & \rho u & \rho v & \rho w & -\dfrac{\rho q^2}{2T} \end{bmatrix} \tag{9.65}$$

式中:θ 同样为预处理参数(稍后将对其进行定义)。预处理矩阵的逆矩阵为

$$\boldsymbol{\Gamma}^{-1}=\begin{bmatrix} a_2\left[\dfrac{c^2}{\gamma-1}-(H-q^2)\right] & -a_2u & -a_2v & -a_2w & a_2 \\ -\dfrac{u}{\rho} & \dfrac{1}{\rho} & 0 & 0 & 0 \\ -\dfrac{v}{\rho} & 0 & \dfrac{1}{\rho} & 0 & 0 \\ -\dfrac{w}{\rho} & 0 & 0 & \dfrac{1}{\rho} & 0 \\ a_3[1-\theta(H-q^2)] & -\theta a_3u & -\theta a_3v & -\theta a_3w & \theta a_3 \end{bmatrix} \tag{9.66}$$

式中

$$a_2=(\gamma-1)\phi$$

$$a_3=\frac{(\gamma-1)\phi T}{\rho} \tag{9.67}$$

$$\phi=\frac{1}{\theta c^2-(\gamma-1)}$$

式(9.64)~式(9.67)中的预处理参数 θ 有几种选择。文献[48]中提供的是一种基于参考速度 u_r 的定义,对于一般流体,θ 为

$$\theta=\frac{1}{u_r^2}-\frac{\rho_T}{\rho h_T} \tag{9.68}$$

而对于完全气体

$$\theta=\frac{1}{u_r^2}+(\gamma-1)\frac{1}{c^2} \tag{9.69}$$

式(9.68)或式(9.69)中的参考速度 u_r 由下面关系式求得:

$$u_r=\min\left[\max\left(\|\boldsymbol{v}\|_2,\frac{v}{\Delta h},\frac{k}{\Delta h},\epsilon\sqrt{\frac{|\Delta p|}{\rho}}\right),c\right] \tag{9.70}$$

式中:Δh 为控制体大小的度量;Δp 为相邻控制体之间的压差;ϵ 为一个小量($\approx 10^{-3}$)。由式(9.70)的定义可以看出,参考速度受当地输运速度的限制,$v/\Delta h$ 和 $k/\Delta h$ 两项在扩散或热传导占主导的边界层内变得非常重要。压力项旨在防止驻点处 u_r 为零。当 $u_r=c$ 时,$\boldsymbol{\Gamma}$ 与 $\boldsymbol{P}$ 完全相同,$\boldsymbol{\Gamma}^{-1}$变为 $\boldsymbol{P}^{-1}$,因此,对于超声速流动,和所期望的一样,预处理被关闭。

另一种针对完全气体而设计的预处理参数 θ 的定义为[123]

$$\theta=\frac{1}{\beta\gamma RT}=\frac{1}{\beta c^2} \tag{9.71}$$

式(9.71)中的参数 β 定义为

$$\beta=\frac{M_r^2}{1+(\gamma-1)M_r^2} \tag{9.72}$$

其中,参考马赫数由下式给出:

$$M_r^2=\max[\min(M^2,1),M_{\min}^2] \tag{9.73}$$

M 为当地马赫数($M^2=\|\boldsymbol{v}\|_2^2/c^2$)。根据文献[123],参数 $M_{\min}^2=KM_\infty^2$,而 $K\approx 3$(也可选择 $K=1$,甚至 $K=0.15$;它的精确值似乎取决于驻点区域或边界层内控制体的数量)。很容易验证,当 $M\geqslant 1$ 时,参数 $\beta=1/\gamma$,矩阵 $\boldsymbol{\Gamma}$ 与 $\boldsymbol{P}$ 相同。

值得注意的是，式(9.69)～式(9.71)中 θ 的定义是等价的。因此，利用式(9.70)可以确定参考马赫数为 $M_r=u_r/c$。

2) 预处理系统的特征值

预处理系统方程组(9.49)，即 $\boldsymbol{P\Gamma}^{-1}\boldsymbol{A}_c=\boldsymbol{P\Gamma}^{-1}\boldsymbol{A}_{c,p}\boldsymbol{P}^{-1}$ 的特征值矩阵为

$$\boldsymbol{\Lambda}_{c,p}=\begin{bmatrix} V & 0 & 0 & 0 & 0 \\ 0 & V & 0 & 0 & 0 \\ 0 & 0 & V & 0 & 0 \\ 0 & 0 & 0 & \dfrac{(a_4+1)}{2}V+c' & 0 \\ 0 & 0 & 0 & 0 & \dfrac{(a_4+1)}{2}V-c' \end{bmatrix} \tag{9.74}$$

式中：$V=\boldsymbol{v}\cdot\boldsymbol{n}$ 表示逆变速度；c'为修正后的声速：

$$c'=\frac{1}{2}\sqrt{V^2(a_4-1)^2+4a_5} \tag{9.75}$$

一般情况下，式(9.74)～式(9.75)中的参数 a_4 和 a_5 为

$$a_4=\frac{a_1}{a_1^\Gamma}$$
$$a_5=\frac{\rho h_T}{a_1^\Gamma} \tag{9.76}$$

其中，a_1 由式(9.58)确定，a_1^Γ 由式(9.64)确定。在完全气体条件下，预处理参数 θ 由式(9.69)确定，参数变为 $a_4=\phi$ 和 $a_5=\phi c^2$，其中 ϕ 由式(9.67)给出。当使用式(9.71)对 θ 的第二种定义时，式(9.76)中的参数分别简化为 $a_4=M_r^2$ 和 $a_5=M_r^2c^2$。可以看出，当 $|M|\to 0$ 时，$c'\approx(V/2)\sqrt{5}$，其特征值如预期一样变成同一量级，因而代表刚性的式(9.40)的特征条件数减小($C_N\approx 1.62$)，时间步进或迭代求解过程的收敛性显著增强。另一方面，当 $|M|\geqslant 1$ 时，$a_4=1$，$c'=c$，重新获得 $\boldsymbol{A}_c$ 的特征值。

根据式(9.74)，预处理系统的谱半径变为

$$\hat{\Lambda}_{c,p}=\left[\frac{(a_4+1)}{2}|V|+c'\right]\Delta S \tag{9.77}$$

其中，ΔS 表示面积。使用该式计算式(6.14)、式(6.18)、式(6.20)或式(6.22)中的时间步长。在中心人工耗散情况下，该式也可代替式(4.53)中的谱半径(见式(5.33)、式(9.52)～式(9.53))。

3）预处理系统的特征矢量

由于使用预处理方法，对流通量雅可比矩阵（见 A. 11 节）的左右特征矢量将被改变，这对于基于特征变量的空间离散非常重要，如 Roe 迎风格式（4. 3. 3 节）或 4. 3. 4 节介绍的迎风 TVD 格式。

式（9. 47）中新的预处理通量雅可比 $\boldsymbol{\Gamma}^{-1}\boldsymbol{A}_{c,p}$ 的左特征矢量矩阵可写成[110]

$$\boldsymbol{T}_p^{-1}=\begin{bmatrix} -\dfrac{\rho_T n_x}{a_1^\Gamma} & 0 & \dfrac{a_6 n_z}{c'} & -\dfrac{a_6 n_y}{c'} & -\dfrac{a_6 n_x}{T} \\ -\dfrac{\rho_T n_y}{a_1^\Gamma} & -\dfrac{a_6 n_z}{c'} & 0 & \dfrac{a_6 n_x}{c'} & -\dfrac{a_6 n_y}{T} \\ -\dfrac{\rho_T n_z}{a_1^\Gamma} & \dfrac{a_6 n_y}{c'} & -\dfrac{a_6 n_x}{c'} & 0 & -\dfrac{a_6 n_z}{T} \\ \dfrac{1}{2}+a_7 & \dfrac{a_6 n_x}{2c'} & \dfrac{a_6 n_y}{2c'} & \dfrac{a_6 n_z}{2c'} & 0 \\ \dfrac{1}{2}-a_7 & -\dfrac{a_6 n_x}{2c'} & -\dfrac{a_6 n_y}{2c'} & -\dfrac{a_6 n_z}{2c'} & 0 \end{bmatrix} \tag{9.78}$$

如式（9. 55）所示，$\boldsymbol{\Gamma}$ 乘以 $\boldsymbol{\Gamma}^{-1}\boldsymbol{A}_{c,p}$ 的右特征矢量矩阵，得[110]

$$\boldsymbol{\Gamma}\boldsymbol{T}_p=\frac{a_1^\Gamma}{\rho h_T}\begin{bmatrix} -a_8 n_x & -a_8 n_y & -a_8 n_z & 1 & 1 \\ -a_8 u n_x & -a_8 u n_y-c' n_z & -a_8 u n_z+c' n_y & u+(a_9+c')n_x & u+(a_9-c')n_x \\ -a_8 v n_x+c' n_z & -a_8 v n_y & -a_8 v n_z-c' n_x & v+(a_9+c')n_y & v+(a_9-c')n_y \\ -a_8 w n_x-c' n_y & -a_8 w n_y+c' n_x & -a_8 w n_z & w+(a_9+c')n_z & w+(a_9-c')n_z \\ a_{11}-a_{10} n_x & a_{12}-a_{10} n_y & a_{13}-a_{10} n_z & H+(a_9+c')V & H+(a_9-c')V \end{bmatrix} \tag{9.79}$$

其中，a_1^Γ 由式（9. 64）定义，式（9. 78）~式（9. 79）中的其他参数为

$$a_6=\rho a_5$$

$$a_7=\frac{V(a_4-1)}{4c'}$$

$$
\begin{aligned}
a_8 &= \frac{\rho_T T}{\rho} \\
a_9 &= -\frac{1}{2}V(a_4-1) \\
a_{10} &= a_8 H + T h_T \\
a_{11} &= c'(v n_z - w n_y) \\
a_{12} &= c'(w n_x - u n_z) \\
a_{13} &= c'(u n_y - v n_x)
\end{aligned} \tag{9.80}
$$

其中，a_4、a_5 由式(9.76)定义，c'的定义见式(9.75)。

以守恒变量表示的预处理 Roe 格式(见式(9.54))所需的矩阵 $\boldsymbol{P\Gamma}^{-1}\boldsymbol{A}_{\mathrm{c}}=\boldsymbol{P\Gamma}^{-1}\boldsymbol{A}_{\mathrm{c,p}}\boldsymbol{P}^{-1}$ 的特征矢量 $\boldsymbol{T}_{\mathrm{c,p}}$ 和 $\boldsymbol{T}_{\mathrm{c,p}}^{-1}$，可以通过对上述特征矢量式(9.78)～式(9.79)的修改得到。可以证明，$\boldsymbol{P\Gamma}^{-1}\boldsymbol{A}_{\mathrm{c}}$ 的右特征矢量矩阵的各列是由 $\boldsymbol{Px}$ 构成的，其中 $\boldsymbol{x}$ 是 $\boldsymbol{\Gamma}^{-1}\boldsymbol{A}_{\mathrm{c,p}}$ 的右特征矢量。因此，$\boldsymbol{P\Gamma}^{-1}\boldsymbol{A}_{\mathrm{c}}$ 的左右特征矢量矩阵可以表示为

$$
\begin{aligned}
\boldsymbol{T}_{\mathrm{c,p}} &= \boldsymbol{P}\boldsymbol{T}_{\mathrm{p}} \\
\boldsymbol{T}_{\mathrm{c,p}}^{-1} &= \boldsymbol{T}_{\mathrm{p}}^{-1}\boldsymbol{P}^{-1}
\end{aligned} \tag{9.81}
$$

将式(9.81)代入用守恒变量表示的预处理 Roe 格式(式(9.54))中，得

$$
\begin{aligned}
(\boldsymbol{F}_{\mathrm{c}})_{I+1/2} = \frac{1}{2}[&\boldsymbol{F}_{\mathrm{c}}(\boldsymbol{W}_{\mathrm{R}}) + \boldsymbol{F}_{\mathrm{c}}(\boldsymbol{W}_{\mathrm{L}}) - \\
&(\boldsymbol{\Gamma T}_{\mathrm{p}}\,|\boldsymbol{\Lambda}_{\mathrm{c,p}}|\,\boldsymbol{T}_{\mathrm{p}}^{-1}\boldsymbol{P}^{-1})_{I+1/2}(\boldsymbol{W}_{\mathrm{R}}-\boldsymbol{W}_{\mathrm{L}})]
\end{aligned} \tag{9.82}
$$

其中，$\boldsymbol{\Gamma T}_{\mathrm{p}}$ 和 $\boldsymbol{T}_{\mathrm{p}}^{-1}$ 分别由式(9.79)和式(9.78)给出。

图9.11～图9.13 给出了式(9.51)(其中 $\boldsymbol{\Gamma}$ 根据式(9.65)计算)的预处理有限体积方法应用于翼型流动的一个实例。由图可以看出，无论是中心格式还是 Roe 迎风格式，都无法得到来流马赫数为 0.01 时的正确解。不采用预处理的格式无法预测压力分布，因而也无法预测升力系数(预测的 $C_{\mathrm{L}}=0.323$ 和 0.324，而正确的升力系数为 0.352)。如图 9.13 所示，预处理有助于得到正确解($C_{\mathrm{L}}=0.353$)，并且显著加快了收敛速度。

请注意，本书附带的 2D 解算器源代码的中心格式以及 Roe 迎风格式都采用了低速预处理技术。

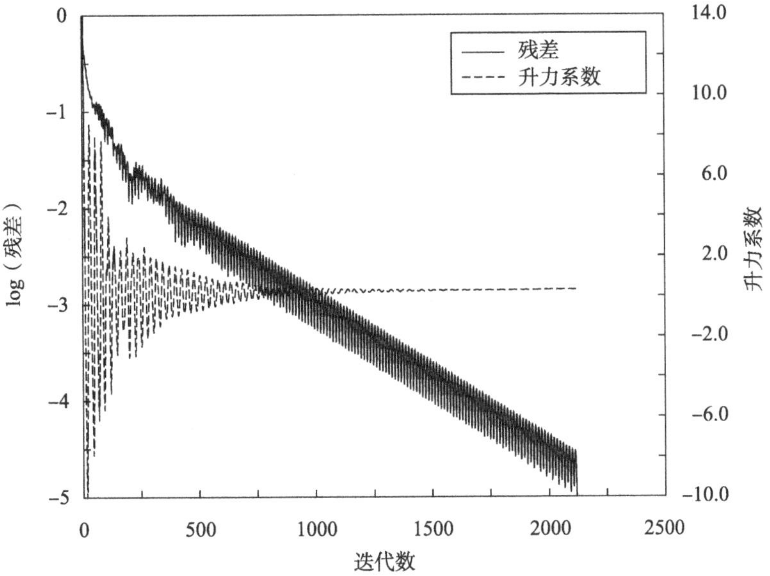

（a）

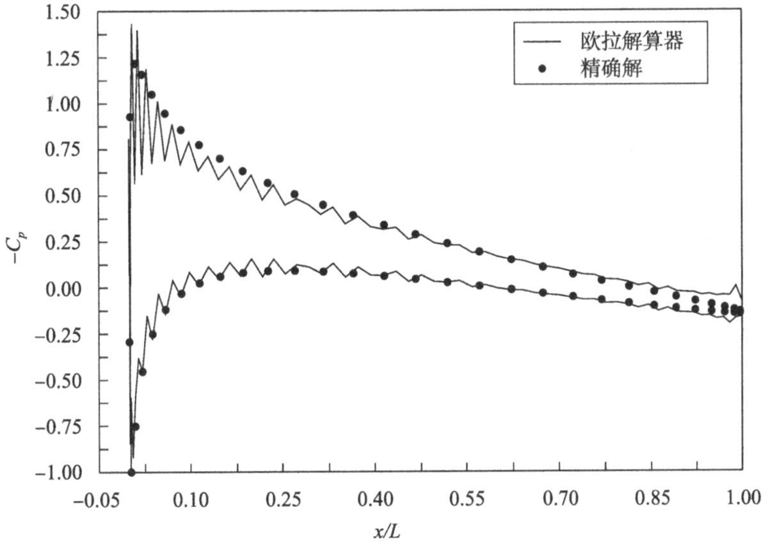

（b）

图 9.11　对称 Joukowsky 翼型（10%厚度）的无黏 2D 绕流。结构网格，$M_\infty=10^{-2}$，$\alpha=3°$，中心空间离散，显式多步时间推进方法，无预处理

（a）收敛历程；（b）欧拉解算器计算的压力系数与精确势流解的比较。

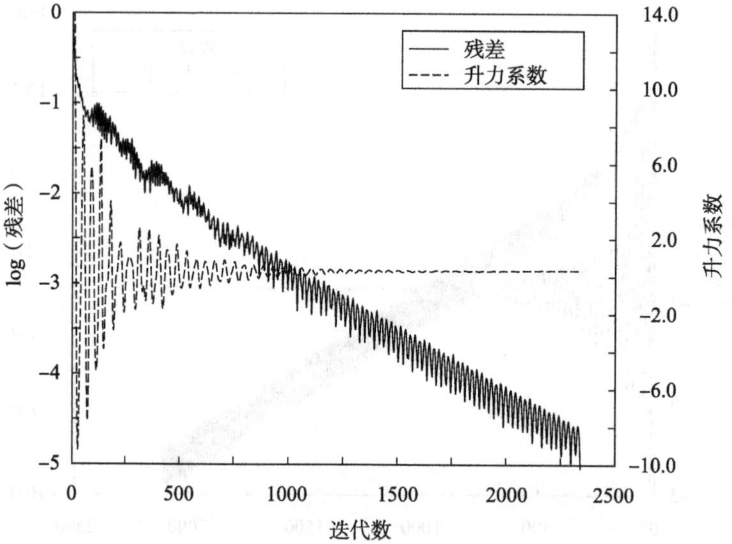

（a）

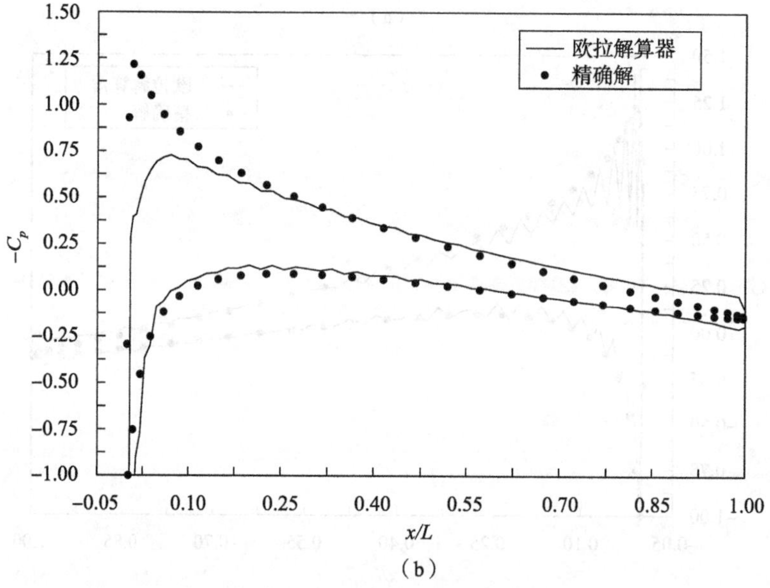

（b）

图 9.12　对称 Joukowsky 翼型（10%厚度）的无黏 2D 绕流。结构网格，$M_\infty = 10^{-2}$，$\alpha = 3°$，二阶 Roe 迎风离散，显式多步时间推进方法，无预处理

（a）收敛历程；（b）欧拉解算器计算的压力系数与精确势流解的比较。

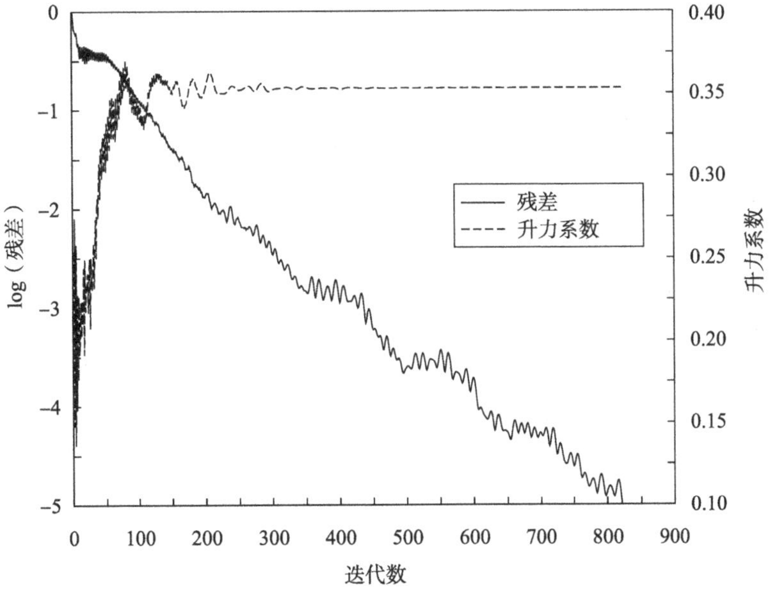

（a）

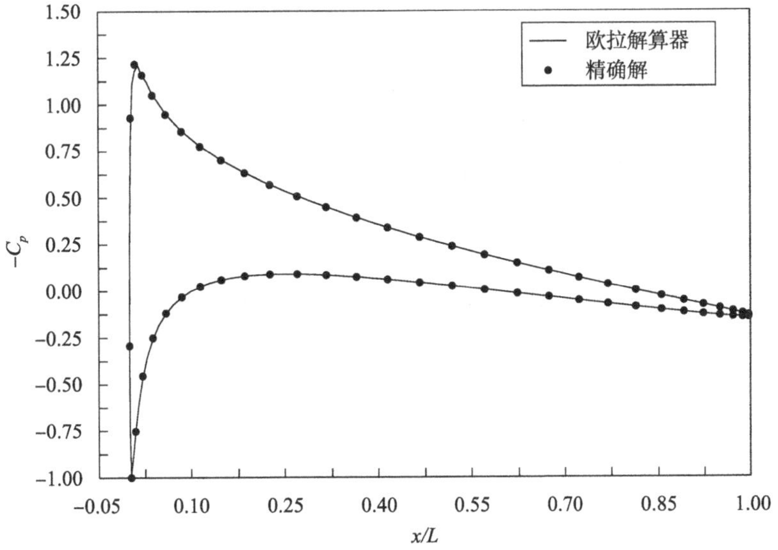

（b）

图9.13 对称Joukowsky翼型(10%厚度)的无黏2D绕流。结构网格，$M_\infty = 10^{-2}$，$\alpha = 3°$，中心空间离散，显式多步时间推进方法，Weiss-Smith预处理

(a) 收敛历程；(b) 欧拉解算器计算的压力系数与精确势流解的比较。

9.6 并行化

如今，包括台式机、笔记本电脑、平板电脑等在内的几乎所有计算机，都有由多核组成的中央处理器(CPU)。此外，现代的图形处理器(GPU)也提供了数百甚至数千个通用可编程处理器。然而，要利用显著提高的计算性能，需要设计适当的软件。虽然并行化与任何特定的数值方法没有直接联系，但它可能会影响数值方法的选择，且必定会影响数值方法的实现。

目前存在数种编程模式可用于对流体力学代码进行并行化。首先是在程序层级上工作的消息传递接口(MPI)，在并行任务之间使用消息传递的还有异步代理库(asynchronous agents library)①，但与 MPI 相比，其命令集受到更多限制[136]。其次，像 OpenMP、CUDA、OpenCL 或 OpenACC[137] 等方法则属于更细粒度的方法，它们在函数或循环层级上工作。由微软提出，于 2012 年底发布的加速大规模并行计算(C++ AMP)编程模型[138]也是如此。可以将诸如 MPI 与其他方法之一相结合，以实现最大加速比，当然，这通常取决于可用的硬件的能力。

下面，将对其中四种方法进行更深入的探讨。为了说明方法的实际用途，我们将使用一个简单的程序，它通过雅可比迭代求解具有固定边值的 2D 拉普拉斯方程。为了展示归约操作，还计算了残差的 2 范数。在附带软件的目录 Parallelization 中提供了示例程序的源代码，源代码用 C 和 C++编写，有关源代码的组织架构和编译的更多细节，请参见第 12 章和 12.8 节。

9.6.1 MPI

MPI 已经成为并行化用 C/C++或 Fortran 编写的数值代码的经典编程模型。基本思想是将计算网格分解成与处理器(或 CPU 核)数量相同的块，将每块网格分配给一个处理器，然后并行计算流场解。用 C 语言编写的一个非常简单的程序如下：

```
#include <mpi.h>
 ...
 MPI_Init( &argc, &argv );                    // 初始化 MPI
 MPI_Comm_rank( MPI_COMM_WORLD, &myProc );// 单独进程号
 MPI_Comm_size( MPI_COMM_WORLD, &nProcs );// 总进程数
 ...
```

① 在适用的情况下，名称是其各自所有者的注册商标。

```
err = ReadData( myProc );
…
err = Solve( myProc );
…
err = SaveSolution( myProc );
…
MPI_Finalize();                          // 完成所有 MPI 运算
…
```

由于一个网格块——称为分区——中的流动通常与其他分区中的流动相关,并且由于每个进程都使用自己的专用内存(分布式内存模型),因此必须在处理器之间交换数据。MPI 为此提供了对进程间通信的显式控制。消息——流场变量、几何数据、残差等——可由分区发送(MPI_Send、MPI_Isend)和接收(MPI_Recv),可以从其他所有处理器收集数据(MPI_Allreduce——例如,残差范数、力和力矩),也可以将数据从一个分区广播至所有其他分区(MPI_Bcast 或 MPI_Scatter),另一种重要的 MPI 通信控制是进程同步(MPI_Barrier)。有关编程接口、库的下载和示例的介绍,请参见文献[139],文献[140]中有非常值得推荐的教程。

尽管 MPI 的概念看似很简单,但还存在一些具体的问题。第一个问题是如何最好地将给定的网格分割成不同的块。为了获得良好的并行效率,至关重要的是,每个处理器的任务量要大致相同,这称为负载平衡。对于结构网格,计算量与控制体数成正比,即与网格单元数(格心格式)或网格节点数(格点格式)成正比。如果网格由多个网格块组成,则这些网格块构成了分区的基础,建议沿块的最大维度来划分,以使相邻分区共享的面(节点)数量最少(以减少通信数据量)。

对于非结构网格,计算量还与分区内的边数(中点-二次格式)或单元的面数(格心格式)有关,对于空间精度阶数大于 2 的格式,计算量还取决于最近的相邻单元(节点)数。由于节点和单元为无序编号,非结构网格的分区会变得非常复杂。幸运的是,有一些库可用于网格分区,(以作者的观点看)最实用的一个是 Metis[141],除了快速可靠之外,Metis 还提供了许多用于优化负载平衡的选项。

网格分区至少沿其公共边界必须进行数据交换。根据划分以及基本的离散方法,可能需要在分区之间进行通信,这些分区在单元边界,甚至是在单个节点处相互连接。结构多块流场解算器需要的相应逻辑(见 8.9 节)在并行化之前已经存在。而非结构网格的解算器为了能够适应 MPI 的执行,需要一个合适的程序设计,或者对现存软件进行大量的重新设计,这与可以持续合并进代码的其他并行方法有很大的不同。

哪些数据必须在分区之间进行通信？下面以 5.2.2 节中的非结构中点-二次格式为例进行说明。首先,在每个分区中单独计算的控制体在跨进程的边界处是不完整的,因此需要对共享给定边界节点的所有分区的(部分)控制体进行求和,如前所述,这可能需要在仅有单个公共节点的分区之间进行通信。MPI 不提供任何搜索相关分区的功能,该搜索功能以及全部必要的逻辑需要由程序开发人员提供。

此外,高阶空间离散和/或纳维-斯托克斯方程组的求解需要计算流场梯度和限制器。在这两种情况下,连接边界节点的所有边的节点处的流动变量都必须为已知的,因此必须从周围的分区进行流动变量的通信(参见示例代码 Unstruct2D 中的周期边界之间的数据传输)。当计算了残差之后,就必须对边界节点处的残差求和,如上文对控制体求和一样。

显式时间推进格式(见 6.1 节)不需要在分区之间进行任何通信。但是,IRS(见 9.3.2 节)与显式格式不同,它与所有隐式格式一样,需要在相邻分区之间至少有近似的耦合形式,否则格式的收敛性和稳定性将受到很大影响。对于应用广泛的 LU-SGS 格式(见 6.2.4 节),一种可能方法是沿边界进行额外的光顺扫描[142]。6.2.5 节讨论的 Newton-Krylov 方法需要修改它的预处理器,因为系统矩阵被分解到了不同的处理器上,最常用的方法是 additive Schwarz 方法,它在分区之间使用重叠区域[143-146]。

PETSc 库(可移植可扩展的科学计算工具包)[147]有助于流场解算器的并行化,它包含一系列并行的线性、非线性方程解算器和时间积分器,使用 MPI,并可由 C/C++或 Fortran 编写的基于 MPI 的应用程序调用。

9.6.2 OpenMP

OpenMP 是为共享内存的、用 C/C++和 Fortran 编写的并行程序设计的[148]。它将计算任务拆分为多个线程,然后各线程并行执行,同时共享应用程序内存的公共部分。OpenMP 提供了实现粗粒度和细粒度并行的能力,与 MPI 的消息传递模式不同,它使程序员能够以增量方式并行化一个串行程序。

OpenMP 提供了一系列简单且有限的编译器指令,因此只需使用几行代码即可实现重要的并行性。此处以附带的软件包中 Parallelization 目录下提供的演示程序 openmp 的(简化)主循环为例:

```
#include <omp.h>
 ...
 #pragma omp parallel default(none) shared(phi, phiOld, Ni, Nj)
                      private(i, j, ij)
```

```
{
  …
  #pragma omp for schedule(static) nowait
  for (j=1; j<(Nj.1); j++)
  {
    for (i=1; i<(Ni-1); i++)
    {
      ij = i + j* Ni;
      phi[ij]= 0.25* (phiOld[ij+1]+phiOld[ij-1]+
                       phiOld[ij+Ni]+phiOld[ij-Ni]);
    }
  }
  …
} // parallel region
```

并行化 j 循环只需要两条编译器指令(#pragma omp)。第一条指令定义了一个并行区域,其中包含了要并行执行的代码,在本示例中,它还声明了所有线程共享的变量。第二条指令建立一个并行 for 循环,循环本身与串行代码保持不变,只有在相应的编译器选项启动时才会执行指令(例如,在 GNU 编译器中的-fopenmp),否则代码将以串行方式执行。

OpenMP 相对简单,因而存在一定的缺陷。例如,没有提供类似于 MPI 的并行 I/O,而是由程序员来处理。正如前面提到的,可以将 OpenMPI 和 MPI 相结合,以实现最优的并行效率。多物理场模拟代码对将两种编程模型相结合的方法尤为感兴趣,如最近演示的粒子流体流动[149]。

9.6.3 CUDA

CUDA 是一个 C/C++和 Fortran 并行编程和计算平台[150],公开发布于 2007 年。CUDA 为代码开发人员提供了可直接访问 NVIDIA GPU 的并行计算单元的虚拟指令集和内存的机会,这样 GPU 就可以用于进行通用处理,这种方法被称为 GPGPU。GPU 有一个并行收集架构,优化后可以相对缓慢地执行多个并发线程,而不是像 CPU 那样快速地执行单个线程。因此,如果一个计算任务可以被分割成许多并行线程,则它可以被大大加速。

CUDA 工具包(可在文献[150]给出的网站免费获得)包括一个编译器、数学库(如用于基本线性代数函数的 CUBLAS),以及用于调试和优化应用程序性能的工具。CFD 程序员感兴趣的还有名为 CUSP 的基于 CUDA 的模板库[151],该库包含了用于稀疏线性系统的各种解算器(如 GMRES),可在文献[152]给出的网站免费下载。文献[153]给出的网站上有使用 CUDA 加速的 CFD 代码的详细介绍。

CUDA 平台可以通过对标准编程语言 C、C++和 Fortran 的拓展来访问。例如，下面的例子来自演示代码 cuda，它位于源代码的 Parallelization 目录中。

```
#include <cuda_runtime.h>
#include <device_launch_parameters.h>


// Body of the kernel function
__global__ void Jacobi(int Ni, int Nj, float * phi, float * phiOld)
{
  int i = blockDim.x * blockIdx.x + threadIdx.x; // i, j indexes
  int j = blockDim.y * blockIdx.y + threadIdx.y; // in phi, phiOld
  ...
  int ij = i + j* Ni;
  phi[ij]= 0.25f* (phiOld[ij+1]+phiOld[ij-1]+
  phiOld[ij+Ni]+phiOld[ij-Ni]);

}
...
// calling kernel function to execute on the GPU
  Jacobi<<<blocksPerGrid,threadsPerBlock>>>(Ni, Nj, phiDev, phiDevOld);
  ...
```

CUDA 允许编程人员定义一种 kernel 函数，然后 kernel 函数由 N 个不同的 CUDA 线程并行执行 N 次。如前面的 Jacobi 函数声明一样，通过使用__global__限定语定义 kernel 函数。在 GPU 上执行特定 kernel 函数的块数和线程数由执行配置指定，执行配置是一个插入到函数名和带括号的参数列表之间、形式为<<<…>>>的运算符。每个执行 kernel 函数的线程都有唯一的索引，可以通过内置的 threadIdx 变量在 kernel 中访问该索引。为了简化 2D 和 3D 网格或矩阵的处理，可以使用分量为 x,y 和 z 的结构 dim3 为每个空间方向指定不同的块数和线程数，它也被用于上面在物理空间和计算空间中使用 2D 网格的例子。

CUDA 的计算效率、CUBLAS 或 CUSP 等优化库的可用性，以及开发和调试工具都非常具有吸引力。该工具包的一大优势还在于可以混合自身和 CUDA 的代码，两者可以一起编译。CUDA 的一个缺点是它对 NVIDIA 的图形硬件具有依赖性，因此，如果你的 GPU 来自 ATI，或者你打算编写一个可移植的软件，将不得不考虑采用 C++ AMP 或下面将要介绍的 OpenCL。

9.6.4 OpenCL

OpenCL的开发由苹果公司发起，它为包括CPU和GPU在内的异构系统的并行编程提供了一个统一的环境[154]。OpenCL将计算系统视为计算设备的一个组成部分，这些设备可能是CPU或GPU之类的“加速器”，它使用一种类似于C的语言（OpenCL C）来编写在计算设备上执行的程序，这种程序称为kernel。每个计算设备通常由许多个单独的处理单元组成（GPU通常有数百或数千个处理器），kernel在处理单元上并行执行。用于编写kernel的编程语言基于C99，但对它进行了调整以适应OpenCL中的设备模型，它省略了函数指针、位字段和可变长度数组，还禁止使用递归。C标准库被一组用于数学编程的自定义函数所取代。

此外，OpenCL定义了一个应用程序编程接口（API），它允许在主机上执行的程序在计算设备上启动kernel并管理设备内存，设备可与主机CPU共享或不共享内存。API提供了设备内存缓冲区的句柄，以及在主机和计算设备之间来回传输数据的函数。OpenCL标准定义了用于C和C++的API，但是其他编程语言（如Python或Java）也有第三方的API。OpenCL C是为运行时编译设计的，因而应用程序在各种设备的实现之间是可移植的，然而，这意味着OpenCL应用程序必须随纯文本格式的内核代码一起安装，这对于专用软件来说是一大难题。

参考文献

[1] Kroll N, Jain RK. Solution of two-dimensional Euler equations—experience with a finite volume code. DFVLR-FB 87-41; 1987.

[2] Mavriplis DJ. On convergence acceleration techniques for unstructured meshes. AIAA Paper 98-2966; 1998.

[3] Jameson A. Iterative solution of transonic flow over airfoil and wings including at Mach 1. Commun Pure Appl Math 1974; 27: 283-309.

[4] Jameson A, Schmidt W, Turkel E. Numerical solutions of the Euler equations by finite volume methods using Runge-Kutta time-stepping schemes. AIAA Paper 81-1259; 1981.

[5] van Leer B, Tai CH, Powell KG. Design of optimally smoothing multi-stage schemes for the Euler equations. AIAA Paper 89-1933; 1989.

[6] Tai C-H, Sheu J-H, van Leer B. Optimal multistage schemes for Euler equations with residual smoothing. AIAA J 1995; 33: 1008-16.

[7] Tai C-H, Sheu J-H, Tzeng P-Y. Improvement of explicit multistage schemes for central spatial discretization. AIAA J 1996; 34: 185-8.

[8] Jameson A, Baker TJ. Solution of the Euler equations for complex configurations. AIAA Paper 83-1929;1983.

[9] Enander R, Karlsson AR. Implicit explicit residual smoothing in multigrid cycle. AIAA Paper 95-0204;1995.

[10] Enander R, Sjöreen B. Implicit explicit residual smoothing for upwind schemes. Uppsala University, Report No. 179;1996.

[11] Martinelli L. Calculation of viscous flows with a multigrid method. PhD Thesis, Dept. of Mechanical and Aerospace Eng. Princeton University;1987.

[12] Turkel E, Swanson RC, Vatsa VN, White JA. Multigrid for hypersonic viscous two- and three-dimensional flows. AIAA Paper 91-1572;1991.

[13] Radespiel R, Kroll N. Multigrid schemes with semicoarsening for accurate computations of hypersonic viscous flows. DLR Internal Report, No. 129-90/19;1991.

[14] Jameson A, Baker TJ, Weatherill NP. Calculation of inviscid transonic flow over a complete aircraft. AIAA Paper 86-0103;1986.

[15] Blazek J, Kroll N, Radespiel R, Rossow C-C. Upwind implicit residual smoothing method for multi-stage schemes. AIAA Paper 91-1533;1991.

[16] Blazek J. Methods to accelerate the solutions of the Euler- and Navier-Stokes equations for steady-state super- and hypersonic flows. Translation of DLR Research Report 94-35, ESA-TT-1331;1995.

[17] Blazek J, Kroll N, Rossow C-C. A comparison of several implicit smoothing methods. In: ICFD conf. numerical methods for fluid dynamics. Reading, England;1992.

[18] Blazek J, Kroll N, Rossow C-C, Swanson RC. A comparison of several implicit residual smoothing methods in combination with multigrid. In: 13th int. conf. on numerical methods in fluid dynamics. Roma, Italy;1992.

[19] Brandt A. Guide to multigrid development. Multigrid methods I. Lecture Notes in Mathematics No. 960. New York: Springer Verlag;1981.

[20] Jameson A. Solution of the Euler equations for two-dimensional transonic flow by a multigrid method. MAE Report No. 1613, Dept. of Mechanical and Aerospace Engineering. Princeton University;1983.

[21] Jameson A. Solution of the Euler equations by a multigrid method. Appl Math Comput 1983; 13:327-56.

[22] Jameson A. Multigrid algorithms for compressible flow calculations. Multigrid Methods II, Lecture Notes in Mathematics No. 1228. New York: Springer Verlag;1985.

[23] Schröer W, Höel D. An unfactored implicit scheme with multigrid acceleration for the solution of the Navier-Stokes equations. Comput Fluids 1987;15:313-20.

[24] Koren B. Multigrid and defect correction for the steady Navier-Stokes equations. J Comput Phys 1990;87:25-46.

[25] Thomas JL. An implicitmultigrid scheme for hypersonic strong-interaction flowfields. In: 5th

Copper Mountain conf. on multigrid methods. Denver, USA; 1991.

[26] Radespiel R, Swanson RC. Progress with multigrid schemes for hypersonic flow problems. ICASE Report, No. 91-89; 1991 (also J Comput Phys 1995; 116: 103-22).

[27] Mavriplis DJ, Venkatakrishnan V. A unified multigrid solver for the Navier-Stokes equations on mixed element meshes. ICASE Report No. 95-53; 1995.

[28] Mavriplis DJ, Venkatakrishnan V. A 3D agglomeration multigrid solver for the Reynolds-averaged Navier-Stokes equations on unstructured meshes. Int J Numer Meth Fluids 1996; 23: 527-44.

[29] Braaten ME, Connell SD. Three-dimensional unstructured adaptive multigrid scheme for the Navier-Stokes equations. AIAA J 1996; 34: 281-90.

[30] Mavriplis DJ. Multigrid strategies for viscous flow solvers on anisotropic unstructured meshes. J Comput Phys 1998; 145: 141-65.

[31] Haselbacher AC. A grid-transparent numerical method for compressible viscous flows on mixed unstructured grids. PhD Thesis, Loughborough University, England; 1999.

[32] Wasserman M, Mor-Yossef Y, Yavneh I, GreenbergJB. A robust implicit multigrid method for RANS equations with two-equation turbulence models. J Comput Phys 2010; 229: 5820-42.

[33] Yoon S, Jameson A. A multigrid LU-SSOR scheme for approximate Newton-iteration applied to the Euler equations. NASA-CR-17954; 1986.

[34] Yoon S, Jameson A. Lower-upper implicit schemes with multiple grids for the Euler equations. AIAA Paper 86-0105; 1986 (also AIAA J 1987; 7: 929-35).

[35] Yoon S, Kwak D. Multigrid convergence of an implicit symmetric relaxation scheme. AIAA Paper 93-3357; 1993.

[36] Blazek J. Investigations of the implicit LU-SSOR scheme. DLR Research Report No. 93-51; 1993.

[37] Blazek J. A multigrid LU-SSOR scheme for the solution of hypersonic flow problems. AIAA Paper 94-0062; 1994.

[38] Gerlinger P, Stoll P, Brüggemann D. An implicit multigrid method for the simulation of chemically reacting flows. J Comput Phys 1998; 146: 322-45.

[39] Langer S. Agglomeration multigrid methods with implicit Runge-Kutta smoothers applied to aerodynamic simulations on unstructured grids. J Comput Phys 2014; 277: 72-100.

[40] Roberts TW, Sidilkover D, Swanson RC. Textbook multigrid efficiency for the steady Euler equations. AIAA Paper 97-1949; 1997.

[41] Thomas JL, Bonhaus DL, Anderson WK, Rumsey CL, Biedron RT. An O(Nm2) plane solver for the compressible Navier-Stokes equations. AIAA Paper 99-0785; 1999.

[42] van Leer B, Darmofal D. Steady Euler solutions in O(N) operations. In: Proc. sixth European multigrid conf. Gent, Belgium; September 27-30, 1999 (Berlin: Springer; 2000).

[43] Nishikawa H, van Leer B. Optimal multigrid convergence by elliptic/hyperbolic splitting. AIAA Paper 2002-2951; 2002 (also J Comput Phys 2003; 190: 52-63).

[44] Ruge JW, Stüben K. Algebraic multigrid. Multigrid methods. In: McCormick SF, editor. SIAM frontiers in applied mathematics. Philadelphia (PA): SIAM; 1987. p. 73-131.

[45] Lonsdale RD. An algebraic multigrid solver for the Navier-Stokes equations on unstructured meshes. Int J Numer Meth Heat Fluid Flow 1993;3:3-14.

[46] Webster R. An algebraic multigrid solver for Navier-Stokes problems. Int J Numer Meth Heat Fluid Flow 1994;18:761-80.

[47] Raw MJ. Robustness of coupled algebraic multigrid for the Navier-Stokes equations. AIAA Paper 96-0297;1996.

[48] Weiss JM, Maruszewski JP, Smith WA. Implicit solution of preconditioned Navier-Stokes equations using algebraic multigrid. AIAA J 1999;37:29-36.

[49] Cleary AJ, Falgout RD, Van Henson E, Jones JE, Manteuffel TA, McCormick SF, et al. Robustness and scalability of algebraic multigrid. SIAM J Sci Comput 2000;21:1886-908.

[50] Stüben K. A review of algebraic multigrid. J Comput Applied Math 2001;128:281-309.

[51] Van Henson E, Yang UM. BoomerAMG: a parallel algebraic multigrid solver and preconditioner. Appl Numer Math 2002;41:155-77.

[52] Haase G, Kuhn M, Reitzinger S. Parallel algebraic multigrid methods on distributed memory computers. SIAM J Sci Comput 2002;24:410-27.

[53] Chang Q, Huang Z. Efficient algebraic multigrid algorithms and their convergence. SIAM J Sci Comput 2002;24:597-618.

[54] Mulder WA. A new multigrid approach to convection problems. J Comput Phys 1989; 83:303-23.

[55] Lynn JF, van Leer B. A semi-coarsened multigrid solver for the Euler and Navier-Stokes equations with local preconditioning. AIAA Paper 95-1667;1995.

[56] Mavriplis DJ. Directional agglomeration multigrid techniques for high Reynolds number viscous flow solvers. AIAA Paper 98-0612;1998.

[57] Mavriplis DJ. Multigrid strategies for viscous flow solvers on anisotropic unstructured meshes. J Comput Phys 1998;145:141-65.

[58] Hackbush W. Multigrid methods and applications. Berlin: Springer Verlag;1985.

[59] Mavriplis DJ. Multigrid techniques for unstructured meshes. ICASE Report No. 95-27;1995.

[60] Mavriplis DJ. Unstructured grid technique. Annu Rev Fluid Mech 1997;29:473-514.

[61] Leclerq MP, Stoufflet B. Characteristic multigrid method application to solve the Euler equations with unstructured and unnested grids. J Comput Phys 1993;104:329-46.

[62] Mavriplis DJ. Three-dimensional multigrid for the Euler equations. AIAA J 1992;30:1753-61.

[63] Peraire J, Peiró J, Morgan K. A 3D finite-element multigrid solver for the Euler equations. AIAA Paper 92-0449;1992.

[64] Morano E, Dervieux A. Looking for O(N) Navier-Stokes solutions on non-structured meshes. In: Proc. 6th Copper Mountain conf. onmultigrid methods;1993. p. 449-64 (also ICASE Report 93-26;1993).

[65] Riemslagh K, Dick E. A multigrid method for steady Euler equations on unstructured adaptive grids. In: Proc. 6th Copper Mountain conf. on multigrid methods; 1993. p. 527-42.

[66] Mavriplis DJ, Martinelli L. Multigrid solution of compressible turbulent flow on unstructured meshes using a two-equation model. Int J Numer Meth Fluids 1994; 18: 887-914.

[67] Crumpton PI, Giles MB. Aircraft computations using multigrid and an unstructured parallel library. AIAA Paper 95-0210; 1995.

[68] Guillard H. Node-nested multigrid with Delaunay coarsening. INRIA Report No. 1898, 1993.

[69] Ollivier-Gooch CF. Multigrid acceleration of an upwind Euler solver on unstructured meshes. AIAA J 1995; 33: 1822-7.

[70] Parthasarathy V, Kallinderis Y. New multigrid approach for three-dimensional unstructured, adaptive grids. AIAA J 1994; 32: 956-63.

[71] Barth TJ. Randomized multigrid. AIAA Paper 95-0207; 1995.

[72] Lawson CL. Software for C1 surface interpolation. In: Rice JR, editor. Mathematical software III. New York: Academic Press; 1977.

[73] Lawson CL. Properties of n-dimensional triangulations. Comput Aided Geomet Design 1986; 3: 231-46.

[74] Crumpton PI, Giles MB. Implicit time accurate solutions on unstructured dynamic grids. AIAA Paper 95-1671; 1995.

[75] Moinier P, Müller J-D, Giles MB. Edge-based multigrid for hybrid grids. AIAA Paper 99-3339; 1999.

[76] Lallemand MH. Schémas décentrés multigrilles pour la résolution des équations d'Euler en éléments finis. PhD Thesis, Université de Provence, France; 1988.

[77] Lallemand MH, Steve H, Dervieux A. Unstructuredmultigridding by volume agglomeration: current status. Comput Fluids 1992; 21: 397-433.

[78] Koobus B, Lallemand MH, Dervieux A. Unstructured volume-agglomeration multigrid: solution of the Poisson equation. Int J Numer Meth Fluids 1994; 18: 27-42.

[79] Smith WA. Multigrid solution of transonic flow on unstructured grids. In: Recent advances and applications in CFD, Proc. ASME winter annual meeting; December 1990. 145-52.

[80] Venkatakrishnan V, Mavriplis DJ. Agglomeration multigrid for the three-dimensional Euler equations. AIAA J 1995; 33: 633-40.

[81] Mavriplis DJ, Venkatakrishnan V. Agglomeration multigrid for viscous turbulent flows. ICASE Report No. 94-62, 1994 (also AIAA Paper 94-2332; 1994).

[82] Elias SR, Stubley GD, Raithby GD. An adaptive agglomeration method for additive correction multigrid. Int J Numer Meth Eng 1997; 40: 887-903.

[83] Okamoto N, Nakahashi K, Obayashi S. A coarse grid generation algorithm for agglomeration multigrid method on unstructured grids. AIAA Paper 98-0615; 1998.

[84] Lassaline JV, Zingg DW. Development of an agglomeration multigrid algorithm with directional coarsening. AIAA Paper 99-3338; 1999.

[85] Nishikawa H, Diskin B, Thomas JL. Critical study of agglomerated multigrid methods for diffusion. AIAA J 2010;48:839-47.
[86] Thomas JL, Diskin B, Nishikawa H. A critical study of agglomerated multigrid methods for diffusion on highly-stretched grids. Comput Fluids 2011;41:82-93.
[87] Nishikawa H, Diskin B, Thomas JL, Hammond DH. Recent advances in agglomerated multigrid. AIAA Paper 2013-863;2013.
[88] Moulitsas I, Karypis G. Multilevel algorithms for generating coarse grids for multigrid methods. In: Proceedings of the 2001 ACM/IEEE Conference on Supercomputing. November 10-16, 2001, Denver, CO, USA. p. 45-55.
[89] http://www-users. cs. umn. edu/~moulitsa/software. html
[90] Volpe G. Performance of compressible flow codes at low Mach number. AIAA J 1993;31:49-56.
[91] Lee WT. Local preconditioning of the Euler equations. PhD Thesis, University of Michigan;1991.
[92] Lee D. Local preconditioning of the Euler and Navier-Stokes equations. PhD Thesis, University of Michigan;1996.
[93] Guillard H, Viozat C. On the behavior of upwind schemes in the low Mach number limit. INRIA Report No. 3160;1997.
[94] Patankar SV. Numerical heat transfer and fluid flow. McGraw-Hill, New York;1980.
[95] Ho Y-H, Lakshminarayana B. Computation of unsteady viscous flow using a pressure-based algorithm. AIAA J 1993;31:2232-40.
[96] Javadia K, Darbandia M, Taeibi-Rahni M. Three-dimensional compressible-incompressible turbulent flow simulation using a pressure-based algorithm. Comput Fluids 2008;37:747-66.
[97] Darwish M, Sraj I, Moukalled F. A coupled finite volume solver for the solution of incompressible flows on unstructured grids. J Comput Phys 2009;228:180-201.
[98] Shterev KS, Stefanov SK. Pressure based finite volume method for calculation of compressible viscous gas flows. J Comput Phys 2010;229:461-80.
[99] Chen ZJ, Przekwas AJ. A coupled pressure-based computational method for incompressible/compressible flows. J Comput Phys 2010;229:9150-65.
[100] Chorin AJ. A numerical method for solving incompressible viscous flow problems. J Comput Phys 1967;2:12-26.
[101] Turkel E. Preconditioned methods for solving the incompressible and low speed compressible equations. J Comput Phys 1987;72:277-98.
[102] Rogers SE, Kwak D. Upwind differencing scheme for the time-accurate incompressible Navier-Stokes equations. AIAA J 1990;28:253-62.
[103] Cabuk H, Sung C-H, Modi V. Explicit Runge-Kutta method for three-dimensional internal incompressible flows. AIAA J 1992;30:2024-31.
[104] Shi J, Toro EF. A Riemann-problem-based approach for steady incompressible flows. Int J Numer Meth Heat Fluid Flow 1996;6:81-93.
[105] Liu C, Zheng X, Sung CH. Preconditioned multigrid methods for unsteady incompressible

flows. J Comput Phys 1998;139:35-57.

[106] Briley WR, Taylor LK, Whitfield DL. High-resolution viscous flow simulations at arbitrary Mach number. J Comput Phys 2003;184:79-105.

[107] Turkel E, Radespiel R, Kroll N. Assessment of preconditioning methods for multidimensional aerodynamics. Comput Fluids 1997;26:613-34.

[108] Rieper F. On the dissipation mechanism of upwind-schemes in the low Mach number regime: a comparison between Roe and HLL. J Comput Phys 2010;229:221-32.

[109] Rieper F. A low-Mach number fix for Roe's approximate Riemann solver. J Comput Phys 2011;230:5263-87.

[110] Mulas M, Chibbaro S, Delussu G, Di Piazza I, Talice M. Efficient parallel computations of flows of arbitrary fluids for all regimes of Reynolds, Mach and Grashof numbers. Int J Numer Meth Heat Fluid Flow 2002;12:637-57.

[111] van Leer B, Lee WT, Roe P. Characteristic time-stepping or local preconditioning of the Euler equations. AIAA Paper 91-1552;1991.

[112] Choi YH, Merkle CL. Time-derivative preconditioning for viscous flows. AIAA Paper 91-1652;1991.

[113] Choi YH, Merkle CL. The application of preconditioning in viscous flows. J Comput Phys 1993;105:207-23.

[114] Turkel E. A review of preconditioning methods for fluid dynamics. Appl Numer Math 1993;12:257-84.

[115] Turkel E, Vatsa VN, Radespiel R. Preconditioning methods for low-speed flows. AIAA Paper 96-2460;1996.

[116] Turkel E. Preconditioning techniques in computational fluid dynamics. Annu Rev Fluid Mech 1999;31:385-416.

[117] Lee D, van Leer B. Progress in local preconditioning of the Euler and Navier-Stokes equations. AIAA Paper 93-3328;1993.

[118] Lee D, van Leer B, Lynn JF. A local Navier-Stokes preconditioner for all Mach and cell Reynolds numbers. AIAA Paper 97-2024;1997.

[119] Jorgenson PC, Pletcher RH. An implicit numerical scheme for the simulation of internal viscous flow on unstructured grids. Comput Fluids 1996;25:447-66.

[120] Weiss J, Smith WA. Preconditioning applied to variable and constant density flows. AIAA J 1995;33:2050-7.

[121] Godfrey AG, Walters RW, van Leer B. Preconditioning for the Navier-Stokes equations with finite-rate chemistry. AIAA Paper 93-0535;1993.

[122] Dailey LD, Pletcher RH. Evaluation of multigrid acceleration for preconditioned time-accurate Navier-Stokes algorithms. Comput Fluids 1996;25:791-811.

[123] Jespersen D, Pulliam T, Buning P. Recent enhancements to OVERFLOW. AIAA Paper 97-0644;1997.

[124] Sharov D, Nakahashi K. Low speed preconditioning and LU-SGS scheme for 3-D viscous flow computations on unstructured grids. AIAA Paper 98-0614; 1998.

[125] Nemec M, Zingg DW. Aerodynamic computations using the convective upstream split pressure scheme with local preconditioning. AIAA Paper 98-2444; 1998.

[126] Merkle CL, Sullivan JY, Buelow PEO, Venkateswaran S. Computation of flows with arbitrary equations of state. AIAA J 1998; 36: 515-21.

[127] Sockol PM. Multigrid solution of the Navier-Stokes equations at low speeds with large temperature variations. J Comput Phys 2003; 192: 570-92.

[128] Puoti V. Preconditioning method for low-speed flows. AIAA J 2003; 41: 817-30.

[129] Lee S-H. Cancellation problem of preconditioning method at low Mach numbers. J Comput Phys 2007; 225: 1199-210.

[130] Lee S-H. Alleviation of cancellation problem of preconditioned Navier-Stokes equations. J Comput Phys 2009; 228: 4970-5.

[131] Park H, Nourgaliev RR, Martineau RC, Knoll DA. On physics-based preconditioning of the Navier-Stokes equations. J Comput Phys 2009; 228: 9131-46.

[132] De Medeiros FEL, De B. Alves LS. Stiffness, sensitiveness and robustness in low Mach preconditioned density-based methods. AIAA Paper 2012-1113; 2012.

[133] Hejranfar K, Kamali-Moghadam R. Preconditioned characteristic boundary conditions for solution of the preconditioned Euler equations at low Mach number flows. J Comput Phys 2012; 231: 4384-402.

[134] Bas O, Tuncer IH, Kaynak U. A Mach-uniform preconditioner for incompressible and subsonic flows. Int J Numer Meth Fluids 2014; 74: 100-12.

[135] Falcã CEG, De Medeiros FEL, De B. Alves LS. Implicit Runge-Kutta physical-time marching in low mach preconditioned density-based methods. AIAA Paper 2014-3085; 2014.

[136] http://msdn. microsoft. com/en-us/library/dd492627. aspx

[137] http://openacc. org/

[138] http://blogs. msdn. com/b/nativeconcurrency

[139] http://www. open-mpi. org

[140] https://computing. llnl. gov/tutorials/mpi

[141] http://www. cs. umn. edu/~metis

[142] Stoll P, Gerlinger P, Brüggemann D. Domain decomposition for an implicit LU-SGS scheme using overlapping grids. AIAA Paper 97-1896; 1997.

[143] Gropp WD, Keyes DE, McInnes LC, Tidriri MD. Globalized Newton-Krylov-Schwarz algorithms and software for parallel implicit CFD. ICASE Report No. 98-24; 1998.

[144] Cai X-C, Sarkis M. A restricted additive Schwarz preconditioner for general sparse linear systems. SIAM J Sci Comput 1999; 21: 792-7.

[145] Knoll DA, Keyes, DE. Jacobian-free Newton-Krylov methods: a survey of approaches and applications. J Comput Phys 2004; 193: 357-97.

[146] Chen R, Wu Y, Yan Z, Zhao Y, Cai X-C. A parallel domain decomposition method for 3D unsteady incompressible flows at high Reynolds number. SIAM J Sci Comput 2014;58:275-89.

[147] http://www.mcs.anl.gov/petsc

[148] http://www.openmp.org

[149] Amritkar A, Deb S, Tafti D. Efficient parallel CFD-DEM simulations using OpenMP. J Comput Phys 2014;256:501-19.

[150] http://www.nvidia.com/object/cuda_home_new.html

[151] https://developer.nvidia.com/cusp

[152] https://github.com/cusplibrary/cusplibrary

[153] Salvadore F, Bernardini M, Botti M. GPU accelerated flow solver for direct numerical simulation of turbulent flows. J Comput Phys 2013;235:129-42.

[154] http://www.khronos.org/opencl

第10章 相容性、精度和稳定性

在有限体积法中,纳维-斯托克斯方程组(2.19)中每个控制体的面积分是通过某种适当的方法近似的,考虑第4章和第5章中各种可能的空间离散方法,在控制体边界处对对流通量和黏性通量进行近似意味着离散方程组和要求解的控制方程组必然存在差异,这种差异引入了一定的空间离散误差,这个误差受计算积分所使用的数值格式和离散分辨率的影响。因此,随着网格的加密,离散方程组的解是否收敛以及以多快的速度收敛到控制方程组精确解是非常重要的问题。我们将在关于相容性和精度的两节之后讨论这个问题。

时间相关控制方程组(2.19)的求解需要对守恒变量的时间导数进行离散,我们在第6章中介绍了几种时间离散方法,每一种方法都有特定的精度阶数,并且对最大时间步长有一定的限制,这些限制可以通过冯·诺依曼稳定性分析来评估,这将在10.3节中介绍。在本章中,我们还将研究显式多步时间推进格式和一类通用隐式格式的时间离散误差衰减特性,衰减特性决定了特定格式的收敛速度和鲁棒性,对多重网格加速(见9.4节)的成败也非常重要。

10.1 相容性要求

如果离散方程组在时间步长和网格尺度均趋近于零的情况下收敛于给定的微分方程,则该离散格式称为相容格式。相容格式确保了求解的是真正的控制方程组,这在文献[1]中称为"正确地求解了方程组",也被称为验证。验证不应与确认混淆,确认的含义为:是否求解了"正确的方程"。因此,验证试图量化数值误差并发现编程错误,而确认关注建模误差,关于这方面详尽的讨论和建议,请参阅文献[2]。

通过将函数展开成泰勒级数可以检验数值格式的相容性。将展开的泰勒级数代入离散方程中,减去微分方程,将得到代表数值误差的截断误差项。对于相容性格式,截断误差应随时间步长和网格尺度的减小而趋于零。

下面用一个简单的例子来说明这个概念。考虑下列 1D 标量方程:

$$\frac{\partial U}{\partial t}+\frac{\partial U}{\partial x}=0 \tag{10.1}$$

一种简单却合理的离散方式是

$$\frac{U_i^{n+1}-U_i^n}{\Delta t}+\frac{U_i^n-U_{i-1}^n}{\Delta x}=0 \tag{10.2}$$

其中,n 表示时间层,i 是节点索引。在时间层 n 上将 U_i^{n+1} 进行展开可以得到

$$U_i^{n+1}=U_i^n+\Delta t\left(\frac{\partial U}{\partial t}\right)_i^n+\frac{(\Delta t)^2}{2}\left(\frac{\partial^2 U}{\partial t^2}\right)_i^n+\cdots \tag{10.3}$$

其中,(…)表示高阶项。U_{i-1}^n 的泰勒展开(见式(3.1))为

$$U_{i-1}^n=U_i^n-\Delta x\left(\frac{\partial U}{\partial x}\right)_i^n+\frac{(\Delta x)^2}{2}\left(\frac{\partial^2 U}{\partial x^2}\right)_i^n+\cdots \tag{10.4}$$

将式(10.3)和式(10.4)代入式(10.2),得到

$$\left(\frac{\partial U}{\partial t}+\frac{\partial U}{\partial x}\right)_i^n=-\frac{\Delta t}{2}\left(\frac{\partial^2 U}{\partial t^2}\right)_i^n+\frac{\Delta x}{2}\left(\frac{\partial^2 U}{\partial x^2}\right)_i^n+\cdots \tag{10.5}$$

与微分方程(式 10.1)比较发现,式(10.5)右端项表示截断误差,量级为 $O(\Delta t,\Delta x)$。随着 $\Delta t\to 0,\Delta x\to 0$,截断误差将消失,因此式(10.2)的离散格式是相容的。

10.2　离散格式的精度

离散格式的精度与其截断误差有关,例如,如果截断误差的首项与 Δx 成正比,称为一阶精度空间格式,如果首项为 $(\Delta x)^2$,则格式为二阶精度,依此类推。从式(10.5)可以看到,式(10.2)的格式在空间和时间方向均为一阶精度。由此可以得到,数值格式相容的条件是它至少为一阶精确,否则随着 Δt 和 Δx 减小,截断误差不可能降低。

现在的问题是,如何评估格式的截断误差。对于具有非线性开关、限制器等的复杂数值格式,泰勒级数展开是不恰当的。以下是几种可能的误差估计方法[1]:

(1) 不同网格上的额外解——使用加密或粗化的网格、无关网格;

(2) 同一网格上的额外解——使用更高阶或更低阶的离散格式;

(3) 同一网格上辅助偏微分方程的解;

(4) 同一网格上的代数估计。

最常用的方法是在一系列不同尺度网格上求解控制方程组,如果恰好知道

精确解,可以很容易地量化每套网格上的误差。截断误差减小的速率决定了离散格式的精度,例如,如果在所有坐标方向上将网格尺度减半,误差降低到 1/4,则该格式是二阶精度的。然而精确解通常是未知的,在这种情况下,可以用下式估计精度的阶数[3]：

$$p=\frac{\ln\left(\dfrac{f_3-f_2}{f_2-f_1}\right)}{\ln(r)} \tag{10.6}$$

其中,p 为精度,r 为网格加密(粗化)因子,f 为数值解,索引 1 表示最细的网格,3 表示最粗的网格。如果网格间的加密因子不是恒定的,可以在文献[1]中找到更一般的关系式。

除了估计截断误差,还可以通过改变网格分辨率得到网格收敛解。如果数值解随着网格进一步加密(在一定的公差范围内)没有改变,则认为该数值解是网格收敛解,这就是网格收敛性研究。尽管网格收敛性研究非常耗时,但建议始终检查数值解是否是网格收敛的。

10.3 冯·诺依曼稳定性分析

在编程实现一种新的离散方法之前,重要的是要掌握,至少是大致了解,该方法将如何影响数值格式的稳定性和收敛性。冯·诺依曼稳定性分析方法被频繁地证实可以可靠地评估求解格式的性质,该方法是 Los Alamos 实验室在第二次世界大战期间发展的,Cranck 和 Nicholson[4] 在 1947 年首次简要介绍了该方法,后来它发表在文献[5]中,文献[6]介绍了冯·诺依曼稳定性分析的细节。

在周期边界条件假设下,冯·诺依曼稳定性分析可应用于离散的线性偏微分方程组。该方法基于傅里叶级数分解,一方面可以研究求解格式的稳定性,另一方面可以详细地考察解在频谱上的表现,准确地说,它对于评估格式的收敛性和鲁棒性是非常重要的,因为解误差的单个分量(傅里叶模态)必须尽可能快地衰减,因此,我们称之为求解格式的衰减特性。

由于冯·诺依曼分析局限于线性问题,欧拉方程组或纳维-斯托克斯方程组必须用一个恰当的模型方程代替。下面将介绍两个常用的 1D 模型问题,第一个描述了扰动的纯对流,模拟欧拉方程组的行为。第二个模型方程包含了额外的扩散项,模拟纳维-斯托克斯方程组的行为。

10.3.1 傅里叶记号和放大因子

在分析模型问题之前,首先对 1D 标量线性方程进行冯·诺依曼分析。在

对通量进行空间离散后,可以得到时间方向的常微分方程组(见式(4.3)):

$$\Delta t\left(\frac{\mathrm{d}}{\mathrm{d}t}U_i\right)=-\frac{\Delta t}{\Delta x}R_i \tag{10.7}$$

其中,U_i 表示点 i 处的标量变量,R_i 表示残差,Δx 表示网格尺度。在周期边界假定下,将式(10.7)中的 U 展开为有限傅里叶级数[6-8]:

$$U_i=\sum_{k=-N}^{N}\hat{U}_k \mathrm{e}^{\mathrm{I}p_k(i\Delta x)} \tag{10.8}$$

其中,点索引 i 从 0 变化到 $N(x_i=i\Delta x)$,p_k 表示波数,I 是虚数单位。由于模型方程是线性的,整体影响可以叠加得到,所以只考虑单个(第 l 个)傅里叶模态就足够了,即

$$\hat{U}\mathrm{e}^{\mathrm{I}p_l(i\Delta x)} \tag{10.9}$$

如果将式(10.9)的傅里叶模态代入到离散模型方程组(10.7),可以得到

$$\Delta t\frac{\mathrm{d}}{\mathrm{d}t}\left(\hat{U}\mathrm{e}^{\mathrm{I}p_l(i\Delta x)}\right)=-\frac{\Delta t}{\Delta x}R_i=-\frac{\Delta t}{\Delta x}z\hat{U}\mathrm{e}^{\mathrm{I}i\Phi} \tag{10.10}$$

定义相位角 $\Phi=p_l\Delta x$。式(10.10)中复函数 z 表示空间算子的傅里叶记号,它的形式取决于离散格式(中心/迎风)和精度阶数。

如果任意谐波的幅值($\hat{U}$)不随时间增长,那么说式(10.7)的显式或隐式时间推进格式是线性稳定的。这意味着下一时间步解的幅值不会大于当前时间步。因此,放大因子定义为

$$g=\frac{\hat{U}^{n+1}}{\hat{U}^{n}} \tag{10.11}$$

如果引入时间推进算子 f,放大因子的一般形式为

$$g=1-fz \tag{10.12}$$

如果 $|g|\leqslant 1$,则时间推进格式是线性稳定的。在空间和时间离散是分离(直线法)的情况下,稳定域取决于时间推进格式(f),而不是空间离散(z)。对任意相位角 Φ,空间算子 z 的傅里叶记号必须都位于稳定域之内。优秀的衰减特性是指 $|g|$ 远小于 1,这意味着数值解的扰动随时间迅速衰减,因此,相比于 $|g|\to 1$,这种时间推进格式能更快地收敛到稳态解。

10.3.2　对流方程

标量线性方程为

$$\frac{\partial U}{\partial t}+\Lambda\frac{\partial U}{\partial x}=0 \tag{10.13}$$

假设对流速度 Λ(也是特征值)为正的常数。下面研究应用于式(10.7)的

几种不同空间离散格式及其傅里叶记号。

1. 带人工耗散的中心格式

根据式(4.48)~式(4.50)，式(10.7)中的残差 R_i 为

$$R_i = \frac{\Lambda}{2}(U_{i+1} - U_{i-1}) + \Lambda\epsilon^{(4)}(U_{i+2} - 4U_{i+1} + 6U_i - 4U_{i-1} + U_{i-2}) \tag{10.14}$$

其中，$\epsilon^{(4)}$ 表示耗散系数。为了简化分析，在式(10.14)中只保留到四阶差分项。根据式(10.10)，将式(10.9)中的谐波代入式(10.14)，可以得到空间算子的傅里叶记号为

$$z = \Lambda[\mathrm{I}\sin\Phi + 4\epsilon^{(4)}(1-\cos\Phi)^2] \tag{10.15}$$

2. 迎风格式

采用一阶迎风格式（见式(4.46)），残差为

$$R_i = \Lambda(U_i - U_{i-1}) \tag{10.16}$$

相应的傅里叶记号为

$$z = \Lambda[\mathrm{I}\sin\Phi + (1-\cos\Phi)] \tag{10.17}$$

二阶迎风格式的残差为

$$R_i = \frac{\Lambda}{2}(3U_i - 4U_{i-1} + U_{i-2}) \tag{10.18}$$

相应的傅里叶记号为

$$z = \Lambda[\mathrm{I}\sin\Phi(2-\cos\Phi) + (1-\cos\Phi)^2] \tag{10.19}$$

10.3.3 对流扩散方程

对流-扩散方程可以写成

$$\frac{\partial U}{\partial t} + \Lambda\frac{\partial U}{\partial x} = \nu\frac{\partial^2 U}{\partial x^2} \tag{10.20}$$

其中，ν 为黏性系数。扩散项 $\partial^2 U/\partial x^2$ 通常用二阶中心差分近似，即

$$\nu\frac{\partial^2 U}{\partial x^2} \approx \frac{\Lambda_{\mathrm{v}}}{\Delta x}(U_{i+1} - 2U_i + U_{i-1}) \tag{10.21}$$

其中

$$\Lambda_{\mathrm{v}} = \frac{\nu}{\Delta x} \tag{10.22}$$

是黏性特征值。

这时，空间算子的傅里叶记号由对流部分和扩散部分组成，即

$$z = z_{\mathrm{c}} - z_{\mathrm{v}} \tag{10.23}$$

其中

$$z_{\mathrm{v}}=2\Lambda_{\mathrm{v}}(\cos\Phi-1) \tag{10.24}$$

z_{c} 是式(10.15)、式(10.17)或式(10.19)中的某一形式。

由式(10.24)可以得出,扩散项只改变傅里叶记号的实部,对于任何类型的扩散或人工耗散(见式(10.15))均是如此。此外,由式(10.17)或式(10.19)可以看出,对流项的迎风离散格式也导致傅里叶记号包含实部($1-\cos\Phi$),这就是迎风耗散。只有中心格式无数值耗散,此时傅里叶记号为 $\mathrm{I}\sin\Phi$,然而中心格式会导致奇偶失联问题。

10.3.4 显式时间推进

离散方程组(10.7)的 m 步显式时间推进格式可以写为(见 6.1.1 节)

$$\begin{aligned}
&U_i^{(0)}=U_i^n\\
&U_i^{(k)}=U_i^{(0)}-\alpha_k\frac{\Delta t}{\Delta x}R_i^{(k-1)},\quad k=1,2,\cdots,m\\
&U_i^{n+1}=U_i^{(m)}
\end{aligned} \tag{10.25}$$

其中,α_k 为各步的系数。如果将变量 U 替换为其傅里叶展开式(10.8),将残差替换为傅里叶记号 z,则上式转换为

$$\begin{aligned}
&\hat{U}^{(0)}=\hat{U}^n\\
&\hat{U}^{(k)}=\hat{U}^{(0)}-\alpha_k\frac{\Delta t}{\Delta x}z\hat{U}^{(k-1)},\quad k=1,2,\cdots,m\\
&U^{n+1}=U^{(m)}
\end{aligned} \tag{10.26}$$

上述 m 步显式格式的放大因子如式(10.12)所示。根据式(10.26)可知,时间推进算子 f 的傅里叶记号为[9]

$$f=\frac{\Delta t}{\Delta x}\left[\alpha_m-\alpha_{m-1}\alpha_m\left(\frac{\Delta t}{\Delta x}z\right)+\cdots-(-1)^m\alpha_1\alpha_2\cdots\alpha_m\left(\frac{\Delta t}{\Delta x}z\right)^{m-1}\right] \tag{10.27}$$

如果在每一步都对 z 的对流和耗散部分进行评估(即(m,m)格式),对于混合多步格式(见 6.1.2 节),f 的推导变得更加复杂。对于((5,3))格式(式(6.7)),时间推进算子的傅里叶记号为[10]

$$\begin{aligned}
f=\frac{\Delta t}{\Delta x}\{&\alpha_5-\alpha_5\alpha_4(z_{\mathrm{I}}+\beta_5z_{\mathrm{R}})(1-\alpha_3z_{\mathrm{I}})-\\
&\alpha_5\alpha_4\alpha_2(1-\alpha_1z_{\mathrm{I}})(z_I+\beta_5z_{\mathrm{R}})[\alpha_3(z_{\mathrm{I}}+\beta_3z_{\mathrm{R}})z_I-\beta_3z_{\mathrm{R}}]-\\
&\alpha_5\alpha_2(1-\beta_5)\beta_3(1-\alpha_1z_I)z_{\mathrm{R}}\}
\end{aligned} \tag{10.28}$$

其中

$$z_{\mathrm{R}}=\frac{\Delta t}{\Delta x}\mathrm{Real}(z)\qquad z_{\mathrm{I}}=\frac{\Delta t}{\Delta x}\mathrm{Imag}(z) \tag{10.29}$$

分别表示 z 的实部和虚部。每步的系数 α_k 和混合系数 β_k 如表 6.2 所示。

时间步长 Δt 可以借助 CFL(Courant-Friedrichs-Lewy)条件[11]确定。对于对流方程(10.13)有如下关系式:

$$\Delta t=\sigma\frac{\Delta x}{|\Lambda|} \tag{10.30}$$

参数 σ 表示 CFL 数,它的大小取决于时间推进格式的类型和每步的系数。式(10.30)的推导见 10.3.6 节。对于对流-扩散方程(10.20)的时间步长,有如下关系式:

$$\Delta t=\sigma\frac{\Delta x}{|\Lambda|+C\Lambda_v} \tag{10.31}$$

式(10.31)中的因子 C 随空间离散方法变化,对于中心格式(式(10.15)),$C=4$ 会产生良好的衰减。对于一阶迎风格式(式(10.17)),$C=2$ 保证了式(10.23)中 $z=z_c-z_v$ 的傅里叶记号仍然以 z_c 为界。实际上,当 $\Phi=\pi$ 时,无论 $\Lambda_v/|\Lambda|$ 比值如何,扩散项的傅里叶记号与 z_c 的傅里叶记号相同。对于二阶迎风格式(式(10.19)),如果设 $C=1$,也会遇到同样的情况。

傅里叶记号和放大因子的示例

接下来考虑冯·诺依曼稳定性分析在对流和对流扩散混合问题中的几个应用。图 10.1 的左半部分显示了空间算子 z 的傅里叶记号的相轨迹(粗实线)以及放大因子 $|g|$ 的幅值等值线。稳定域的边界用 $|g|=1$ 表示。右半部分显示了 $|g|$ 相对于相位角的变化,据此可以评估衰减特性。

对流方程迎风离散的傅里叶记号和衰减特性如图 10.1 所示,两种情况均使用了最优系数的三步时间推进格式(表 6.1)。一阶迎风格式(式(10.16))的特性如图 10.1(a)和(b)所示。可以看到,傅里叶记号的相轨迹通过了 $|g|$ 的所有三个最小值(等值线图上的小圆区域),这导致相位角 $\Phi\geqslant\pi/2$ 时放大因子幅值很小,它对于高效的多重网格格式尤其重要(粗网格上 $\Phi\geqslant\pi/2$ 对应于细网格上的 $\Phi<\pi/2$)。还可以观察到,对于 $\Phi<\pi/2$,衰减性能较差,这种特性对多步格式是典型的,它解释了这些格式渐近收敛速度较低的原因,但可以采用多重网格改善。需要注意的是,$0\leqslant\Phi<\pi$ 表示图 10.1 左部分中相轨迹的前半部分。

图 10.1(b)表明了 CFL 数 σ 增加太多的影响。与预期一致,傅里叶记号扩展到了稳定边界之外,因此 g 的幅值大于 1(虚线),这意味着对于相应的相位角(频率),解的误差被放大,这种时间推进格式显然会发散。图 10.1(c)给出了二阶迎风格式(式(10.18))的性质,其表现基本上与一阶迎风格式相似。

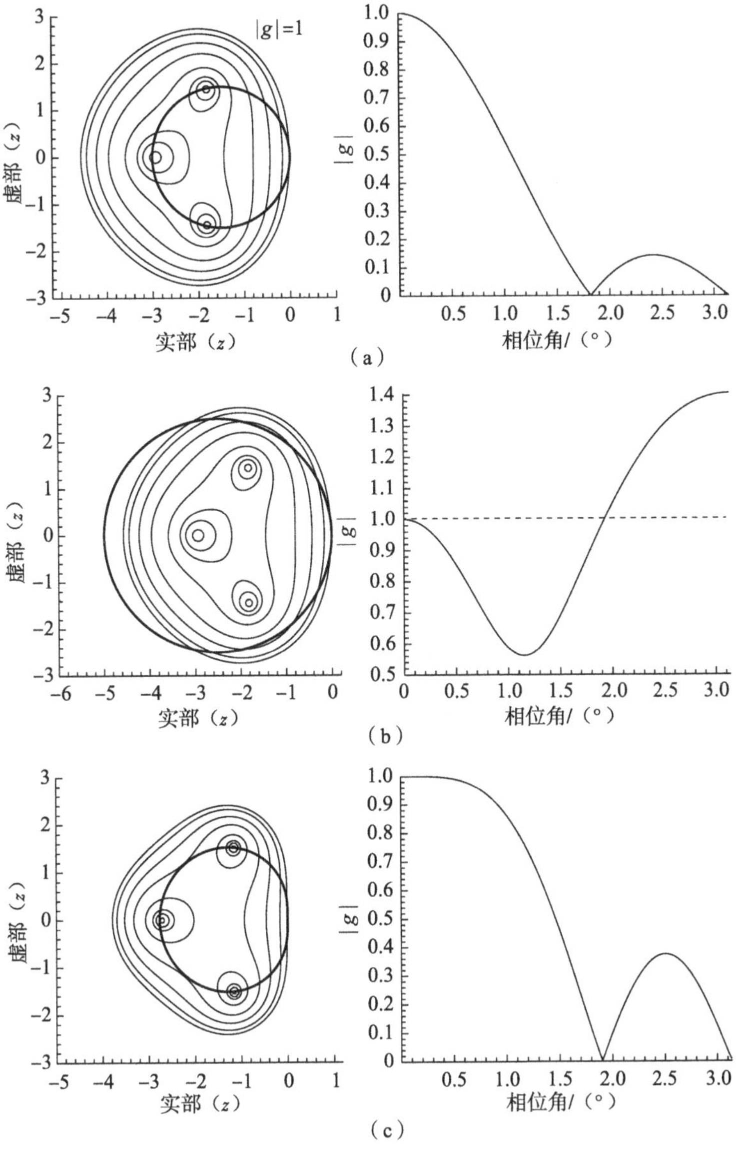

图 10.1　对流方程-显式(3,3)格式的傅里叶记号(z)和放大因子幅值($|g|$)

(a) 一阶迎风离散,$\sigma=1.5$,各级系数:0.1481,0.4,1.0;(b) 跟(a)一致,除了 $\sigma=2.5$;(c) 二阶迎风;$\sigma=0.69$;各级系数 0.1918,0.4929,1.0。

图 10.2 展示了使用(5,3)混合格式(见 6.1.2 节)求解对流方程时,傅里叶记号的相轨迹和格式的衰减特性。空间离散使用了添加人工耗散的中心格式(式(10.14)),为了揭示耗散系数 $\epsilon^{(4)}$ 的影响,其值由 1/16(图 10.2(a))变化到

1/256(图 10.2(c))。由结果可以总结出,随着人工耗散减小,相轨迹沿着实轴收缩,这种影响是由式(10.15)中的项$(1-\cos\Phi)^2$缩小引起的,衰减特性随着人工耗散减小而恶化,这意味着在实际应用中格式的收敛速度和鲁棒性会随着耗散的降低而变差。

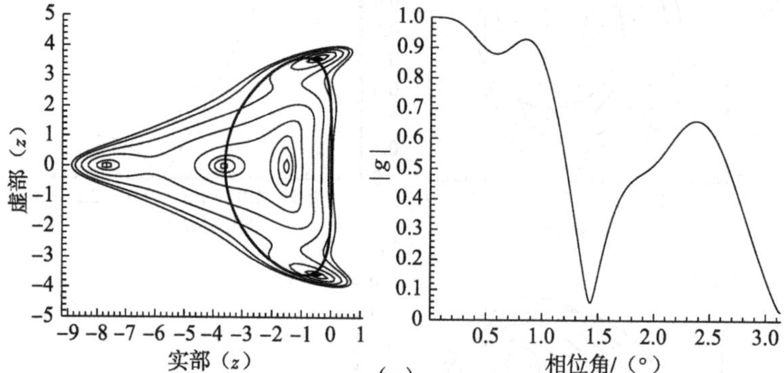

(a)

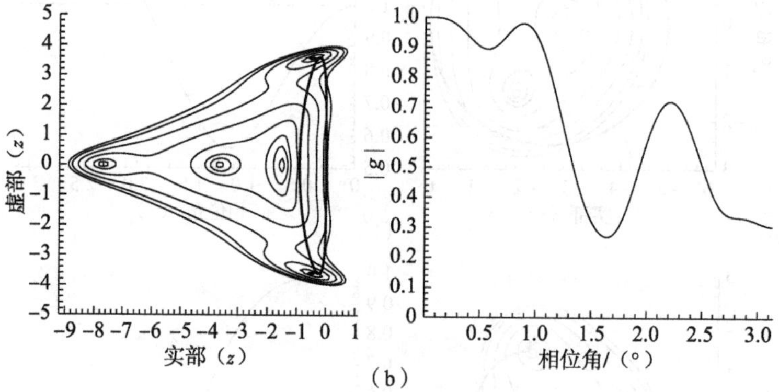

(b)

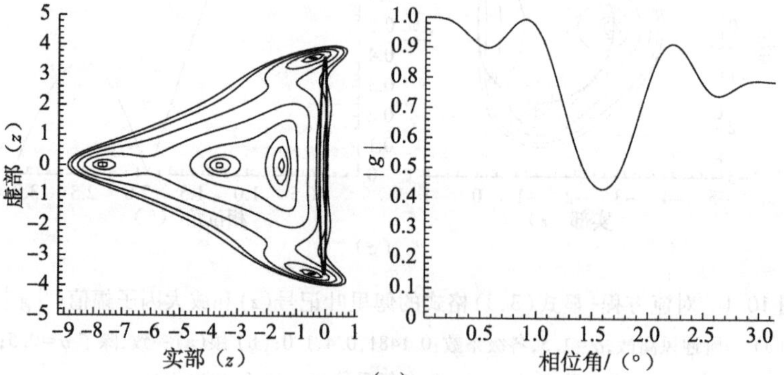

(c)

图 10.2　对流方程-显式(5,3)格式的傅里叶记号(z)和放大因子幅值($|g|$),中心格式空间离散,$\sigma=3.6$,各级系数如表 6.2 所示

(a) $\epsilon^{(4)}=1/16$;(b) $\epsilon^{(4)}=1/64$;(c) $\epsilon^{(4)}=1/256$。

接下来研究了使用(5,3)混合格式求解对流扩散方程(10.20)的特性。首先,该格式和一阶迎风空间离散(式(10.16))耦合使用,在图 10.3 中,对比了 $\Lambda_v=0$(纯对流)和 $\Lambda_v/\Lambda=2$(虚线表示)的情况。可以看到,傅里叶记号的相轨迹沿着虚轴收缩,这是由于扩散项 z_v 的影响,它只有一个实分量。如果根据式(10.31)使用 $C=2$ 计算时间步长,傅里叶记号沿实轴的范围保持不变,该格式的衰减特性保持在与纯对流方程大致相等的不错水平上,这是由于在 z_c 相轨迹环绕的区域,$|g|$ 具有最优的、平缓的极小值(图 10.3(b))。

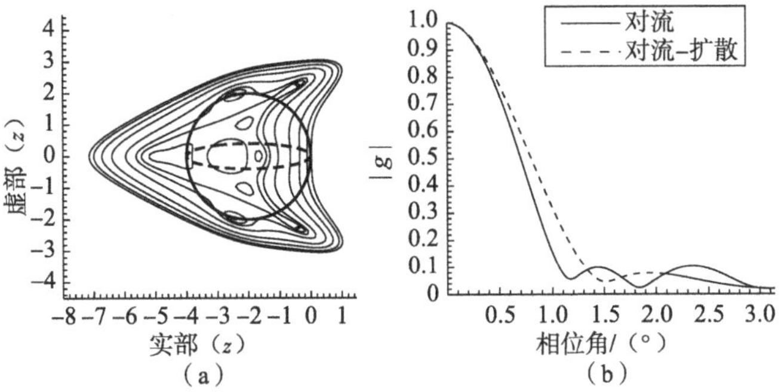

图 10.3 对流扩散方程-显式(5,3)格式的傅里叶记号(z)和放大因子幅值($|g|$),一阶迎风空间离散格式,$\sigma=2.0$,各级系数和混合系数如表 6.2 所示

图 10.4 给出了基于中心格式空间离散的(5,3)混合格式的性质。耗散系数设为 $\epsilon^{(4)}=1/64$,特征值比值为 $\Lambda_v/\Lambda=2$。可以观察到,与对流方程相比,傅里叶记号的相轨迹完全改变了形态,在 $\Lambda_v\to\infty$ 极限情况下,相轨迹退化成一条直线,因此对时间推进算法进行优化使得 $|g|$ 沿着实轴有平缓的极小值是重要的。本算例中时间步长根据式(10.31)计算,$C=4$。

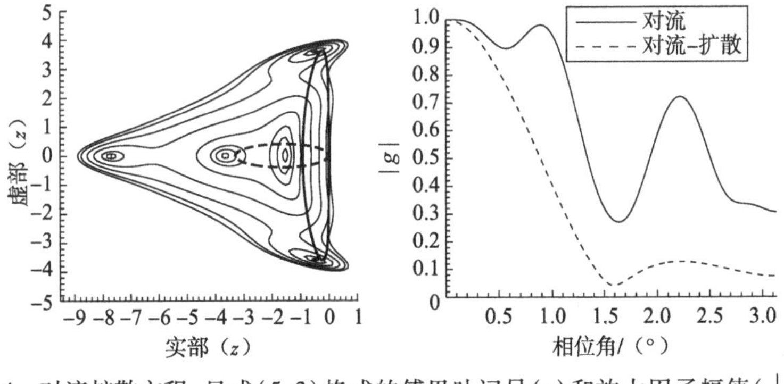

图 10.4 对流扩散方程-显式(5,3)格式的傅里叶记号(z)和放大因子幅值($|g|$),中心离散格式,$\sigma=3.6$,$\epsilon^{(4)}=1/64$,各级系数和混合系数如表 6.2 所示

冯·诺依曼稳定性分析的更多例子可以查阅文献[9,12-18]。本书还提供了一个用于分析显式多级格式(含隐式残差光滑)稳定性的源代码程序。

10.3.5 隐式时间推进

根据式(6.28),用以处理模型方程积分的通用隐式格式可表示为

$$\left[\frac{\Delta x}{\Delta t}+\beta\left(\frac{\partial R}{\partial U}\right)_i\right]\Delta U^n=-R_i^n \tag{10.32}$$

将通量雅可比 $\partial R/\partial U$ 分解成两个微分算子之和,分别是式(10.20)中的对流项和扩散项,即

$$\left\{\frac{\Delta x}{\Delta t}+\beta\left[(D_x^{\mathrm{I}})_{\mathrm{c}}-(D_x^{\mathrm{I}})_{\mathrm{v}}\right]_i\right\}\Delta U^n=-R_i^n \tag{10.33}$$

对流差分算子可以采取多种形式,例如添加人工耗散的中心格式的算子为(式(6.47))

$$(D_x^{\mathrm{I}})_{\mathrm{c}}\Delta U^n=\frac{\Lambda}{2}(\Delta U_{i+1}^n-\Delta U_{i-1}^n)+\Lambda\epsilon^{\mathrm{I}}(\Delta U_{i+1}^n-2\Delta U_i^n+\Delta U_{i-1}^n) \tag{10.34}$$

其中,ϵ^{I} 表示隐式耗散系数。这里没有包含四阶差分项,因为在实际中很少使用(计算量大)。一阶或二阶迎风格式的隐式算子分别为

$$(D_x^{\mathrm{I}})_{\mathrm{c}}\Delta U^n=\Lambda(\Delta U_i^n-\Delta U_{i-1}^n) \tag{10.35}$$

或

$$(D_x^{\mathrm{I}})_{\mathrm{c}}\Delta U^n=\frac{\Lambda}{2}(3\Delta U_i^n-4\Delta U_{i-1}^n+\Delta U_{i-2}^n) \tag{10.36}$$

扩散差分算子通常使用中心类格式:

$$(D_x^{\mathrm{I}})_{\mathrm{v}}\Delta U^n=\frac{\Lambda_{\mathrm{v}}}{2}(\Delta U_{i+1}^n-2\Delta U_i^n+\Delta U_{i-1}^n) \tag{10.37}$$

显式算子的离散可以采用式(10.14)、式(10.16)或式(10.18)中的任何一个。

将式(10.9)的傅里叶模态代入隐式格式(式(10.33)),可以得到

$$\left[\frac{\Delta x}{\Delta t}+\beta z^{\mathrm{I}}\right]\Delta\hat{U}^n=-z^{\mathrm{E}}\hat{U}^n \tag{10.38}$$

其中,$z^{\mathrm{I}}=z_{\mathrm{c}}^{\mathrm{I}}-z_{\mathrm{v}}^{\mathrm{I}}$ 表示通量雅可比的傅里叶记号,$z^{\mathrm{E}}=z_{\mathrm{c}}^{\mathrm{E}}-z_{\mathrm{v}}^{\mathrm{E}}$ 表示显式算子的傅里叶记号。傅里叶记号的形式如10.3.2节和10.3.3节中的推导。基于式(10.12)的放大因子为

$$g=1-\frac{z^{\mathrm{E}}}{\left[\dfrac{\Delta x}{\Delta t}+\beta z^{\mathrm{I}}\right]} \tag{10.39}$$

因此,时间推进算子的傅里叶记号为 $f=1/[\cdots]$。从式(10.39)可以看出,对于 $\Delta t\to\infty$,$\beta=1$ 且 $z^{\mathrm{I}}=z^{\mathrm{E}}$,相位角 $\Phi>0$ 时的放大因子为零。如果通量雅可比是精确的,牛顿格式(见 6.2.5 节)会出现这种情况,这也是精确牛顿法收敛非常快的原因。

放大因子的示例

下面,将研究一些隐式格式在不同显式和隐式离散算子下的衰减特性。感兴趣的读者可以查阅文献[19]了解更多关于 2D 问题中流行的 LU-SGS 方法的冯·诺依曼分析(单重和多重网格),本书还提供了隐式格式分析程序的源代码。

在第一个例子中,考虑将中心格式空间离散应用于显式算子(式(10.14))和隐式算子(式(10.34)),这类似于 6.2.3 节中介绍的标准 ADI 格式,我们希望研究隐式算子的参数设置对放大因子的影响。图 10.5 比较了三个算例结果。第一条曲线(实线)的参数设置为 $\beta=1$ 和 $\epsilon^{\mathrm{I}}=1/20$,$\epsilon^{\mathrm{I}}$ 取值使得在 $\Phi=\pi$ 处 $|g|=0$。可以用下式计算:

$$\epsilon^{\mathrm{I}}=\frac{1}{\beta}\left(4\epsilon^{(4)}-\frac{\Delta x}{4\,|\Lambda|\,\Delta t}\right) \tag{10.40}$$

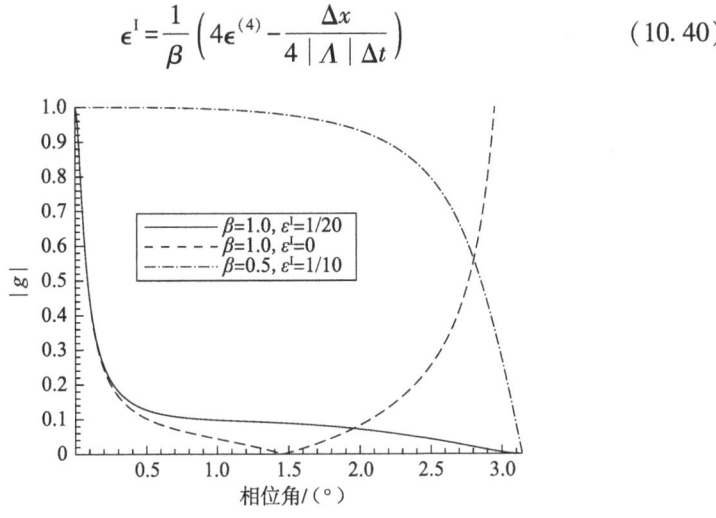

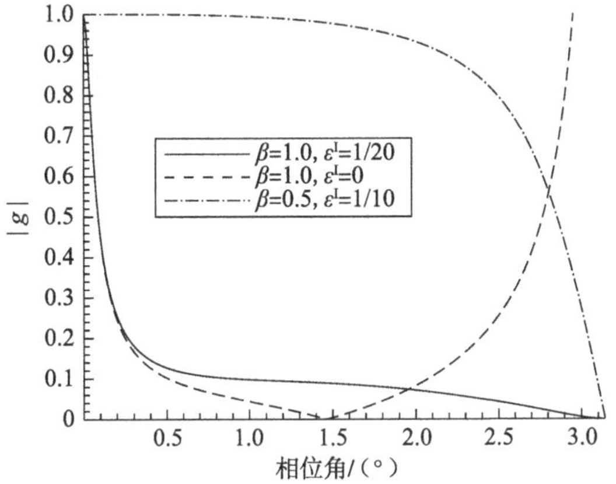

图 10.5　对流方程-隐式格式的放大因子幅值($|g|$)。使用中心离散的显式和隐式算子:$\sigma=20$,$\epsilon^{(4)}=1/64$

第二条曲线(虚线)表明,隐式算子须包含人工耗散,对于 $\epsilon^{\mathrm{I}}=0$ 的情况,在高频处($\Phi\to\pi$)格式变得不稳定。最后的曲线(点线)显示了 $\beta=1/2$(二阶时间格式)情况下 g 的幅值,可以看到此时的衰减特性比 $\beta=1$ 时差很多,因此应该首选 $\beta=1$。使用 6.3.2 节所述的双时间步推进能够更有效地求解非定常问题。

增大 CFL 数产生的影响如图 10.6 所示。隐式算子和显式算子都使用了一阶迎风格式(式(10.16)和式(10.35))。由于两种算子的相似性,格式的衰减性质非常好,这证实了我们对牛顿格式的看法。

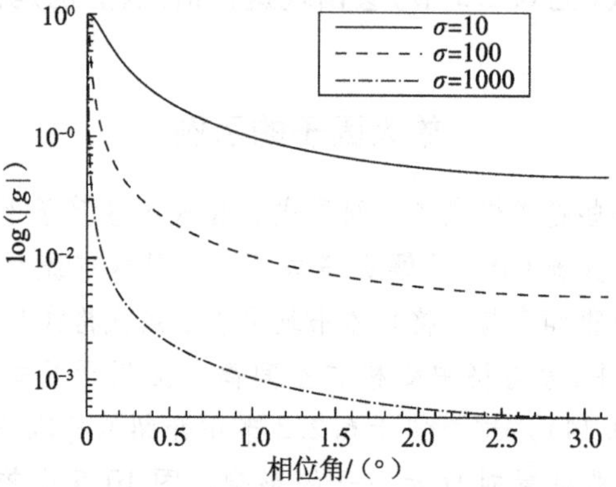

图 10.6　对流方程-隐式格式下放大因子的幅值($|g|$);使用一阶迎风格式离散的显式和隐式算子,$\beta=1$。注意 y 轴使用了对数尺度

图 10.7 比较了隐式算子(左端项)的两种不同离散格式的放大因子。两种离散格式下,显式算子都使用二阶迎风格式(式(10.18))离散。很明显,左端项显式算子的精确表达带来更好的衰减特性,从而收敛更快,这点常在实践中得到证实。一阶/二阶混合离散的衰减特性可以通过增加超松弛来改善,这点在 LU-SGS 格式中得到了证实[19-20]。

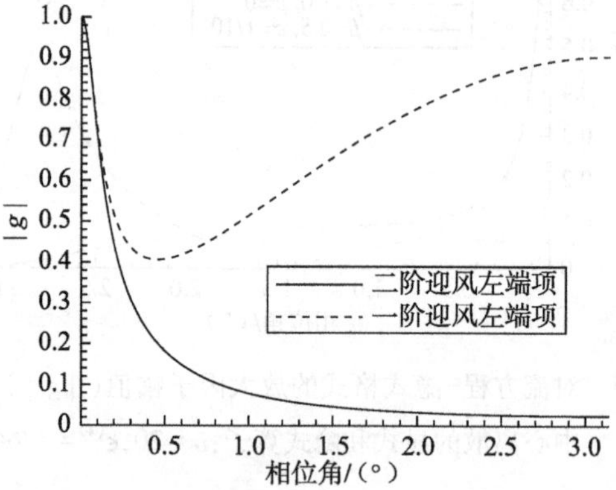

图 10.7　对流方程-隐式格式下放大因子的幅值($|g|$);使用二阶迎风格式离散的显式算子,$\sigma=10$,$\beta=1.0$

最后一个例子是对流扩散方程(10.20),如图 10.8 所示。在此比较了两种情况。第一种情况下,隐式算子中也包含了扩散项的离散(实线),可以观察到

衰减特性是非常有利的,这点与对流方程情况类似。然而,如果将扩散项从隐式算子中移除(图 10.8 中的虚线),即使黏性特征值与对流的特征值的比值较小($\Lambda_v/\Lambda=2$),格式也会变得不稳定。因此,为了得到鲁棒的格式,在通量雅可比的近似中需要包含黏性通量的影响。

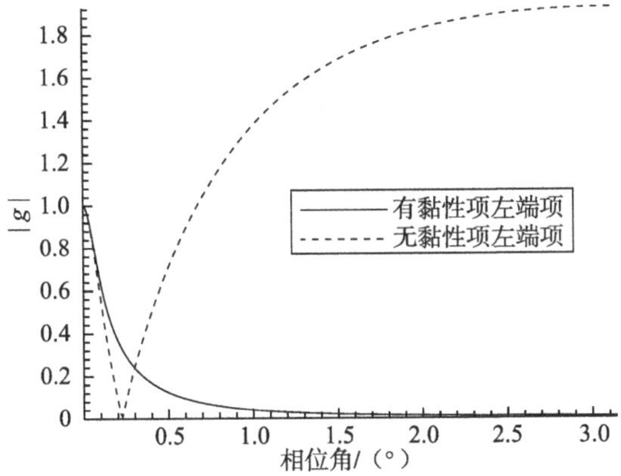

图 10.8　对流扩散方程-隐式格式下放大因子的幅值($|g|$);使用二阶迎风格式离散的显式和隐式算子,$\sigma=10$,$\beta=1.0$,$\Lambda_v/\Lambda=2$

总之,隐式算子至少应该包括空间离散和物理问题最重要的特征,对通量雅可比 $\partial R/\partial W$ 做过于粗糙的近似会很容易导致格式不稳定,因此在通量雅可比的计算量和准确度之间寻求合理的折中是非常重要的。

与显式多步格式相比,合理设计的隐式格式的衰减性能明显更好,在低相位角(频率)处尤其如此,因此隐式格式会比多步格式的渐近收敛速度更快。

10.3.6　CFL 条件的推导

每个显式时间推进格式只有在时间步长 Δt 位于一定范围内才能保持稳定,Courant 等给出了时间推进格式稳定的必要非充分条件[11],所谓 CFL 条件是指数值格式的依赖域必须包括偏微分方程的依赖域。为了说明这一观点,考虑图 10.9 中的 x-t 图。对流方程(10.13)的依赖域由特征关系 $\mathrm{d}x/\mathrm{d}t=\Lambda$ 给定,这意味任何的信息都以该速度在域内传播,因此 $t_0+\Delta t$ 时刻的精确解与 t_0 时刻、$x^*=x_i-\Lambda\Delta t$ 位置处的解相同。为了正确模拟精确解,空间离散格式的模板应该包含 x^*,(若 $\Lambda>0$)至少应包括点 x_{i-1},该数值格式的依赖域由图 10.9 中的阴影区域表示,因此得到下列条件:

$$\Lambda\Delta t\leqslant\Delta x\quad\text{或}\quad\Delta t\leqslant\frac{\Delta x}{\Lambda}\tag{10.41}$$

这将产生 CFL 条件:

$$\sigma = |\Lambda| \frac{\Delta t}{\Delta x} \leqslant 1 \tag{10.42}$$

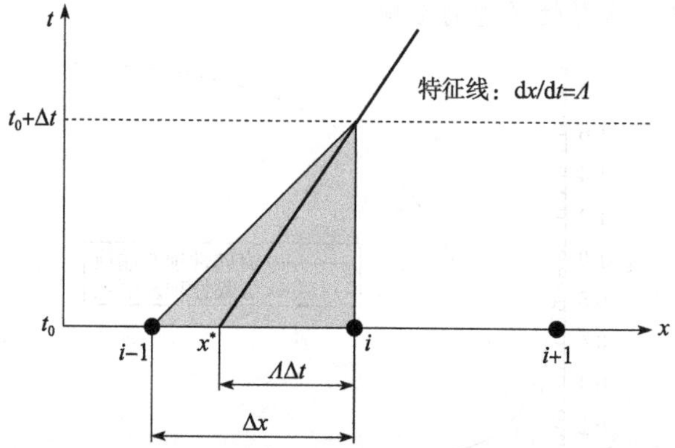

图 10.9 显式时间推进格式的依赖域(阴影区域)和对流方程的依赖域(实线)

由于在显式多步格式中,$t_0+\Delta t$ 时刻的解由不止一个时间步获得,所以该格式允许 CFL 数 $\sigma>1$。

基于以上论证,可以很容易地证明,像式(10.32)这样的隐式格式总是满足 CFL 条件,因为依赖的数值域扩展到了所有网格点。

冯·诺依曼分析和 CFL 条件

式(10.42)中的 CFL 条件也可以通过冯·诺依曼稳定性分析得到。考虑使用一阶迎风格式(式(10.16))和单步显式格式求解对流方程(10.13)。该数值格式的依赖域如图 10.9 中的阴影区域所示。时间推进算子的傅里叶记号简化为 $f=\Delta t/\Delta x$,根据式(10.12),通过关系式:

$$g=1-\frac{\Delta t}{\Delta x}z \tag{10.43}$$

可以得到放大因子 g。如果将式(10.17)z 的表达式代入式(10.43),放大因子将变为

$$g=1-\frac{\Delta t}{\Delta x}\Lambda[\,\mathrm{I}\sin\Phi+(1-\cos\Phi)\,] \tag{10.44}$$

因此,放大因子 g 的幅值为(假定 $\Lambda>0$)

$$|g|^2=2\frac{\Delta t}{\Delta x}\Lambda\left[\frac{\Delta t}{\Delta x}\Lambda-1\right](1-\cos\Phi)+1 \tag{10.45}$$

很容易证明,$|g|^2$ 的最大值出现在相位角 $\Phi=\pi$ 处。如果在式(10.45)中

假定 $\Phi=\pi$,可以得到

$$|g(\Phi=\pi)|^2=4\frac{\Delta t}{\Delta x}\Lambda\left[\frac{\Delta t}{\Delta x}\Lambda-1\right]+1 \tag{10.46}$$

为了使时间推进格式稳定,必须满足 $|g|^2\leqslant 1$,这个条件只有在

$$\frac{\Delta t}{\Delta x}\Lambda\leqslant 1 \tag{10.47}$$

才成立。因此有

$$\Delta t\leqslant\frac{\Delta x}{\Lambda} \tag{10.48}$$

这与式(10.41)完全对应。

需要指出的是,CFL 条件(式(10.42))并不足以保证数值格式的稳定性(但却是必要的),因此,还应该进行冯·诺依曼分析。

参考文献

[1] Roache PJ. Quantification of uncertainty in computational fluid dynamics. Ann Rev Fluid Mech 1997;29:123-60.

[2] AIAA. Guide for the verification and validation of computational fluid dynamics simulations. AIAA G-077-1998(2002);1998 and 2002.

[3] De Vahl DG. Natural convection of air in a square cavity: a bench mark numerical solution. Int J Num Method Fluids 1983;3:249-64.

[4] Cranck J,Nicholson P. A practical method for numerical evaluation of solutions of partial differential equations of the heat conduction type. Proc Camb Philos Soc 1947;43:50-67.

[5] Charney JG,Fjortoft R,von Neumann J. Numerical integration of the barotropic vorticity equation. Tellus 1950;2:237-54.

[6] Hirsch C. Numerical computation of internal and external flows,vol. 1. New York: John Wiley and Sons;1988.

[7] Roache PJ. Computational fluid dynamics. Albuquerque,USA: Hermosa Publishers;1972.

[8] Roache PJ. Fundamentals of computational fluid dynamics. Albuquerque,USA: Hermosa Publishers;1998.

[9] Kroll N,Jain RK. Solution of two-dimensional Euler equations—experience with a finite volume code. DLR Research Report,No. 87-41;1987.

[10] Radespiel R,Swanson RC. Progress with multigrid schemes for hypersonic flow problems. ICASE Report No. 91-89;1991;also J Comput Phys 1995;116:103-22.

[11] Courant R,Friedrichs KO,Lewy H. Über die partiellen differenzengleichungen der mathematischen Physik. Math Ann 1928;100:32-74. Translation: On the partial difference equations of

mathematical physics. IBM J 1967;11:215-34.

[12] Jameson A. Multigrid algorithms for compressible calculations. multigrid methods II, Lecture notes in mathematics, vol. 1228. New York: Springer Verlag;1985.

[13] van Leer B, Tai C-H, Powell KG. Design of optimally smoothing multi-stage schemes for the Euler equations. AIAA Paper 89-1933;1989.

[14] Lötstedt P, Gustafsson B. Fourier analysis of multigrid methods for general systems of PDE. Report No. 129/1990. Sweden: Department of Scientific Computing, Uppsala University;1990.

[15] Blazek J, Kroll N, Radespiel R, Rossow, C-C. Upwind implicit residual smoothing method for multi-stage schemes. AIAA Paper 91-1533;1991.

[16] Blazek J. Methods to accelerate the solution of the Euler- and the Navier-Stokes equations for steady-state super- and hypersonic flows. Translation of DLR Research Report, No. 94-35, ESA-TT-1331;1995.

[17] Tai C-H, Sheu J-H, van Leer B. Optimal multistage schemes for Euler equations with residual smoothing. AIAA J 1995;33:1008-16.

[18] Tai C-H, Sheu J-H, Tzeng P-Y. Improvement of explicit multistage schemes for central spatial discretization. AIAA J 1996;34:185-8.

[19] Blazek J. Investigations of the implicit LU-SSOR scheme. DLR Research Report, No. 93-51;1993.

[20] Blazek J. A multigrid LU-SSOR scheme for the solution of hypersonic flow problems. AIAA Paper 94-0062;1994.

第 11 章
网格生成原理

在对控制方程组进行数值求解之前，必须对所有边界面进行离散，并在流动区域内生成体网格。正如在第 3 章开始时所讨论的（见 3.1 节），基本上有两种网格类型可供选择：

(1) 结构网格；

(2) 非结构网格。

图 11.1 和图 11.2 是一个民用飞机结构网格的例子[1-2]，图 11.3 和图 11.4 为类似构型的非结构表面网格和体网格[3-6]。

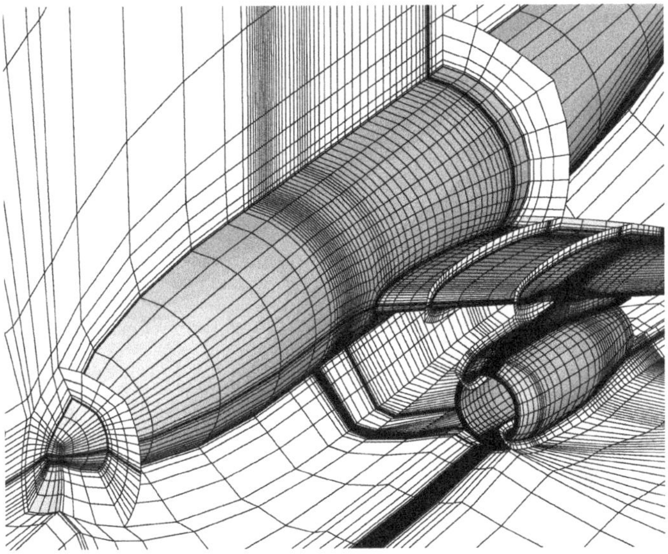

图 11.1　翼身组合体构型的结构面网格和体网格（由德国 DLR 的 O. Brodersen 提供）

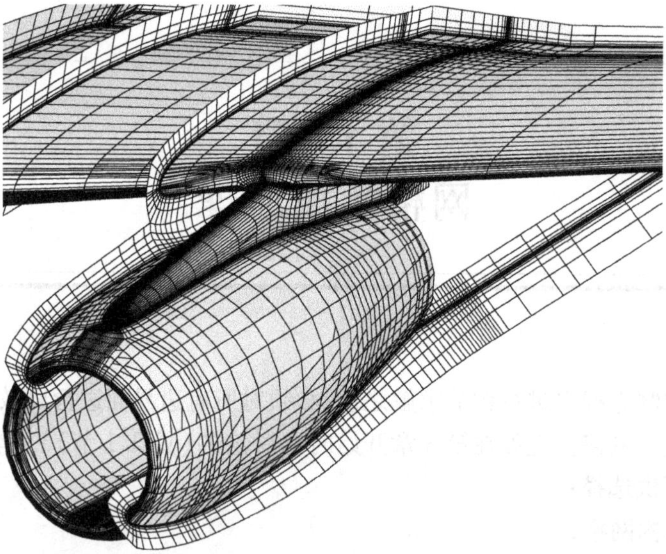

图 11.2 翼身组合体构型的结构面网格和体网格：挂架和发动机短舱局部放大图（由德国 DLR 的 O. Brodersen 提供）

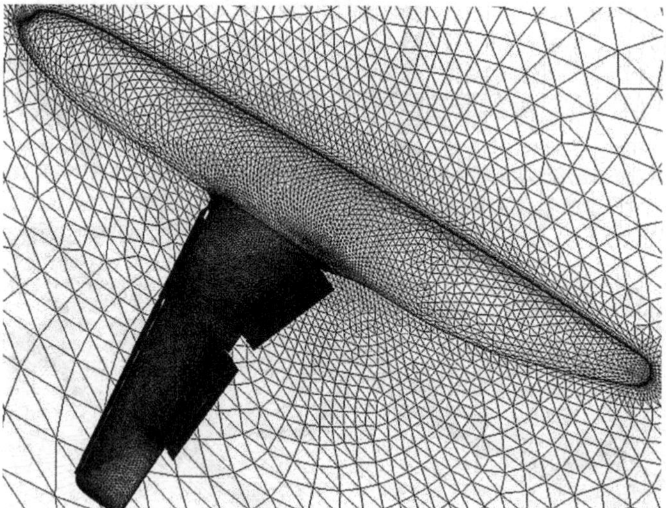

图 11.3 翼身组合体构型的非结构面网格（由美国怀俄明大学 D. Mavriplis 提供）

图 11.4　翼身组合体构型的非结构网格:缝隙和体网格剖面局部放大图
(由美国怀俄明大学 D. Mavriplis 提供)

结构网格和非结构网格都有各自的优缺点(3.1 节中进行了介绍)。然而,不管是何种网格类型,通常遇到的瓶颈是从 CAD(计算机辅助设计)系统导入到网格生成程序中的数据质量。分析的曲面定义通常通过某种标准格式进行转换,如 IGES(initial graphics exchange specification,初始图形交换规范[7-8])、STEP(standard for the exchange of product model data,产品模型数据交换标准[9-11])或 DXF(drawing exchange format,绘图交换格式[12])。另一种常用的方法是由 STL(stereolithography)格式直接转换为三角形曲面网格[13-15],所有现代 CAD 程序都提供这种格式。直接利用原始 CAD 数据的情况非常少[16]。经验表明,转换过程会影响数据的准确性,此外,在 CAD 系统中,曲面的表达本身往往就不精确,大多数情况会使相邻曲面之间产生缝隙,重叠或不连续,在离散曲面之前或者用来生成体网格之前必须消除这些缺陷,这个过程称为“CAD 修复”[17-18]。

下面将介绍生成结构网格(11.1 节)和非结构网格(11.2 节)的基本方法。由于篇幅所限,在此仅作简要说明,更详细的介绍请参阅文献[19-20],这两篇文献对曲面建模、结构与非结构面、体网格生成等问题进行了更深入的探讨。关于网格生成方法的发展回顾可参见文献[21]。

11.1　结构网格

结构网格的显著特征是,物理空间中的网格点以一种独特的方式映射到由

三个整数 i,j,k（每个坐标方向对应一个整数）组成的连续集合上，该整数集定义了一个“计算空间”（图 3.2）。在计算空间中，ξ,η,ζ 与 i,j,k 的关系如下：

$$
\begin{aligned}
\xi&=i/i_{\max}, \quad i=0,1,2,\cdots,i_{\max}\\
\eta&=j/j_{\max}, \quad j=0,1,2,\cdots,j_{\max}\\
\zeta&=k/k_{\max}, \quad k=0,1,2,\cdots,k_{\max}
\end{aligned}
\tag{11.1}
$$

该映射关系意味着 $0\leqslant\xi\leqslant1$，$0\leqslant\eta\leqslant1$ 和 $0\leqslant\zeta\leqslant1$。连接相邻网格点，在计算空间形成立方体，在物理空间形成六面体（2D 为四边形）。结构网格生成方式是用四边形（称为面网格）离散流场区域的边界面，流场区域内部用六面体填充。流场区域内的网格称为体网格。

生成结构网格首先沿着边界线（面的边界）分布网格点，通常在曲率大的区域布置的节点密度大，然后利用边界曲线上的点分布，生成面网格，最后基于包围物理域的面网格构建体网格。常见的问题是如何以已知边界离散为基础生成体网格，这可以通过两种不同的方法来解决：

（1）代数网格生成方法；

（2）求解偏微分方程组（partial differential equations，PDEs）的网格生成方法。

应用 PDEs 生成网格需要一个有效的初始（面或体）网格，初始网格大多数采用代数方法生成。一般应用两种不同类型的 PDEs：

（1）椭圆型方程组；

（2）双曲型方程组。

代数网格生成（见 11.1.2 节）使用直接函数描述计算空间和物理空间之间的坐标转换，应用最广泛的代数技术是超限插值（transfinite interpolation，TFI），给定所有边界上的点分布，通过插值生成物理区域内的网格点，TFI 方法的特殊公式可以在边界处控制角度和网格间距。

基于椭圆型 PDEs（见 11.1.3 节）的方法是应用最广的方法之一，允许用户指定网格线与边界之间的角度，控制表面附近网格的间距和增长比。椭圆型网格生成方法还能保证整个区域内网格的光滑性，可以生成高质量的边界正交网格。与代数或双曲型网格生成方法相比，椭圆型网格生成方法的缺点是计算时间长，并且存在数值实现困难的问题。

双曲型 PDEs 也可以用于网格生成（见 11.1.4 节）。该方法在两个面之间沿着某个特定的计算坐标方向推进生成体网格，推进过程非常直观，即基于已知的表面点分布生成新层。双曲型网格生成方法固有的缺陷是不能完全控制外部网格边界的形状，但这并不一定表示存在问题，例如在外流情况下。双曲型网格生成可以使整个区域的网格都基本正交，而且计算量也很小。

11.1.1　C 型、H 型和 O 型网格拓扑结构

开始生成任何网格之前,必须考虑网格的拓扑结构,即决定需要多少个网格块,以及这些块之间应该如何排列(对于复杂的几何外形,这项工作可能需要几周到几个月的时间)。对于每个网格块,必须将计算区域中的边界(或其部分边界)指定到物理空间中的特定边界(如固壁、远场等),在物理空间中网格的形状与边界的指定有很大关系。在实际应用中,构建了三种标准的单块网格拓扑结构,因为在平面视图中,网格线类似于相应的大写字母,因此将它们称为 C 型、H 型或 O 型网格拓扑结构。3D 网格可以描述为两种网格拓扑结构的组合,如机翼周围的网格通常由流向的 C 型网格(见图 11.2)和展向的 O 型网格(或 H 型网格)组成,这种情况称为 C-O 型网格。下面将详细讨论 C 型、H 型和 O 型三种网格拓扑结构。

1. C 型网格拓扑结构

对于 C 型网格拓扑结构,气动外形由一族网格线包围,这组网格线还构成尾迹区(如果存在的话),具体情况如图 11.5 所示,由图可见,η = 常数的网格线从远场($\xi=0$)开始,然后是尾迹,通过后缘(节点 b),顺时针方向围绕物体,最后再次延续到远场($\xi=1$)。另一组网格线(ξ = 常数)从物体和尾迹的法向开始。$\eta=0$ 网格线的 $a-b$ 部分(段)表示坐标切割边界,这意味着将物理空间的线段 $a-b$ 映射到计算空间的两段,即线段 $b\leqslant\xi\leqslant a$ 和线段 $b'\leqslant\xi\leqslant a'$,因此,坐标切割的上部(线段 $b'-a'$)和下部(线段 $a-b$)的节点分别保存在计算机内存中,关于坐标切割边界条件的介绍见 8.7 节(图 11.6)。

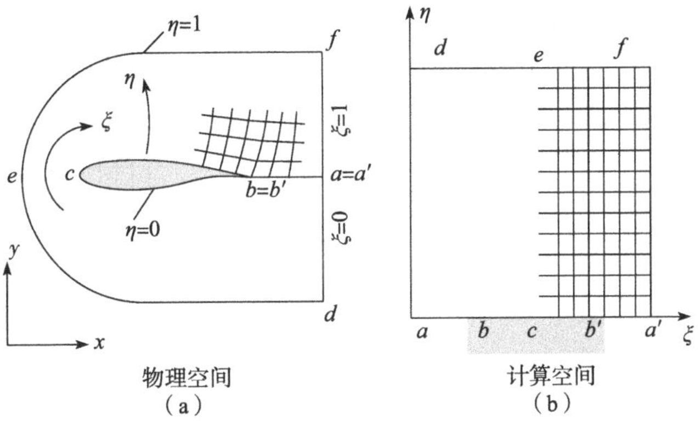

图 11.5　2D 情况下 C 型网格的拓扑结构

2. H 型网格拓扑结构

H 型网格拓扑结构是涡轮机械中叶片流道网格生成常用的拓扑结构,如

图 11.7 所示。由图可见，物体表面用 $\eta=0$ 和 $\eta=1$ 两条不同的网格线来描述。与 C 型网格不同，一族网格线（η= 常数）与流线非常相近（入口位于 $\xi=0$ 处，出口位于 $\xi=1$ 处）。

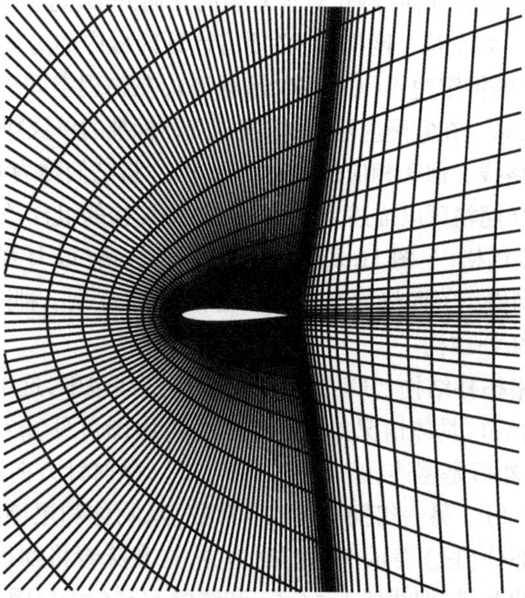

图 11.6　围绕 NACA0012 翼型的 C 型网格的局部视图

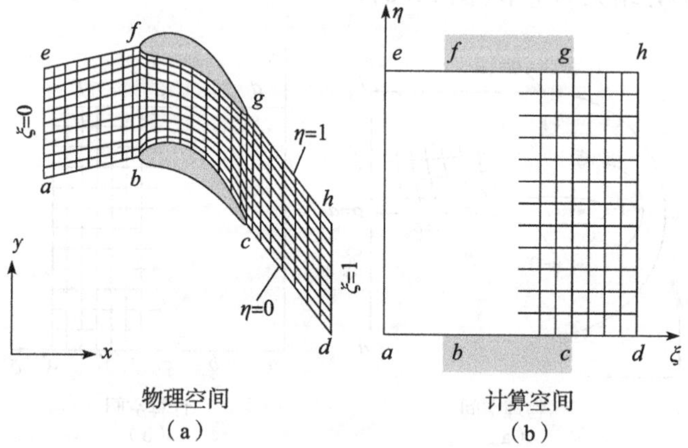

图 11.7　2D 情况下 H 型网格的拓扑结构

初看一眼，网格线没有明显的坐标切割。然而，在涡轮机械中，a–b 段和 e–f 段互为周期（3D 情况为旋转周期），c–d 段和 g–h 段同样如此。这类边界条件将在 8.8 节中讨论。图 11.8 给出了涡轮叶片之间非正交 H 型网格的一个实例。

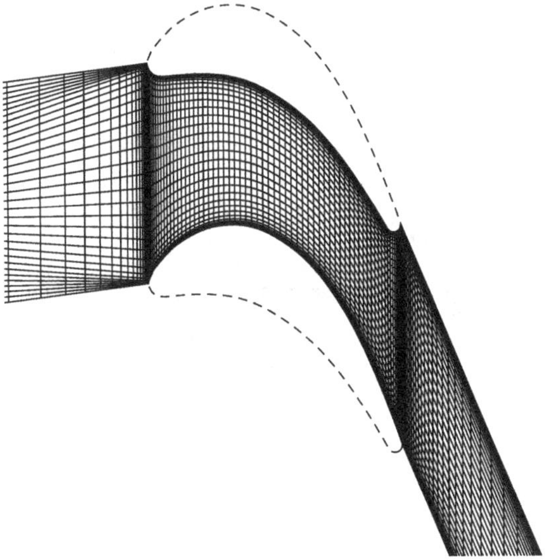

图 11.8　涡轮叶片(虚线)之间的 H 型网格的局部视图

3. O 型网格拓扑结构

从图 11.9 的 O 型拓扑结构示意图可以看出,一族网格线(η=常数)在物体周围形成闭合曲线,另一族网格线(ξ=常数)沿物体和外边界(在此为远场)之间的径向方向。整个边界线 $\eta=0$ 表示物体的轮廓($a-a'$)。坐标切割由计算空间中的边界 $\xi=0$(节点 $a-c$)和 $\xi=1$(节点 $a'-c'$)确定。图 11.10 显示了一个用于模拟翼型无黏绕流的标准 O 型网格。O 型拓扑结构的缺点是在尖锐后缘处网格质量差。

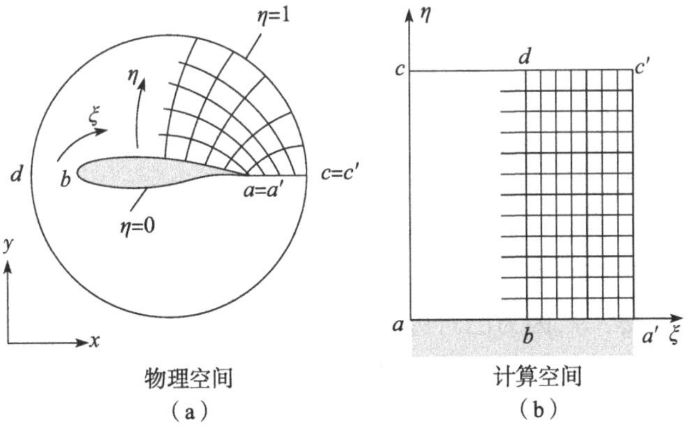

图 11.9　2D 情况下 O 型网格拓扑结构图

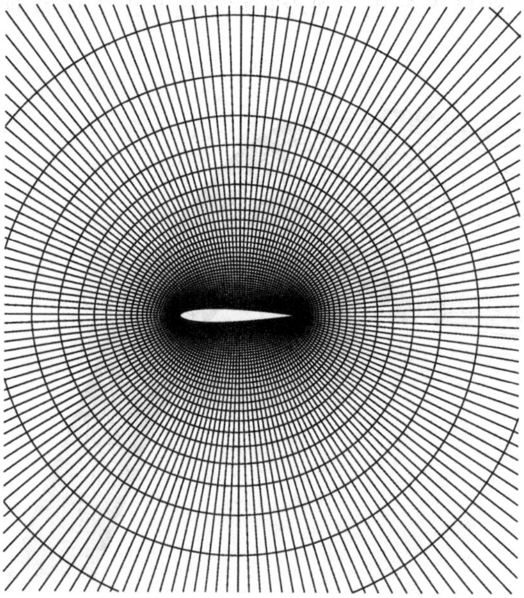

图 11.10　围绕 NACA0012 翼型的 O 型网格的局部视图

11.1.2　代数网格生成

由给定的边界点分布生成面或体网格使用最广泛的代数技术是 TFI 方法，该方法由 Gordon 和 Hall 于 1973 年首次提出[22]。TFI 方法在计算空间的每个坐标方向上采用 1D 单变量插值。单变量插值函数的一般形式如下：

$$
\begin{aligned}
\boldsymbol{U} &= \sum_{i=1}^{L}\sum_{n=0}^{P}\alpha_i^n(\xi)\frac{\partial^n \boldsymbol{r}(\xi_i,\eta,\zeta)}{\partial\xi^n}\\
\boldsymbol{V} &= \sum_{j=1}^{M}\sum_{m=0}^{Q}\beta_j^m(\eta)\frac{\partial^m \boldsymbol{r}(\xi,\eta_j,\zeta)}{\partial\eta^m}\\
\boldsymbol{W} &= \sum_{k=1}^{N}\sum_{l=0}^{R}\gamma_k^l(\zeta)\frac{\partial^l \boldsymbol{r}(\xi,\eta,\zeta_k)}{\partial\zeta^l}
\end{aligned}
\tag{11.2}
$$

式中：$\boldsymbol{U},\boldsymbol{V},\boldsymbol{W}$ 分别表示 ξ,η,ζ 方向的单变量插值函数。$\alpha_i^n(\xi)$，$\beta_j^m(\eta)$，$\gamma_k^l(\zeta)$ 为混合函数，$\boldsymbol{r}$ 表示网格点在物理空间中的位置。为了计算插值函数，必须指定位置 $\boldsymbol{r}$ 和导数 $\partial^n\boldsymbol{r}/\partial\xi^n$ 等，因为已对边界曲线或边界面进行了离散，所以可以将上述指定的值代入式(11.2)，$\boldsymbol{U},\boldsymbol{V},\boldsymbol{W}$ 已知后，使用插值函数的布尔和运算就可以生成流场区域内的网格：

$$
\boldsymbol{r}=\boldsymbol{U}\oplus\boldsymbol{V}\oplus\boldsymbol{W}=\boldsymbol{U}+\boldsymbol{V}+\boldsymbol{W}-\boldsymbol{UV}-\boldsymbol{UW}-\boldsymbol{VW}+\boldsymbol{UVW}
\tag{11.3}
$$

式(11.3)中的方法保证 2D 中所有四条边界曲线和 3D 中所有六个边界面都匹配,式中的张量积计算为

$$
\begin{aligned}
\boldsymbol{UV} &= \sum_{i=1}^{L}\sum_{j=1}^{M}\sum_{m=0}^{Q}\sum_{n=0}^{P}\alpha_i^n\beta_j^m\frac{\partial^{mn}\boldsymbol{r}(\xi_i,\eta_j,\zeta)}{\partial\eta^m\partial\xi^n}\\
\boldsymbol{UW} &= \sum_{i=1}^{L}\sum_{k=1}^{N}\sum_{l=0}^{R}\sum_{n=0}^{P}\alpha_i^n\gamma_k^l\frac{\partial^{ln}\boldsymbol{r}(\xi_i,\eta,\zeta_k)}{\partial\zeta^l\partial\xi^n}\\
\boldsymbol{VW} &= \sum_{j=1}^{M}\sum_{k=1}^{N}\sum_{l=0}^{R}\sum_{m=0}^{Q}\beta_j^m\gamma_k^l\frac{\partial^{lm}\boldsymbol{r}(\xi,\eta_j,\zeta_k)}{\partial\zeta^l\partial\eta^m}\\
\boldsymbol{UVW} &= \sum_{i=1}^{L}\sum_{j=1}^{M}\sum_{k=1}^{N}\sum_{l=0}^{R}\sum_{m=0}^{Q}\sum_{n=0}^{P}\alpha_i^n\beta_j^m\gamma_k^l\frac{\partial^{lmn}\boldsymbol{r}(\xi_i,\eta_j,\zeta_k)}{\partial\zeta^l\partial\eta^m\partial\xi^n}
\end{aligned}
\tag{11.4}
$$

与网格生成相关的布尔算子和张量积的更多细节可参见文献[23-24]。

可使用线性插值、拉格朗日插值、埃尔米特插值、样条插值等各种类型的插值函数,线性 TEI 是使用最广的插值方法。设式(11.2)和式(11.4)中的 $L=M=N=2$ 和 $P=Q=R=0$,就得到线性 TFI,采用这种方法,可以仅根据给定的 6 个边界面(设边界面处 $\xi_1=0,\xi_2=1,\eta_j,\zeta_k$ 类似)的点分布来生成体网格。线性 TFI 的混合函数为[25]

$$
\begin{aligned}
\alpha_1^0(\xi) &= 1-\xi,\quad \alpha_2^0(\xi)=\xi\\
\beta_1^0(\eta) &= 1-\eta,\quad \beta_2^0(\eta)=\eta\\
\gamma_1^0(\zeta) &= 1-\zeta,\quad \gamma_2^0(\zeta)=\zeta
\end{aligned}
\tag{11.5}
$$

线性 TFI 计算效率非常高。图 11.8 为使用线性 TFI 代数生成网格的一个实例。

采用三次混合函数的埃尔米特 TFI 还可以在边界上指定网格线的斜率,设式(11.2)和式(11.4)中的 $L=M=N=2$ 和 $P=Q=R=1$,就得到埃尔米特 TFI,也可以在不同计算坐标中将线性插值和埃尔米特插值相结合。关于埃尔米特 TFI 技术以及对 TFI 技术(如网格间距控制)的进一步拓展可参见文献[20]的第 3 章。

11.1.3 椭圆型网格生成

众所周知,基于椭圆型 PDEs 的网格生成方法可以生成单元尺寸和网格线斜率光滑变化的网格,而且椭圆型网格生成方法能够控制边界附近网格的正交性和间距,这对于黏性流动的模拟尤为重要。椭圆型 PDEs 网格生成法最早是 Thompson 等[26]在 1974 年提出的,如文献[27-28]对该方法的 2D 数值实现进行了详细介绍。

在 3D 中,网格点的未知笛卡儿坐标 $\boldsymbol{r}=[x,y,z]^{\mathrm{T}}$ 的泊松方程组可写成

$$\alpha_{11}\frac{\partial^2 \boldsymbol{r}}{\partial \xi^2}+\alpha_{22}\frac{\partial^2 \boldsymbol{r}}{\partial \eta^2}+\alpha_{33}\frac{\partial^2 \boldsymbol{r}}{\partial \zeta^2}+2\left(\alpha_{12}\frac{\partial^2 \boldsymbol{r}}{\partial \xi\partial \eta}+\alpha_{13}\frac{\partial^2 \boldsymbol{r}}{\partial \xi\partial \zeta}+\alpha_{23}\frac{\partial^2 \boldsymbol{r}}{\partial \eta\partial \zeta}\right)$$
$$=-\frac{1}{J^2}\left(P\frac{\partial \boldsymbol{r}}{\partial \xi}+Q\frac{\partial \boldsymbol{r}}{\partial \eta}+R\frac{\partial \boldsymbol{r}}{\partial \zeta}\right) \tag{11.6}$$

式中：P,Q,R 表示控制函数，度量系数 α 由下面关系式给出：

$$\alpha_{11}=\left(\frac{\partial \boldsymbol{r}}{\partial \eta}\cdot\frac{\partial \boldsymbol{r}}{\partial \eta}\right)\left(\frac{\partial \boldsymbol{r}}{\partial \zeta}\cdot\frac{\partial \boldsymbol{r}}{\partial \zeta}\right)-\left(\frac{\partial \boldsymbol{r}}{\partial \eta}\cdot\frac{\partial \boldsymbol{r}}{\partial \zeta}\right)^2$$
$$\alpha_{22}=\left(\frac{\partial \boldsymbol{r}}{\partial \xi}\cdot\frac{\partial \boldsymbol{r}}{\partial \xi}\right)\left(\frac{\partial \boldsymbol{r}}{\partial \zeta}\cdot\frac{\partial \boldsymbol{r}}{\partial \zeta}\right)-\left(\frac{\partial \boldsymbol{r}}{\partial \xi}\cdot\frac{\partial \boldsymbol{r}}{\partial \zeta}\right)^2$$
$$\alpha_{33}=\left(\frac{\partial \boldsymbol{r}}{\partial \xi}\cdot\frac{\partial \boldsymbol{r}}{\partial \xi}\right)\left(\frac{\partial \boldsymbol{r}}{\partial \eta}\cdot\frac{\partial \boldsymbol{r}}{\partial \eta}\right)-\left(\frac{\partial \boldsymbol{r}}{\partial \xi}\cdot\frac{\partial \boldsymbol{r}}{\partial \eta}\right)^2 \tag{11.7}$$
$$\alpha_{12}=\left(\frac{\partial \boldsymbol{r}}{\partial \xi}\cdot\frac{\partial \boldsymbol{r}}{\partial \zeta}\right)\left(\frac{\partial \boldsymbol{r}}{\partial \eta}\cdot\frac{\partial \boldsymbol{r}}{\partial \zeta}\right)-\left(\frac{\partial \boldsymbol{r}}{\partial \xi}\cdot\frac{\partial \boldsymbol{r}}{\partial \eta}\right)\left(\frac{\partial \boldsymbol{r}}{\partial \zeta}\cdot\frac{\partial \boldsymbol{r}}{\partial \zeta}\right)$$
$$\alpha_{13}=\left(\frac{\partial \boldsymbol{r}}{\partial \xi}\cdot\frac{\partial \boldsymbol{r}}{\partial \eta}\right)\left(\frac{\partial \boldsymbol{r}}{\partial \eta}\cdot\frac{\partial \boldsymbol{r}}{\partial \zeta}\right)-\left(\frac{\partial \boldsymbol{r}}{\partial \xi}\cdot\frac{\partial \boldsymbol{r}}{\partial \zeta}\right)\left(\frac{\partial \boldsymbol{r}}{\partial \eta}\cdot\frac{\partial \boldsymbol{r}}{\partial \eta}\right)$$
$$\alpha_{23}=\left(\frac{\partial \boldsymbol{r}}{\partial \xi}\cdot\frac{\partial \boldsymbol{r}}{\partial \zeta}\right)\left(\frac{\partial \boldsymbol{r}}{\partial \xi}\cdot\frac{\partial \boldsymbol{r}}{\partial \eta}\right)-\left(\frac{\partial \boldsymbol{r}}{\partial \eta}\cdot\frac{\partial \boldsymbol{r}}{\partial \zeta}\right)\left(\frac{\partial \boldsymbol{r}}{\partial \xi}\cdot\frac{\partial \boldsymbol{r}}{\partial \xi}\right)$$

式(11.7)中，括号中的项表示标量积，坐标变换雅可比矩阵行列式的倒数(J^{-1})可根据式(A.19)求出。需要注意的是，设 $P=Q=R=0$，则式(11.6)退化为拉普拉斯方程，利用拉普拉斯方程固有的光滑特性，可提高代数生成网格的质量，但无法控制内部点的分布。

略去关于 ζ 方向的导数，可以很容易地由式(11.6)推导出用于生成 2D 网格或表面网格的椭圆型方程组：

$$\alpha_{11}\frac{\partial^2 x}{\partial \xi^2}-2\alpha_{12}\frac{\partial^2 x}{\partial \xi\partial \eta}+\alpha_{22}\frac{\partial^2 x}{\partial \eta^2}=-\frac{1}{J^2}\left(P\frac{\partial x}{\partial \xi}+Q\frac{\partial x}{\partial \eta}\right)$$
$$\alpha_{11}\frac{\partial^2 y}{\partial \xi^2}-2\alpha_{12}\frac{\partial^2 y}{\partial \xi\partial \eta}+\alpha_{22}\frac{\partial^2 y}{\partial \eta^2}=-\frac{1}{J^2}\left(P\frac{\partial y}{\partial \xi}+Q\frac{\partial y}{\partial \eta}\right) \tag{11.8}$$

式(11.8)中的度量系数 α 为

$$\alpha_{11}=\left(\frac{\partial x}{\partial \eta}\right)^2+\left(\frac{\partial y}{\partial \eta}\right)^2$$
$$\alpha_{12}=\frac{\partial x}{\partial \xi}\frac{\partial x}{\partial \eta}+\frac{\partial y}{\partial \xi}\frac{\partial y}{\partial \eta} \tag{11.9}$$
$$\alpha_{22}=\left(\frac{\partial x}{\partial \xi}\right)^2+\left(\frac{\partial y}{\partial \xi}\right)^2$$

式(11.8)中坐标变换雅可比矩阵行列式的倒数 J^{-1} 由式(A.25)确定。

边界正交网格的生成需要在求解方程组(11.6)或方程组(11.8)过程中指定合适的边界条件,从根本上说我们可以应用 Neumann 条件或 Dirichlet 条件。Neumann 边界条件可直接指定网格线与边界之间的夹角,在这种情况下,不需要控制函数($P=Q=R=0$),但无法控制边界上的点分布和间距,事实上,边界点将自动重新配置,以匹配给定的网格线偏斜度,因此该方法不适用于黏性壁面边界。关于椭圆型网格生成的 Neumann 条件的更多细节,请参见文献[20]的第 6 章。

Dirichlet 边界条件用于边界点位置必须保持固定不变的情况,应用控制函数来达到所需的夹角和间距。2D 情况下控制函数对网格的影响如图 11.11 所示,由图可见,因为边界点保持固定不变,所以式(11.8)中的负 P 值使 $\xi=$常数的网格线向左旋转;而 $Q<0$ 使 $\eta=$常数的网格线移动到离边界更近的位置。首先,根据偏斜度和网格间距的规定值与实际值之差,计算在各自边界上的控制函数的值;然后,将 P,Q,R 的边界值插值到区域内,并求解方程组(11.6)或方程组(11.8)。重复上述过程,直到达到所需的网格特性,这种基于 Dirichlet 条件的方法是由 Sorenson[27]、Thomas 和 Middlecoff[29] 在 2D 中开发的,后来由 Sorenson[30] 和 Thompson[31] 将其推广到 3D。

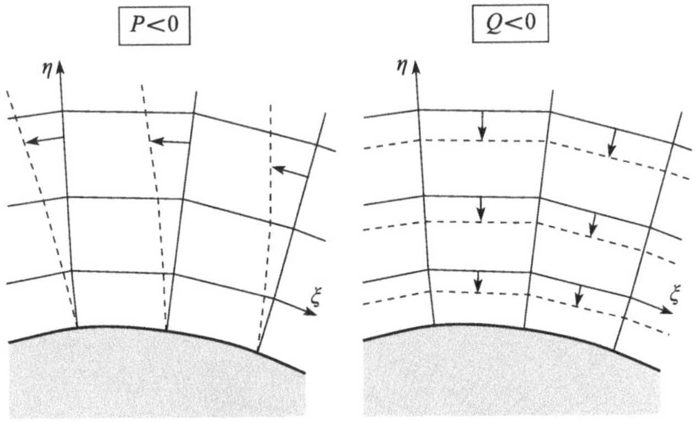

图 11.11　2D 椭圆型网格生成中,在 $\eta=$常数边界处控制函数的影响。P 控制偏斜度,Q 控制网格间距

椭圆型方程组(11.6)或椭圆型方程组(11.8)通常用二阶中心有限差分格式离散,所得到的线性代数方程组可以用任何标准方法求解,例如用多重网格技术加速的高斯-赛德尔松弛格式,控制函数的值在外迭代中更新。为了提高迭代过程的鲁棒性,建议用拉普拉斯方程组($P=Q=R=0$)进行几次迭代,以光顺初始网格(通常为代数网格)[32]。

采用椭圆型方程组生成网格的效果如图 11.12 所示,首先,使用 11.1.2 节中介绍的 TFI 方法生成围绕翼型的网格,生成的网格如图 11.12(a)所示;然后,用拉普拉斯方程（即设式(11.8)中的 $P=Q=R=0$）对代数网格进行光顺，由图可见，拉普拉斯光顺改善了翼型前缘周围的网格，但无法控制垂直于表面的网格间距；只有借助控制函数 $P(\xi,\eta)$ 和 $Q(\xi,\eta)$ 才能实现网格间距的控制，如图 11.12(c)所示，控制函数还能确保网格线与机翼表面垂直。图 11.1、图 11.2 和图 11.6 显示了采用椭圆型方法生成网格的其他实例。

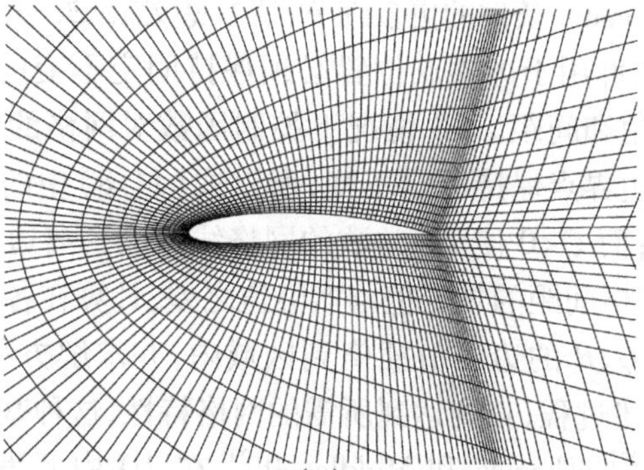

(a)

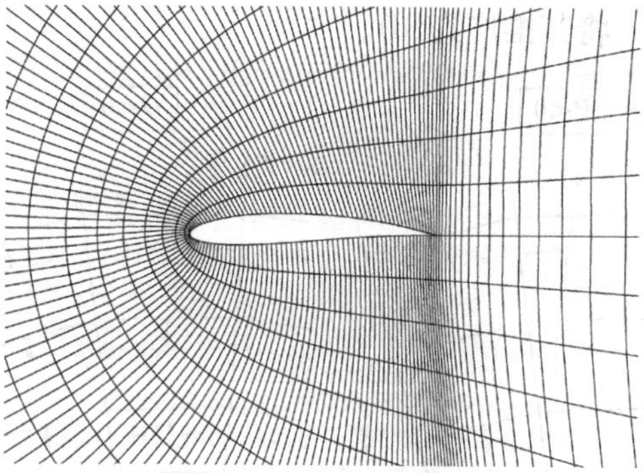

(b)

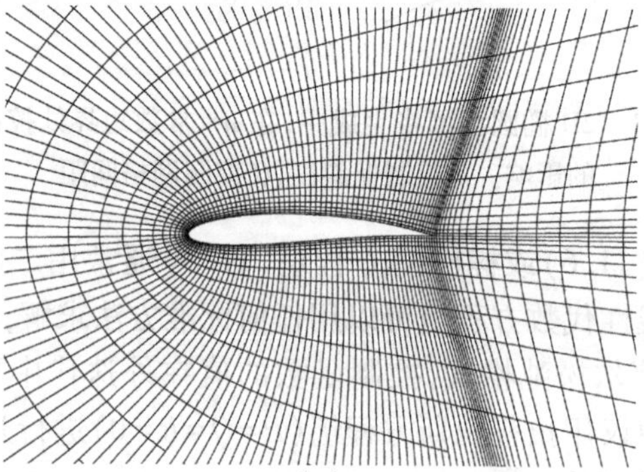

(c)

图 11.12　同一翼型的网格序列：从上到下分别为采用 TFI 代数方法，拉普拉斯方程（$P=Q=0$）和椭圆型方程组(11.8)生成的网格

11.1.4　双曲型网格生成

双曲型 PDEs 适合于不需要精确控制外边界形状的网格生成。为了生成一套网格，首先必须指定一个初始点分布；然后，沿法线方向，由上一层网格点推进一给定的距离得到新一层的网格点，从而构建网格。Starius[33] 以及 Steger 和 Chaussee[34] 提出了双曲型 PDEs 在网格生成中的应用，文献[20]的第 5 章中有关于双曲型网格生成方法的最新探讨，文献[35-42]介绍了对该方法的各种推广和改进。

生成体网格的双曲型方程组为

$$\begin{cases}\dfrac{\partial \boldsymbol{r}}{\partial \xi}\cdot\dfrac{\partial \boldsymbol{r}}{\partial \zeta}=0\\[2mm]\dfrac{\partial \boldsymbol{r}}{\partial \eta}\cdot\dfrac{\partial \boldsymbol{r}}{\partial \zeta}=0\\[2mm]\dfrac{\partial \boldsymbol{r}}{\partial \zeta}\cdot\left(\dfrac{\partial \boldsymbol{r}}{\partial \xi}\times\dfrac{\partial \boldsymbol{r}}{\partial \eta}\right)=\Omega\end{cases}\tag{11.10}$$

式中：$\boldsymbol{r}=[x,y,z]^{\mathrm{T}}$ 表示网格点的笛卡儿坐标；ξ,η,ζ 表示计算坐标（式(11.1)）；Ω 为用户指定的网格单元体积，此外，假定式(11.10)中 $\zeta=$常数的面表示初始状态。式(11.10)中的前两个关系式表示正交条件（$\xi=$常数和 $\eta=$常数的面垂直于 $\zeta=$常数的面）。式(11.10)中的最后一个关系式确保网格单元体积等于 Ω，这为控制网格层之间的间距提供了可能。

双曲型方程组的 2D 公式可写成

$$\begin{cases}\dfrac{\partial x}{\partial \xi}\dfrac{\partial x}{\partial \eta}+\dfrac{\partial y}{\partial \xi}\dfrac{\partial y}{\partial \eta}=0\\[2mm]\dfrac{\partial x}{\partial \xi}\dfrac{\partial y}{\partial \eta}-\dfrac{\partial y}{\partial \xi}\dfrac{\partial x}{\partial \eta}=\Omega\end{cases}\tag{11.11}$$

其中，Ω 表示指定的网格单元面积，初始点分布在曲线 $\eta=0$ 上给定。

对双曲型网格生成方程组(11.10)或方程组(11.11)，采用二阶中心差分格式，在 ξ 和 η 坐标(3D)或 ξ 坐标(2D)上进行离散，采用一阶隐式格式（ADI 格式，见 6.2.3 节）沿 ζ 方向（式(11.10)）或 η 方向（式(11.11)）推进。如此一来，可以仅根据所需的网格间距来选择推进步长，用标准三对角解算器对隐式算子求逆。为了使推进过程稳定，必须在离散方程中添加人工耗散项（二阶差分）。双曲型方程组(11.10)或方程组(11.11)的求解比椭圆型方程组的求解更快，通常也更容易。

11.2　非结构网格

典型的非结构网格由三角形（2D 情况）和四面体（3D 情况）组成，然而，由各种单元类型组合构建的非结构网格变得越来越流行,例如用六面体或三棱柱离散边界层,然后流场区域的其余部分用四面体填充,用金字塔作为六面体或三棱柱和四面体之间的过渡单元,称为混合网格。与四面体网格相比,结构的六面体网格的优点是,保持了高度拉伸的黏性网格在壁面法向上的精度,并且减少了单元、边和面的数量。另外,非结构四面体网格的优点是能够快速离散复杂外形(图 11.3 和图 11.4),并且用户干预少。混合网格试图将结构网格和非结构网格的优点结合起来。

对于非结构网格,节点和网格单元是准随机排序的,即相邻单元或网格点不能通过其索引直接识别(图 3.3)。因为非结构网格不需要遵循任何预定的拓扑结构,使得其具有极大的几何灵活性。此外,对物理解的网格自适应——网格细化或粗化——在非结构网格上要比在结构网格上更容易实现。

应用于 CFD 的非结构网格生成方法主要有:

(1) Delaunay 方法;

(2) 阵面推进法。

也可以将这两种方法相结合,根据所采用的基本方法,称之为阵面推进 Delaunay 方法[43-45]或阵面 Delaunay 方法[46-48]。除了这两种标准技术,还有气泡填充法[49-54]等其他方法。本章仅介绍 Delaunay 方法和阵面推进方法的基本特征,文献[20,55-56]对这两种方法进行了概述。

Delaunay 方法主要是指连接网格点以形成三角形(2D)和四面体(3D)的特定方法。Delaunay 三角化最重要的特征是任何三角形(四面体)的外接圆(或 3D 中的外接球)不包含其他网格点。空外接圆准则使得 2D 情况下所有三角形的最小角度最大化(最大-最小三角化),因此可以生成正则性高的网格。

阵面推进法的思路是从已知的边界离散(表面网格)开始,逐单元依次生成网格,单元的开放曲面构成阵面,新的三角形(四面体)通过在阵面前放置点来构建,阵面以这种方式向计算域推进,直到所有的空间被填满。点的配置由背景网格控制。阵面推进法的优点是点分布光滑,且能无条件地保证边界的完整性,但阵面推进法生成网格的速度比 Delaunay 方法要慢。

11.2.1 Delaunay 三角化

Delaunay 三角化基于 Dirichlet 在 1850 年提出的一种将空间划分成唯一的凸区域集的理论[57]。对于给定的点集,每个凸区域表示特定点周围的空间,该空间内的点到该特定点的距离小于到其他点的距离,这些凸区域形成的多边形(3D 为多面体),称为 Dirichlet 棋格或 Voronoj 图[58]。如果用直线连接 Voronoj 图中共享某段(面)的点对,则得到 Delaunay 三角化[59]。Delaunay 三角化定义了一组包含点的凸包的三角形(3D 为四面体),如图 11.13 所示,它是 Voronoj 图的二次图。Voronoj 多边形的顶点,在 2D 情况是三角形的外接圆圆心,在 3D 情况为四面体的外接球球心,这表明每个三角形的外接圆(每个四面体的外接球)的内部不包含其他点。

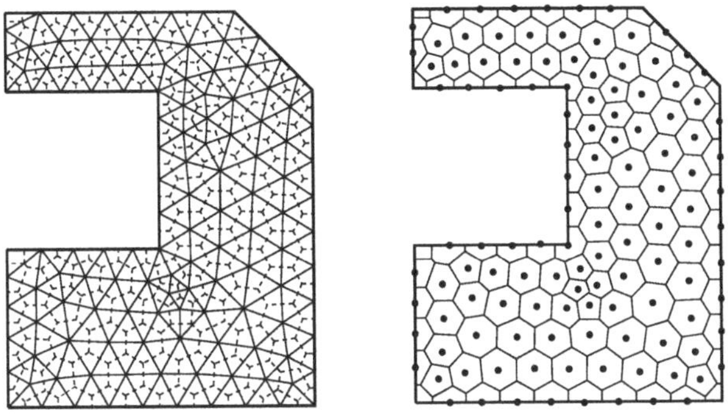

(a)　　　　　　　　(b)

图 11.13　Delaunay 三角化((a)用虚线表示)和 Voronoj 图((b)用实线表示)

如前所述,Delaunay 法是一种连接网格点的特殊方法,网格点位置必须采用其他技术来确定。因此,构建 Delaunay 网格的一种常用方法是将点逐点插入初始三角剖分中,然后对网格进行局部重新划分三角形,以满足空外接圆(外接球)准则。递增点插入算法的步骤描述如下:

(1) 离散物理域的边界。

(2) 生成包含所有边界点的初始 Delaunay 网格,可以是边界点本身的三角化(图 11.14(a)),也可以是如图 11.14(b)所示的围绕物理域的可分成三角形(四面体)的四边形(3D 为六面体)。

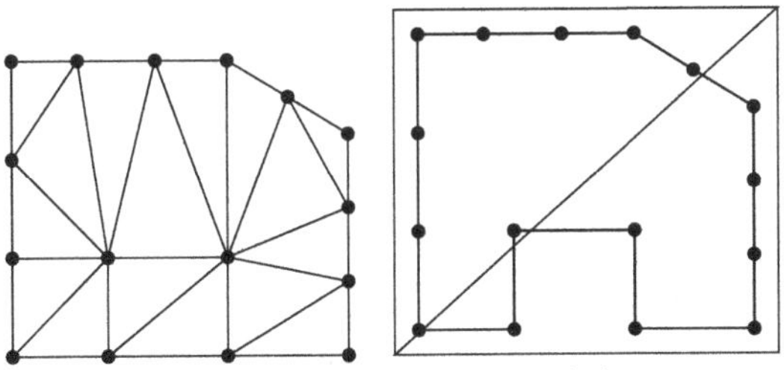

(a)　　　　　　　　(b)

图 11.14　2D 情况不同的初始 Delaunay 网格

(a) 边界点的三角化;(b) 分成两个三角形的四边形

(3) 如果尚未完成剖分,则使用约束 Delaunay 技术将所有边界点插入初始三角剖分中。

(4) 将网格中所有不符合尺寸或质量标准的三角形(四面体)建立一个列表,列表从最差单元开始排序。

(5) 在列表第一个单元的外接圆圆心[60-62](图 11.15)或 Voronoj 图线段[63-64]处放置一个新点,并对网格进行局部重新三角化。检查每个新单元,如果不符合尺寸/质量标准,则将其添加到列表中。

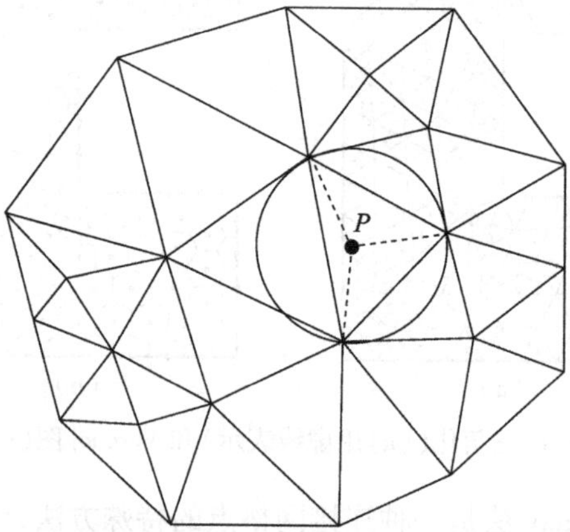

图 11.15 将新点 P 插入有效的 Delaunay 三角剖分内,点位于外接圆圆心

(6) 如果列表中仍有单元,继续步骤(5);

(7) 删除区域外的单元并恢复边界;

(8) 检查网格质量(见 11.2.5 节),如有必要,对网格进行光顺和/或交换边。

也可以在步骤(3)之后进行边界恢复(步骤(7))。

由上可见,Delaunay 网格生成的核心任务是将新点插入到有效的三角剖分中(步骤(3)和步骤(5)),为此提出了几种算法,应用最广泛的是 Green 和 Sibson[65]、Bowyer[66]以及 Watson[67]提出的算法。后面将介绍 Watson 算法。

执行步骤(4)和步骤(5),必须用一些合适的度量标准对单元进行评价。度量标准可以是单元质量(最小角度、长宽比,见文献[68-69]),或单元尺寸(体积、边长)。可直接由其几何外形对单元质量进行评价,但单元尺寸度量必须表示为区域内空间位置的函数,存在下面几种单元尺寸度量方法:

(1) 指定分析函数(例如,单元尺寸与到物体的距离成正比);

(2) 使用初始三角剖分从边界进行尺寸分布插值(见文献[20]的第 1 章,17~20 页);

(3) 基于四叉树(3D 为八叉树)结构的背景网格插值[70-75];

(4) 指定区域内的源(点、线等)(见文献[20]的第 1 章,20~22 页)。

此外,可以借助阵面推进技术布点[45-46]。

1. Watson 算法

点插入和再三角化的 Watson 方法包括以下步骤[67]：

(1) 定位包含插入点 P 的单元(图 11.15)；

(2) 找出外接圆(3D 为外接球)包含点 P 的所有单元，如图 11.16(a)所示；

(3) 删除三角剖分的所有相交单元；

(4) 将凸腔边界上的顶点与点 P 连接，形成新的单元(图 11.16(b))。

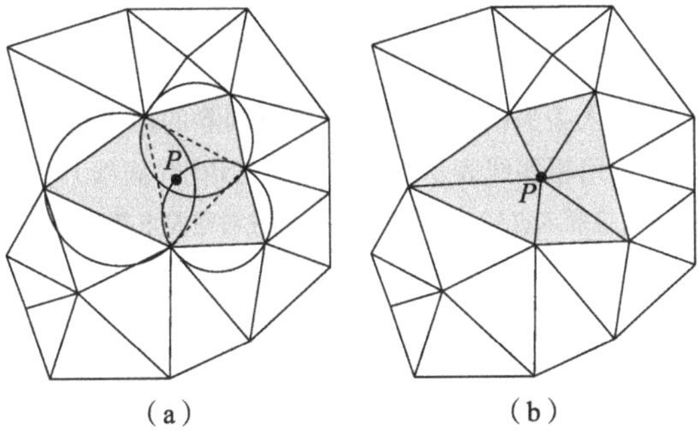

图 11.16　Watson 算法

(a) 剔除不符合要求的三角形；(b) 凸腔重新三角剖分。

当存储每个单元的信息时，特别适合于 Watson 算法的数据结构是：

(1) 组成单元的节点的索引；

(2) 与该单元共享一个面的各相邻单元的指针；

(3) 外接圆(球)的圆心和半径。

这种数据结构可以在步骤(2)中高效搜索相交单元，从包含插入点的单元的相邻单元开始搜索，然后继续搜索相邻单元的相邻单元，依此类推，如果单元的外接圆(外接球)不相交，则停止在该方向上的搜索。Delaunay 三角化的几何特性确保了该单元的相邻单元不相交，还保证所涉及的所有单元可采用简单的局部优化策略[67,76]。Watson 算法的计算量为 $N\log(N)$ 量级，其中 N 为网格点数，因此它是一种非常快速的网格生成方法。

2. 约束 Delaunay 三角化

Delaunay 三角化不能自动处理指定的边和面，比如物理区域的边界边和边界面。约束 Delaunay 三角剖分旨在解决这个问题[77]。2D 的边界边恢复如图 11.17 所示，根据具体情况采用边交换或者重新三角剖分。约束 Delaunay 三角化在 2D 中总是能得到一个有效的网格，但在 3D 中却无法保证。3D 边界离散的恢复必须分两步进行：第一步进行边界边恢复，第二步进行边界面恢复[56]。在一些情况下，必须在边界上插入额外的点(称为 Steiner 点)[78-81]，否则空腔无法重新进行四面体剖分。恢复边界边(面)后,可以剔除流动区域之外的单元。

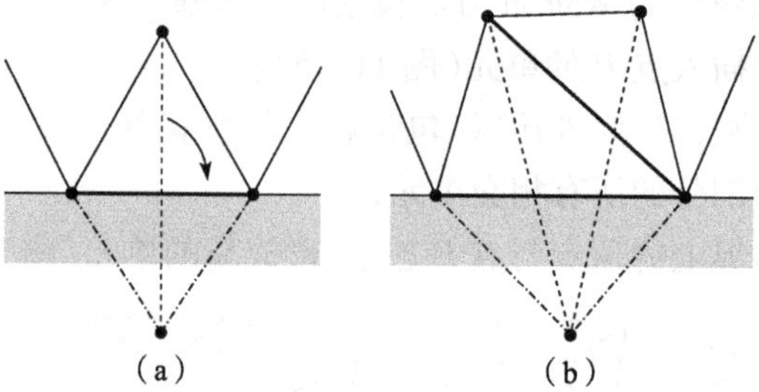

图 11.17 插入丢失的边界边。交换或剔除的边用虚线表示,新边用粗实线表示。区域外的三角形(虚线)被剔除
(a) 通过边交换插入;(b) 剔除相交三角形并对空腔重新进行三角剖分。

11.2.2 阵面推进法

Peraire 等[82-83]以及 Löhner 和 Parikh[84]在 20 世纪 80 年代末首次提出阵面推进法,该网格生成方法的步骤可归纳如下:

(1) 离散物理域边界(生成面网格);

(2) 生成一个表示阵面的边(3D 情况下为面)列表,起始时为边界的边(面),按照边(面)尺寸递增顺序对列表进行排序[55],此策略有助于生成平滑变化的单元;

(3) 选择列表的第一条边(面),并在阵面边(面)中心上方的法线方向上放置一个新点 P_0,如图 11.18 所示,与域的距离 d 由尺寸分布函数的当地值决定(参见文献[56]);

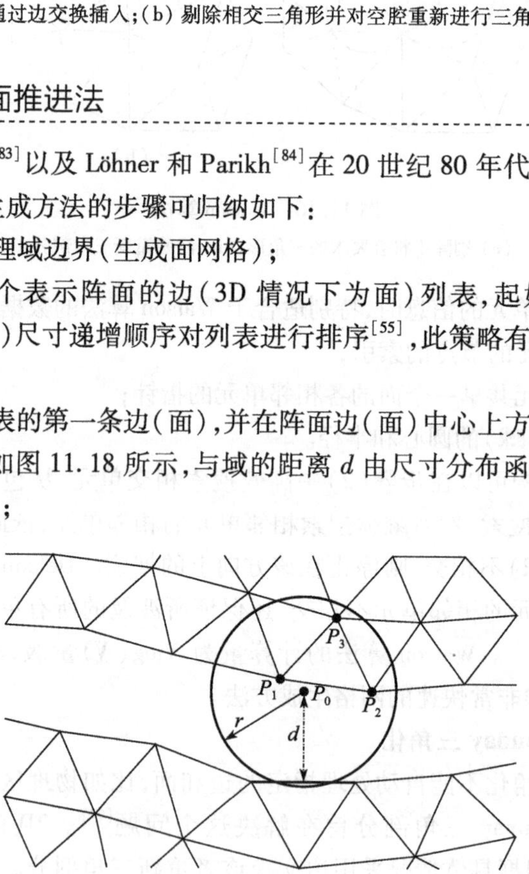

图 11.18 2D 阵面推进法中新节点 P_0 的插入。活动边用粗线表示

(4) 定义一个半径为 r、圆心为点 P_0 的圆(球),半径 r 取决于当地网格尺寸;

(5) 确定位于圆(球)内的所有点;

(6) 如果没有交点,利用点 P_0 生成一个新的单元,否则,根据交点到点 P_0 的距离对交点进行排序(即 P_1,P_2,P_3),用这些点形成单元,并接受第一个不与任何其他单元相交且满足给定质量标准的单元;

(7) 删除当前的阵面边(面),并将新形成的边(面)添加到列表中,再次对列表进行排序;

(8) 继续执行步骤(3),直到列表为空;

(9) 检查网格质量(见 11.2.5 节),如果有必要,光顺网格以及交换边。

典型 2D 构型推进阵面过程的各个阶段如图 11.19 所示。

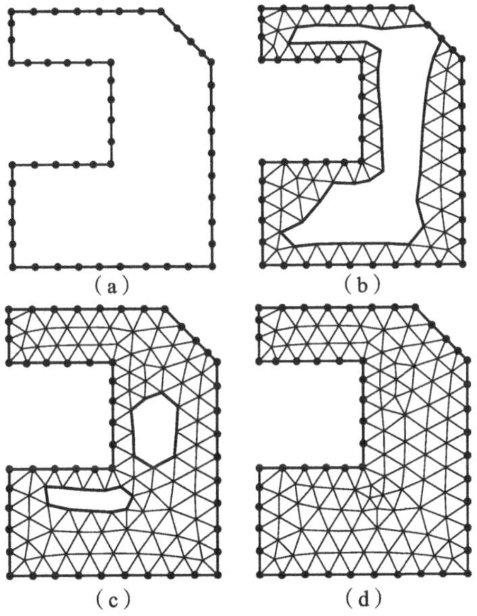

图 11.19　使用阵面推进技术生成网格的不同阶段。实际阵面用粗实线表示

域内单元尺寸的分布主要由边界上的点的密度决定。此外,步骤(3)中的距离 d 可以通过在区域内放置如点、线、圆柱等各种类型的源来控制(参见文献[20]的第 1 章,20~22 页)。当地单元尺寸可以从基于四叉树(3D 为八叉树)数据结构的背景网格中得到[70-75]。另外的方法是使用边界节点的 Delaunay 三角剖分[46-48,85]进行单元尺寸插值,或使用笛卡儿背景网格[86]。四叉树(八叉树)数据结构还可用于附近点的高效搜索(步骤 5),以及检查单元之间可能的交点。文献[87]设计了一个简单的基于弹簧近似的阵面相交测试,文献[88]对不同的搜索算法进行了详细比较。

与 Delaunay 三角化方法相比,阵面推进法的优点在于无条件地保留边界离散,以及更好地控制单元尺寸和网格平滑度。此外,Delaunay 方法比阵面推进法对舍入误差更敏感,但点搜索和交叉检查算法需要大量数值计算。在 CFD 领域,通常采用阵面推进技术和 Delaunay 算法相结合的方法,特别是在 3D 情况下[43-48,89]。文献[90]中对阵面推进法进行了详细介绍。

11.2.3 各向异性网格生成

在 11.2.1 节和 11.2.2 节中,描述了生成各向同性三角形或四面体网格单元的两种方法。各向同性网格适用于无黏流动的模拟,但 RANS 方程组的精确求解需要在边界层内和尾迹中具有很高的空间分辨率,因此在高 *Re* 流动情况下,各向同性网格将导致单元数量过多。显然,必须在黏性流动区域内采用各向异性的大长宽比单元,可采用非常容易进行拉伸的四边形(3D 情况为六面体或三棱柱),从而出现了如今非常流行的混合单元网格(见 11.2.4 节)。当然,对六面体或三棱柱进行剖分也能获得纯四面体网格,另外,可以直接生成拉伸的三角形(四面体)网格。在任何情况下,网格生成方法都应防止生成图 11.20(a)所示形式的钝三角形(四面体),文献[91]证实,此类单元会导致空间离散的截断误差非常大,因此建议生成如图 11.20(b)和图 11.20(c)所示的直角单元,但根据控制体类型的不同,直角单元也可能引起离散误差[92-93](见 5.2.3 节)。

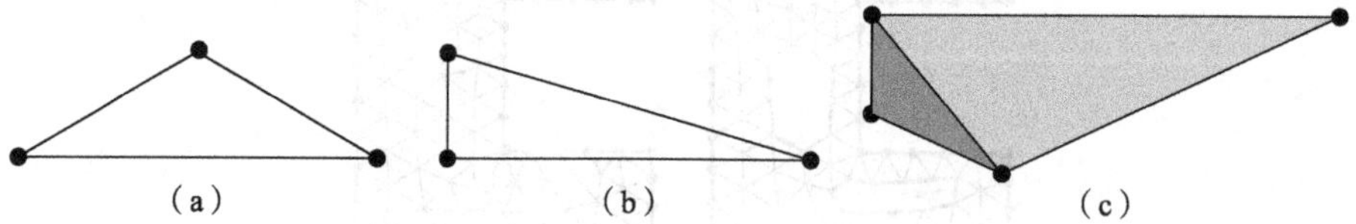

图 11.20 大长宽比单元

(a) 钝角三角形;(b) 直角三角形;(c) 直角四面体。

1. 拉伸 Delaunay 三角化

生成拉伸三角形(四面体)单元需要定义物理域中拉伸的大小和方向,在拉伸 Delaunay 三角剖分中使用背景网格来定义它。可以采用与生成各向同性 Delaunay 网格相同的点插入技术,但外接圆(外接球)准则被外接椭圆(外接椭球)条件所代替,椭圆(椭球)定义拉伸矢量的当地方向,拉伸量通过轴的比值来反映[94-96]。另一种方法是用椭球代替球的气泡填充(bubblepack)算法[51-53]。

2. 层推进法

对于几何复杂的区域,各向异性 Delaunay 三角剖分中拉伸矢量的指定非常复杂,因此,Pirzadeh[3,4,97-98]及其他人[99-100]提出的层推进法(advancing-layer method)得到了更广泛的应用。层推进法与双曲型结构网格生成技术(见 11.1.4 节)类似,也可以看作一种改进的阵面推进技术。

层推进法生成拉伸三角形(四面体)网格的步骤如下(图 11.21):

(1) 将所有边界面剖分成三角形;

(2) 计算边界点处的近似法向;

(3) 沿法线布置网格点,并形成厚度不断增加的四边形(3D 为三棱柱)单元层;

(4) 将每个四边形(三棱柱)剖分成两个三角形(三个四面体),需要特别注意 3D 情况的连接模式(connectivity pattern);

(5) 继续生长每个单元层,直到前沿自交或与其他前沿相交,或单元长宽比接近于 1。

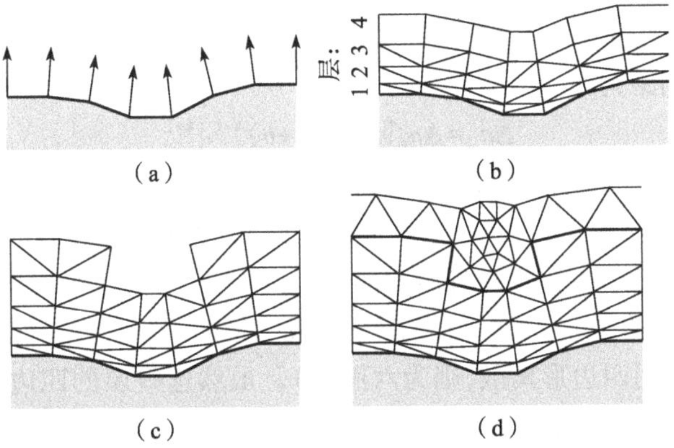

图 11.21　层推进法的步骤

(a) 计算边界法向;(b) 生成新层并剖分单元;(c) 完成拉伸网格;(d) 区域其他部分的各向同性单元生成。

然后用各向同性四面体填充流动区域的其余部分,通常采用阵面推进法,该方法从最后一层边界单元所表示的表面开始。图 5.2 和图 11.4 是使用上述过程生成的网格实例。

在理论上,计算边界点法向(步骤 2)有两种方法。对于第一种方法,如果边界面是用如 NURBS(非均匀有理 B 样条)等解析函数描述[101],则法向导数很容易计算,文献[102]介绍了这类网格生成方法,但在边界相交处(如后缘处)必须谨慎。

第二种更广泛的方法是使用面三角化。这种方法在 2D 情况很容易实现,只需将与特定节点连接的边界边的法向进行平均即可。然而,这种简单平均的方法对 3D 情形并不适用,原因是边界点可以由任意数量的三角形面共享,因此法向会因三角剖分不同而有偏差,在尖角和弯曲处,这个问题尤为严重。最优法向的必要条件由可见性准则(visibility criterion)[103-104]给出,该准则指出,沿法向放置的点从共享边界点的所有三角形面看必须均是可见的,基于可见性准则设计了各种方法来构造法向，详见文献[103-106]。为防止推进方向的突变和网格线交叉，需对法向进行光滑处理，通常采用加权拉普拉斯类光顺法[104-105]，即

$$\boldsymbol{n}_i = \omega \boldsymbol{n}_i^{(0)} + \frac{(1-\omega)}{\sum_j 1/|\boldsymbol{r}_{ij}|}\sum_j \frac{\boldsymbol{n}_j}{|\boldsymbol{r}_{ij}|} \tag{11.12}$$

式中：$\boldsymbol{n}_i$ 和 $\boldsymbol{n}_i^{(0)}$ 分别表示新节点的法向和初始节点的法向；$\boldsymbol{n}_j$ 表示相邻节点的法向；$|\boldsymbol{r}_{ij}|$ 为节点 i 与节点 j 之间的距离。权重系数 ω 与面曲率相关，凹区域取小值，凸区域取大值。分别对每层进行光顺，常见的做法是对开始几层使用初始节点的法向，然后逐渐对法向进行光顺，这样处理的原因是边界正交网格有助于减少空间离散格式的离散误差。

根据用户指定的第一层高度和法向拉伸比计算推进步长的大小，例如由文献[3]可得层间距为

$$\Delta n_k = \Delta n_0 [1+a(1+b)^{k-1}]^{k-1} \tag{11.13}$$

式中：Δn_k 表示第 k 层的间距；Δn_0 为给定的第一层厚度；a 和 b 表示拉伸参数，其范围是 $0.04 \leqslant a \leqslant 0.2$ 和 $0 \leqslant b \leqslant 0.07$。此外，可修改凹凸角处的推进步长 Δn_k[106]，也可以沿边界层生长方向增加 Δn_k 以保持 y^+ 不变。

层推进算法的每一步中，2D 情况下生成一层四边形，3D 情况下生成一层三棱柱（如果表面用四边形离散，则为六面体）。虽然这些层的四边形可以很容易地剖分成两个三角形（沿着最短对角线），但将三棱柱剖分成三个四面体必须特别小心，三棱柱之间的面必须以相同的方式进行剖分。文献[3]提出了一种基于整数的三棱柱一致剖分方法，文献[107]介绍了一种更简单的方法，文献[108]报道了另一种剖分方法。

3. 法向推进与 Delaunay 法的结合

生成拉伸三角形（四面体）网格的另一种方法是通过改进的阵面推进法（在此称为法向布点推进法）与 Delaunay 三角剖分相结合而得到的方法。Müller[109]提出了针对 2D 网格的相应方法，Marcum[110-111] 以及 Sharov 和 Nakahashi[106] 开发了该算法的 3D 形式。这种网格生成方法首先进行边界节点的体三角剖分，将剖分的网格用作背景网格，也可用作高效的搜索结构，在此可以恢复边界[106]，但对这一步是非必要的[110-111]。然后，使用类似于层推进法的技术，将点插入拉伸网格区域内，背景三角剖分用于相邻阵面的确定和相交检查。最后一步，扩大后的面被用作标准 Delaunay 法生成各向同性单元的新边界。

4. 拉伸的表面网格

在流动呈现很强方向性的区域，在不影响求解精度的情况下，各向异性网格可以大幅减少三角形面的数量，从而减少空间单元的数量（图 11.22）。例如使

用 RANS 进行模拟时,在机翼或叶片的展向不需要具有很高网格分辨率,对于机身的轴向同样如此。根据文献[112-113],使用各向异性网格后,表面三角形网格量减少了 1/6~1/2。然而,如图 11.22 所示,正确确定表面单元的方向非常重要,特别是曲率大的区域,否则,因为无法正确地表达真实的表面轮廓,流场解将变为伪解。可以使用上述任何一种各向异性生成方法生成拉伸表面网格(有关表面网格生成的简要讨论,参见文献[16,75,114]),拉伸比和方向通常通过一个线源指定。另一种方法则是首先在参数坐标中生成各向同性网格,然后利用拉伸函数将网格转换到物理空间。

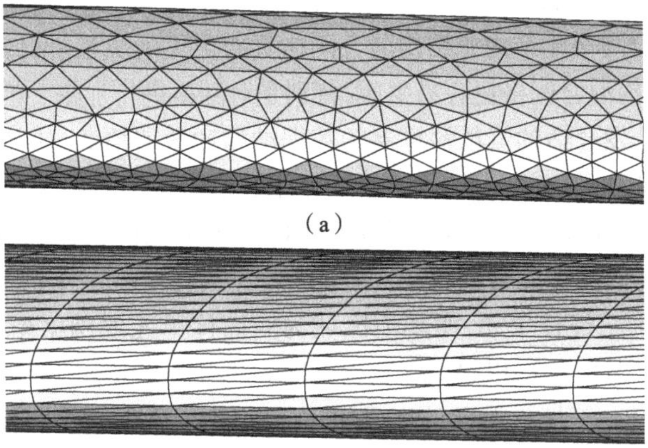

(a)

(b)

图 11.22　前缘非结构表面网格

(a) 无特定方向的各向同性网格;(b) 和展向拉伸的各向异性网格。

11.2.4　非结构混合网格和结构非结构混合网格

当前,使用由不同单元类型(三棱柱、四面体、金字塔等)组成的网格对流动区域进行离散变得非常流行,此类网格被称为非结构混合网格或单元类型混合网格(mixed grid)。另一种思路是网格由结构区和非结构区的网格构成[115-122],这种情况称为结构非结构混合网格(hybrid grid)。值得注意的是,有些文献作者使用了“hybrid grid”一词,但实际上他们指的是“mixed grid”。使用非结构混合网格或结构非结构混合网格都旨在使每个流动区域采用最合适的网格单元或网格拓扑关系,其目的是提高模拟精度和降低计算成本。但其限制在于,必须以自动化程度较高的方式生成此类网格,且生成网格过程也必须足够快。

事实上,结构非结构混合网格的应用相对较少,其困难在于需要两种不同的流场解算器,一种是结构的,另一种是非结构的,两种解算器必须进行耦合,两种

CFD 代码不仅编写工作量大,更主要的是代码的升级和维护所需的工作量也很大,因此有必要开发一种特殊的数据和程序结构。结构非结构混合网格的经典应用是,使用结构网格(六面体或可能是半结构的棱柱网格)对近壁区域进行离散,这使得更容易实现高阶空间格式,隐式方法[123]或多重网格[124-125];流场其余部分用非结构网格填充,这在几何外形复杂的区域具有相当大的优势。结构非结构混合网格的理念也可用于如叶轮机械中的转子-定子相互作用的情况,在这种情况下,转子和定子周围的结构网格之间的交界面由一片非结构网格组成(类似于文献[126]的概念),随着每次转子运动,非结构网格重新生成,通过这种方法,交界面之间可以保持离散格式的精度和守恒特性。

非结构混合网格比结构非结构混合网格具有更大的灵活性,特别是对于复杂几何外形,非结构混合网格更容易生成,特别是对网格进行加密。但这种网格面临的挑战是需要开发合适的流场解算器的数据结构和数值格式,使其能以准确且高效的方式处理各种网格单元类型,在第 5 章中我们讨论了一些相关问题。下面简要介绍三棱柱/四面体混合网格和三棱柱/四面体/笛卡儿混合网格的生成方法。

1. 三棱柱/四面体混合网格

近壁区域为三棱柱单元,而其他区域为四面体单元的网格生成方法最初由 Nakahashi[127-129]提出。Kallinderis[130-131]、Connell 和 Braaten[107]对这种方法进行了进一步的研究。三棱柱单元的生成使用与前面介绍的层推进或法向推进方法相同的技术,四面体则是直接从最后一层三棱柱生长而来。在三棱柱的一个四边形面仍然暴露的情况下,采用金字塔作为四面体的过渡单元。

2. 三棱柱/四面体/笛卡儿混合网格

笛卡儿网格由与笛卡儿坐标轴对齐的正方形(3D 为立方体)组成。即使对于几何外形复杂的区域,笛卡儿网格的生成也很容易,且计算量很小。但笛卡儿方法的缺点在于固体边界处流场求解的精度,固体边界有的是弯曲的,有的不是沿笛卡儿坐标方向,这个问题在进行 RANS 模拟时变得特别严重,因为需要使用高度拉伸和与边界垂直的单元。因此,Melton 等[132]、Karman[133]、Smith 和 Leschziner[134]、Wang 等[135]、Delanaye 等[136]、Hitzel 等[137]等学者提出,在壁面附近插入贴体四边形(三棱柱)单元,通过金字塔和四面体作为过渡单元,在内层和外部笛卡儿网格之间创建连续的交界面。另外,通过允许悬挂节点和悬挂线来简化网格生成过程,这种情况如图 11.23 所示。如果处理得当,交界面处通量的交换可保持守恒[136,138]。关于笛卡儿混合网格方法的详细叙述见文献[139]。Karman[140]提出的一种有趣的思路,将笛卡儿体网格投影到边界上,从而自动对表面进行离散,然后沿固体边界插入黏性层。

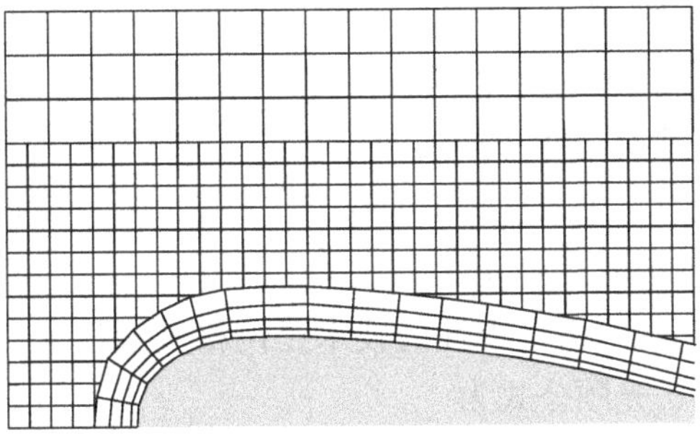

图 11.23　物面附近笛卡儿/拉伸四边形混合网格示例

11.2.5　网格质量评价与改善

网格质量对模拟精度有很大的影响,特别是对于非结构网格,第 5 章多处提到这个问题(如 5.3.3 节关于解的重构)。因此网格尽可能由规则单元组成,这一点很重要,而且单元尺寸(以及黏性区域的单元拉伸)在区域内应平滑变化,网格也应足够精细,以分辨相关的几何和流动特征。

图 11.24 显示了应避免的四面体形状:钝单元(图 11.20(a))、长薄单元(单元由四个近似共面的点组成)、针状单元和楔形单元。唯一的例外是黏性区域内的楔形拉伸四面体(图 11.20),但它可以使用三棱柱来替代,由于流动的最大梯度位于壁面法向(楔的短边),数值误差不会过度振荡。

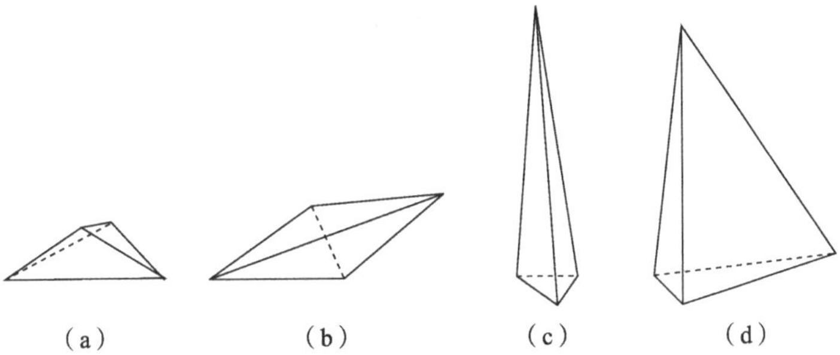

图 11.24　不理想的四面体单元形状

(a) 钝角;(b) 长薄片;(c) 针状;(d) 楔形。

除了单元角[111]之外,还可以使用以下参数作为四面体网格的质量度量[20,141](以下数值是指对应等边四面体的值):

(1) 外接球半径/内切球半径 = 3.0;

(2) 最大边长/内切球半径 = 4.8990;

(3) 外接球半径/最大边长 = 0.6125;

(4) 最大边长/最小边长 = 1.0；

(5) (平均单元边长)^3/体积 = 8.4797；

(6) (体积)^4/(所有三角形面的面积和)^3 = 4.585×10^{-4}。

为了提高网格质量，可以使用 Lawson[142-143] 提出的边交换算法，边交换特别适合用于剔除网格中的薄片单元[48]。

另一种常用于改善网格正则性的技术是网格光顺技术，它使用近似拉普拉斯算子剔除网格节点，光顺公式为

$$\boldsymbol{r}_i^{n+1} = \boldsymbol{r}_i^n + \frac{\omega}{N_A}\sum_{j=1}^{N_A}(\boldsymbol{r}_j - \boldsymbol{r}_i) \tag{11.14}$$

式中：N_A 为 i 附近的节点数；ω 为松弛因子（内部取 0.5~1.0，与边界相邻的点处取 0.25，边界点处取 0.0，参见文献[48]）。式(11.14)中的关系式需要迭代求解，文献[47]中采用点高斯-赛德尔格式，而不是采用点雅可比格式，因为点高斯-赛德尔格式能更好地抑制负体积。图 11.25 是使用拉普拉斯算子（式(11.14)）光顺的非结构化网格的例子。

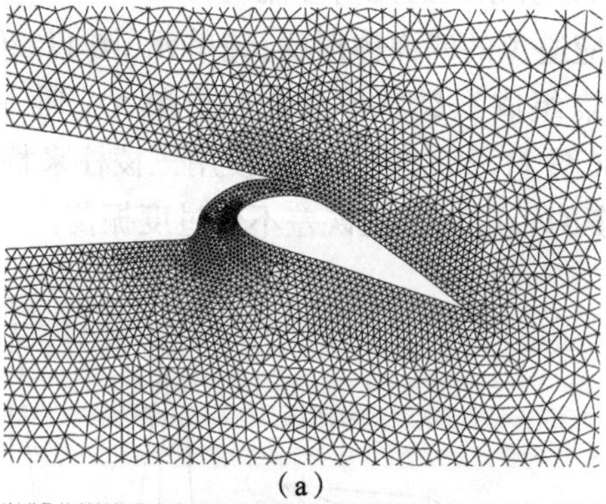

(a)

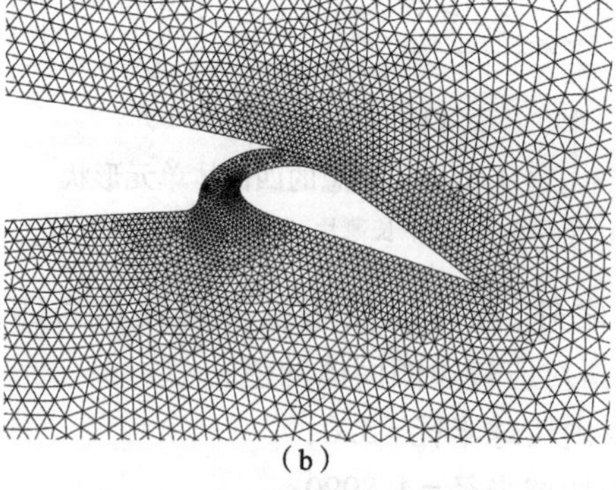

(b)

图 11.25　非结构三角形网格

(a) 光顺前；(b) 光顺后。

参考文献

[1] BrodersenO, Hepperle M, Ronzheimer A, Rossow CC, Schöning B. The parametric grid generation system MegaCads. Proceeding of the 5th international conference on numerical grid generation in computing field simulations. Mississippi: NSF Engineering Center; 1996. p. 353-62.

[2] Brodersen O, Ronzheimer A, Ziegler R, Kunert T, Wild J, Hepperle M. Aerodynamic applications using MegaCads. Proceedings of the 6th international conference on numerical grid generation in computing field simulations. London, England: University of Greenwich; 1998. p. 793-802.

[3] Pirzadeh S. Three-dimensional unstructured viscous grids by the advancing-layers method. AIAA J 1996; 34: 43-9.

[4] Pirzadeh S. Progress toward a user-oriented unstructured viscous grid generator. AIAA Paper 96-0031; 1996.

[5] Mavriplis DJ, Pirzadeh S. Large-scale parallel unstructured mesh computations for 3D high-lift analysis. AIAA Paper 99-0537; 1999.

[6] Pirzadeh S. Unstructured grid generation for complex 3D high-lift configurations. Paper No. 1999-01-5557. San Francisco: World Aviation Congress and Exposition, October 1999.

[7] Digital Representation for Communication of Product Definition Data: IGES 5. 2 (Initial Graphics Exchange Specification Version 5. 2). US Product Data Association, November 1993, ISBN 978-1-88538900-8.

[8] http://www. wiz-worx. com/iges5x/.

[9] ISO 10303-21: 2002 Industrial automation systems and integration—product data representation and exchange—part 21: implementation methods: clear text encoding of the exchange structure.

[10] STEP application handbook: ISO 10303(Version 3 ed.). North Charleston, SC: SCRA, 30 June 2006.

[11] http://www. wikistep. org/index. php/Main_Page.

[12] http://images. autodesk. com/adsk/files/autocad_2014_pdf_dxf_reference_enu. pdf.

[13] Burns M. Automated fabrication. Prentice-Hall, New Jersey; 1993, ISBN 978-0-13-119462-5.

[14] Chua CK, Leong KF, Lim CS. Rapid prototyping: principles and applications (2nd ed.). Singapore: World Scientific Publishing Co. ; 2003, ISBN 981-238-117-1.

[15] http://www. ennex. com/~fabbers/StL. asp.

[16] Aftosmis MJ, Delanaye M, Haimes R. Automatic generation of CFD-ready surface triangulations from CAD geometry. AIAA Paper 99-0776; 1999.

[17] Bohn JW, Zozny, MJ. Automatic CAD-model repair: shell-closure. Proceedings symposium on freeform fabrication, Department of Mechanical Engineering, University of Texas at Austin; 1992.

[18] Guéziec A, Taubin G, Lazarus F, Horn W. Cutting and stitching: efficient conversion of a non-manifold polygonal surface to a manifold. IBM-RC-20935, IBM Research Division, Yorktown Heights; 1997.

[19] Carey GF. Computational grids: generation, adaption, and solution strategies. Washington, DC:

Taylor & Francis; 1997.

[20] Thompson JF, Soni BK, Weatherill NP, editors. Handbook of grid generation. Boca Raton: CRC Press; 1999.

[21] Baker TJ. Three decades of meshing; a retrospective view. AIAA Paper 2003-3563; 2003.

[22] Gordon WN, Hall CA. Construction of curvilinear coordinate systems and application to mesh generation. Int J Numer Methods Eng 1973; 7: 461-77.

[23] Gordon WJ, Thiel LC. Transfinite mappings and their application to grid generation. In: Thompson JF, editor. Numerical grid generation, North-Holland, p. 171; 1982.

[24] Thompson JF, Warsi ZUA, Mastin CW. Numerical grid generation: foundations and applications. Amsterdam, Netherlands: Elsevier Science; 1985.

[25] Soni BK. Two- and three-dimensional grid generation for internal flow applications of computational fluid dynamics. AIAA Paper 85-1526; 1985.

[26] Thompson JF, Thames FC, Mastin CW. Automatic numerical generation of body-fitted curvilinear coordinate system for field containing any number of arbitrary two-dimensional bodies. J Comput Phys 1974; 15: 299-319.

[27] Sorenson RL. A computer program to generate two-dimensional grids about airfoils and other shapes by the use of Poisson's equation. NASA TM-81198; 1980.

[28] Hsu K, Lee SL. A numerical technique for two-dimensional grid generation with grid control at all of the boundaries. J Comput Phys 1991; 96: 451-69.

[29] Thomas PD, Middlecoff JF. Direct control of the grid point distribution in meshes generated by elliptic equations. AIAA J 1980; 18: 652-6.

[30] Sorenson RL. Three-dimensional elliptic grid generation about fighter aircraft for zonal finite difference computations. AIAA Paper 86-0429; 1986.

[31] Thompson JF. A general three-dimensional elliptic grid generation system based on a composite block structure. Comp Method Appl Mech Eng 1987; 64: 377-411.

[32] Sonar T. Grid generation using elliptic partial differential equations. DFVLR-FB 89-15; 1989.

[33] Starius G. Constructing orthogonal curvilinear meshes by solving initial value problems. Numer Math 1977; 28: 25-48.

[34] Steger JL, Chaussee DS. Generation of body-fitted coordinates using hyperbolic partial differential equations. SIAM J Sci Stat Comput 1980; 1: 431-7.

[35] Steger JL, Rizh YM. Generation of three-dimensional body-fitted coordinates using hyperbolic partial differential equations. NASA TM-86753; 1985.

[36] Steger JL. Generation of three-dimensional body-fitted grids by solving hyperbolic partial differential equations. NASA TM-101069; 1989.

[37] Dwyer HA. A geometric interpretation of hyperbolic grid generation. AIAA Paper 94-0315; 1994.

[38] Sethian JA. Curvature flow and entropy conditions applied to grid generation. J Comput Phys 1994; 115: 440-54.

[39] Jeng YN, Shu YL. The grid combination method for the hyperbolic grid solver in regions with

enclosed boundaries. AIAA Paper 95-0857;1995.

[40] Jeng YN, Liou Y-C. Hyperbolic equation method of grid generation for enclosed regions. AIAA J 1996;34:1293-5.

[41] Tai CH, Yin SL, Soong CY. A novel hyperbolic grid generation procedure with inherent adaptive dissipation. J Comput Phys 1995;116:173-9.

[42] Brakhage K-H, Müller S. Algebraic-hyperbolic grid generation with precise control of intersection of angles. Int J Numer Method Fluids 2000;33:89-123.

[43] Mavriplis DJ. An advancing front Delaunay triangulation algorithm designed for robustness. ICASE Report No. 92-49, 1992; also AIAA Paper 93-0671; 1993; also J Comput Phys 1995; 117:90-101.

[44] Frey PJ, Borouchaki H, George P-L. Delaunay tetrahedralization using an advancing-front approach. Proceedings of the 5th international meshing roundtable; October 1996. p. 31-46.

[45] Fleischmann P, Selberherr S. Three-dimensional Delaunay mesh generation using a modified advancing front approach. Proceedings of the 6th international meshing roundtable; October 1997. p. 31-46.

[46] Müller JD, Roe PL, Deconinck H. A frontal approach for internal node generation in Delaunay triangulations. Int J Numer Method Fluids 1993;17:241-56.

[47] Müller JD. On triangles and flow [PhD Thesis]. The University of Michigan;1996.

[48] Marcum DL, Weatherill NP. Unstructured grid generation using iterative point insertion and local reconnection. AIAA J 1995;33:1619-25.

[49] Bern M, Eppstein D. Quadrilateral meshing by circle packing. Proceedings of the 6th international meshing roundtable; October 1997. p. 7-19.

[50] Liu J, Tang R. Ball-packing method: a new approach for quality automatic triangulation of arbitrary domains. Proceedings of the 6th international meshing roundtable; October 1997. p. 85-96.

[51] Shimada K, Yamada A, Itoh T. Anisotropic triangular meshing of parametric surfaces via close packing of ellipsoidal bubbles. Proceedings of the 6th international meshing roundtable; October 1997. p. 375-90.

[52] Li X-Y, Teng S-H, Ungor A. Biting spheres in 3D. Proceedings of the 8th international meshing roundtable; October 1999. p. 85-95.

[53] Soji Y, Shimada K. High quality anisotropic tetrahedral mesh generation via ellipsoidal bubble packing. Proceedings of the 9th international meshing roundtable; October 2000. p. 263-73.

[54] Benabbou A, Borouchaki H, Laug P, Lu J. Sphere packing and applications to granular structure modeling. Proceedings of the 17th international meshing roundtable; October 2008. p. 1-18.

[55] Mavriplis DJ. Unstructured mesh generation and adaptivity. ICASE Report No. 95-26;1995.

[56] Weatherill NP, Hassan O, Morgan K, Peraire J, Peiro J, Marchant MJ. Unstructured grid generation. COSMASE: short course on grid generation, automation and parallel utilisation, Lausanne, Switzerland, September 23-27;1996.

[57] Dirichlet GL. über die Reduktion der positiven quadratischen Formen mit drei unbestimmten ganzen Zahlen. Z Reine Angew Math 1850;40:209-27.

[58] Voronoj G. Novelles applications de paramètres continus à la théorie des formes quadratiques.

Z Reine Angew Math 1908;133:97-178.
[59] Delaunay B. Sur la sphère vide. Bull Acad Sci USSR, Class Sci Mat Nat 1934;7:793-800.
[60] Holmes DG, Snyder DD. The generation of unstructured meshes using Delaunay triangulation. Proceedings of the 2nd international conference on numerical grid generation in CFD, Pineridge, Swansea, Wales, UK; 1988. p. 643-52.
[61] George PL, Hecht F, Saltel E. Fully automatic mesh generator for 3-D domains of any shape. Imp Comput Sci Eng 1990;2:187-218.
[62] Weatherill NP, Hassan O, Marcum DL. Compressible flowfield solutions with unstructured grids generated by Delaunay triangulation. AIAA J 1995;33:1196-204.
[63] Rebay S. Efficient unstructured mesh generation by means of Delaunay triangulation and Bowyer-Watson algorithm. J Comput Phys 1993;106:125-38.
[64] Baker TJ. Triangulations, mesh generation and point placement strategies. Proc. conf. frontiers of computational fluid dynamics 1994, Ithaca, New York, November 1994.
[65] Green PJ, Sibson R. Computing the Dirichlet tessellation in the plane. Comput J 1977;21:168-73.
[66] Bowyer A. Computing Dirichlet tessellations. Comput J 1981;24:162-6.
[67] Watson DF. Computing the n-dimensional Delaunay tessellation with application to Voronoj polytopes. Comput J 1981;24:167-72.
[68] Ruppert J. A Delaunay refinement algorithm for quality 2-dimensional mesh generation. J Algorithms 1995;18:548-85.
[69] Shewchuk JR. Triangle: engineering a 2D quality mesh generator and Delaunay triangulator. Proceedings of the 1st workshop application computational geometry, Philadelphia, USA; May 1996. p. 124-33.
[70] Finkel RA, Bentley JL. Quad trees, a data structure for retrieval of composite keys. Acta Inform 1974;4:1-9.
[71] Samet H. The quadtree and related hierarchical data structures. Comput Surv 1984; 16:188-260.
[72] Samet H. The design and analysis of spatial data structures. Boston: Addison-Wesley; 1990.
[73] Yerry MA, Shephard MS. A modified quadtree approach to finite element mesh generation. IEEE Comput Graph Appl 1983;3:39-46.
[74] Yerry MA, Shephard MS. Automatic three-dimensional mesh generation by the modified-octree technique. Int J Numer Methods Eng 1984;20:1965-90.
[75] Miranda ACO, Martha LF. Mesh generation on high-curvature surfaces based on a background quadtree structure. Proceedings of the 11th international meshing roundtable; September 2002. p. 333-42.
[76] Baker TJ. Three dimensional mesh generation by triangulation of arbitrary point sets. AIAA Paper 87-1124; 1987.
[77] Chew LP. Constrained Delaunay triangulations. Algorithmica 1989;4:97-108.
[78] George PL, Hecht F, Saltel E. Automatic mesh generator with specified boundary. Comput Methods Appl Mech Eng 1991;33:975-95.
[79] Weatherill NP, Hassan O, Marcum DL. Calculation of steady compressible flowfields with the

finite-element method. AIAA Paper 93-0341;1993.

[80] Sharov D,Nakahashi K. A boundary recovery algorithm for Delaunay tetrahedral meshing. Proceedings of the 5th international conference on numerical grid generation in computing field simulations,Mississippi State University;1996. p. 229-38.

[81] Joe B,Construction of three-dimensional constrained triangulations. Technical Report ZCS2008-06. Zhou Computing Services,Inc. ;2008.

[82] Peraire J,Vahdati M,Morgan K,Zienkiewicz OC. Adaptive remeshing for compressible flow computations. J Comput Phys 1987;72:449-66.

[83] Peraire J,Peiro J,Formaggia L,Morgan K,Zienkiewicz OC. Finite element Euler computations in three dimensions. Int J Numer Method Eng 1988;26:2135-59.

[84] Löhner R,Parikh P. Three-dimensional grid generation by the advancing front method. Int J Numer Method Fluids 1988;8:1135-49;also AIAA Paper 88-0515;1988.

[85] Löhner R. Progress in grid generation via the advancing front technique. Eng Comput 1996; 12:186-210.

[86] Pirzadeh S. Structured background grids for generation of unstructured grids by advancing-front method. AIAA J 1993;31:257-65.

[87] Pirzadeh S. Unstructured viscous grid generation by advancing-front method. NASA CR-191449;1993.

[88] Barth TJ. On unstructured grids and solvers. VKI lecture series 1990-03;1990. p. 1-65.

[89] Marcum DL. Advancing-front/local-reconnection (AFLR) unstructured grid generation. In: Hafez M,Oshima K,editors. CFD review 1998,Part I. Singapore: World Scientific Publishing Co. ;1998.

[90] Ito Y,Shih AM,Soni BK. Reliable isotropic tetrahedral mesh generation based on an advancing front method. Proceedings of the 13th international meshing roundtable; September 2004. p. 95-106.

[91] Babuška I,Aziz AK. On the angle condition in the finite-element method. SIAM J Numer Anal 1976;13:214-26.

[92] Baker TJ. Discretization of the Navier-Stokes equations and mesh induced errors. Proceedings of the 5th international conference on numerical grid generation in CFD. Mississippi: Mississippi State University;April 1996.

[93] Baker TJ. Irregular meshes and the propagation of solution errors. Proceedings of the 15th international conference on numerical methods in fluid dynamics,Monterey,CA;June 1996.

[94] Mavriplis DJ. Adaptive mesh generation for viscous flows using Delaunay triangulation. J Comput Phys 1990;90:271-91.

[95] Vallet MG,Hecht F,Mantel B. Anisotropic control of mesh generation based upon a Voronoj type method. In: Arcilla AS,et al. ,editors. Proceedings of the 3rd international conference on numerical grid generation in computing fluid dynamics and related fields,Barcelona,Spain. North-Holland;1991. p. 93-103.

[96] Irmisch S,Schwolow R,Erzeugung unstrukturierter Dreiecknetze mittels der Delaunay-triangulierung. Zeitschrift Wundheilung 1994;18:361-8.

[97] Pirzadeh S. Unstructured viscous grid generation by the advancing-layers method. AIAA J 1994;32:1735-7;also AIAA Paper 93-3453;1993.

[98] Pirzadeh S. Viscous unstructured three-dimensional grids by the advancing-layers method. AIAA Paper 94-0417;1994.

[99] Hassan O, Probert EJ, Weatherill NP, Marchant MJ, Morgan K, Marcum DL. The numerical simulation of viscous transonic flow using unstructured grids. AIAA Paper 94-2346;1994.

[100] Marchant MJ, Weatherill NP. Unstructured grid generation for viscous flow simulations. In: Weatherill NP, et al. editors. Proceedings of the 4th international conference on numerical grid generation in computing fluid dynamics and related fields, Swansea, UK. Pineridge Press;1994. p. 151-62.

[101] Piegl L, Tiller W. The NURBS book. 2nd ed. Berlin: Springer Verlag;1997.

[102] WhitmireJB, Dollar TW. Prismatic grid generation for NURBS bounded triangulations. AIAA Paper 98-0219;1998.

[103] Kallinderis Y, Ward S. Prismatic grid generation with an efficient algebraic method for aircraft configurations. AIAA Paper 92-2721;1992.

[104] Kallinderis Y, Ward S. Prismatic grid generation for 3-D complex geometries. AIAA J 1993; 31:1850-6.

[105] Kallinderis Y, Khawaja A, McMorris H. Hybrid prismatic/tetrahedral grid generation for viscous flows around complex geometries. AIAA J 1996;34:291-8.

[106] Sharov D, Nakahashi K. Hybrid prismatic/tetrahedral grid generation for viscous flow applications. AIAA Paper 96-2000;1996;also AIAA J 1998;36:157-62.

[107] Connell SD, Braaten ME. Semistructured mesh generation for three-dimensional Navier-Stokes calculations. AIAA J 1995;33:1017-24.

[108] Dompierre J, Labbé P, Vallet M-G, Camarero R. How to subdivide pyramids, prisms and hexahedra into tetrahedra. Report CERCA R99-78;1999; also 8th international meshing roundtable, Lake Tahoe, CA;1999.

[109] Müller JD. Quality estimates and stretched meshes based on Delaunay triangulations. AIAA J 1994;32:2372-9.

[110] Marcum DL. Generation of unstructured grids for viscous flow applications. AIAA Paper 95-0212;1995.

[111] Marcum DL, Gaither JA. Mixed element type unstructured grid generation for viscous flow applications. AIAA Paper 99-3252;1999.

[112] McMorris H, Kallinderis Y. A combined octree-advancing front method for tetrahedra and anisotropic surface meshes. AIAA Paper 96-2442;1996.

[113] McMorris H, Kallinderis Y. Octree-advancing front method for generation of unstructured surface and volume meshes. AIAA J 1997;35:976-84.

[114] Chen H, Bishop J. Delaunay triangulation for curved surfaces. Proceedings of the 6th international meshing roundtable, Park City, Utah, USA, October 13-15;1997. p. 115-27.

[115] Nakahashi K. FDM-FEM zonal approach for computations of compressible viscous flows. Lecture notes in physics, vol. 264. Heidelberg: Springer Verlag;1986. p. 494-8.

[116] Nakahashi K, Obayashi S. FDM-FEM zonal approach for viscous flow computations over multiple bodies. AIAA Paper 87-0604; 1987.

[117] Nakahashi K, Obayashi S. Viscous flow computations using a composite grid. AIAA Paper 87-1128; 1987.

[118] Weatherill NP. Mixed structured-unstructured meshes for aerodynamic flow simulations. Aeronaut J 1990; 94: 111-23.

[119] Shaw JA, Georgala JM, Peace AJ, Childs PN. The construction, application and interpretation of three-dimensional hybrid meshes. In: Arcilla AS, et al., editors. Proceedings of the 3rd international conference on numeric grid generation in computing fluid dynamics an related fields, Barcelona, Spain. Amsterdam: North-Holland; 1991.

[120] Michal T, Johnson J. A hybrid structured/unstructured grid multi-block flow solver for distributed parallel processing. AIAA Paper 97-1895; 1997.

[121] Shaw JA, Peace AJ. Simulating three-dimensional aeronautical flows on mixed block-structured/semi-unstructured/unstructured grids. Technical Memorandum 428. Bedford: Aircraft Research Association; 1998.

[122] Blazek J. Comparison of two conservative coupling algorithms for structured-unstructured grid interfaces. AIAA Paper 2003-3536; 2003.

[123] Pandya SA, Hafez MM. A semi-implicit finite-volume scheme for the solution of Euler equations on 3-D prismatic grids. AIAA Paper 93-3431; 1993.

[124] Parthasarathy V, Kallinderis Y, Nakajima K. Hybrid adaption method and directional viscous multigrid with prismatic tetrahedral meshes. AIAA Paper 95-0670; 1995.

[125] Parthasarathy V, Kallinderis Y. Directional viscous multigrid using adaptive prismatic meshes. AIAA J 1995; 33: 69-78.

[126] Kao KH, Liou MS. Direct replacement of arbitrary grid-overlapping by nonstructured grid. AIAA Paper 95-0346; 1995.

[127] Nakahashi K. Computations of three-dimensional Navier-Stokes equations by using a prismatic mesh. Proceedings of the 6th NAL symposium on aircraft computational aerodynamics, Tokyo, Japan; June 1988.

[128] Nakahashi K. A finite-element method on prismatic elements for the three-dimensional Navier-Stokes equations. Lecture notes in physics, vol. 323. Heidelberg: Springer Verlag; 1989. p. 434-8.

[129] Nakahashi K. Marching grid generation for external viscous flow problems. Trans Jpn Soc Aerospace Sci 1992; 35: 88-102.

[130] Kallinderis Y. A new finite-volume Navier-Stokes scheme on three-dimensional semi-structured prismatic elements. AIAA Paper 92-2697; 1992.

[131] Kallinderis Y. Adaptive hybrid prismatic/tetrahedral grids. Int J Numer Method Fluids 1995; 20: 1023-37.

[132] Melton JE, Pandya SA, Steger JL. 3-D Euler flow solutions using unstructured Cartesian and prismatic grids. AIAA Paper 93-0331; 1993.

[133] Karman SL. SPLITFLOW: a 3-D unstructured Cartesian/prismatic grid CFD code for com-

plex geometries. AIAA Paper 95-0343, 1995.

[134] Smith RJ, Leschziner MA. Automatic grid generation for complex geometries. Aeronaut J 1996; 100: 7-14.

[135] Wang ZJ, Hufford GS, Przekwas AJ. Adaptive Cartesian/adaptive prism (ACAP) grid generation for complex geometries. AIAA Paper 97-0860; 1997.

[136] Delanaye M, Aftosmis M, Berger MJ, Liu Y, Pulliam TH. Automatic hybrid-Cartesian grid generation for high-Reynolds number flows around complex geometries. AIAA Paper 99-0777; 1999.

[137] Hitzel SM, Tremel U, Deister F, Rieger H. Complex configuration meshing—an industrial view and approach. AIAA Paper 2003-4130; 2003.

[138] Rai MM. A conservative treatment of zonal boundaries for Euler equation calculations. J Comput Phys 1986; 62: 472-503.

[139] Shaw JA, Stokes S, Lucking MA. The rapid and robust generation of efficient hybrid grids for RANS simulations over complete aircraft. Int J Numer Method Fluids 2003; 43: 785-821.

[140] Karman SL. Hierarchical unstructured mesh generation. AIAA Paper 2004-0613; 2004.

[141] Parthasarathy VN, Graichen CM, Hathaway AF. A comparison of tetrahedron quality measures. Finite Elements Anal Des 1993; 15: 255-61.

[142] Lawson CL. Software for C1 surface interpolation. In: Rice JR, editor. Mathematical software III. New York: Academic Press; 1977. p. 161-94.

[143] Lawson CL. Properties of n-dimensional triangulations. Comput Aided Geom Des 1986; 3: 231-46.

第12章 软件应用

本软件包包含了数个 1D 和 2D 流场解算器、网格生成器的源代码和可执行程序,还提供了各种 1D 和 2D 流动算例的输入数据和网格。此外,还给出了对显式和隐式时间推进格式进行冯·诺依曼稳定性分析的两套程序。最后,给出了使用不同方法进行代码并行化的示例。

该软件的目的是演示如何将前几章中介绍的 CFD 理论原理转换成计算机代码,这些程序可作为进一步试验和改进的基础。源代码是在 GNU 通用公共许可证的条款下提供的,这意味着它们可以免费复制和重新发布。更多细节请参阅其中一个目录中的 LICENSE. txt 文件。

流场解算器和网格生成器的源代码是用 Fortran-77 和 Fortran-90 编写的,2D 非结构解算器也提供了现代 C++语言的版本。这些程序不包含任何外部库的调用,源代码尽量保持简单,但仍然足够灵活,作者没有对软件的速度或内存消耗进行深入优化。大部分源代码的每个文件只有一个子程序或函数,可使用 Doxygen① 工具对 2D 流场解算器代码完整溯源。

所有网格、解和收敛文件都以纯 ASCII 格式存储。收敛文件和流场解文件的输出形式适合使用 Vis2D② 程序进行可视化,两者格式相同,每个变量占有一列,在文件头中给出变量的名称。对于非结构流场解算器生成的流场解文件,在坐标和流场变量之后存储每个单元节点的索引,第一列表示节点 1,第二列表示节点 2,依此类推。文件格式详见 Vis2D 的相关文档,由于其形式简单,必要时可以很容易地改为其他可视化工具需要的格式。

该软件包三个最高级别的目录为:C++、Fortran 和 Parallelization。这些目录包含以下子目录:

① http://www.doxygen.org.

② 可在 CFD 咨询和分析网查阅(http://www/cfd-ca.de)。

（1）C++。

unstructured_2d：2D 非结构三角形网格欧拉/纳维－斯托克斯方程组的流场解。

（2）Fortran。

①analysis：1D 模型方程组冯·诺依曼稳定性分析；

②grid_1d：1D 网格生成（拉瓦尔喷管）；

③grid_struct_2d：内外流的 2D 结构网格生成；

④grid_unstr_2d：将 2D 结构网格转换为非结构三角形网格；

⑤structured_1d：准 1D 欧拉方程组求解（拉瓦尔喷管流动），多重网格演示；

⑥structured_2d：2D 结构网格欧拉/纳维－斯托克斯方程组求解；

⑦unstructured_2d：非结构三角形网格 2D 欧拉/纳维－斯托克斯方程组求解。

（3）Parallelization。

①agents：异步代理库；

②cuda：CUDA；

③mpi：MPI；

④openmp：OpenMP；

⑤serial：串行版本。

上述子目录的内容将在下面的章节中进一步解释。通常，每个模拟程序的子目录包含一个 README. txt 文件，它描述了特定的文件、如何编译和运行应用程序以及如何存储结果，此外，还解释了主要变量和输入参数的含义。对于 2D 流场解算器，doc 子目录包含 HTML 格式的文档（index. html）。每个流场解算器的子目录也有一个 run 文件夹，包含各个演示算例的输入文件、网格、流场解和图像。

最后一点是关于收敛历程，这里提供的所有流场解算器的收敛性都是通过两个连续时间步密度差的 2 范数来衡量，即

$$\|\Delta\rho\|_2 = \sqrt{\sum_{I=1}^{N}(\rho_I^{n+1} - \rho_I^n)^2} \tag{12.1}$$

其中，求和是对所有 N 个控制体进行累加。为了方便，常用第一次迭代的值对式（12.1）中的收敛测度进行归一化。程序直接用对数尺度存储归一化的收敛历程。

12.1　稳定性分析程序

analysis 目录包含对线性 1D 方程组进行冯·诺依曼稳定性分析(见 10.3 节)的两个程序。第一个程序计算显式多步时间推进格式(见 6.1.1 节)和混合格式(见 6.1.2 节)的傅里叶记号和放大因子幅值,子目录 mstage 包含了源代码。第二个程序分析了一般隐式格式(见 6.2 节)的衰减特性,位于 implicit 子目录中。这两个程序既可以处理 1D 对流问题,也可以处理 1D 对流扩散方程(见 10.3.2 节和 10.3.3 节)。空间离散可以采用添加人工耗散的中心格式、一阶迎风格式和二阶迎风格式。同样的方法也适用于隐式算子。在显式格式情况下,也可以研究(中心或迎风)隐式残差光滑(见 9.3 节)的影响。

12.2　一维结构网格生成器

grid_1d 目录包含了生成 1D 结构网格的程序。每个网格节点都与特定面积关联,沿 x 轴的面积分布与拉瓦尔喷管相对应,喷管面积使用下式进行计算:

$$
\begin{aligned}
A(x) &= 1+\frac{1}{2}(A_1-1)\left\{1+\cos\left(\frac{\pi x}{x_{\text{thr}}}\right)\right\}, && 0\leqslant x\leqslant x_{\text{thr}} \\
A(x) &= 1+\frac{1}{2}(A_2-1)\left\{1-\cos\left(\frac{\pi(x-x_{\text{thr}})}{1-x_{\text{thr}}}\right)\right\}, && x_{\text{thr}}<x\leqslant 1
\end{aligned}
\tag{12.2}
$$

其中,x_{thr} 表示喉部位置,喷管长度假定是 1。在式(12.2)中,A_1 表示进口面积,A_2 表示出口面积(图 12.1)。根据式(12.2)得到喉道面积为 $A^*(x=x_{\text{thr}})=1$。该程序生成的网格作为 12.5 节中的准 1D 欧拉解算器的输入。

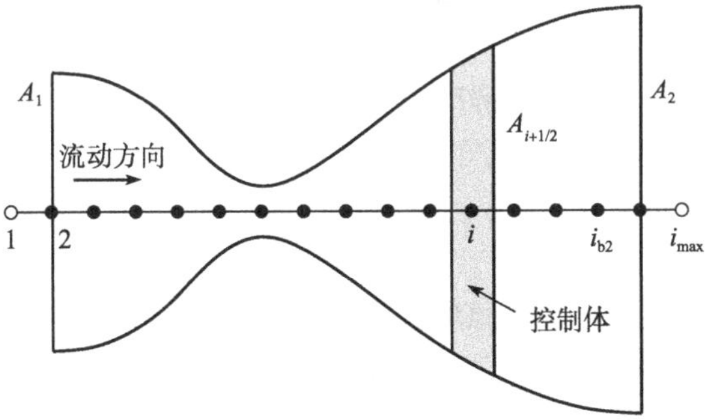

图 12.1　1D 欧拉解算器的网格和控制体;$i=1$ 和 $i=i_{max}$ 均为虚拟单元

12.3 二维结构网格生成器

grid_struct_2d 目录包含三个不同的程序,用于生成内外流的 2D 结构网格。第一个程序的源代码在子目录 cgrid 中,生成绕翼型的 C 型网格(见 11.1.1 节),如图 11.6 所示。初始网格由线性 TFI 方法(式(11.5))代数地生成,然后使用椭圆 PDEs(见 11.1.3 节)生成具有指定壁面距离的边界正交网格。翼型外形使用 Bézier 样条近似(关于 Bézier 样条见文献[1-2])。

第二个程序用以在 2D 带圆形鼓包的槽道内生成代数网格,网格在鼓包的前部和后部加密,源代码存储在 channel 子目录中。

第三个程序位于子目录 hgrid 中,可为叶栅外形生成 H 型网格(见 11.1.1 节),图 11.8 给出了该程序生成的网格示例。本例采用基于线性 TFI 技术(式(11.5))的代数网格生成方法,叶片外形同样由 Bézier 样条对给定点进行插值。

程序 cgrid 和 hgrid 依赖于 srccom 子目录中提供的库,该库包含样条插值、线性 TFI、椭圆网格生成和网格拉伸的程序。

12.4 结构到非结构网格转换器

grid_unstr_2d 目录包含将 2D 结构网格转换为非结构三角形网格的程序。该程序通过连接距离最短的对角节点,将结构网格的每个四边形单元划分为两个三角形。三角形网格可以作为 12.7 节中非结构流场解算器的输入。

12.5 准一维欧拉解算器

目录 structured_1d 包含求解准 1D 欧拉方程组的程序,该方程组是喷嘴内或管道中无黏流动的控制方程,其守恒微分形式为[3]

$$\frac{\partial \boldsymbol{W}}{\partial t}+\frac{\partial \boldsymbol{F}_{\mathrm{c}}}{\partial x}=\boldsymbol{Q} \tag{12.3}$$

其中,的守恒变量、对流通量和源项分别为

$$\boldsymbol{W}=\begin{bmatrix}\rho A\\ \rho uA\\ \rho EA\end{bmatrix} \quad \boldsymbol{F}_{\mathrm{c}}=\begin{bmatrix}\rho uA\\ (\rho u^2+p)A\\ \rho HuA\end{bmatrix} \quad \boldsymbol{Q}=\begin{bmatrix}0\\ p\mathrm{d}A/\mathrm{d}x\\ 0\end{bmatrix} \tag{12.4}$$

这里 A 表示喉道的面积。总焓 H 由式(2.12)给出,压力由式(2.29)得到

$$p=(\gamma-1)\rho\left(E-\frac{u^2}{2}\right) \tag{12.5}$$

控制方程组(12.3)在结构网格上使用二次控制体方法离散(见 4.2.3 节)，图 12.1 显示了网格和控制体的示意图,空间离散采用了三种不同的方法:

(1) 具有标量人工耗散的中心格式(见 4.3.1 节);

(2) CUSP 矢通量分裂格式(见 4.3.2 节);

(3) Roe 通量差分裂格式(见 4.3.3 节)。

离散方程组(12.3)采用显式多步格式(6.1.1 节或 6.1.2 节)进行时间推进。该程序使用局部时间步长(见 9.1 节)和中心 IRS 技术(见 9.3.1 节)来加速收敛。源代码存储在 src 子目录中,输入文件和运行结果存储在 run 子目录中。

第二个版本的流场解算器采用了 9.4 节中的多重网格方法来加速收敛,空间和时间离散格式与上述相同。多重网格在粗网格上采用 FAS 策略求解,在最细网格上采用 FMG 策略初始化。源代码包含在 srcmg 子目录中,输入文件示范存储在 runmg 子目录中。

入口和出口平面的边界条件基于 8.4 节的特征变量实现,为此采用了 8.1 节中描述的虚拟网格点概念(图 12.1)。

运行示例

作为示例,考虑拉瓦尔喷管流动,其中进口面积 $A_1=1.5$,出口面积 $A_2=2.5$,边界条件如下:

(1) 进口总压 $p_{t,in}=1.0\times10^5$Pa;

(2) 进口总温 $T_{t,in}=288.0$K;

(3) 出口静压 $p_{out}=7.0\times10^4$Pa。

模拟采用 Roe 迎风格式进行空间离散和五重的多重网格技术,参数设置为:$\sigma=4.5$,$\epsilon=0.8$,限制器系数 $K=1.5$,熵修正系数 $\delta=0.05c$。

可以使用 12.2 节中介绍的程序生成计算网格。子目录 runmg 中提供了网格文件 laval.grd。流场解算器还需要各种输入参数来设置边界条件、数值格式、收敛历程和绘图文件输出,这些数据存在名为 input_r 的用户输入文件中。假设解算器源代码已经编译过(详情见 README.txt),可以通过如下方式开始模拟:打开终端窗口(Windows 上的命令提示符),进入 runmg 目录,然后输入:

```
% eul1dmg < input_r
```

这将呈现如图 12.2 所示的收敛历程,图 12.3 显示了沿喷管径向的马赫数分布。

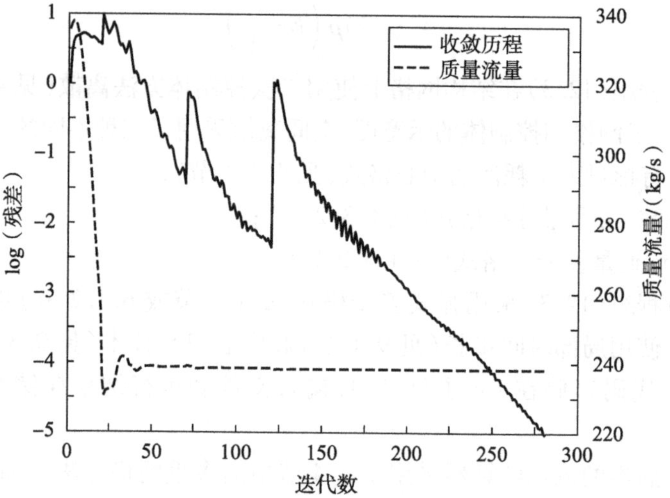

图 12.2　密度残差和质量流量的收敛历程

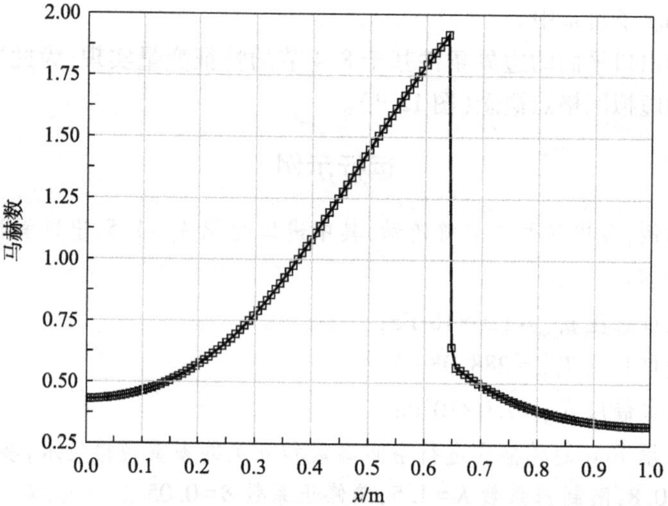

图 12.3　沿喷管径向的马赫数分布

12.6　二维结构欧拉/纳维-斯托克斯解算器

目录 structured_2d 包含了在贴体结构网格上求解 2D 欧拉和纳维-斯托克斯方程组的程序。空间离散使用 4.2.1 节所述的格心有限体积方法，采用带有标量人工耗散的中心离散格式（见 4.3.1 节），以及 4.3.3 节给出的 Roe 迎风格式。控制方程组使用显式多步格式(见 6.1.1 节或 6.1.2 节)进行时间积分,通过局部时间步长(见 9.1 节)和中心 IRS 技术(见 9.3.1 节)进行加速。速度分量和温度的梯度采用 4.4.1 节描述的二次控制体计算。该程序还能在极低马赫数下模拟不可压缩流动(见 9.5 节)。为了使代码易于理解,没有实现湍流模型。

流场解算器的源代码存储在 src 子目录中,输入文件、网格和结果在 run 子目录中,doc 子目录包含了 HTML 格式的文档(文件、模块、函数、变量)。

该程序利用虚拟单元(见 8.1 节)来处理边界条件,计算中使用了两层虚拟单元,图 12.4 给出了计算域网格的示意图(亦可见图 8.1)。该程序只能处理单块网格,但处理边界条件的规格和类型非常灵活。程序提供了以下八种边界类型:

(1) 坐标切割(C 型或 O 型网格);

(2) 远场(可选择涡修正、亚声速和超声速);

(3) 入流(亚声速);

(4) 出流(亚声速和超声速);

(5) 流体注入(或提取);

(6) 线周期;

(7) 固壁(黏性或无黏);

(8) 对称(沿 x 方向为常数或 y 方向为常数的线)。

上述边界条件的实现与第 8 章的讨论密切相关。

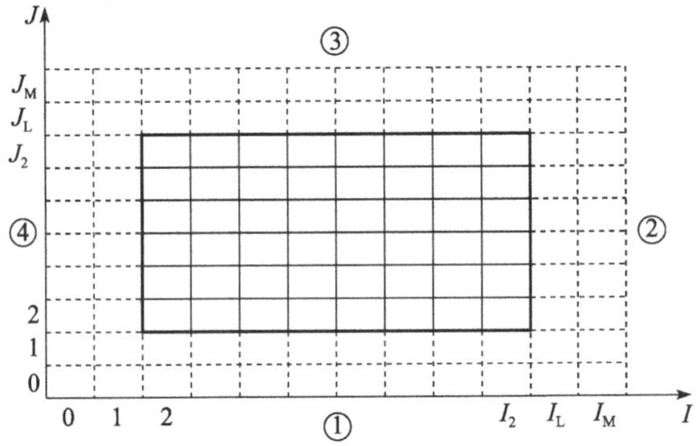

图 12.4　2D 结构欧拉/纳维-斯托克斯解算器计算空间的网格示意。每个边界都有两层虚拟单元(虚线)。带圈数字表示计算域的边界

计算空间中的四个边界可以划分为任意数量的段,每个段都可以与不同的边界条件相关联,这种方法与 8.9 节中描述的块界面非常相似。实际上,该程序可以相对容易地推广到多块网格。段的定义与拓扑文件中的网格分开存储(文件扩展名为 top)。

解算器中面矢量 $\boldsymbol{SI}$ 和 $\boldsymbol{SJ}$(即 $\boldsymbol{n}\Delta S$)的定义如图 12.5 所示,面矢量与控制体 VOL(I,J)(对应于 $\Omega_{I,J}$)的左面和底面相关联,指向控制体的外侧。控制体剩余两个面的面矢量为 $-\boldsymbol{SI}(I+1,J)$ 和 $-\boldsymbol{SJ}(I,J+1)$。这样,每个控制体只需存储两个面矢量。

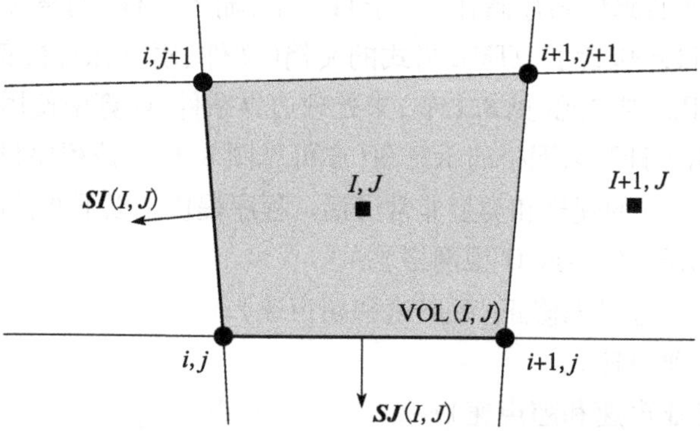

图 12.5　2D 结构欧拉/纳维-斯托克斯解算器的控制体和面矢量。索引(I,J)表示单元,索引(i,j)表示网格节点

运行示例

作为示例,考虑下列远场边界条件下 NACA0012 对称翼型的跨声速绕流:

(1) 马赫数 $M_\infty=0.8$;

(2) 迎角 $\alpha=1.25°$;

(3) 静压 $p_\infty=1.0\times10^5\text{Pa}$;

(4) 静态温度 $T_\infty=288.0\text{K}$。

空间离散采用 Roe 迎风格式,时间离散采用 $\sigma=5.0$、$\epsilon=1.2$ 的显式时间推进格式,限制器系数 $K=30.0$,熵修正系数 $\delta=0.05c$。

可以使用 12.3 节中介绍的程序生成示例网格,run 子目录包含了网格文件 n0012.grd。流场解算器还需要各种输入参数来设置边界条件、数值格式、收敛历程和绘图文件输出,这些数据存在名为 n0012_input_r 的用户输入文件中。假设解算器源代码已经编译过(详情见 README.txt),可以通过如下方式开始模拟:打开终端窗口(Windows 上的命令提示符),进入 run 目录,然后输入:

```
% Struct2D n0012_input_r
```

将呈现如图 12.6 所示的收敛历程,图 12.7 展示了翼型附近的马赫数云图。

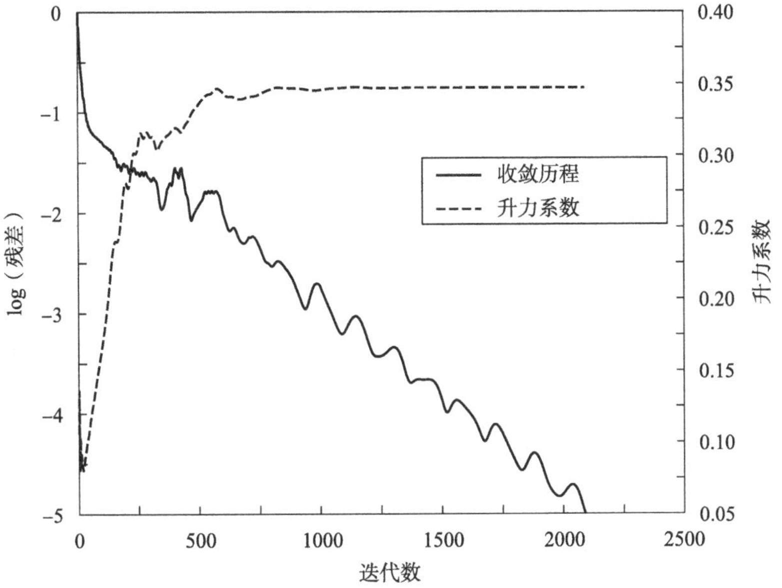

图 12.6　密度残差和升力系数的收敛历程

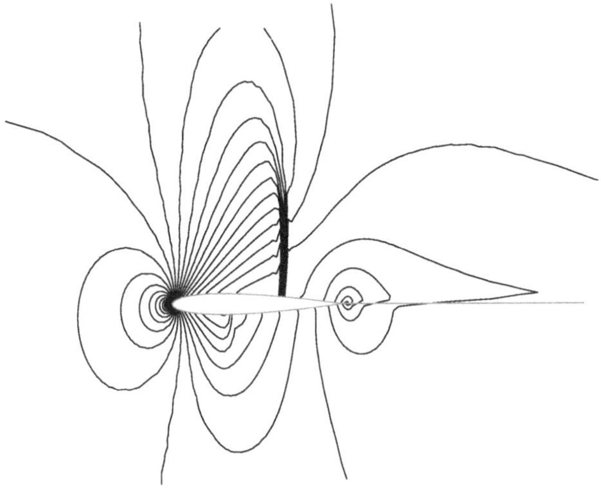

图 12.7　机翼附近的马赫数分布

12.7　二维非结构欧拉/纳维-斯托克斯解算器

unstructured_2d 和 C++/unstructured_2d 目录包含非结构三角形网格上求解

2D 欧拉和纳维-斯托克斯方程组的程序。该程序使用 5.2.2 节中的中点-二次格点格式进行空间离散，对流通量、黏性通量以及梯度计算采用基于边的数据结构。该程序还能模拟极低马赫数下的不可压缩流动（见 9.5 节）。

流场解算器的源代码存储在 src 子目录中，输入文件、网格和结果存储在 run 子目录中，doc 子目录存储了 HTML 格式的文档（文件、模块、函数、变量）。

采用 Roe 通量差分裂格式（见 4.3.3 节和 5.3.2 节）计算对流通量，根据式(5.42)，通过分段线性重构获得控制体界面左右两侧的状态，采用 Green-Gauss 方法（式(5.49)~式(5.50)）计算流动变量的梯度，采用 5.3.5 节（式(5.67)）介绍的 Venkatakrishnan 限制器对重构解进行限制，计算黏性通量时需要控制体界面处的流场梯度，其计算方法见 5.4.2 节，采用显式多步格式（见 6.1.1 节或 6.1.2 节）对离散控制方程进行时间积分，并采用局部时间步长（见 9.1 节）和中心隐式残值光滑方法（见 9.3.2 节）加速收敛。

该程序包含了与前面介绍的结构网格流场解算器相同的边界条件（除了注入和坐标切割），数据结构按照每个边界面可以有不同的边界条件的形式组织，远场、入口和出口边界条件采用了虚拟节点（一层）的概念实现，所有与边界相关的拓扑数据都与网格坐标和节点索引一起存储在一个公共文件中（扩展名为 ugr）。

运行示例

作为示例，考虑 VKI-1 涡轮叶栅跨声速流动，边界条件为[4]：

(1) 进口总压 $p_{t,in}=1.0\times10^5$Pa；

(2) 进口总温 $T_{t,in}=300.0$K；

(3) 进口气流角 $\alpha_{inl}=30°$；

(4) 出口静压 $p_{out}=5.283\times10^4$Pa。

空间离散采用 Roe 迎风格式，时间离散采用 $\sigma=5.5$、$\epsilon=0.4$ 的显式时间推进格式，限制器系数 $K=1.0$，熵修正系数 $\delta=0.05\cdot c$。

示例网格可以使用 12.3 节和 12.4 节中介绍的程序生成，run 子目录包含使用阵面推进法（见 11.2.2 节）生成的名为 vkil.ugr 的非结构网格。流场解算器还需要各种输入参数来设置边界条件、数值格式、收敛历史和绘图文件输出，这些数据存在名为 vki1_input 的用户输入文件中。假设解算器的源代码已经编译过（详细信息请参阅 README.txt），可以通过如下方式开始模拟：打开终端窗口（Windows 上的命令提示符），进入 run 目录，然后输入：

```
% Unstruct2D vki1_input
```

这将呈现如图 12.8 所示的收敛历程，图 12.9 展示了叶栅内部的压力云图。

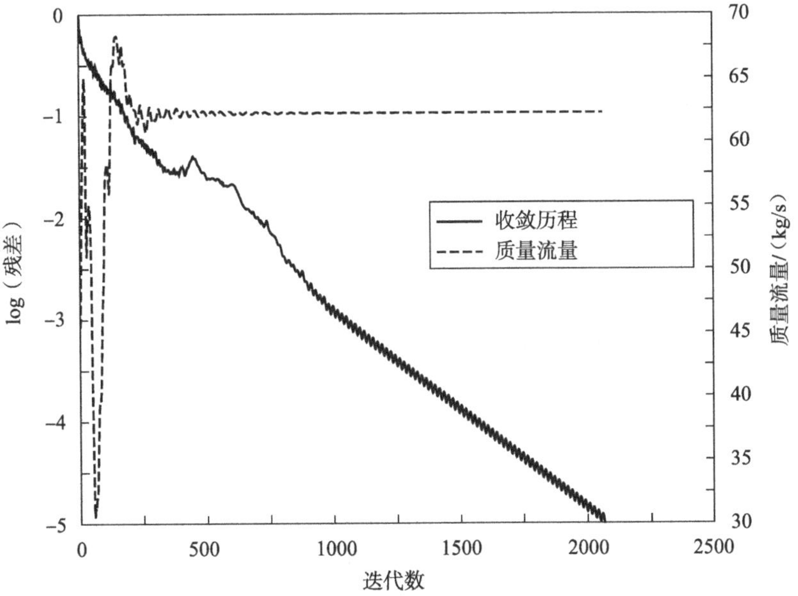

图 12.8　密度残差和质量流量的收敛历程

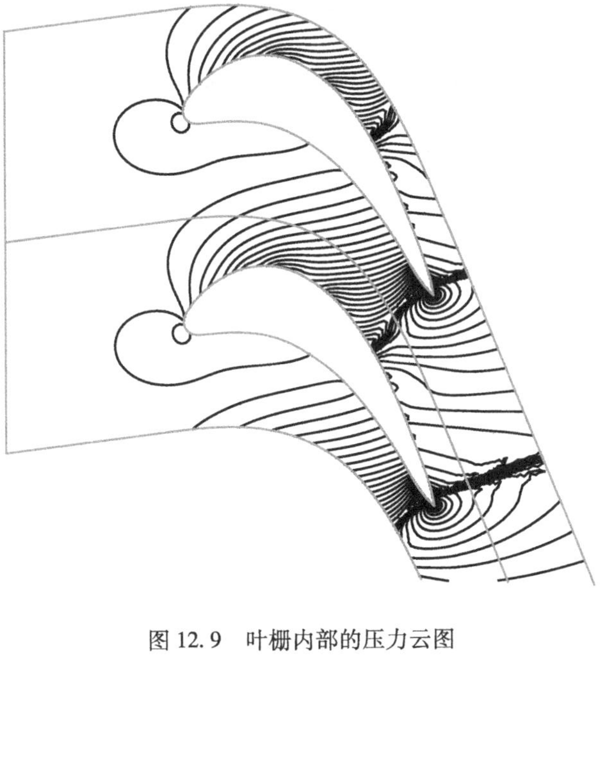

图 12.9　叶栅内部的压力云图

12.8 并行化

Parallelization 目录包含使用异步代理库、CUDA、MPI 和 OpenMP(9.6 节)进行代码并行化的示例。并行化程序在结构矩形网格上使用雅可比迭代求解 2D 拉普拉斯方程:

$$\phi_{xx}+\phi_{yy}=0 \tag{12.6}$$

边界值在迭代过程中保持固定。

示例使用 C(CUDA 和 MPI)和 C++语言编写。使用 MPI 时,编译由 Makefile 文件控制,在其他情况下,提供了 Visual Studio(VS2012)的项目文件。注意,除了 OpenMP 之外,所有示例都需要相应的库和编译工具(见 9.6 节中的参考文献)。OpenMP 程序无须任何特殊准备,任何最新的 C++编译器都可以满足。

参考文献

[1] Farin GE. Curves and surfaces for computer aided geometric design—a practical guide. 3rd ed. Boston:Academic Press;1993.

[2] Hoschek J,Lasser D. Fundamentals of computer aided geometric design. Wellesley,MA: A. K. PetersLtd;1993.

[3] Shapiro AH. The dynamics and thermodynamics of compressible fluid flow. New York: Ronald Press;1953.

[4] Kiock R,Lehthaus F,Baines NC,Sieverding CH. The transonic flow through a plane turbine cascadeas measured in four European wind tunnels. ASME J Eng Gas Turbines Power 1986;108:277-84.

附　录

A.1　微分形式的控制方程组

假设对流通量和黏性通量是连续的，则控制方程组(2.19)可以由积分形式转化为微分形式。首先应用高斯定理，将式(2.19)变为

$$\frac{\partial}{\partial t}\int_{\Omega}\boldsymbol{W}\mathrm{d}\Omega+\int_{\Omega}\nabla\cdot(\boldsymbol{F}_{\mathrm{c}}-\boldsymbol{F}_{\mathrm{v}})\mathrm{d}\Omega=\int_{\Omega}\boldsymbol{Q}\mathrm{d}\Omega \tag{A.1}$$

其中，$\boldsymbol{F}_{\mathrm{c}}$ 和 $\boldsymbol{F}_{\mathrm{v}}$ 分别为对流通量和黏性通量的张量。对于任意控制体，式(A.1)可写成微分形式

$$\frac{\partial \boldsymbol{W}}{\partial t}+\nabla\cdot(\boldsymbol{F}_{\mathrm{c}}-\boldsymbol{F}_{\mathrm{v}})=\boldsymbol{Q} \tag{A.2}$$

考虑到任意贴体网格，我们引入笛卡儿坐标系(x,y,z)和曲线坐标系(ξ,η,ζ)之间的映射函数(见图3.2)

$$\begin{aligned}\xi&=\xi(x,y,z,t)\\ \eta&=\eta(x,y,z,t)\\ \zeta&=\zeta(x,y,z,t)\end{aligned} \tag{A.3}$$

据此，3D纳维-斯托克斯方程组(A.2)的微分形式转化为

$$\frac{\partial \boldsymbol{W}^*}{\partial t}+\frac{\partial \boldsymbol{F}_{\mathrm{c},1}}{\partial \xi}+\frac{\partial \boldsymbol{F}_{\mathrm{c},2}}{\partial \eta}+\frac{\partial \boldsymbol{F}_{\mathrm{c},3}}{\partial \zeta}=\frac{\partial \boldsymbol{F}_{\mathrm{v},1}}{\partial \xi}+\frac{\partial \boldsymbol{F}_{\mathrm{v},2}}{\partial \eta}+\frac{\partial \boldsymbol{F}_{\mathrm{v},3}}{\partial \zeta}+\boldsymbol{Q}^* \tag{A.4}$$

其中，守恒变量为

$$\boldsymbol{W}^*=J^{-1}\begin{bmatrix}\rho\\ \rho u\\ \rho v\\ \rho w\\ \rho E\end{bmatrix} \tag{A.5}$$

对流通量由下面的关系式得到：

$$\boldsymbol{F}_{c,1}=J^{-1}\begin{bmatrix}\rho V_1\\ \rho V_1 u+\xi_x p\\ \rho V_1 v+\xi_y p\\ \rho V_1 w+\xi_z p\\ \rho V_1 H\end{bmatrix},\quad \boldsymbol{F}_{c,2}=J^{-1}\begin{bmatrix}\rho V_2\\ \rho V_2 u+\eta_x p\\ \rho V_2 v+\eta_y p\\ \rho V_2 w+\eta_z p\\ \rho V_2 H\end{bmatrix},\quad \boldsymbol{F}_{c,3}=J^{-1}\begin{bmatrix}\rho V_3\\ \rho V_3 u+\zeta_x p\\ \rho V_3 v+\zeta_y p\\ \rho V_3 w+\zeta_z p\\ \rho V_3 H\end{bmatrix}\tag{A.6}$$

黏性通量定义为

$$\boldsymbol{F}_{v,1}=J^{-1}\begin{bmatrix}0\\ \xi_x\tau_{xx}+\xi_y\tau_{xy}+\xi_z\tau_{xz}\\ \xi_x\tau_{yx}+\xi_y\tau_{yy}+\xi_z\tau_{yz}\\ \xi_x\tau_{zx}+\xi_y\tau_{zy}+\xi_z\tau_{zz}\\ \xi_x\Theta_x+\xi_y\Theta_y+\xi_z\Theta_z\end{bmatrix}$$

$$\boldsymbol{F}_{v,2}=J^{-1}\begin{bmatrix}0\\ \eta_x\tau_{xx}+\eta_y\tau_{xy}+\eta_z\tau_{xz}\\ \eta_x\tau_{yx}+\eta_y\tau_{yy}+\eta_z\tau_{yz}\\ \eta_x\tau_{zx}+\eta_y\tau_{zy}+\eta_z\tau_{zz}\\ \eta_x\Theta_x+\eta_y\Theta_y+\eta_z\Theta_z\end{bmatrix}\tag{A.7}$$

$$\boldsymbol{F}_{v,3}=J^{-1}\begin{bmatrix}0\\ \zeta_x\tau_{xx}+\zeta_y\tau_{xy}+\zeta_z\tau_{xz}\\ \zeta_x\tau_{yx}+\zeta_y\tau_{yy}+\zeta_z\tau_{yz}\\ \zeta_x\tau_{zx}+\zeta_y\tau_{zy}+\zeta_z\tau_{zz}\\ \zeta_x\Theta_x+\zeta_y\Theta_y+\zeta_z\Theta_z\end{bmatrix}$$

最后，源项变换为

$$\boldsymbol{Q}^{*}=J^{-1}\begin{bmatrix}0\\ \rho f_{e,x}\\ \rho f_{e,y}\\ \rho f_{e,z}\\ \rho\boldsymbol{f}_e\cdot\boldsymbol{v}+\dot{q}_h\end{bmatrix}\tag{A.8}$$

其中，$\boldsymbol{f}_e$ 为外体力矢量。

对于静止网格,逆变速度在ξ,η,ζ坐标方向上的形式为

$$\begin{aligned}V_1&=\xi_x u+\xi_y v+\xi_z w\\V_2&=\eta_x u+\eta_y v+\eta_z w\\V_3&=\zeta_x u+\zeta_y v+\zeta_z w\end{aligned}\tag{A.9}$$

黏性应力张量的分量和热通量的分量为

$$\begin{aligned}\tau_{xx}&=2\mu\frac{\partial u}{\partial x}+\lambda\left(\frac{\partial u}{\partial x}+\frac{\partial v}{\partial y}+\frac{\partial w}{\partial z}\right)\\\tau_{yy}&=2\mu\frac{\partial v}{\partial y}+\lambda\left(\frac{\partial u}{\partial x}+\frac{\partial v}{\partial y}+\frac{\partial w}{\partial z}\right)\\\tau_{zz}&=2\mu\frac{\partial w}{\partial z}+\lambda\left(\frac{\partial u}{\partial x}+\frac{\partial v}{\partial y}+\frac{\partial w}{\partial z}\right)\\\tau_{xy}&=\tau_{yx}=\mu\left(\frac{\partial u}{\partial y}+\frac{\partial v}{\partial x}\right)\\\tau_{xz}&=\tau_{zx}=\mu\left(\frac{\partial u}{\partial z}+\frac{\partial w}{\partial x}\right)\\\tau_{yz}&=\tau_{zy}=\mu\left(\frac{\partial v}{\partial z}+\frac{\partial w}{\partial y}\right)\\\Theta_x&=u\tau_{xx}+v\tau_{xy}+w\tau_{xz}+k\frac{\partial T}{\partial x}\\\Theta_y&=u\tau_{yx}+v\tau_{yy}+w\tau_{yz}+k\frac{\partial T}{\partial y}\\\Theta_z&=u\tau_{zx}+v\tau_{zy}+w\tau_{zz}+k\frac{\partial T}{\partial z}\end{aligned}\tag{A.10}$$

式中:k为热传导系数;μ为动力学黏性系数;λ为第二黏性系数(根据斯托克斯假设$\lambda=-(2/3)\mu$),热扩散通量也可以写成下面的变形:

$$\begin{aligned}k\frac{\partial T}{\partial x}&=\frac{\mu}{(\gamma-1)Pr}\frac{\partial c^2}{\partial x}\\k\frac{\partial T}{\partial y}&=\frac{\mu}{(\gamma-1)Pr}\frac{\partial c^2}{\partial y}\\k\frac{\partial T}{\partial z}&=\frac{\mu}{(\gamma-1)Pr}\frac{\partial c^2}{\partial z}\end{aligned}\tag{A.11}$$

其中,c为声速,速度分量u,v,w,以及温度T在笛卡儿坐标系x,y,z下的导数可以借助链式法则表示为ξ,η和ζ的导数,例如:

$$\frac{\partial u}{\partial x}=\xi_x\frac{\partial u}{\partial \xi}+\eta_x\frac{\partial u}{\partial \eta}+\zeta_x\frac{\partial u}{\partial \zeta}$$
$$\frac{\partial u}{\partial y}=\xi_y\frac{\partial u}{\partial \xi}+\eta_y\frac{\partial u}{\partial \eta}+\zeta_y\frac{\partial u}{\partial \zeta} \tag{A.12}$$

坐标变换雅可比矩阵 $\partial(\xi,\eta,\zeta)/\partial(x,y,z)$ 的行列式的倒数定义为

$$J^{-1}=\frac{\partial x}{\partial \xi}\frac{\partial y}{\partial \eta}\frac{\partial z}{\partial \zeta}+\frac{\partial x}{\partial \eta}\frac{\partial y}{\partial \zeta}\frac{\partial z}{\partial \xi}+\frac{\partial x}{\partial \zeta}\frac{\partial y}{\partial \xi}\frac{\partial z}{\partial \eta}-\frac{\partial x}{\partial \xi}\frac{\partial y}{\partial \zeta}\frac{\partial z}{\partial \eta}-\frac{\partial x}{\partial \eta}\frac{\partial y}{\partial \xi}\frac{\partial z}{\partial \zeta}-\frac{\partial x}{\partial \zeta}\frac{\partial y}{\partial \eta}\frac{\partial z}{\partial \xi} \tag{A.13}$$

最后，通过下列关系式定义度量项

$$\begin{aligned}
\xi_x&=J\left(\frac{\partial y}{\partial \eta}\frac{\partial z}{\partial \zeta}-\frac{\partial y}{\partial \zeta}\frac{\partial z}{\partial \eta}\right)\\
\xi_y&=J\left(\frac{\partial z}{\partial \eta}\frac{\partial x}{\partial \zeta}-\frac{\partial z}{\partial \zeta}\frac{\partial x}{\partial \eta}\right)\\
\xi_z&=J\left(\frac{\partial x}{\partial \eta}\frac{\partial y}{\partial \zeta}-\frac{\partial x}{\partial \zeta}\frac{\partial y}{\partial \eta}\right)\\
\eta_x&=J\left(\frac{\partial y}{\partial \zeta}\frac{\partial z}{\partial \xi}-\frac{\partial y}{\partial \xi}\frac{\partial z}{\partial \zeta}\right)\\
\eta_y&=J\left(\frac{\partial z}{\partial \zeta}\frac{\partial x}{\partial \xi}-\frac{\partial z}{\partial \xi}\frac{\partial x}{\partial \zeta}\right)\\
\eta_z&=J\left(\frac{\partial x}{\partial \zeta}\frac{\partial y}{\partial \xi}-\frac{\partial x}{\partial \xi}\frac{\partial y}{\partial \zeta}\right)\\
\zeta_x&=J\left(\frac{\partial y}{\partial \xi}\frac{\partial z}{\partial \eta}-\frac{\partial y}{\partial \eta}\frac{\partial z}{\partial \xi}\right)\\
\zeta_y&=J\left(\frac{\partial z}{\partial \xi}\frac{\partial x}{\partial \eta}-\frac{\partial z}{\partial \eta}\frac{\partial x}{\partial \xi}\right)\\
\zeta_z&=J\left(\frac{\partial x}{\partial \xi}\frac{\partial y}{\partial \eta}-\frac{\partial x}{\partial \eta}\frac{\partial y}{\partial \xi}\right)
\end{aligned} \tag{A.14}$$

理解式(A.13)和式(A.14)中的度量项如何与体积或面矢量等几何量相关联具有重要意义。为此，我们考虑一个2D结构网格，如图A.1所示，假设 ξ 坐标对应于 i 方向，η 坐标对应于 j 方向，则式(A.14)中的笛卡儿坐标关于曲线坐标的导数在点 (i,j) 可以近似为

$$\left(\frac{\partial x}{\partial \xi}\right)_{i,j}\approx\frac{\frac{1}{2}[(x_{i,j}+x_{i+1,j})-(x_{i,j}+x_{i-1,j})]}{(i+1/2)-(i-1/2)}=\frac{1}{2}(x_{i+1,j}-x_{i-1,j})$$

$$\left(\frac{\partial y}{\partial \xi}\right)_{i,j} \approx \frac{\frac{1}{2}[(y_{i,j}+y_{i+1,j})-(y_{i,j}+y_{i-1,j})]}{(i+1/2)-(i-1/2)} = \frac{1}{2}(y_{i+1,j}-y_{i-1,j}) \tag{A.15}$$

和

$$\left(\frac{\partial x}{\partial \eta}\right)_{i,j} \approx \frac{\frac{1}{2}[(x_{i,j}+x_{i,j+1})-(x_{i,j}+x_{i,j-1})]}{(j+1/2)-(j-1/2)} = \frac{1}{2}(x_{i,j+1}-x_{i,j-1})$$

$$\left(\frac{\partial y}{\partial \eta}\right)_{i,j} \approx \frac{\frac{1}{2}[(y_{i,j}+y_{i,j+1})-(y_{i,j}+y_{i,j-1})]}{(j+1/2)-(j-1/2)} = \frac{1}{2}(y_{i,j+1}-y_{i,j-1}) \tag{A.16}$$

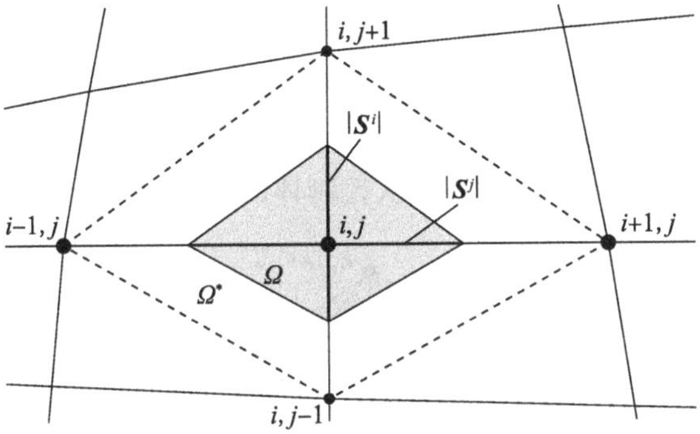

图 A.1　2D 度量项的几何表示

因此,使用式(A.15)和式(A.16)的近似,式(A.14)中的公式变为

$$\begin{aligned}
\xi_x &= J\frac{\partial y}{\partial \eta} \approx JS_x^I \\
\xi_y &= -J\frac{\partial x}{\partial \eta} \approx JS_y^I \\
\eta_x &= -J\frac{\partial y}{\partial \xi} \approx JS_x^J \\
\eta_y &= J\frac{\partial x}{\partial \xi} \approx JS_y^J
\end{aligned} \tag{A.17}$$

其中,$\boldsymbol{S}^I=[S_x^I,S_y^I]^{\mathrm{T}}$ 为 i 方向上的平均面矢量,$\boldsymbol{S}^J=[S_x^J,S_y^J]^{\mathrm{T}}$ 为 j 方向的平均面矢量(图 A.1)。式(A.17)中的关系也可表示为

$$\boldsymbol{S}^I = J^{-1}\begin{bmatrix}\xi_x \\ \xi_y\end{bmatrix} \quad \boldsymbol{S}^J = J^{-1}\begin{bmatrix}\eta_x \\ \eta_y\end{bmatrix} \tag{A.18}$$

由上述关系可以看到,度量项对应于有限体积格式中的面矢量分量除以坐

标变换雅可比矩阵的行列式 J。

那么，J^{-1} 的几何解释又是什么？式(A.13)在2D情况下表示为

$$J^{-1}=\frac{\partial x}{\partial \xi}\frac{\partial y}{\partial \eta}-\frac{\partial x}{\partial \eta}\frac{\partial y}{\partial \xi} \tag{A.19}$$

将式(A.17)的度量项代入式(A.19)，可得

$$J^{-1}=2\Omega=\frac{1}{2}\Omega^* \tag{A.20}$$

图A.1显示了体积 Ω 和 Ω^*（厚度 $b=1$ 保持不变）。Ω 由点 $(i+1/2,j)$，$(i,j+1/2)$，$(i-1/2,j)$，$(i,j-1/2)$ 定义，Ω^* 由网格节点 $(i,j-1)$，$(i+1,j)$，$(i,j+1)$，$(i-1,j)$ 定义。需要注意的是，在矩形网格上 J^{-1} 等于4.2.3节中二次控制体的体积。

上述度量项的推导同样可以在3D空间中进行，结果如图A.2所示，变换雅可比矩阵(J^{-1})行列式的倒数等于网格节点 $(i,j,k-1/2)$，$(i+1/2,j,k)$，$(i,j,k+1/2)$，$(i-1/2,j,k)$，$(i,j-1/2,k)$ 和 $(i,j+1/2,k)$ 定义的体积的两倍，在长方体网格上，它恰好又对应于4.2.3节的二次控制体的体积。

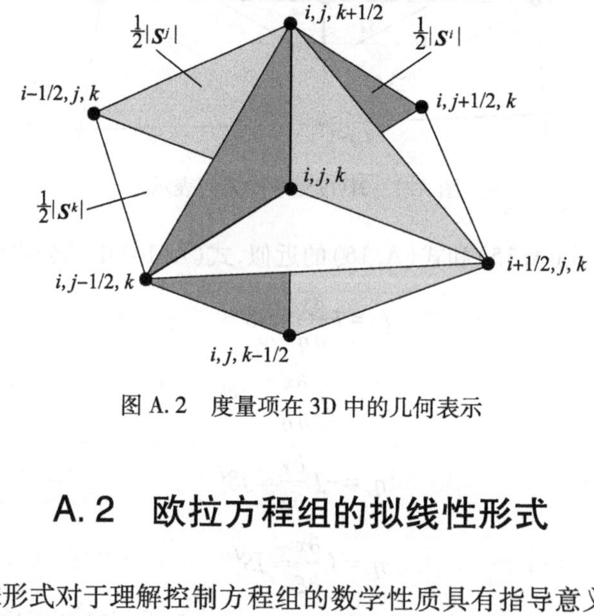

图A.2　度量项在3D中的几何表示

A.2　欧拉方程组的拟线性形式

这种特殊形式对于理解控制方程组的数学性质具有指导意义，也有助于数值格式的发展和分析。

拟线性形式由微分形式的控制方程组(A.4)推导出。在此我们仅考虑欧拉方程组，即

$$\frac{\partial \boldsymbol{W}^*}{\partial t}+\frac{\partial \boldsymbol{F}_{c,1}}{\partial \xi}+\frac{\partial \boldsymbol{F}_{c,2}}{\partial \eta}+\frac{\partial \boldsymbol{F}_{c,3}}{\partial \zeta}=0 \tag{A.21}$$

其中，$W^* = J^{-1}W$，对流通量 $F_{c,1/2/3}$ 在式(A.6)中定义。线性化过程首先将通量的空间导数替换为守恒变量的导数，例如：

$$\frac{\partial F_{c,1}}{\partial \xi} = \frac{\partial F_{c,1}}{\partial W}\frac{\partial W}{\partial \xi} \tag{A.22}$$

这样空间导数现在变为守恒变量的线性函数，利用式(A.65)对流通量雅可比矩阵的定义，可以得到欧拉方程组的拟线性形式为

$$J^{-1}\frac{\partial W}{\partial t} + A_{c,1}\frac{\partial W}{\partial \xi} + A_{c,2}\frac{\partial W}{\partial \eta} + A_{c,3}\frac{\partial W}{\partial \zeta} = 0 \tag{A.23}$$

对 $A_{c,1}$ 由式(A.68)(或2D情况下的式(A.66))给出，其中用 $J^{-1}(\xi_x,\xi_y,\xi_z)$ 代替式中的(n_x,n_y,n_z)，$A_{c,2}$ 和 $A_{c,3}$ 分别用类似的方法定义关于 η 和 ζ 的导数。雅可比矩阵的特征值决定了波(具有对流模态和声波模态)在流域中传播的速度和方向，因此，特征值决定了显式时间格式的时间步长，雅可比矩阵的特征值和特征矢量在迎风空间离散格式中也起着重要的作用，在迎风空间离散格式中，模板偏向于波传播的方向。

A.3 控制方程组的数学性质

通过下面的拟线性二阶方程的经典例子，可以很好地说明 PDEs 的数学性质，方程为

$$a\frac{\partial^2 U}{\partial x^2} + b\frac{\partial^2 U}{\partial x \partial y} + c\frac{\partial^2 U}{\partial y^2} = d \tag{A.24}$$

其中，U 表示一般标量函数，系数 a,b,c 和 d 是坐标、U 及其一阶导数的非线性函数，但不是 U 的二阶导数的函数。根据判别式函数(b^2-4ac)的符号，可以定义三种不同类型的方程[1-2]：如果判别式为正，则方程是双曲型的；如果$(b^2-4ac)<0$，则方程变成椭圆型；如果(b^2-4ac)为零，则方程表示抛物型。

纳维-斯托克斯方程组不能用这样简单的方式来判别，事实上，它们通常是这三种类型的混合，具体取决于流动条件和问题的几何形状。然而，对双曲型、抛物型和椭圆型 PDEs 的物理性质进行说明是有价值的，流体力学求解方法的正确数学表述必须考虑这些性质。

A.3.1 双曲型方程组

如果方程(A.24)是双曲型的，则它有两个实特征值，情形如图 A.3 所示。从理论可知，点 P 的信息仅影响前行的特征线之间的区域，另外，点 P 仅从特征线 AP 与 BP 之间的区域接收信息。此外，点 P 的解仅取决于通过点 P 的两条特征线所

截断的边界部分，以及它们的交点（即 AB 区间），例如，点 P 不从点 C 获得信息，因为 P 不在从点 C 传播的特征线的封闭区内。因此，为了确定给定区域的解，我们只需要在一部分边界上指定条件，从而，双曲型方程组表示一个初值问题。

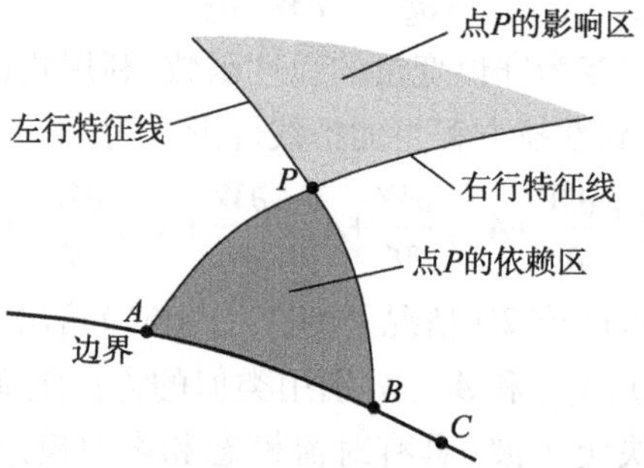

图 A.3　每个点具有两条特征线的双曲型 PDE 的依赖域和影响域[3]

在流体力学中，以下是由双曲型 PDEs 控制流动的例子：

(1) 定常无黏超声速流动：如果流动是 2D 的，则其性质如上所述，对于 3D 流动，在(x,y,z)空间中具有特征面，点 P 则影响前行的（下游）特征面包围的区域，在线化位势方程中，特征面对应于马赫锥。

(2) 非定常流：在这种情况下，控制方程组在时间上是双曲型的，在空间上可以是不同的类型，因此，局部流动是亚声速还是超声速并不重要，由于具有（部分）双曲型的性质，有必要指定一个初始解，然后在时间上进行推进。

A.3.2　抛物型方程组

在这种情况下，方程（A.24）只有一个实特征值，图 A.4 显示了特征线和影响域。对于抛物型方程组，点 P 的信息影响特征线（图中的 BP）一侧的整个区域，从而包括点 C 和 D，另外，点 A 处的解完全独立于点 P，因此，像双曲型 PDEs 一样，信息只向一个方向传播。此外，从图 A.4 中可以看出，点 D 的解既依赖于整个特征线 BP 的条件，也依赖于边界段 BC 的条件。因此，抛物型方程组代表了一个混合的初值和边值问题。

一种特殊形式的简化控制方程组，即抛物型纳维-斯托克斯方程组（PNS）（见 2.4.3 节）表现出所描述的抛物型性质，在这种情况下，黏性应力项中涉及流向的导数在控制方程组中被忽略。PNS 方程组的求解从入口边界指定的（初始）数据开始，然后以向下游推进的方式进行模拟，每个流向站位代表一个垂直于主流方向的平面（2D 情况下为线），并以一种显式方式与一个或两个上游平面耦合，在每个平面上，方程组都需要迭代求解。

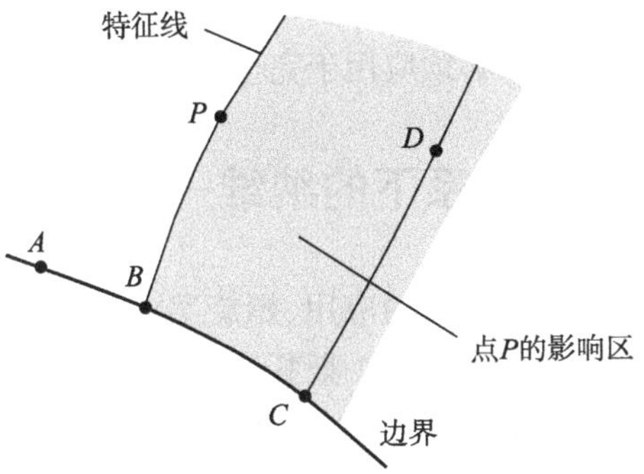

图 A.4 抛物型 PDE 的影响域[3]

对高 *Re* 数情况下的纳维-斯托克斯方程组的进一步简化产生了著名的边界层方程组，它也是抛物型的，可以用类似的显式空间推进来求解。

A.3.3 椭圆型方程组

如果方程(A.24)为椭圆型，则它具有两个复特征值，从理论可知，在某一点 P 上的信息会影响区域内的所有其他分区，另外，P 点的解依赖于周围的整个区域，情况如图 A.5 所示，这意味着 P 点处的解受整个封闭边界(A,B,C)的影响。因此，椭圆型 PDEs 表示一个边值问题，其中 P 点的解必须与域内所有其他点的解同时进行求解，这与应用于抛物型和双曲型方程组的空间或时间推进过程形成了强烈的对比。由图 A.5 可知，必须在整个边界(A,B,C)上指定边界条件，边界条件可以为以下两种形式之一：

(1) 在边界上指定因变量，这种类型的边界条件称为 Dirichlet 条件。

(2) 在边界上指定因变量的导数，这种类型的边界条件被称为 Neumann 条件。

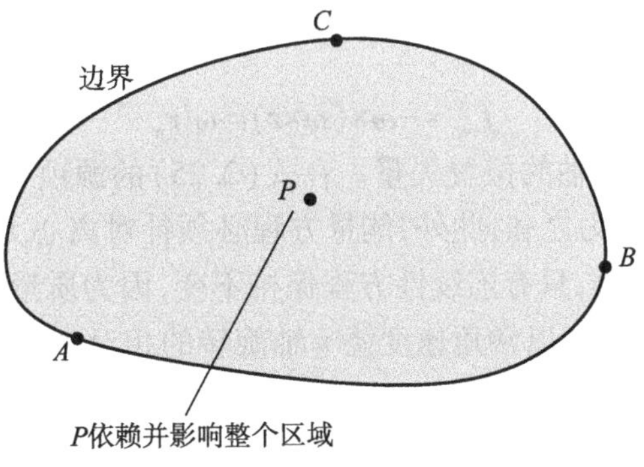

图 A.5 椭圆型 PDE 的影响域和依赖域[3]

例如，在流体力学中，定常亚声速/不可压缩无黏流动由椭圆型方程组控制，对于这些流动，物理边界条件必须应用于完全包围模拟流场的面上。

A.4 旋转坐标系下的纳维-斯托克斯方程组

在某些情况下，例如涡轮机械的应用、螺旋桨或地球物理等，计算域绕某个轴恒定地旋转，在这些应用中，将纳维-斯托克斯方程组转化到旋转坐标系下是很实用的。考虑图 A.6 所示的情况，点 P 绕固定轴以恒定角速度 $\boldsymbol{\omega}$ 旋转，为了简单起见，我们假设旋转轴与 x 轴重合，因此角速度为 $\boldsymbol{\omega}=[\omega_1,0,0]^{\mathrm{T}}$，绝对速度 $\boldsymbol{v}_{\mathrm{a}}$ 是相对速度 $\boldsymbol{v}_{\mathrm{r}}$ 与牵连速度 $\boldsymbol{v}_{\mathrm{e}}$ 之和，即

$$\boldsymbol{v}_{\mathrm{a}}=\boldsymbol{v}_{\mathrm{r}}+\boldsymbol{v}_{\mathrm{e}}=\boldsymbol{v}_{\mathrm{r}}+\boldsymbol{\omega}\times\boldsymbol{r} \tag{A.25}$$

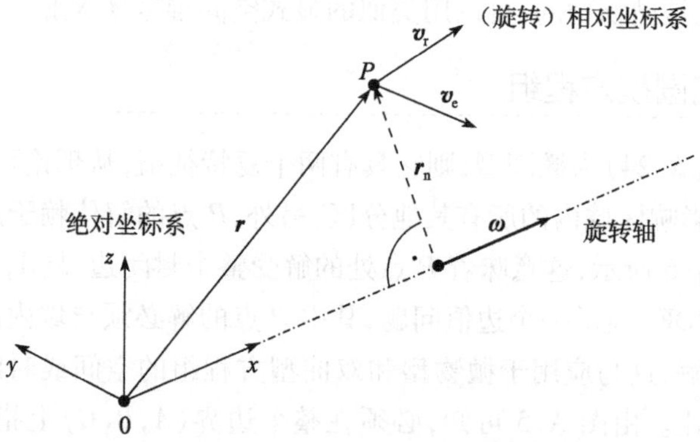

图 A.6　绝对坐标系和旋转坐标系。旋转轴与 x 轴重合

如果我们在相对坐标系下重写纳维-斯托克斯方程组(2.19)，必须考虑科里奥利力和离心力的影响，单位质量的科氏力定义为

$$\boldsymbol{f}_{\mathrm{Corr}}=-2(\boldsymbol{\omega}\times\boldsymbol{v}_{\mathrm{r}}) \tag{A.26}$$

单位质量的离心力为

$$\boldsymbol{f}_{\mathrm{cen}}=-\boldsymbol{\omega}\times(\boldsymbol{\omega}\times\boldsymbol{r})=\omega_1^2\boldsymbol{r}_{\mathrm{n}} \tag{A.27}$$

其中，$\boldsymbol{r}_{\mathrm{n}}$ 为垂直于旋转轴的位置矢量。在式(2.25)的源项 $\boldsymbol{Q}$ 中，必须为动量方程增加科氏力和离心力之和，此外，能量方程必须针对离心力(科氏力对能量平衡没有贡献)进行修改，只有连续性方程保持不变，因为质量平衡对系统的旋转是保持不变的。因此，在以常角速度绕 x 轴旋转的相对坐标系下，纳维-斯托克斯方程组的形式为

$$\frac{\partial}{\partial t}\int_{\Omega}\boldsymbol{W}\mathrm{d}\Omega+\oint_{\partial\Omega}(\boldsymbol{F}_{\mathrm{c}}-\boldsymbol{F}_{\mathrm{v}})\mathrm{d}S=\int_{\Omega}\boldsymbol{Q}\mathrm{d}\Omega \tag{A.28}$$

守恒变量为

$$\boldsymbol{W}=\begin{bmatrix}\rho\\ \rho u\\ \rho v\\ \rho w\\ \rho E\end{bmatrix} \tag{A.29}$$

其中,ρ 表示密度,u,v,w 是相对坐标系下的笛卡儿速度分量,E 是单位质量的相对总能,由下式给定:

$$E=e+\frac{|\boldsymbol{v}_{\mathrm{r}}|^2}{2}-\frac{|\boldsymbol{v}_{\mathrm{e}}|^2}{2}=e+\frac{u^2+v^2+w^2}{2}-\frac{\omega_1^2|\boldsymbol{r}_{\mathrm{n}}|^2}{2} \tag{A.30}$$

其中,$|\boldsymbol{r}_{\mathrm{n}}|^2=y^2+z^2$,对流通量 $\boldsymbol{F}_{\mathrm{c}}$ 定义为

$$\boldsymbol{F}_{\mathrm{c}}=\begin{bmatrix}\rho V\\ \rho uV+n_x p\\ \rho vV+n_y p\\ \rho wV+n_z p\\ \rho IV\end{bmatrix} \tag{A.31}$$

其中,p 是静压,n_x,n_y,n_z 分别表示面 $\partial\Omega$ 的指向外的单位法向矢量的分量。此外:

$$I=h+\frac{|\boldsymbol{v}_{\mathrm{r}}|^2}{2}-\frac{|\boldsymbol{v}_{\mathrm{e}}|^2}{2}=H-\frac{\omega_1^2|\boldsymbol{r}_{\mathrm{n}}|^2}{2} \tag{A.32}$$

$$V=n_x u+n_y v+n_z w$$

其中,I 表示滞止焓,H 表示相对总焓,V 表示逆变速度。滞止焓代表恒定旋转坐标系中的总能。

黏性通量 $\boldsymbol{F}_{\mathrm{v}}$ 与式(2.23)保持相同的形式。黏性应力张量的分量在形式上仍保持如式(2.15)所示,然而,源项 $\boldsymbol{Q}$ 增加了关于科氏力和离心力的扩展:

$$\boldsymbol{Q}=\begin{bmatrix}0\\ \rho f_{\mathrm{e},x}\\ \rho\omega_1(y\omega_1+2w)+\rho f_{\mathrm{e},y}\\ \rho\omega_1(z\omega_1-2v)+\rho f_{\mathrm{e},z}\\ \rho\boldsymbol{f}_{\mathrm{e}}\cdot\boldsymbol{v}_{\mathrm{r}}+\dot{q}_{\mathrm{h}}\end{bmatrix} \tag{A.33}$$

$\rho\boldsymbol{f}$ 是单位体积的(除科氏力和离心力之外的)体力,$\dot{q}_{\mathrm{h}}$ 表示单位质量的传热率。

转换形式的控制方程组(A.28)利用状态变量之间的热力学关系进行封闭。在理想气体的情况下,压力由如下公式计算:

$$p=(\gamma-1)\rho\left[E-\frac{u^2+v^2+w^2-\omega_1^2\,|\boldsymbol{r}_n|^2}{2}\right] \tag{A.34}$$

式中：γ 为比热比。此外，黏性系数和热传导系数必须以流动状态的函数的形式提供。

A.5 运动网格的纳维-斯托克斯方程组

在某些情况下，例如研究流体-结构的相互作用或外挂物分离投放的模拟时，有必要在运动的，同时也可能存在变形的网格上求解控制方程组，用于解决这类问题的两种最流行的方法是任意拉格朗日欧拉公式[4-6]和运动网格[7]，这两种方法密切相关，并产生对控制方程组的相同的修改形式，用于解释网格对流体的相对运动。

对于面元为 $\mathrm{d}S$ 的运动和/或变形的控制体 Ω，纳维-斯托克斯方程组(2.19)的时间相关积分形式为

$$\frac{\partial}{\partial t}\int_{\Omega}\boldsymbol{W}\mathrm{d}\Omega+\oint_{\partial\Omega}(\boldsymbol{F}_c^{\mathrm{M}}-\boldsymbol{F}_v)\mathrm{d}S=\int_{\Omega}\boldsymbol{Q}\mathrm{d}\Omega \tag{A.35}$$

守恒变量为

$$\boldsymbol{W}=\begin{bmatrix}\rho\\ \rho u\\ \rho v\\ \rho w\\ \rho E\end{bmatrix} \tag{A.36}$$

其中，ρ 表示密度，u,v,w 是笛卡儿速度分量，E 是单位质量的总能，运动网格下对流通量 $\boldsymbol{F}_c^{\mathrm{M}}$ 变为

$$\boldsymbol{F}_c^{\mathrm{M}}=\boldsymbol{F}_c-V_t\boldsymbol{W} \tag{A.37}$$

其中，$\boldsymbol{F}_c$ 由式(2.21)给出，V_t 为控制体的面的逆变速度，因此：

$$V_t=\boldsymbol{g}\cdot\boldsymbol{n}=n_x\frac{\partial x}{\partial t}+n_y\frac{\partial y}{\partial t}+n_z\frac{\partial z}{\partial t} \tag{A.38}$$

其中，$\boldsymbol{g}$ 表示网格的速度，n_x,n_y,n_z 分别表示面 $\partial\Omega$ 的指向外的单位法向矢量的分量，利用式(2.21)，对流通量可以写成

$$\boldsymbol{F}_c^{\mathrm{M}}=\begin{bmatrix}\rho V_r\\ \rho uV_r+n_xp\\ \rho vV_r+n_yp\\ \rho wV_r+n_zp\\ \rho HV_r+V_tp\end{bmatrix} \tag{A.39}$$

其中,H 表示总焓,p 表示静压。V_r 表示相对于运动网格的逆变速度,即

$$V_r = n_x u + n_y v + n_z w - V_t = V - V_t \tag{A.40}$$

黏性通量 $\boldsymbol{F}_v$ 与源项 $\boldsymbol{Q}$ 分别与式(2.23)和式(2.25)保持相同的形式,黏性应力张量的分量与式(2.15)也相同,另外,对流通量的雅可比矩阵(见 A.9 节)、其特征值(式(A.84)或式(A.88))以及谱半径都因网格运动而发生改变(在式(4.53)、式(5.33)或式(9.77)中将 V 替换为 V_r)。

Thomas 和 Lombard[8] 首先指出,除了质量守恒、动量守恒和能量守恒外,还必须满足 GCL,以避免由控制体变形引起的误差[7,9-11]。GCL 源于如下要求:在控制体体积或网格速度的计算中,要求数值格式能够保持均匀流动的状态而不受网格变形的影响[11],GCL 是由连续性方程推导出的,对于运动网格,质量守恒方程(2.3)的公式为

$$\frac{\partial}{\partial t}\int_{\Omega} \rho \mathrm{d}\Omega + \oint_{\partial\Omega} \rho (V - V_t) \mathrm{d}S = 0 \tag{A.41}$$

式(A.41)在匀速流动(即 V 不变)和密度恒定的情况也必须成立,假设控制体总是封闭的,则对逆变速度 V 的积分为零,由此,我们得到了积分形式的 GCL 为

$$\frac{\partial}{\partial t}\int_{\Omega} \mathrm{d}\Omega + \oint_{\partial\Omega} V_t \mathrm{d}S = 0 \tag{A.42}$$

如我们所见,GCL 将控制体体积的变化与它的面的运动联系起来,很明显,当控制体的形状不随时间变化时,这种运动网格自动满足 GCL。为了获得自洽的求解方法,必须使用与应用于控制方程组(A.35)相同的格式对式(A.42)的 GCL 进行时间离散,此外,GCL 必须与流体方程组同时求解,我们将通过两个典型的时间离散来说明这一点。

1. 显式时间格式

让我们从式(6.4)给出的一个基本的显式时间格式(前向欧拉格式)开始,对于这种特定的时间离散,式(A.42)变为

$$\Omega^{n+1} - \Omega^{n} = \int_{t}^{t+\Delta t} \oint_{\partial\Omega} (\boldsymbol{g} \cdot \boldsymbol{n}) \mathrm{d}S \mathrm{d}t \tag{A.43}$$

由于控制体 Ω 的面由凸面组成,因此式(A.43)中体积的总变化等于每个小面的局部变化之和,考虑控制体的一个特定小面 m,如图 A.7 所示,我们可以将当地的 GCL 表示为

$$\Delta\Omega_m = \Omega_m^{n+1} - \Omega_m^{n} = \Delta t [(\boldsymbol{g} \cdot \boldsymbol{n}) \Delta S]_m^{n+1/2} \tag{A.44}$$

假设网格运动在时间和空间上均为线性的,式(A.44)中 $\Delta\Omega_m$ 项表示小面

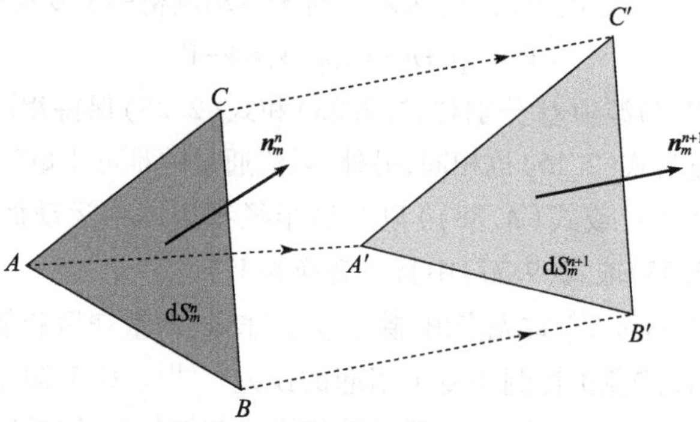

图 A.7　控制体的小面 m 在时间层 n 和 $n+1$ 之间的运动

在时间段 Δt 扫过的体积，在图 A.7 中由 $ABCA'B'C'$ 表示，由于控制体体积的当地变化为已知量（在 3D 情况下一般有一定的近似），我们可以直接从式（A.44）计算出式（A.35）中求解对流通量 $\boldsymbol{F}_c^M$ 所需的 $V_t\Delta S$，式（A.44）中的法向矢量定义为 t 和 $t+\Delta t$ 处法向矢量的算术平均，即

$$(\boldsymbol{n}\Delta S)_m^{n+1/2}=\frac{1}{2}[(\boldsymbol{n}\Delta S)_m^{n+1}+(\boldsymbol{n}\Delta S)_m^n] \tag{A.45}$$

它被用于计算式（A.35）中的对流通量和黏性通量。

当使用 6.1.1 节中的显式多步格式对式（A.35）进行时间离散时，第 k 步的流动解将由如下关系式给出：

$$\boldsymbol{W}^{(k)}=\boldsymbol{W}^{(0)}\frac{\Omega^n}{\Omega^{n+1}}-\frac{\alpha_k\Delta t}{\Omega^{n+1}}\left[\sum_{m=1}^{N_F}(\boldsymbol{F}_c^M-\boldsymbol{F}_v)_m\Delta S_m^{n+1/2}-\boldsymbol{Q}\Omega^n\right] \tag{A.46}$$

式中：N_F 表示控制体的小面的数量①，对流通量和黏性通量以及源项都是前一步的解 $\boldsymbol{W}^{(k-1)}$ 的函数，此外，为了与式（A.46）的多步格式相一致，网格速度由下式得到：

$$[(\boldsymbol{g}\cdot\boldsymbol{n})\Delta S]_m^{n+1/2}=\frac{\Omega_m^{n+1}-\Omega_m^n}{\alpha_k\Delta t} \tag{A.47}$$

最后，法向矢量由式（A.45）计算。

2. 隐式时间格式

时间离散的第二个例子，考虑式（6.79）中的三点后向欧拉格式，省略索引 I，式（6.79）的右端项在运动网格下变为

---

① 注意，小面的数量可能与面的数量不同，对于非结构网格的中点-二次格式尤其如此。

$$\boldsymbol{R}^{n+1}=\sum_{m=1}^{N_{\mathrm{F}}}(\boldsymbol{F}_{\mathrm{c}}^{\mathrm{M}}-\boldsymbol{F}_{\mathrm{v}})_m\Delta S_m^{n+1}-\boldsymbol{Q}\Omega^{n+1} \tag{A.48}$$

在这种情况下,对流通量和黏性通量以及源项都是新解 $\boldsymbol{W}^{n+1}$ 的函数,采用相同的时间积分格式计算网格速度:

$$[(\boldsymbol{g}\cdot\boldsymbol{n})\Delta S]_m^{n+1}=\frac{3\Omega_m^{n+1}-4\Omega_m^n+\Omega_m^{n-1}}{2\Delta t} \tag{A.49}$$

计算对流通量和黏性通量所需要的法向矢量,必须由前两个时间层外插计算:

$$(\boldsymbol{n}\Delta S)_m^{n+1}=\frac{3}{2}(\boldsymbol{n}\Delta S)_m^{n+1/2}-\frac{1}{2}(\boldsymbol{n}\Delta S)_m^{n-1/2} \tag{A.50}$$

其中,时间层($n+1/2$)和($n-1/2$)的项由式(A.45)得到,与文献[12]给出的条件相比较,式(A.49)和式(A.50)中的定义确保了在运动网格上的时间离散仍然保持二阶精度。

3. 物面边界条件

正如在 8.2 节中讨论的,第一个条件是在垂直于物面的方向上没有流动,对于运动边界 $V_r=0$,应用于式(A.39),$(\boldsymbol{F}_{\mathrm{c}}^{\mathrm{M}})_{\mathrm{w}}$ 中只保留 $\boldsymbol{n}p_{\mathrm{w}}$ 项和 $V_t p_{\mathrm{w}}$ 项。物面压力 p_{w} 的计算方法与 8.2 节相同,物面速度由式(A.47)或式(A.49)给出。第二个条件,在黏性流动的情况下,流体以与物面速度相同的速度运动,即 $\boldsymbol{v}_{\mathrm{w}}=\boldsymbol{g}_{\mathrm{w}}$,这影响到 8.2.2 节中讨论的虚拟单元的外插。

关于运动和/或变形网格的控制方程的数值实现以及 GCL 的离散的更多细节可以在上述引用的文献中找到,也可以参考文献[12-20],文献[15,20]对非结构网格格式进行了非常详细的描述。

A.6 薄层近似

在高 Re 数的情况下,流动只在物面周围的狭窄区域内受黏性应力的影响,因此,可以合理地假设物面法向的梯度占主导地位,而忽略其他方向的梯度[21-22],在这种情况下,我们称之为纳维-斯托克斯方程组的薄剪切层近似(简称薄层近似)。

考虑一般曲线网格下的纳维-斯托克斯方程组的微分形式(式(A.4)),如图 2.4 所示,如果我们进一步假设物面位于 η 为常数的方向,则扩散项中流向方向($\partial/\partial\xi$)和横向方向($\partial/\partial\zeta$)的所有导数都被忽略,微分形式的 3D TSL 纳维-斯托克斯方程组在曲线坐标系(ξ,η,ζ)中的形式如下:

$$\frac{\partial \boldsymbol{W}^*}{\partial t}+\frac{\partial \boldsymbol{F}_{c,1}}{\partial \xi}+\frac{\partial \boldsymbol{F}_{c,2}}{\partial \eta}+\frac{\partial \boldsymbol{F}_{c,3}}{\partial \zeta}=\frac{\partial \boldsymbol{F}_{v,2}}{\partial \eta}+\boldsymbol{Q}^* \tag{A.51}$$

守恒变量($\boldsymbol{W}^*$)、对流通量($\boldsymbol{F}_{c,1/2/3}$)和扩散通量($\boldsymbol{F}_{v,2}$)分别由式(A.5)~式(A.7)给出。此外,源项($\boldsymbol{Q}^*$)也保持式(A.8)给出的形式,然而,黏性应力张量(式(A.10))和热通量的分量变为

$$\begin{aligned}
\tau_{xx}&=2\mu\eta_x\frac{\partial u}{\partial \eta}+\lambda\left(\eta_x\frac{\partial u}{\partial \eta}+\eta_y\frac{\partial v}{\partial \eta}+\eta_z\frac{\partial w}{\partial \eta}\right)\\
\tau_{yy}&=2\mu\eta_y\frac{\partial v}{\partial \eta}+\lambda\left(\eta_x\frac{\partial u}{\partial \eta}+\eta_y\frac{\partial v}{\partial \eta}+\eta_z\frac{\partial w}{\partial \eta}\right)\\
\tau_{zz}&=2\mu\eta_z\frac{\partial w}{\partial \eta}+\lambda\left(\eta_x\frac{\partial u}{\partial \eta}+\eta_y\frac{\partial v}{\partial \eta}+\eta_z\frac{\partial w}{\partial \eta}\right)\\
\tau_{xy}&=\tau_{yx}=\mu\left(\eta_y\frac{\partial u}{\partial \eta}+\eta_x\frac{\partial v}{\partial \eta}\right)\\
\tau_{xz}&=\tau_{zx}=\mu\left(\eta_z\frac{\partial u}{\partial \eta}+\eta_x\frac{\partial w}{\partial \eta}\right)\\
\tau_{yz}&=\tau_{zy}=\mu\left(\eta_z\frac{\partial v}{\partial \eta}+\eta_y\frac{\partial w}{\partial \eta}\right)\\
\Theta_x&=u\tau_{xx}+v\tau_{xy}+w\tau_{xz}+k\eta_x\frac{\partial T}{\partial \eta}\\
\Theta_y&=u\tau_{yx}+v\tau_{yy}+w\tau_{yz}+k\eta_y\frac{\partial T}{\partial \eta}\\
\Theta_z&=u\tau_{zx}+v\tau_{zy}+w\tau_{zz}+k\eta_z\frac{\partial T}{\partial \eta}
\end{aligned} \tag{A.52}$$

在式(A.52)中,根据 TSL 假设,笛卡儿坐标的偏导数近似为

$$\begin{aligned}
\frac{\partial u}{\partial x}&\approx\eta_x\frac{\partial u}{\partial \eta}\\
\frac{\partial u}{\partial y}&\approx\eta_y\frac{\partial u}{\partial \eta}\\
\frac{\partial u}{\partial z}&\approx\eta_z\frac{\partial u}{\partial \eta}
\end{aligned} \tag{A.53}$$

A.7 PNS 方程组

如果满足以下三个条件:

(1) 流动是定常的(即 $\partial \boldsymbol{W}/\partial t=0$);

（2）流体主要向一个方向运动（不存在边界层分离）；

（3）横向流动分量可以忽略不计。

则纳维-斯托克斯方程组（2.19）可以简化为 PNS 方程组[23]，于是，速度分量和温度关于流向的导数可以从黏性应力项中去掉，此外，式（2.23）的黏性通量中的黏性应力张量 $\boldsymbol{\tau}$ 和它的功（$\boldsymbol{\tau}\cdot\boldsymbol{v}$）在流向方向上的分量也可以忽略不计。

为了说明 PNS 方法，让我们考虑一般曲线网格中的微分形式的纳维-斯托克斯方程组，如式（A.4）所示，假定流向方向对应于坐标 ξ，横向流动方向分别为坐标 η 和 ζ，于是，在曲线坐标系（ξ,η,ζ）下微分形式的 3D PNS 方程组的形式为

$$\frac{\partial \boldsymbol{W}^*}{\partial t}+\frac{\partial \boldsymbol{F}_{c,1}}{\partial \xi}+\frac{\partial \boldsymbol{F}_{c,2}}{\partial \eta}+\frac{\partial \boldsymbol{F}_{c,3}}{\partial \zeta}=\frac{\partial \boldsymbol{F}_{v,2}}{\partial \eta}+\frac{\partial \boldsymbol{F}_{v,3}}{\partial \zeta}+\boldsymbol{Q}^* \tag{A.54}$$

因此，流向方向黏性通量（$\boldsymbol{F}_{v,1}$）为零，在 PNS 方法中，不再使用黏性应力张量和热通量（式（A.10））中的笛卡儿速度分量和温度的偏导数的精确形式（式（A.12）），而是近似为

$$\begin{aligned}
\frac{\partial u}{\partial x}&\approx\eta_x\frac{\partial u}{\partial \eta}+\zeta_x\frac{\partial u}{\partial \zeta}\\
\frac{\partial u}{\partial y}&\approx\eta_y\frac{\partial u}{\partial \eta}+\zeta_y\frac{\partial u}{\partial \zeta}\\
\frac{\partial u}{\partial z}&\approx\eta_z\frac{\partial u}{\partial \eta}+\zeta_z\frac{\partial u}{\partial \zeta}
\end{aligned} \tag{A.55}$$

守恒变量（式（A.5））、对流通量（式（A.6））、其余两个黏性通量分量（式（A.7））以及源项（式（A.8））都保持不变。

A.8 纳维-斯托克斯方程组的轴对称形式

在模拟圆柱体的外流或内流时，通常可以假设它关于纵轴对称，这意味着没有旋流，同时流动变量沿圆周方向的导数为零，这样就可以从纳维-斯托克斯方程组（2.19）中去掉一个动量方程，从而控制方程组从 3D 降到 2D，大大减少了计算量。

轴对称形式的纳维-斯托克斯方程组为

$$\frac{\partial}{\partial t}\int_{\Omega}\boldsymbol{W}\mathrm{d}\Omega+\oint_{\partial\Omega}(\boldsymbol{F}_c-\boldsymbol{F}_v)r\mathrm{d}S=\int_{\Omega}\boldsymbol{Q}\mathrm{d}\Omega \tag{A.56}$$

其中，r 是径向坐标。守恒变量由四个分量组成：

$$\boldsymbol{W}=\begin{bmatrix}\rho\\ \rho u\\ \rho v\\ \rho E\end{bmatrix} \tag{A.57}$$

在此，u 和 v 分别是轴向和径向的速度分量，对流通量为

$$\boldsymbol{F}_{\mathrm{v}}=\begin{bmatrix}\rho V\\ \rho uV+n_x p\\ \rho vV+n_r p\\ \rho HV\end{bmatrix} \tag{A.58}$$

其中，n_x 是轴向的单位法向矢量，n_r 是径向的单位法向矢量，逆变速度 $V=n_x u+n_r v$，黏性通量定义为

$$\boldsymbol{F}_{\mathrm{v}}=\begin{bmatrix}0\\ n_x\tau_{xx}+n_r\tau_{xr}\\ n_x\tau_{rx}+n_r\tau_{rr}\\ n_x\Theta_x+n_r\Theta_r\end{bmatrix} \tag{A.59}$$

其中

$$\begin{aligned}\Theta_x&=u\tau_{xx}+v\tau_{xr}+k\frac{\partial T}{\partial x}\\ \Theta_r&=u\tau_{rx}+v\tau_{rr}+k\frac{\partial T}{\partial r}\end{aligned} \tag{A.60}$$

表示流体中黏性应力和热传导的做功，最后，源项为

$$\boldsymbol{Q}=\frac{1}{r}\begin{bmatrix}0\\ 0\\ p-\tau_{\theta\theta}\\ 0\end{bmatrix} \tag{A.61}$$

黏性应力张量的分量由下列关系给出：

$$\begin{aligned}\tau_{xx}&=-\frac{2\mu}{3}\nabla\cdot\boldsymbol{v}+2\mu\frac{\partial u}{\partial x}\\ \tau_{rr}&=-\frac{2\mu}{3}\nabla\cdot\boldsymbol{v}+2\mu\frac{\partial v}{\partial r}\\ \tau_{\theta\theta}&=-\frac{2\mu}{3}\nabla\cdot\boldsymbol{v}+2\mu\frac{v}{r}\\ \tau_{xr}&=\tau_{rx}=\mu\left(\frac{\partial u}{\partial r}+\frac{\partial v}{\partial x}\right)\end{aligned} \tag{A.62}$$

速度的散度为

$$\nabla \cdot \boldsymbol{v}=\frac{\partial u}{\partial x}+\frac{1}{r}\frac{\partial(rv)}{\partial r}=\frac{\partial u}{\partial x}+\frac{\partial v}{\partial r}+\frac{v}{r} \tag{A.63}$$

对称轴上的边界条件为

$$\begin{aligned}&\rho v=0\\&\frac{\partial u}{\partial r}=\frac{\partial T}{\partial r}=\frac{\partial p}{\partial r}=0\end{aligned} \tag{A.64}$$

需要注意的是，体积 Ω 是控制体的面的面积与平均半径 r 的乘积，我们可以看到，轴对称方程组（A.56）类似于 2D 纳维-斯托克斯方程组，不同之处是需要乘以半径，增加源项 $(p-\tau_{\theta\theta})/r$，以及改变 $\nabla \cdot \boldsymbol{v}$ 的定义。

A.9 对流通量雅可比矩阵

对流通量雅可比矩阵表示对流通量关于守恒变量的梯度，即

$$\boldsymbol{A}_{\mathrm{c}}=\frac{\partial \boldsymbol{F}_{\mathrm{c}}}{\partial \boldsymbol{W}} \tag{A.65}$$

其中，$\boldsymbol{F}_{\mathrm{c}}$ 由式（2.21）给出。

1. 二维形式

对流通量雅可比矩阵在 2D 情况下表示为

$$\boldsymbol{A}_{\mathrm{c}}=\begin{bmatrix}-V_{\mathrm{t}} & n_x & n_y & 0\\ n_x\phi-uV & V-V_{\mathrm{t}}-a_3n_xu & n_yu-a_2n_xv & a_2n_x\\ n_y\phi-vV & n_xv-a_2n_yu & V-V_{\mathrm{t}}-a_3n_yv & a_2n_y\\ V(\phi-a_1) & n_xa_1-a_2uV & n_ya_1-a_2vV & \gamma V-V_{\mathrm{t}}\end{bmatrix} \tag{A.66}$$

其中

$$\begin{aligned}&a_1=\gamma E-\phi\\&a_2=\gamma-1\\&a_3=\gamma-2\\&V=n_xu+n_yv\\&\phi=\frac{1}{2}(\gamma-1)(u^2+v^2)\end{aligned} \tag{A.67}$$

n_x,n_y 表示单位法向矢量 $\boldsymbol{n}$ 的笛卡儿坐标分量（图 2.1），γ 为比热比，V 为逆变速度。对于静止网格，控制体的面的逆变速度（V_{t}，见式（A.38））为零。

2. 三维形式

对流通量雅可比矩阵在 3D 情况下表示为

$$
\boldsymbol{A}_{\mathrm{c}}=\begin{bmatrix}
-V_{\mathrm{t}} & n_x & n_y & n_z & 0 \\
n_x\phi-uV & V-V_{\mathrm{t}}-a_3n_xu & n_yu-a_2n_xv & n_zu-a_2n_xw & a_2n_x \\
n_y\phi-vV & n_xv-a_2n_yu & V-V_{\mathrm{t}}-a_3n_yv & n_zv-a_2n_yw & a_2n_y \\
n_z\phi-wV & n_xw-a_2n_zu & n_yw-a_2n_zv & V-V_{\mathrm{t}}-a_3n_zw & a_2n_z \\
V(\phi-a_1) & n_xa_1-a_2uV & n_ya_1-a_2vV & n_za_1-a_2wV & \gamma V-V_{\mathrm{t}}
\end{bmatrix} \tag{A.68}
$$

其中

$$
\begin{aligned}
a_1&=\gamma E-\phi \\
a_2&=\gamma-1 \\
a_3&=\gamma-2 \\
V&=n_xu+n_yv+n_zw \\
\phi&=\frac{1}{2}(\gamma-1)(u^2+v^2+w^2)
\end{aligned} \tag{A.69}
$$

其中，n_x,n_y,n_z 分别表示单位法向矢量 $\boldsymbol{n}$ 的笛卡儿坐标分量（图 2.1），对于静止网格 $V_{\mathrm{t}}=0$。

A.10 黏性通量雅可比矩阵

黏性通量雅可比矩阵表示黏性通量关于守恒变量的梯度，即

$$
\boldsymbol{A}_{\mathrm{v}}=\frac{\partial \boldsymbol{F}_{\mathrm{v}}}{\partial \boldsymbol{W}} \tag{A.70}
$$

其中，$\boldsymbol{F}_{\mathrm{v}}$ 由式(2.23)给出，为了简单起见，下面只考虑纳维-斯托克斯方程组的 TSL 近似（参见 2.4.3 节和 A.6 节），此外，还假定动力学黏性系数和热传导系数不随守恒变量及空间的变化而变化。

1. 二维形式

2D 情况下，在曲线坐标系中 TSL 的黏性通量雅可比矩阵为[24]

$$
\boldsymbol{A}_{\mathrm{v}}=\frac{\mu}{J}\begin{bmatrix} 0 & 0 & 0 & 0 \\ b_{21} & a_1\partial_\psi(\rho^{-1}) & a_2\partial_\psi(\rho^{-1}) & 0 \\ b_{31} & a_2\partial_\psi(\rho^{-1}) & a_3\partial_\psi(\rho^{-1}) & 0 \\ b_{41} & b_{42} & b_{43} & a_4\partial_\psi(\rho^{-1}) \end{bmatrix} \tag{A.71}
$$

其中，J 表示对应于式(A.13)的坐标变换雅可比矩阵的行列式，系数 a 和 b 为

$$
\begin{aligned}
& b_{21}=-a_1\partial_\psi(u/\rho)-a_2\partial_\psi(v/\rho) \\
& b_{31}=-a_2\partial_\psi(u/\rho)-a_3\partial_\psi(v/\rho) \\
& b_{41}=a_4\partial_\psi[(u^2+v^2)/\rho-E/\rho]-a_1\partial_\psi(u^2/\rho)- \\
& \qquad 2a_2\partial_\psi(uv/\rho)-a_3\partial_\psi(v^2/\rho) \\
& b_{42}=-a_4\partial_\psi(u/\rho)-b_{21} \\
& b_{43}=-a_4\partial_\psi(v/\rho)-b_{31} \\
& a_1=(4/3)\psi_x^2+\psi_y^2 \\
& a_2=(1/3)\psi_x\psi_y \\
& a_3=\psi_x^2+(4/3)\psi_y^2 \\
& a_4=(\gamma/Pr)(\psi_x^2+\psi_y^2)
\end{aligned} \tag{A.72}
$$

在以上关系式中，μ 为动力学黏性系数，γ 为比热比，Pr 表示普朗特数。

在此需要将 $\partial_\psi(\,)=\partial(\,)/\partial\psi$ 视为一个运算符，以式(A.71)中的 $a_1\partial_\psi(\rho^{-1})$ 为例，在许多隐式格式中，存在黏性通量雅可比矩阵 $\boldsymbol{A}_{\mathrm{v}}$ 乘以守恒变量的变化量 $\Delta\boldsymbol{W}=\boldsymbol{W}^{n+1}-\boldsymbol{W}^{n}$，因此，完整项为

$$
[J^{-1}\mu a_1\partial_\psi(\rho^{-1})]\Delta W_2=A_{22}\Delta W_2 \tag{A.73}
$$

如果在结构网格上使用后向差分离散上述项，在中点$(i+1/2)$处有

$$
[A_{22}\Delta W_2]_{i+1/2}=J_{i+1/2}^{-1}\mu_{i+1/2}(a_1)_{i+1/2}\frac{\left(\dfrac{\Delta W_2}{\rho}\right)_{i+1}-\left(\dfrac{\Delta W_2}{\rho}\right)_i}{\Delta\psi} \tag{A.74}
$$

当 $\psi=\xi$ 或 $\psi=\eta$ 时，得到特定坐标方向的雅可比矩阵。

2. 三维形式

3D 情况下，在曲线坐标系中 TSL 的黏性通量雅可比矩阵为[24]

$$
\boldsymbol{A}_{\mathrm{v}}=\frac{\mu}{J}\begin{bmatrix} 0 & 0 & 0 & 0 & 0 \\ b_{21} & a_1\partial_\psi(\rho^{-1}) & a_2\partial_\psi(\rho^{-1}) & a_3\partial_\psi(\rho^{-1}) & 0 \\ b_{31} & a_2\partial_\psi(\rho^{-1}) & a_4\partial_\psi(\rho^{-1}) & a_5\partial_\psi(\rho^{-1}) & 0 \\ b_{41} & a_3\partial_\psi(\rho^{-1}) & a_5\partial_\psi(\rho^{-1}) & a_6\partial_\psi(\rho^{-1}) & 0 \\ b_{51} & b_{52} & b_{53} & b_{54} & a_7\partial_\psi(\rho^{-1}) \end{bmatrix} \tag{A.75}
$$

其中

$$
\begin{aligned}
b_{21} &= -a_1\partial_\psi(u/\rho)-a_2\partial_\psi(v/\rho)-a_3\partial_\psi(w/\rho)\\
b_{31} &= -a_2\partial_\psi(u/\rho)-a_4\partial_\psi(v/\rho)-a_5\partial_\psi(w/\rho)\\
b_{41} &= -a_3\partial_\psi(u/\rho)-a_5\partial_\psi(v/\rho)-a_6\partial_\psi(w/\rho)\\
b_{51} &= a_7\partial_\psi[(u^2+v^2+w^2)/\rho-E/\rho]-\\
&\quad a_1\partial_\psi(u^2/\rho)-a_4\partial_\psi(v^2/\rho)-a_6\partial_\psi(w^2/\rho)-\\
&\quad 2a_2\partial_\psi(uv/\rho)-2a_3\partial_\psi(uw/\rho)-2a_5\partial_\psi(vw/\rho)\\
b_{52} &= -a_7\partial_\psi(u/\rho)-b_{21}\\
b_{53} &= -a_7\partial_\psi(v/\rho)-b_{31}\\
b_{54} &= -a_7\partial_\psi(w/\rho)-b_{41}
\end{aligned}
\tag{A.76}
$$

以及

$$
\begin{aligned}
a_1 &= (4/3)\psi_x^2+\psi_y^2+\psi_z^2\\
a_2 &= (1/3)\psi_x\psi_y\\
a_3 &= (1/3)\psi_x\psi_z\\
a_5 &= (1/3)\psi_y\psi_z\\
a_4 &= \psi_x^2+(4/3)\psi_y^2+\psi_z^2\\
a_6 &= \psi_x^2+\psi_y^2+(4/3)\psi_z^2\\
a_7 &= (\gamma/Pr)(\psi_x^2+\psi_y^2+\psi_z^2)
\end{aligned}
\tag{A.77}
$$

同样,需要将 $\partial_\psi(\,)=\partial(\,)/\partial\psi$ 视为一个运算符,特定坐标方向的雅可比矩阵分别由 $\psi=\xi$ 或 $\psi=\eta$ 或 $\psi=\zeta$ 得到。

采用有限体积离散时,度量项 ψ_x,ψ_y 和 ψ_z 可根据式(A.17)转化为面矢量 $\boldsymbol{S}=\boldsymbol{n}\Delta S$ 的分量,此外,坐标变换雅可比矩阵的行列式 J 替换为如图 A.1 和式(A.20)所示的围绕控制体的面所形成体积的倒数,再次以式(A.73)为例,对于有限体积格式,式(A.73)中的乘积 $J^{-1}a_1$ 将变为

$$
J^{-1}a_1=\frac{2}{\Omega^*}\left(\frac{4}{3}S_x^2+S_y^2\right)
\tag{A.78}
$$

需要注意的是,式(A.78)中包含了面的面积,以反映残差计算时黏性通量与 ΔS 的乘积——$\boldsymbol{F}_v$ 本身只包含单位法向矢量的分量,则在中点$(i+1/2)$处的离散变为

$$
[A_{22}\Delta W_2]_{i+1/2}=\frac{2\mu_{i+1/2}}{\Omega_{i+1/2}^*}\left(\frac{4}{3}S_x^2+S_y^2\right)\left[\left(\frac{\Delta W_2}{\rho}\right)_{i+1}-\left(\frac{\Delta W_2}{\rho}\right)_i\right]
\tag{A.79}
$$

其中,根据 $\boldsymbol{S}^I$ 和 $\boldsymbol{S}^J$ 的定义(见图 A.1 和式(A.15)和式(A.16)),我们假定 $\Delta\psi=1$。

A.11 从守恒变量到特征变量的转换

对流通量雅可比矩阵(见 A.9 节)具有实特征值和一组完整的特征矢量,因此可以对角化为[25]

$$\boldsymbol{A}_{\mathrm{c}}=\boldsymbol{T}\boldsymbol{\Lambda}_{\mathrm{c}}\boldsymbol{T}^{-1} \tag{A.80}$$

其中,$\boldsymbol{T}^{-1}$ 表示左特征矢量矩阵,$\boldsymbol{T}$ 表示右特征矢量矩阵,$\boldsymbol{\Lambda}_{c}$ 表示特征值对角矩阵。对于任意的变分 δ(类似于 ∂_t 或∇),从守恒变量($\boldsymbol{W}$)到特征变量($\boldsymbol{C}$)的转换为

$$\delta\boldsymbol{C}=\boldsymbol{T}^{-1}\delta\boldsymbol{W} \tag{A.81}$$

对流通量雅可比矩阵的对角化可以看作将其分解为不同的波,右特征矢量代表波,特征变量是波幅,特征值是相应的波速。在下面介绍对角化的内容中,我们区分两种类型的波:第一种是对流波或线性波,它与类型为 $\Lambda_{\mathrm{c}}=\boldsymbol{v}\cdot\boldsymbol{n}$ 的特征值相关;第二种是声波,它与类型为 $\Lambda_{\mathrm{c}}=\boldsymbol{v}\cdot\boldsymbol{n}\pm c$ 的特征值相关。需要注意的是,不能同时在多个空间方向上进行对角化[26],这就是通量差分裂格式和 TVD 格式只能使用 1D 波分解的原因,只有波动分裂格式(见 3.1.5 节)使用了近似多维分解。

1. 二维形式

2D 情况下,$\boldsymbol{\Lambda}_{\mathrm{c}}$ 的左特征矢量矩阵为[24]

$$\boldsymbol{T}^{-1}=\begin{bmatrix}(1-\phi c^{-2}) & a_1uc^{-2} & a_1vc^{-2} & -a_1c^{-2}\\ -(n_xu-n_yv)\rho^{-1} & n_y\rho^{-1} & -n_x\rho^{-1} & 0\\ a_2(\phi-cV) & a_2(n_xc-a_1u) & a_2(n_yc-a_1v) & a_1a_2\\ a_2(\phi+cV) & -a_2(n_xc+a_1u) & -a_2(n_yc+a_1v) & a_1a_2\end{bmatrix} \tag{A.82}$$

右特征矢量矩阵为[24]

$$\boldsymbol{T}=\begin{bmatrix}1 & 0 & a_3 & a_3\\ u & n_y\rho & a_3(u+n_xc) & a_3(u-n_xc)\\ v & -n_x\rho & a_3(v+n_yc) & a_3(v-n_yc)\\ \phi a_1^{-1} & \rho(n_yu-n_xv) & a_3(a_4+cV) & a_3(a_4-cV)\end{bmatrix} \tag{A.83}$$

实特征值组成的对角矩阵为

$$\boldsymbol{\Lambda}_{\mathrm{c}}=\begin{bmatrix}\Lambda_1 & 0 & 0 & 0\\ 0 & \Lambda_2 & 0 & 0\\ 0 & 0 & \Lambda_3 & 0\\ 0 & 0 & 0 & \Lambda_4\end{bmatrix} \tag{A.84}$$

以上各式中的参数定义为

$$
\begin{aligned}
&a_1=\gamma-1\\
&a_2=\frac{1}{\rho c\sqrt{2}}\\
&a_3=\frac{\rho}{c\sqrt{2}}\\
&a_4=\frac{\phi+c^2}{\gamma-1}\\
&V=n_x u+n_y v\\
&\phi=\frac{1}{2}(\gamma-1)(u^2+v^2)\\
&\Lambda_1=\Lambda_2=V-V_t\\
&\Lambda_3=V-V_t+c\\
&\Lambda_4=V-V_t-c
\end{aligned}
\tag{A.85}
$$

其中,γ 为比热比,c 为声速,$\boldsymbol{n}=[n_x,n_y]^{\mathrm{T}}$ 为单位法向矢量,V 为逆变速度,最后,V_t 为控制体的面的逆变速度,如式(A.38)所示。在静止网格中,V_t 为零。

2. 三维形式

3D 情况下,$\boldsymbol{\Lambda}_c$ 的左特征矢量矩阵为[24]

$$
\boldsymbol{T}^{-1}=\begin{bmatrix}
n_x a_5-(n_z v-n_y w)\rho^{-1} & n_x a_1 u c^{-2} & n_x a_1 v c^{-2}+n_z\rho^{-1} & n_x a_1 w c^{-2}-n_y\rho^{-1} & -n_x a_1 c^{-2}\\
n_y a_5-(n_x w-n_z u)\rho^{-1} & n_y a_1 u c^{-2}-n_z\rho^{-1} & n_y a_1 v c^{-2} & n_y a_1 w c^{-2}+n_x\rho^{-1} & -n_y a_1 c^{-2}\\
n_z a_5-(n_y u-n_x v)\rho^{-1} & n_z a_1 u c^{-2}+n_y\rho^{-1} & n_z a_1 v c^{-2}-n_x\rho^{-1} & n_z a_1 w c^{-2} & -n_z a_1 c^{-2}\\
a_2(\phi-cV) & -a_2(a_1 u-n_x c) & -a_2(a_1 v-n_y c) & -a_2(a_1 w-n_z c) & a_1 a_2\\
a_2(\phi+cV) & -a_2(a_1 u+n_x c) & -a_2(a_1 v+n_y c) & -a_2(a_1 w+n_z c) & a_1 a_2
\end{bmatrix}
\tag{A.86}
$$

右特征矢量矩阵为[24]

$$
\boldsymbol{T}=\begin{bmatrix}
n_x & n_y & n_z & a_3 & a_3 \\
n_x u & n_y u-n_z\rho & n_z u+n_y\rho & a_3(u+n_x c) & a_3(u-n_x c) \\
n_x v+n_z\rho & n_y v & n_z v-n_x\rho & a_3(v+n_y c) & a_3(v-n_y c) \\
n_x w-n_y\rho & n_y w+n_x\rho & n_z w & a_3(w+n_z c) & a_3(w-n_z c) \\
n_x a_6+\rho(n_z v-n_y w) & n_y a_6+\rho(n_x w-n_z u) & n_z a_6+\rho(n_y u-n_x v) & a_3(a_4+cV) & a_3(a_4-cV)
\end{bmatrix} \tag{A.87}
$$

实特征值组成的对角矩阵为

$$
\boldsymbol{\Lambda}_{\mathrm{c}}=\begin{bmatrix}
\Lambda_1 & 0 & 0 & 0 & 0 \\
0 & \Lambda_2 & 0 & 0 & 0 \\
0 & 0 & \Lambda_3 & 0 & 0 \\
0 & 0 & 0 & \Lambda_4 & 0 \\
0 & 0 & 0 & 0 & \Lambda_5
\end{bmatrix} \tag{A.88}
$$

以上各式中使用的缩写为

$$
\begin{aligned}
& a_1=\gamma-1 \\
& a_2=\frac{1}{\rho c\sqrt{2}} \\
& a_3=\frac{\rho}{c\sqrt{2}} \\
& a_4=\frac{\phi+c^2}{\gamma-1} \\
& a_5=1-\frac{\phi}{c^2} \\
& a_6=\frac{\phi}{\gamma-1} \\
& V=n_x u+n_y v+n_z w \\
& \phi=\frac{1}{2}(\gamma-1)(u^2+v^2+w^2) \\
& \Lambda_1=\Lambda_2=\Lambda_3=V-V_{\mathrm{t}}
\end{aligned} \tag{A.89}
$$

$$\Lambda_4 = V - V_t + c$$
$$\Lambda_5 = V - V_t - c$$

在关系式(A.86)~式(A.89)中，γ 为比热比，c 为声速，$\boldsymbol{n}=[n_x,n_y,n_z]^{\mathrm{T}}$ 为单位法向矢量，V 为逆变速度，最后，V_t 为控制体的面的逆变速度，如式(A.38)所示，在静止网格的情况下，V_t 为零。

A.12 GMRES 算法

考虑线性方程组：

$$\boldsymbol{Ax}=\boldsymbol{b} \tag{A.90}$$

寻找形式如下的近似解：

$$\boldsymbol{x}=\boldsymbol{x}_0+\boldsymbol{z} \tag{A.91}$$

其中，$\boldsymbol{x}_0$ 表示初始猜测值，$\boldsymbol{z}$ 是 Krylov 子空间的成员之一：

$$\boldsymbol{z}\in K_m,\quad K_m\equiv \text{span}\{\boldsymbol{r}_0,\boldsymbol{Ar}_0,\boldsymbol{A}^2\boldsymbol{r}_0,\cdots,\boldsymbol{A}^{m-1}\boldsymbol{r}_0\} \tag{A.92}$$

其中，$\boldsymbol{r}_0=\boldsymbol{b}-\boldsymbol{Ax}_0$，$m$ 是 K 的维数，也称为搜索方向的个数，GMRES 算法[27]通过残差的 2 范数来确定 $\boldsymbol{z}$，即

$$\|\boldsymbol{b}-\boldsymbol{A}(\boldsymbol{x}_0+\boldsymbol{z})\| \tag{A.93}$$

的最小化。下面，我们给出了 GMRES 算法的具体步骤。

A.12.1 K_m 正交基的计算

采用改进的 Gram-Schmidt 过程：

$$\boldsymbol{r}_0=\boldsymbol{b}-\boldsymbol{Ax}_0$$
$$\boldsymbol{v}_1=\boldsymbol{r}_0/\|\boldsymbol{r}_0\|$$

DO $j=1,m$

  $\tilde{\boldsymbol{v}}_{j+1}=\boldsymbol{Av}_j$

  DO $i=1,j$

    $h_{i,j}=\tilde{\boldsymbol{v}}_{j+1}\cdot\boldsymbol{v}_i$

    $\tilde{\boldsymbol{v}}_{j+1}=\tilde{\boldsymbol{v}}_{j+1}-h_{i,j}\boldsymbol{v}_i$

  ENDDO

  $h_{j+1,j}=\|\tilde{\boldsymbol{v}}_{j+1}\|$

  $\boldsymbol{v}_{j+1}=\tilde{\boldsymbol{v}}_{j+1}/h_{j+1,j}$

ENDDO

式中：$h_{i,j}$ 为上 Hessenberg 矩阵的系数(i 为行，j 为列)，然而，矩阵还包括了元素 $h_{j+1,j}$。因此，维数变成$(m+1)\times m$。

A. 12. 2　上 Hessenberg 矩阵的生成

这个矩阵具有如下接近三角的形式：

$$\boldsymbol{H}_m^* = \begin{bmatrix} h_{1,1} & h_{1,2} & \cdots & h_{1,m-1} & h_{1,m} \\ h_{2,1} & h_{2,2} & \cdots & h_{2,m-1} & h_{2,m} \\ 0 & h_{3,2} & \ddots & \vdots & \vdots \\ \vdots & 0 & \ddots & h_{m-1,m-1} & h_{m-1,m} \\ \vdots & \vdots & \ddots & h_{m,m-1} & h_{m,m} \\ 0 & 0 & \cdots & 0 & h_{m+1,m} \end{bmatrix}^{(m+1)\times m} \quad (A.94)$$

下面进一步用它来表述和解决残差（式（A. 93））的最小化问题。

A. 12. 3　残差最小化

初始解 $\boldsymbol{x}_0$ 的修正定义为[27]

$$\boldsymbol{z} = \sum_{j=1}^{m} y_j \boldsymbol{v}_j \quad (A.95)$$

其中，y_j 表示如下矢量的分量：

$$\boldsymbol{y} = [y_1, y_2, \cdots, y_m]^{\mathrm{T}} \quad (A.96)$$

此外，可以证明：

$$\boldsymbol{A}\boldsymbol{V}_m = \boldsymbol{V}_{m+1}\boldsymbol{H}_m^* \quad (A.97)$$

其中

$$\boldsymbol{V}_m = [\boldsymbol{v}_1, \boldsymbol{v}_2, \cdots, \boldsymbol{v}_m] \quad (A.98)$$

为矩阵，$\boldsymbol{v}_j$ 是它的列矢量。我们引入符号

$$\boldsymbol{e} = [\|\boldsymbol{r}_0\|, 0, \cdots, 0]^{\mathrm{T}} \quad (A.99)$$

其中，$\boldsymbol{e}$ 有（$m+1$）个元素。使用式（A. 99）的定义，我们发现 $\boldsymbol{r}_0 = \boldsymbol{b} - \boldsymbol{A}\boldsymbol{x}_0 = \boldsymbol{V}_{m+1}\boldsymbol{e}$，因此，可以得到式（A. 93）中的残差为

$$\begin{aligned} \|\boldsymbol{b} - A(\boldsymbol{x}_0 + \boldsymbol{z})\| &= \left\| \boldsymbol{r}_0 - \boldsymbol{A}\left(\sum_{j=1}^{m} y_j \boldsymbol{v}_j\right) \right\| \\ &= \|\boldsymbol{r}_0 - \boldsymbol{A}\boldsymbol{V}_m\boldsymbol{y}\| \\ &= \|\boldsymbol{V}_{m+1}(\boldsymbol{e} - \boldsymbol{H}_m^*\boldsymbol{y})\| \\ &= \|\boldsymbol{e} - \boldsymbol{H}_m^*\boldsymbol{y}\| \end{aligned} \quad (A.100)$$

最后，使用 $\boldsymbol{V}_{m+1}$ 的标准正交化（这意味着对于 $i \neq j$，$\boldsymbol{v}_i^{\mathrm{T}} \cdot \boldsymbol{v}_j = 0$，对于 $i = j$，$\boldsymbol{v}_i^{\mathrm{T}} \cdot \boldsymbol{v}_j = 1$），残差最小化问题可以简化为

$$\min_{z\in K_m}\|\boldsymbol{b}-\boldsymbol{A}(\boldsymbol{x}_0+\boldsymbol{z})\|=\min_{\boldsymbol{y}\in R^m}\|\boldsymbol{e}-\boldsymbol{H}_m^*\boldsymbol{y}\| \tag{A. 101}$$

最小化问题的解可以借助 Q-R 算法得到，下面将介绍 Q-R 算法。

A. 12. 4 Q-R 算法

定义 $\boldsymbol{R}_m=\boldsymbol{Q}_m\boldsymbol{H}_m^*$，其中

$$\boldsymbol{Q}_m\overset{\text{def}}{=}\boldsymbol{F}_m\boldsymbol{F}_{m-1}\cdots\boldsymbol{F}_1 \tag{A. 102}$$

是下面给定的旋转矩阵的乘积：

$$\boldsymbol{F}_j=\begin{bmatrix}\boldsymbol{I}_{j-1} & & \\ & \begin{bmatrix}c_j & s_j\\ -s_j & c_j\end{bmatrix} & \\ & & \boldsymbol{I}_{m-j}\end{bmatrix}^{(m+1)\times(m+1)} \tag{A. 103}$$

在式（A. 103）中，$\boldsymbol{I}_j$ 表示维数为 j 的单位矩阵，c_j 和 s_j 表示旋转角度的正弦/余弦值（$c_j^2+s_j^2=1$），选择旋转使得 $\boldsymbol{H}_m^*$ 转换为上三角矩阵 $\boldsymbol{R}_m$，其维度为 $(m+1)\times m$，最后一行只包含 0。由于 $\boldsymbol{Q}_m^{\mathrm{T}}\boldsymbol{Q}_m=\boldsymbol{I}$，我们可以改写式（A. 101）中的 $\|\boldsymbol{e}-\boldsymbol{H}_m^*\boldsymbol{y}\|$ 为

$$\|\boldsymbol{e}-\boldsymbol{H}_m^*\boldsymbol{y}\|=\|\boldsymbol{Q}_m^{\mathrm{T}}(\boldsymbol{Q}_m\boldsymbol{e}-\boldsymbol{Q}_m\boldsymbol{H}_m^*\boldsymbol{y})\|=\|\boldsymbol{g}-\boldsymbol{R}_m\boldsymbol{y}\| \tag{A. 104}$$

其中，$\boldsymbol{g}=\boldsymbol{Q}_m\boldsymbol{e}$ 表示矢量 $\boldsymbol{e}$（式（A. 99））的变换，$\boldsymbol{R}_m$ 的最后一行由零组成，因此矢量 $(\boldsymbol{g}-\boldsymbol{R}_m\boldsymbol{y})$ 只有第 $(m+1)$ 行的项 g_{m+1} 是非零的，如果我们将 $(\boldsymbol{g}-\boldsymbol{R}_m\boldsymbol{y})$ 的前 m 个分量表示为 $p_j(j=1,2,\cdots,m)$，则式（A. 104）中的范数为

$$\|\boldsymbol{g}-\boldsymbol{R}_m\boldsymbol{y}\|=\sqrt{g_{m+1}^2+\sum_{j=1}^{m}p_j^2} \tag{A. 105}$$

如果选择 $\boldsymbol{y}$ 的分量 y_j，使 $j=1,2,\cdots,m$ 时，$p_j^2=0$，我们得到式（A. 101）中的最小化问题：

$$\min_{y\in R^m}\|\boldsymbol{g}-\boldsymbol{R}_m\boldsymbol{y}\|=|g_{m+1}| \tag{A. 106}$$

分量 y_j 由下列线性方程组通过回代得到：

$$\begin{bmatrix}R_{1,1} & \cdots & R_{1,m-1} & R_{1,m}\\ 0 & \ddots & \vdots & \vdots\\ \vdots & \ddots & R_{m-1,m-1} & R_{m-1,m}\\ 0 & \cdots & 0 & R_{m,m}\end{bmatrix}\begin{bmatrix}y_1\\ \vdots\\ y_{m-1}\\ y_m\end{bmatrix}=\begin{bmatrix}g_1\\ \vdots\\ g_{m-1}\\ g_m\end{bmatrix} \tag{A. 107}$$

由已知的 y_j，根据式（A. 95），就可得到方程组（A. 90）的解。

需要重点强调的是

$$
\begin{aligned}
\|\boldsymbol{r}_m\| &= \|\boldsymbol{b}-\boldsymbol{A}(\boldsymbol{x}_0+\boldsymbol{z})\| \\
&= \|\boldsymbol{e}-\boldsymbol{H}_m^*\boldsymbol{y}\| \\
&= \|\boldsymbol{g}-\boldsymbol{R}_m\boldsymbol{y}\| \\
&= |g_{m+1}|
\end{aligned} \tag{A.108}
$$

这意味着实际残差很容易确定为$|g_m+1|$。

A.13 张量符号

像坐标(x_i)或速度分量(v_i)这样的式子表示一阶张量,它们有三个分量,从而对应矢量,即

$$
\begin{aligned}
x_i &= [x_1, x_2, x_3] = [x, y, z] = \boldsymbol{r} \\
v_i &= [v_1, v_2, v_3] = [u, v, w] = \boldsymbol{v}
\end{aligned} \tag{A.109}
$$

二阶张量由 9 个分量组成,可以写成 3×3 的矩阵,例如:

$$
v_i v_j \equiv \begin{bmatrix} v_1v_1 & v_1v_2 & v_1v_3 \\ v_2v_1 & v_2v_2 & v_2v_3 \\ v_3v_1 & v_3v_2 & v_3v_3 \end{bmatrix} \tag{A.110}
$$

同样,黏性应力张量 $\boldsymbol{\tau}=\tau_{ij}$ 为

$$
\tau_{ij} \equiv \begin{bmatrix} \tau_{11} & \tau_{12} & \tau_{13} \\ \tau_{21} & \tau_{22} & \tau_{23} \\ \tau_{31} & \tau_{32} & \tau_{33} \end{bmatrix} \tag{A.111}
$$

Kronecker 符号 δ_{ij} 是一个特殊的二阶张量,它对应 3×3 的单位矩阵,因此,存在如下关系:

$$
\delta_{ij} = \begin{cases} 1, & i=j \\ 0, & i\neq j \end{cases} \tag{A.112}
$$

最后一个重要的规则是爱因斯坦求和约定,它指出,每当表达式中出现两个相同的下标时,它意味着对所有三个坐标方向求和,这样,矢量 $\boldsymbol{u}$ 和 $\boldsymbol{v}$ 的标量积可以表示为

$$
u_i v_i = u_1v_1 + u_2v_2 + u_3v_3 = \boldsymbol{u} \cdot \boldsymbol{v} \tag{A.113}
$$

此外,矢量 $\boldsymbol{v}$ 的散度使用张量符号可表示为

$$
\frac{\partial v_i}{\partial x_i} = \frac{\partial v_1}{\partial x_1} + \frac{\partial v_2}{\partial x_2} + \frac{\partial v_3}{\partial x_3} = \nabla \cdot \boldsymbol{v} \tag{A.114}
$$

参考文献

[1] Sommerfeld AJW. Partial differential equations in physics. New York: Academic Press;1949.

[2] Courant R, Hilbert D. Methods of mathematical physics. New York: Interscience;1962.

[3] Hirsch C. Numerical computation of internal and external flows, vol. 1. Chichester: John Wiley and Sons;1988.

[4] Hirt CW, Amsden AA, Cook JL. An arbitrary Lagrangian-Eulerian computing method for all flow speeds. J Comput Phys 1974;14:227-53.

[5] Pracht WE. Calculating three-dimensional fluid flows at all speeds with an Eulerian-Lagrangian computing mesh. J Comput Phys 1975;17:132-59.

[6] Donea J. An arbitrary Lagrangian-Eulerian finite element method for transient fluid-structure interactions. Comput Methods Appl Mech Eng 1982;33:689-23.

[7] Batina JT. Unsteady Euler airfoil solutions using unstructured dynamic meshes. AIAA Paper 89-0115;1989.

[8] Thomas PD, Lombard CK. Geometric conservation law and its application to flow computations on moving grids. AIAA J 1979;17:1030-37.

[9] Tamura Y, Fujii K. Conservation law for moving and transformed grids. AIAA Paper 93-3365;1993.

[10] Lesoinne M, Farhat C. Geometric conservation laws for flow problems with moving boundaries and deformable meshes and their impact on aeroelastic computations. AIAA Paper 95-1709; 1995 [also in Comput Methods Appl Mech Eng 1996;134:71-90].

[11] Guillard H, Farhat C. On the significance of the GCL for flow computations on moving meshes. AIAA Paper 99-0793;1999.

[12] Geuzaine P, Farhat C. Design and time-accuracy analysis of ALE schemes for inviscid and viscous flow computations on moving meshes. AIAA Paper 2003-3694;2003.

[13] Nkonga B, Guillard H. Godunov type method on non-structured meshes for three-dimensional moving boundary problems. Comput Methods Appl Mech Eng 1994;113:183-204.

[14] Singh KP, Newman JC, Baysal O. Dynamic unstructured method for flows past multiple objects in relative motion. AIAA J 1995;33:641-49.

[15] Venkatakrishnan V, Mavriplis DJ. Implicit method for the computation of unsteady flows on unstructured grids. J Comput Phys 1996;127:380-97.

[16] Slater JW, Freund D, Sajben M. Study of CFD methods applied to rapidly deforming boundaries. AIAA Paper 97-2041;1997.

[17] Vierendeels J, Riemslagh K, Dick E. Flow calculation in complex shaped moving domains. AIAA Paper 97-2044;1997.

[18] Koobus B., Farhat C. Second-order time-accurate and geometrically conservative implicit schemes for flow computations on unstructured dynamic meshes. Comput Methods Appl Mech Eng 1999;

170:103-30.

[19] Dubuc L, Cantariti F, Woodgate M, Gribben B, Badcock KJ, Richards BE. A grid deformation technique for unsteady flow computations. Int J Numer Methods Fluids 2000;32:285-11.

[20] Sørensen KA, Hassan O, Morgan K, Weatherill NP. A multigrid accelerated time-accurate inviscid compressible fluid flow solution algorithm employing mesh movement and local remeshing. Int J Numer Methods Fluids 2003;43:517-36.

[21] Steger JL. Implicit finite difference simulation of flows about arbitrary geometries. AIAA J 1978;17:679-86.

[22] Pulliam TH, Steger JL. Implicit finite difference simulations of 3-D compressible flows. AIAA J 1980;18:159-67.

[23] Patankar SV, Spalding DB. A calculation procedure for heat, mass and momentum transfer in three-dimensional parabolic flows. Int J Heat Mass Transfer 1972;15:1787-806.

[24] Pulliam TH, Steger JL. Recent improvements in efficiency, accuracy, and convergence for implicit approximate factorization algorithms. AIAA Paper 85-0360;1985.

[25] Warming RF, Beam R, Hyett BJ. Diagonalization and simultaneous symmetrization of the gas-dynamic matrices. Math Comput 1975;29:1037.

[26] Whitham GB. Linear and nonlinear waves. New York: Wiley;1974.

[27] Saad Y, Schulz MH. GMRES: a generalized minimum residual algorithm for solving nonsymmetric linear systems. SIAM J Sci Stat Comput 1986;7:856-69.

内容简介

本书详细讲述了 CFD 的基本原理和常用的数值方法，并给出了大量实例程序供读者学习 CFD 编程和应用。本书的特点是理论联系实际，着重介绍工程实用的 CFD 方法，侧重于具体算法的实现，在介绍各种数值方法的同时讨论了方法的优缺点，每个理论部分都有相应的实践部分来增强对知识的理解，并展示了这些方法如何应用到实际 CFD 程序中。

本书可供相关专业本科生及研究生作为教材使用，也可作为研究人员开展 CFD 领域工作的入门书和参考手册。